AF905476

Studies in Computational Intelligence

Volume 912

Series Editor
Janusz Kacprzyk, Polish Academy of Sciences, Warsaw, Poland

The series "Studies in Computational Intelligence" (SCI) publishes new developments and advances in the various areas of computational intelligence—quickly and with a high quality. The intent is to cover the theory, applications, and design methods of computational intelligence, as embedded in the fields of engineering, computer science, physics and life sciences, as well as the methodologies behind them. The series contains monographs, lecture notes and edited volumes in computational intelligence spanning the areas of neural networks, connectionist systems, genetic algorithms, evolutionary computation, artificial intelligence, cellular automata, self-organizing systems, soft computing, fuzzy systems, and hybrid intelligent systems. Of particular value to both the contributors and the readership are the short publication timeframe and the world-wide distribution, which enable both wide and rapid dissemination of research output.

The books of this series are submitted to indexing to Web of Science, EI-Compendex, DBLP, SCOPUS, Google Scholar and Springerlink.

More information about this series at http://www.springer.com/series/7092

Aboul Ella Hassanien · Roheet Bhatnagar ·
Ashraf Darwish
Editors

Artificial Intelligence for Sustainable Development: Theory, Practice and Future Applications

Springer

Editors
Aboul Ella Hassanien
Information Technology Department,
Faculty of Computers and Information
Cairo University
Giza, Egypt

Roheet Bhatnagar
Department of Computer Science
and Engineering, Faculty of Engineering
Manipal University
Jaipur, Rajasthan, India

Ashraf Darwish
Faculty of Science
Helwan University
Cairo, Egypt

ISSN 1860-949X ISSN 1860-9503 (electronic)
Studies in Computational Intelligence
ISBN 978-3-030-51919-3 ISBN 978-3-030-51920-9 (eBook)
https://doi.org/10.1007/978-3-030-51920-9

© The Editor(s) (if applicable) and The Author(s), under exclusive license to Springer Nature Switzerland AG 2021
This work is subject to copyright. All rights are solely and exclusively licensed by the Publisher, whether the whole or part of the material is concerned, specifically the rights of translation, reprinting, reuse of illustrations, recitation, broadcasting, reproduction on microfilms or in any other physical way, and transmission or information storage and retrieval, electronic adaptation, computer software, or by similar or dissimilar methodology now known or hereafter developed.
The use of general descriptive names, registered names, trademarks, service marks, etc. in this publication does not imply, even in the absence of a specific statement, that such names are exempt from the relevant protective laws and regulations and therefore free for general use.
The publisher, the authors and the editors are safe to assume that the advice and information in this book are believed to be true and accurate at the date of publication. Neither the publisher nor the authors or the editors give a warranty, expressed or implied, with respect to the material contained herein or for any errors or omissions that may have been made. The publisher remains neutral with regard to jurisdictional claims in published maps and institutional affiliations.

This Springer imprint is published by the registered company Springer Nature Switzerland AG
The registered company address is: Gewerbestrasse 11, 6330 Cham, Switzerland

Preface

Currently, AI is influencing the larger trends in global sustainability; it could play a significant role where humanity co-exists harmoniously with machines, or portend a dystopian world filled with conflict, poverty and suffering. It would also accelerate our progress on the United Nations Sustainable Development Goals (SDGs). In this context, the SDGs are defining the development goals for the countries of the world, and AI is rapidly opening up a new ways in the fields of industry, health, business, education, environment and space industry. As a result, AI has been incorporated in various forms into the SDGs through experimentation and in sustainable management and leadership programmes. Therefore, there are many countries all over the world started to establish nation AI strategies.

AI, when is applied for sustainability inducing applications and projects, will present massive and geographically wide-ranging opportunities, enable more efficient and effective public policy for sustainability, and will enhance connectivity, access and the efficiency in many different sectors.

The academic community has an important role to play to be ready for this AI revolution, in preparing the future generations of national and international policy-makers and decision-makers in addressing the opportunities and the challenges presented by AI and the imperative to advance the global goals. In this regard, AI for sustainability is challenged by large amount of data in machine learning models, increased cybersecurity risks and adverse impacts of AI applications.

This book includes future studies and researches of AI for sustainable development and to show how AI can deliver immediate solutions without introducing long-term threats to environmental sustainability.

This book aims at emphasizing the latest developments and achievements in the field AI and related technologies with a special focus on sustainable development and eco-friendly AI applications. The book describes theory, applications and conceptualization of ideas and critical surveys which are covering most of aspects of AI for SDGs. This book will help to identify those aspects of connected smarter world that are key enablers of AI sustainable applications and its sustenance as a futuristic technology.

The content of this book is divided into four parts: first part presents the role and importance of AI technology in agriculture sector as one of the main SDGs. Healthcare sector is considered as one of the important goals of SDGs. Therefore, second part describes and analyses the effective role of AI in healthcare industry to enable countries to overcome the developing of diseases and in crisis times of pandemic such as COVID-19 (Coronavirus). Third part introduces the machine and deep learning as the most important branches of AI and their impact in many areas of applications for SDGs. There are other emerging technologies such as Internet of Things, sensor networks and cloud computing which can be integrated with AO for the future of SDGs. As a result, the fourth part presents the applications of the most merging technologies and smart networking as integrated technologies with AI to fulfil the SDGs.

Finally, editors of this book would like to acknowledge all the authors for their studies and contributions. Editors also would like to encourage the reader to explore and expand the knowledge in order to create their implementations according to their necessities.

Book Editors

Giza, Egypt — Aboul Ella Hassanien
Jaipur, India — Roheet Bhatnagar
Cairo, Egypt — Ashraf Darwish

Contents

Artificial Intelligence in Sustainability Agricultures

Optimization of Drip Irrigation Systems Using Artificial Intelligence Methods for Sustainable Agriculture and Environment 3
Dmitriy Klyushin and Andrii Tymoshenko

Artificial Intelligent System for Grape Leaf Diseases Classification 19
Kamel K. Mohammed, Ashraf Darwish, and Aboul Ella Hassenian

Robust Deep Transfer Models for Fruit and Vegetable Classification: A Step Towards a Sustainable Dietary 31
Nour Eldeen M. Khalifa, Mohamed Hamed N. Taha, Mourad Raafat Mouhamed, and Aboul Ella Hassanien

The Role of Artificial Neuron Networks in Intelligent Agriculture (Case Study: Greenhouse) 45
Abdelkader Hadidi, Djamel Saba, and Youcef Sahli

Artificial Intelligence in Smart Health Care

Artificial Intelligence Based Multinational Corporate Model for EHR Interoperability on an E-Health Platform 71
Anjum Razzaque and Allam Hamdan

Predicting COVID19 Spread in Saudi Arabia Using Artificial Intelligence Techniques—Proposing a Shift Towards a Sustainable Healthcare Approach 83
Anandhavalli Muniasamy, Roheet Bhatnagar, and Gauthaman Karunakaran

Machine Learning and Deep Learning Applications

A Comprehensive Study of Deep Neural Networks for Unsupervised Deep Learning 101
Deepti Deshwal and Pardeep Sangwan

An Overview of Deep Learning Techniques for Biometric Systems 127
Soad M. Almabdy and Lamiaa A. Elrefaei

Convolution of Images Using Deep Neural Networks in the Recognition of Footage Objects 171
Varlamova Lyudmila Petrovna

A Machine Learning-Based Framework for Efficient LTE Downlink Throughput 193
Nihal H. Mohammed, Heba Nashaat, Salah M. Abdel-Mageid, and Rawia Y. Rizk

Artificial Intelligence and Blockchain for Transparency in Governance 219
Mohammed AlShamsi, Said A. Salloum, Muhammad Alshurideh, and Sherief Abdallah

Artificial Intelligence Models in Power System Analysis 231
Hana Yousuf, Asma Y. Zainal, Muhammad Alshurideh, and Said A. Salloum

Smart Networking Applications

Internet of Things for Water Quality Monitoring and Assessment: A Comprehensive Review 245
Joshua O. Ighalo, Adewale George Adeniyi, and Goncalo Marques

Contribution to the Realization of a Smart and Sustainable Home 261
Djamel Saba, Youcef Sahli, Rachid Maouedj, Abdelkader Hadidi, and Miloud Ben Medjahed

Appliance Scheduling Towards Energy Management in IoT Networks Using Bacteria Foraging Optimization (BFO) Algorithm 291
Arome Junior Gabriel

Artificial Intelligence in Sustainability Agricultures

Optimization of Drip Irrigation Systems Using Artificial Intelligence Methods for Sustainable Agriculture and Environment

Dmitriy Klyushin and Andrii Tymoshenko

Abstract An AI system of optimal control and design of drip irrigation systems is proposed. This AI system is based on simulation of the water transport process described by Richards-Klute equation. Using the Kirchhoff transformation the original nonlinear problem is reduced to a linear problem of optimal control of non-stationary moisture transport in an unsaturated soil providing the desirable water content. For minimization of a cost functional a variational algorithm is used. Also, the finite-difference method is used to solve direct and conjugate problems. The optimization is achieved by minimizing the mean square deviation of the moisture content from the target distribution at a given moment in time. The chapter describes the optimization of drip irrigation system with buried sources in a dry soil. Results demonstrate high accuracy and effectiveness of the method. The purpose of the chapter is to describe an AI system for optimal control and design of drip irrigation system based on modern mathematical methods. The novelty of the proposed approach is that it is the first attempt to optimize drip irrigation system using the linearization via the Kirchhoff transformation. The contribution of the chapter is that it describes the effectiveness of the holistic AI approach to the design and control of drip irrigation systems for sustainable agriculture and environment. The scope of the future work it to introduce the impulse control in time and optimization the pipe functioning in the scale of an irrigation module at whole.

Keywords Sustainable environment · Sustainable agriculture · Drip irrigation · Optimal control · Artificial intelligence · Richards–Klute equation

D. Klyushin (✉) · A. Tymoshenko
Faculty of Computer Science and Cybernetics, Taras Shevchenko National University of Kyiv, Akademika Glushkova Avenu, 4D, Kyiv 03680, Ukraine
e-mail: dokmed5@gmail.com

© The Editor(s) (if applicable) and The Author(s), under exclusive license to Springer Nature Switzerland AG 2021

A. E. Hassanien et al. (eds.), *Artificial Intelligence for Sustainable Development: Theory, Practice and Future Applications*, Studies in Computational Intelligence 912,
https://doi.org/10.1007/978-3-030-51920-9_1

1 Introduction

Drip irrigation is one of the most effective watering methods which provide sustainability of the agriculture and environment [1]. These systems save water and allow optimal control of soil water content and plant growing. They open wide possibilities for using smart technologies including various sensors for measuring moisture content of soil and pressure in pipes [2–4]. The schematic design of an irrigation module is presented in Fig. 1.

The structure of the optimizing system for automatic control of drip irrigation consists of the following components:

1. Managed object which is an irrigation module or a group of irrigation modules which are turned on at the same time.
2. Sensors measuring soil moisture at the module location site.
3. A device for generating control commands (processor).
4. Actuators (valves on-off modules).

The purpose of the drip irrigation design and control system is to determine the optimal parameters, as well as to generate and transfer control actions to actuators to optimize the watering regime based on the operational control of soil moisture using sensors and a priori information on the intensity of moisture extraction by plant roots (Fig. 2).

The main tool for determining the optimal parameters of the drip irrigation system is the emitter discharge optimization algorithm, which provides the required irrigation mode depending on the level of moisture.

To work out the optimal control action, an algorithm for calculating the optimal irrigation schedule [5] which determines the order of switching on the modules,

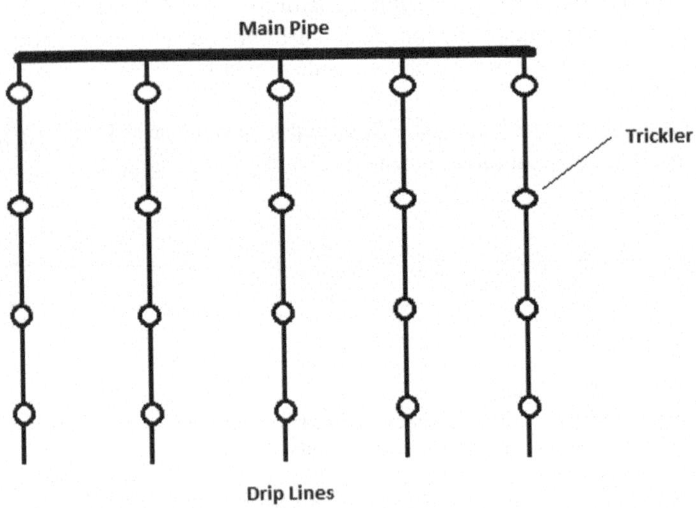

Fig. 1 Design of irrigation module

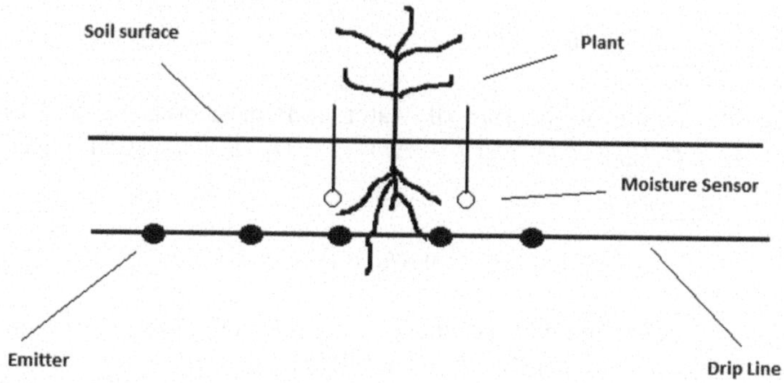

Fig. 2 Buried sources and sensors

the duration of the next watering and the planned date for the next watering for each module, and an algorithm for optimization of the emitter discharge are used. The criterion for optimizing the irrigation schedule is to minimize economic losses by minimizing the total delay of irrigation relative to the planned dates taking into account the priorities of the modules and maximizing the system utilization factor (minimizing the total downtime).

Based on this, the problem of controlling the drip irrigation system at a given point in time t can be formulated as follows:

1. Determine the priority irrigation module (or group of modules).
2. Determine the start and end moment of watering on the priority module (group of modules).
3. Determine planned time for the start of the next watering on each of the modules.
4. Determine the total delay of the schedule.
5. Determine the economic efficiency of the schedule (the value of yield losses due to watering delays and system downtimes).

At the initial time, the control command generation device (processor) generates a polling signal for sensors that measure soil moisture in irrigated areas. As a result of the survey, a vector $\Theta(t_0) = (\theta_1(t_0), \theta_2(t_0), \ldots, \theta_N(t_0))$ is formed containing information about the moisture values in N sections at a time t_0, and a vector $W(t_0) = (w_1(t_0), w_2(t_0), \ldots, w_N(t_0))$ consisting of the values of the volume of moisture reserve in each cite. The volume of moisture is calculated for every cites.

After a specified period of time Δt, which determines the discreteness of the survey of sensors, the water content vector $\Theta(t_0 + \Delta t) = (\theta_1(t_0 + \Delta t), \theta_2(t_0 + \Delta t), \ldots, \theta_N(t_0 + \Delta t))$ and the moisture storage volume vector $W(t_0 + \Delta t) = (w_1(t_0 + \Delta t), w_2(t_0 + \Delta t), \ldots, w_N(t_0 + \Delta t))$ are formed similarly, containing the measurement results at a time $t_0 + \Delta t$. This information is enough to calculate the vector $V(t_0 + \Delta t) = (v_1(t_0 + \Delta t), v_2(t_0 + \Delta t), \ldots, v_N(t_0 + \Delta t))$ of the rate of decreasing of water content at ith cite, where is the rate $v_i(t_0 + \Delta t)$ of decreasing water content at the ith cite, is determined by the formula

$$v_i(t_0 + \Delta t) = \frac{\theta_i(t_0 + \Delta t) - \theta_i(t_0)}{\Delta t}. \tag{1}$$

Knowing the vector of the rate of water content decreasing and the vector $D(t_0 + \Delta t) = (D_1(t_0 + \Delta t), D_2(t_0 + \Delta t), \ldots, D_N(t_0 + \Delta t))$ of water balance deficit where

$$D_i(t_0 + \Delta t) = w_i(t_0 + \Delta t) - w_i(t_0), \tag{2}$$

we can calculate the estimated time of water content decreasing at the ith cite to a critical level (planned term for the next watering) by formula

$$T_i^* = \frac{\theta_i(t_0 + \Delta t) - \theta_i^*(t_0)}{v_i(t_0 + \Delta t)}, \tag{3}$$

as well as the duration of irrigation required to compensate for the water balance deficit

$$P_i^* = \frac{D_i(t_0 + \Delta t)}{Q_i}, \tag{4}$$

where Q_i is the discharge rate at the ith cite, which we will consider to be a constant value obtained by the algorithm of emitter discharge optimization.

Now, on the basis of the available information, it is possible to determine the optimal irrigation schedule that determines the order of inclusion of irrigation modules, taking as a quality criterion the minimum delay in irrigation relative to the planned period, taking into account the priorities of irrigated crops.

Moisture transfer in unsaturated soil with point sources is an object of various researches. This process is simulated using either computer models or analytical solutions. The point source is assumed dimensionless, the atmosphere pressure and temperature of soil matrix are considered be constant. The effectiveness of the methods is estimated by their accuracy, flexibility and complexity [6, 7]. Computer simulation allows using of real attributes of porous media based on Richards-Klute equation [8]. But the stability of the methods is not guaranteed because of quasi-linearity of the problem [9].

As a rule, computer simulations are based on the finite differences [10] or finite elements methods [11, 12]. To reduce the problem to a linear case, the Kirchhoff transformation is used [13, 14]. This allows using method for solving a linearized problem. However, the optimization problem is still unsolved and considered only in the context of the identification of distributed soil parameters [15] but not discharge of a point source.

For simulation and optimal control of fluid flow through the soil, a variational method is proposed [16, 17]. The Kirchhoff transformation allows reducing the model to a linear initial-boundary problem and use the finite difference method. Mathematical correctness of this approach is considered in [18]. Algorithmic aspects of the used computational methods described in [19].

The purpose of the chapter is to provide a holistic view of the problem of optimal control for a drip irrigation model based on the variational algorithm of determination the optimal point source discharge for in a porous medium. This approach demonstrates the advantages of AI approach to design and control of a drip irrigation system for sustainable agriculture and environment.

2 Mathematical Model

Consider a two-dimensional problem of fluid flow filtration throughout unsaturated soil with zero initial water content and a given water content at the target time moment. This problem is described by the Richards–Klute equation:

$$\frac{\partial \omega}{\partial t} = \frac{\partial}{\partial x}\left[K_x(\omega)\frac{\partial H}{\partial x}\right] + \frac{\partial}{\partial y}\left[K_y(\omega)\frac{\partial H}{\partial y}\right] + \sum_{j=1}^{N} Q_j(t)\delta(x-x_j)\times\delta(y-y_j), \quad (5)$$

$$\begin{aligned} &(x, y, t) \in \Omega_0 \times (0, T),\\ &\omega|_{x=0} = 0;\ \omega|_{x=L_1} = 0;\\ &\omega|_{y=0} = 0;\ \omega|_{y=L_2} = 0;\\ &\omega(x, y, 0) = 0,\ (x, y) \in \overline{\Omega}_0. \end{aligned} \quad (6)$$

Here, $H = \psi(\omega) - y$ is the piezometric head, $\psi(\omega)$ is the hydrodynamic potential, $D_y(\omega) = K_y(\omega)\frac{d\psi}{d\omega}$ is the diffusiveness along the axis y, $\Omega_0 = [(x, y) : 0 < x < L_1, 0 < y < L_2]$ is the rectangular region, $y = y_0$ is the plane at the ground surface level (Oy is the vertical axis taken downward). It is assumed that $K_x(\omega) = k_1 k(\omega)$, $K_y(\omega) = k_2 k(\omega)$, where k_1, k_2 are filtration coefficients along Ox, Oy, $k(\omega)$ is the function depending of the water content in the soil. Let us simplify the evaluations, setting $k_1 = k_2$, $L_1 = L_2 = 1$. To make the differential equation dimensionless, let us add variables [20]:

$$\beta_2 = 0,5\ell,\ \beta_1 = \sqrt{\tfrac{k_2}{k_1}}\beta_2,\ \alpha = \tfrac{\langle D_y\rangle\beta_2^2}{T},$$
$$\xi = \tfrac{\beta_1}{L_1}x,\ \zeta = \tfrac{\beta_2}{L_2}y,\ \tau = \alpha t.$$

where $\langle D_y\rangle$ is the average value of D_y, ξ represent dimensionless width coordinate and ζ stands for dimensionless depth coordinate.

Then, we apply the Kirchhoff transformation [20]:

$$\Theta = \frac{4\pi k_1}{Q^* k_2 \beta_2}\int_{\omega_0}^{\omega} D_y(\omega)d\omega$$

where Q^* is the scale multiplier. It is supposed that the following conditions are met:

- $\Theta(\omega)$ and $K_y(\omega)$ have linear relationship: $D_y^{-1}(\omega)\frac{\mathrm{d}K_y(\omega)}{\mathrm{d}\omega} = \ell = \text{const}$ and
- $\frac{\partial\omega}{\partial t} = \frac{k_2\beta_2 Q^*}{4\pi k_1}\frac{1}{D_y(\omega)}\frac{\partial\Theta}{\partial t} \simeq \frac{k_2\beta_2^3 Q^*}{4\pi k_1}\frac{\partial\Theta}{\partial\tau}$.

To make the Eq. (5) dimensionless we need for the additional variable $q_j = \frac{Q_j}{Q^*}$ that is the scaled source point discharge. Hereinafter, Ω, Γ are dimensionless equivalents of Ω_0, Γ_0, where Γ_0 is the boundary of Ω_0.

In this case, we may to reformulate the problem (5), (6) as

$$\frac{\partial\Theta}{\partial\tau} = \frac{\partial^2\Theta}{\partial\xi^2} + \frac{\partial^2\Theta}{\partial\zeta^2} - 2\frac{\partial\Theta}{\partial\zeta} + 4\pi\sum_{j=1}^{N} q_j(\tau)\delta(\xi-\xi_j)\times\delta(\zeta-\zeta_j),\ (\xi,\zeta,\tau)\in\Omega\times(0,1], \quad (7)$$

$$\begin{gathered}\Theta\big|_{\xi=0} = 0;\ \Theta\big|_{\xi=1} = 0;\\ \Theta\big|_{\zeta=0} = 0;\ \Theta\big|_{\zeta=1} = 0;\ (\xi,\zeta,\tau)\in\Gamma\times[0,1].\\ \Theta(\xi,\zeta,0) = 0,\ (\xi,\zeta)\in\Omega.\end{gathered} \quad (8)$$

The points r_j, $j = \overline{1,N}$, define the location of the point sources with discharges $q_j(\tau)$. The target water content values $\varphi_m(\tau)$ are averaged values of $\Theta(\xi,\zeta,\tau)$ in the small area ω_m around the given points $(\xi_m,\zeta_m)\in\Omega, m = \overline{1,M}$ (sensors). The purpose is to find $q_j(\tau)$, $j = \overline{1,N}$, minimizing the mean square deviation of $\Theta(\xi_m,\zeta_m,\tau)$ from $\varphi_m(\tau)$ by the norm of $L_2(0,1)$.

Assume that the optimal control belongs to the Hilbert space $(L_2(0,1))^N$ with the following inner product

$$\langle X,Y\rangle = \sum_{j=1}^{N}\int_0^1 x_j(\tau)y_j(\tau)d\tau.$$

Then the cost functional is

$$J_\alpha\left(\overline{Q}\right) = \sum_{m=1}^{M}\int_0^1\left(\varphi_m(\tau) - \int_\Omega g_m(\xi,\zeta)\Theta(\xi,\zeta,\tau)d\Omega\right)^2 d\tau + \alpha\left\|\overline{Q}\right\|^2, \quad (9)$$

where $\overline{Q}(\tau) = (q_1(\tau),\ldots,q_N(\tau))^T$ is the control vector, $g_m(\xi,\zeta) = \frac{\chi_{\omega_m}}{\mathrm{diam}\,\omega_m}$ is the averaging core in ω_m, χ_{ω_m} is the indicator function, $\alpha > 0$ is the regularization parameter. The vector of optimal discharges of the point sources $\overline{Q}^*$ minimizes the cost functional:

$$J_\alpha\left(\overline{Q}^*\right) = \min_{q\in(L_2(0,1))^N} J_\alpha\left(\overline{Q}\right). \quad (10)$$

The existence and uniqueness of the solution of the similar problem were proved in [21–25]. The conditions providing existence and uniqueness of the problem (7)–(10) are established in [18].

3 Algorithm

The problem (7)–(10) is solved using the following iterative algorithm [17].

1. Solve the direct problem.

$$L\Theta^{(k)} \equiv \frac{\partial \Theta^k}{\partial \tau} - \frac{\partial^2 \Theta^k}{\partial \xi^2} - \frac{\partial^2 \Theta^k}{\partial \zeta^2} + 2\frac{\partial \Theta^k}{\partial \zeta} = 4\pi \sum_{j=1}^{N} q_j(\tau)\delta(\xi - \xi_j) \times \delta(\zeta - \zeta_j);$$

$$0 < \tau \leq 1;\ \Theta^k(0) = 0; \tag{11}$$

2. Solve the conjugate problem.

$$L^*\Psi^{(k)} \equiv -\frac{\partial \Psi^k}{\partial \tau} - \frac{\partial^2 \Psi^k}{\partial \xi^2} - \frac{\partial^2 \Psi^k}{\partial \zeta^2} - 2\frac{\partial \Psi^k}{\partial \zeta} = 2\left(\Theta^k - \varphi(\tau)\right);\ 0 \leq \tau < 1,\ \Psi^{(k)}(1) = 0; \tag{12}$$

3. Evaluate the new point source discharge approximation.

$$\frac{Q^{(k+1)} - Q^{(k)}}{\tau_{k+1}} + \Psi^{(k)} + \alpha Q^{(k)} = 0, \quad k = 0, 1, \ldots.$$

For solving the direct problem, the implicit numerical scheme was used. The region $0 \leq \xi, \zeta \leq 1$ was partitioned with a step $h = \frac{1}{30}$ and the time interval was partitioned using time steps $\tilde{\tau} = \frac{1}{100}$ for $0 \leq \tau \leq 1$.

The resulting system of linear finite-difference equations which approximate the problem (7), (8) has the following form:

$$\Lambda(\xi, \zeta) = \Lambda_1(\xi) + \Lambda_2(\zeta) = \frac{\partial \Theta}{\partial \tau} - \left(\varphi(\xi, \zeta) + \frac{1}{h}\varphi_1(\xi) + \frac{1}{h}\varphi_2(\zeta)\right)$$

Accounting the boundary conditions, we have

$$\Lambda_1(\xi) = \begin{cases} 0, \ \xi = 0 \\ \left(\Theta_{\bar{\xi}}\right)_{\xi}, \ 0 < \xi < 1 \\ 0, \ \xi = 1 \end{cases}, \Lambda_2(\zeta) = \begin{cases} 0, \ \zeta = 0 \\ \left(\Theta_{\bar{\zeta}}\right)_{\zeta} - 2\Theta_{\hat{\zeta}}, \ 0 < \zeta < 1 \\ 0, \ \zeta = 1 \end{cases}. \quad (13)$$

Here $\hat{y}$ is the central finite-difference derivative. Due to the boundary conditions, $\varphi_1(\xi) = \varphi_2(\zeta) = 0$. To solve this system the Jacobi method is used.

At the last step, the accuracy of the approximation of the optimal discharge depends on the regularization parameter, which is chosen according to computational errors. The error order of this approximation is $O(h^2)$ for space steps and $O(\tau)$ for time steps.

The iterative algorithm stops according to the following conditions.

1. The average absolute value of the difference between the current and previous approximations of Θ is less 10^{-7}.
2. The number of iterations is more that a threshold established in advance (for example 1000).

4 Simulation

The target function is taken as a result of modelling with the initially chosen optimal emitter discharge 10. For models with several sources we also assume $q = 10$ and calculate the function according to that value. Iterations start from zero source discharge approximation. There are different positions for the point source: near the top left corner, near the middle of the top boundary, near the middle of the left boundary and in the center are considered (Figs. 3, 4, 5 and 6). The right-hand side of equation was the following:

$$\varphi(\xi,\zeta) = \begin{cases} 4\pi q, \ \xi = \frac{7}{30}, \ \zeta = \frac{7}{30}; \\ 0, \quad \text{else,} \end{cases} \quad \varphi(\xi,\zeta) = \begin{cases} 4\pi q, \ \xi = 0,5, \ \zeta = \frac{7}{30}; \\ 0, \quad \text{else,} \end{cases}$$

$$\varphi(\xi,\zeta) = \begin{cases} 4\pi q, \ \xi = \frac{7}{30}, \ \zeta = 0,5; \\ 0, \quad \text{else} \end{cases} \quad \varphi(\xi,\zeta) = \begin{cases} 4\pi q, \ \xi = 0,5, \ \zeta = 0,5; \\ 0, \quad \text{else.} \end{cases}$$

The deviation of the computed source discharge from the optimal was less than 2% when the regularization parameter was chosen equal to 10^{-7}. The isolines of dimensionless water content for these four tests are shown below. As ζ we denoted depth, so the top of our area is $\zeta = 0$ and the bottom is $\zeta = 1$ in order not to get confused, the figures are named according to space coordinates. Table 1 demonstrates necessary number of iterations of the variational algorithm to achieve 98% accuracy of the optimal discharge for various finite-difference schemes (two- and three-layered) and step sizes. The optimization was done either by comparing dimensionless water content during all time, or by minimizing the difference at the last time moment.

Also, three possible source locations were tested (Figs. 7, 8 and 9). In case of horizontal symmetry, two point sources were used (Fig. 7):

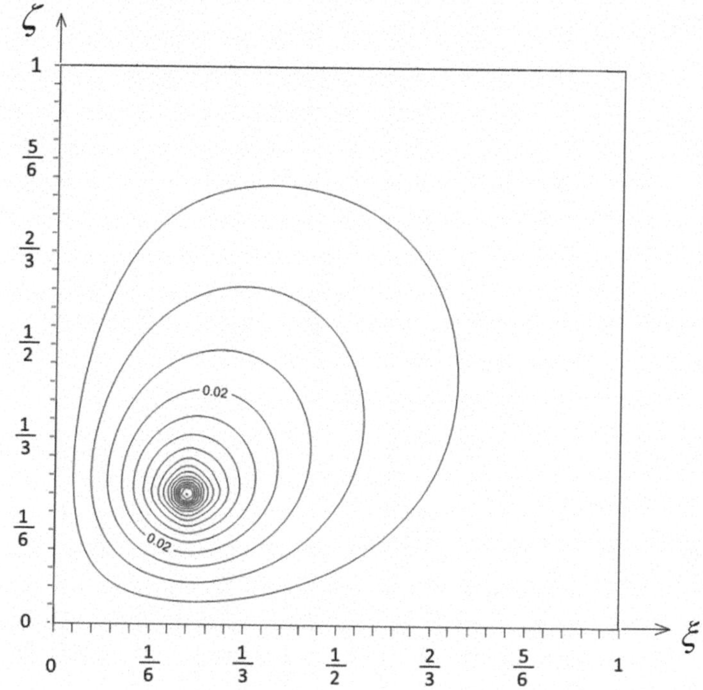

Fig. 3 Corner source

$$\varphi(\xi,\zeta)=\begin{cases}4\pi q,\ \xi=\frac{7}{30},\ \zeta=\frac{7}{30}\text{ AND }\xi=\frac{23}{30},\ \zeta=\frac{23}{30};\\ 0, \qquad\qquad \text{else,}\end{cases}$$

providing humidification with central priority. The optimal discharge was taken constant to guarantee symmetry.

In case of vertical placement, one source was placed near the top and another at the center (Fig. 8):

$$\varphi(\xi,\zeta)=\begin{cases}4\pi q,\ \xi=\frac{1}{2},\ \zeta=\frac{7}{30}\text{ AND }\xi=\frac{1}{2},\ \zeta=\frac{1}{2};\\ 0, \qquad\qquad \text{else,}\end{cases}$$

This placement lead to more humidification for the top-central part.

Finally, the triangular humidification model was tested with the central emitter having half of usual discharge (Fig. 9).

$$\varphi(\xi,\zeta)=\begin{cases}4\pi q,\xi=\frac{7}{30},\zeta=\frac{7}{30}\text{ and }\xi=\frac{23}{30},\zeta=\frac{7}{30},\\ 4\pi\frac{q}{2},\xi=\frac{1}{2},\zeta=\frac{1}{2},\\ 4\pi\frac{q}{2},\xi=\frac{1}{2},\zeta=\frac{1}{2}.\end{cases}$$

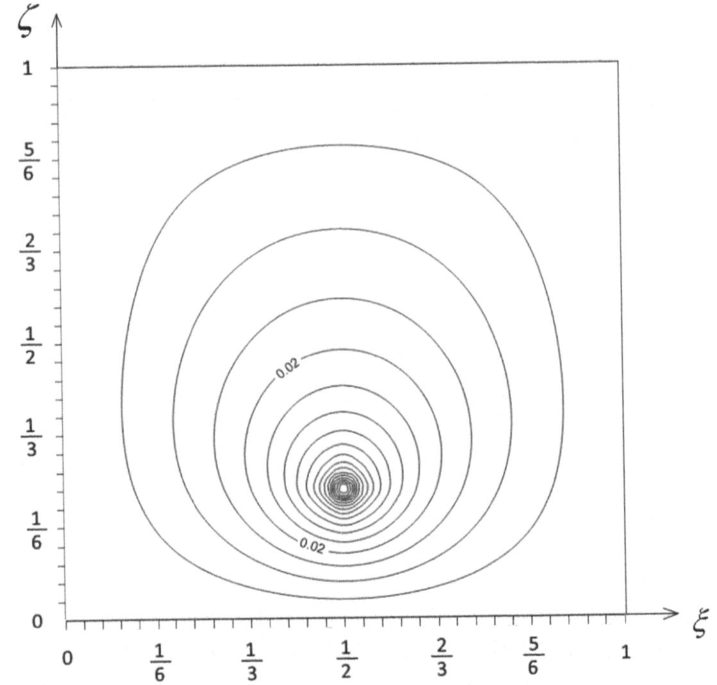

Fig. 4 Top central source

In all these cases, the variational algorithm led to accuracy improvement from the initial approximation of discharge to new received values for each source. Thus, the offered method demonstrated high accuracy and stability in defining the optimal source discharge for several options of source placement. The regularization parameter was chosen with respect to calculation errors and received values of theta.

Thus, in all cases the minimum of the cost functional was achieved with the precision not less than 98%. The rate of convergence is defined by the number of iterations required for such accuracy (Table 1). Therefore, this mathematical approach may be successfully used as a base for development of an AI system for design and optimal control of drip irrigation systems providing sustainable agriculture and environment.

5 Conclusion

The AI approach for design and optimal control of drip irrigation system is proposed. It is based on simulation of the water transport process described by Richards-Klute equation. The simulation shows the effectiveness of the Kirchhoff transformation for reducing the original quasi-linear problem to the linear problem of optimal control of non-stationary moisture transport in unsaturated soil. It is demonstrated the accuracy

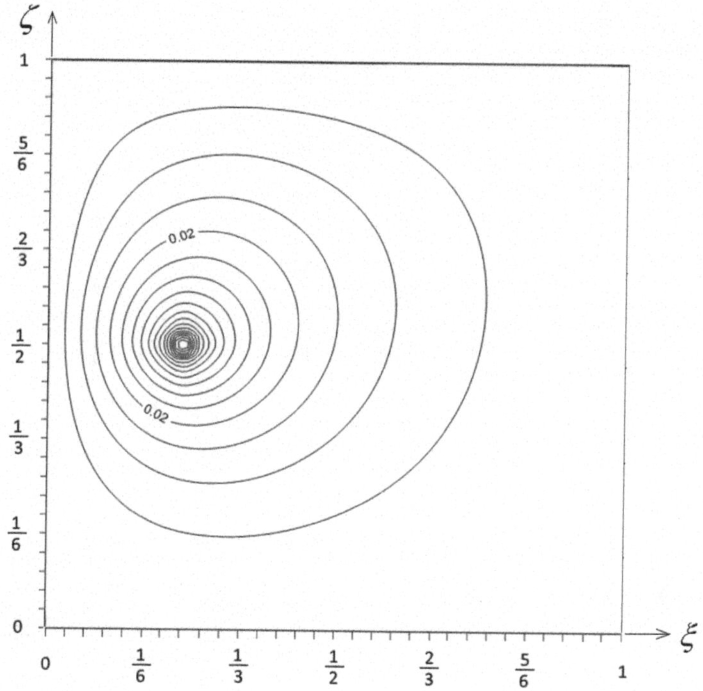

Fig. 5 Source near the left boundary

and effectiveness of the proposed variational algorithm for minimization of a cost functional based on the finite-difference method used to solve direct and conjugate problems. The proposed approach may be used as a base for effective system for design and optimal control of drip irrigation systems for sustainable agriculture and environment. In the future, this approach may be expanded for the impulse control in time and optimization the pipe functioning in the scale of an irrigation module at a whole.

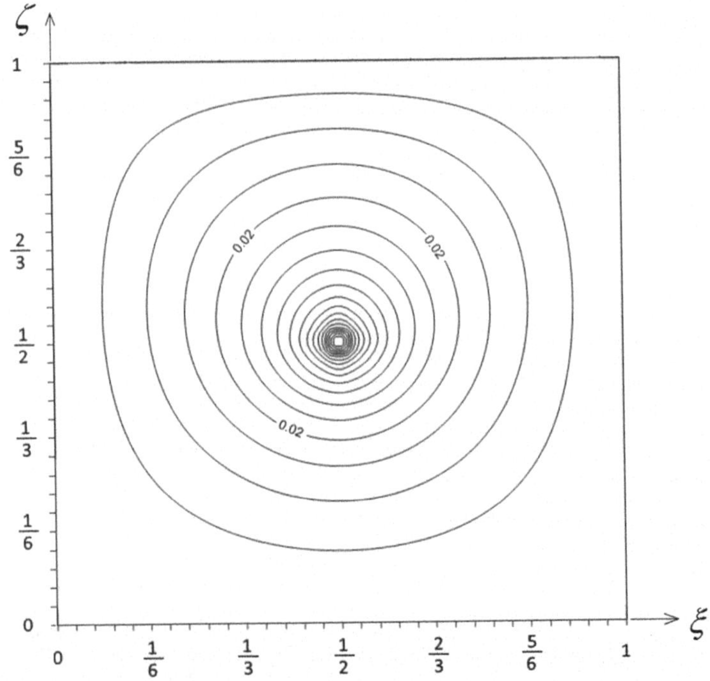

Fig. 6 Source in the center

Table 1 Number of iterations

Scheme	Step	Comparisons	Iterations
Two-layered	10	All time	2552
Two-layered	100	All time	254
Three-layered	100	At last moment	219

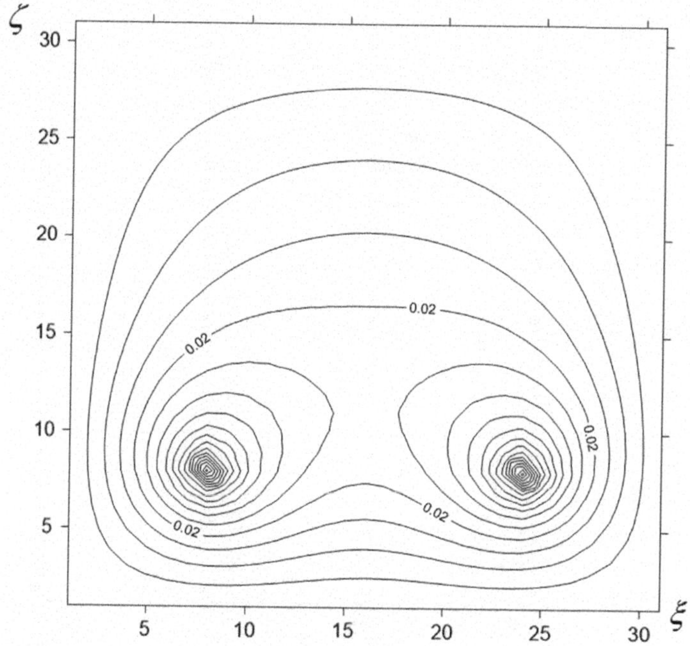

Fig. 7 Sources granting horizontal symmetry

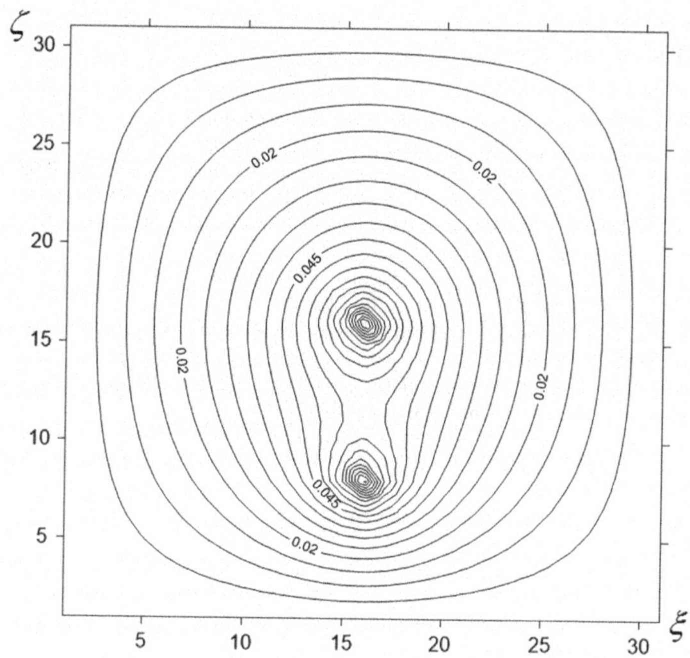

Fig. 8 Sources focused on central humidification

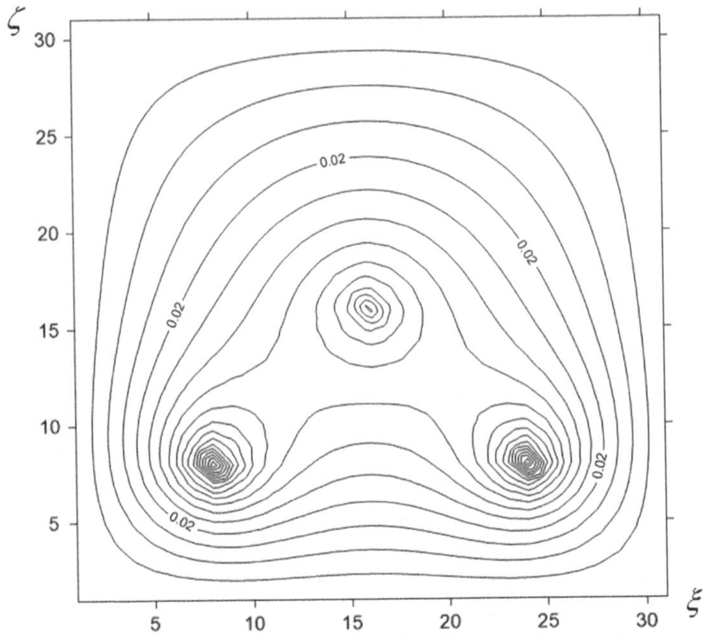

Fig. 9 Sources forming a triangle

References

1. M.R. Goyal, P. Panigrahi, *Sustainable Micro Irrigation Design Systems for Agricultural Crops: Methods and Practices* (Apple Academic Press, Oakville, ON, 2016)
2. J. Kirtan, D. Aalap, P. Poojan, Intelligent irrigation system using artificial intelligence and machine learning: a comprehensive review. Int. J. Adv. Res. **6**, 1493–1502 (2018)
3. A. Gupta, S. Mishra, N. Bokde, K. Kulat, Need of smart water systems in India. Int. J. Appl. Eng. Res. **11**(4), 2216–2223 (2016)
4. M. Savitha, O.P. UmaMaheshwari, Smart crop field irrigation in IOT architecture using sensors. Int. J. Adv. Res. Comput. Sci. **9**(1), 302–306 (2018)
5. R.W. Conway, W.L. Maxwell, L.W. Miller, *Theory of Scheduling* (Dover Publications, Mineola, New York, 2003)
6. S.P. Friedman, A. Gamliel, Wetting patterns and relative water-uptake rates from a ring-shaped water source. Soil Sci. Soc. Am. J. **83**(1), 48–57 (2019)
7. M. Hayek, An exact explicit solution for one-dimensional, transient, nonlinear Richards equation for modeling infiltration with special hydraulic functions. J. Hydrol. **535**, 662–670 (2016)
8. M. Farthing, F.L. Ogden, Numerical solution of Richards' equation: a review of advances and challenges. Soil Sci. Soc. Am. J. **81**(6), 1257–1269 (2017)
9. Y. Zha et al., A modified picard iteration scheme for overcoming numerical difficulties of simulating infiltration into dry soil. J. Hydrol. **551**, 56–69 (2017)
10. F. List, F. Radu, A study on iterative methods for solving Richards' equation. Comput. Geosci. **20**(2), 341–353 (2015)
11. D.A. Klyushin, V.V. Onotskyi, Numerical simulation of 3D unsaturated infiltration from point sources in porous media. J. Coupled Syst. Multiscale Dyn. **4**(3), 187–193 (2016)

12. Z.-Y. Zhang et al., Finite analytic method based on mixed-form Richards' equation for simulating water flow in vadose zone. J. Hydrol. **537**, 146–156 (2016)
13. H. Berninger, R. Kornhuber, O. Sander, Multidomain discretization of the Richards equation in layered soil. Comput. Geosci. **19**(1), 213–232 (2015)
14. I.S. Pop, B. Schweizer, Regularization schemes for degenerate Richards equations and outflow conditions. Math. Model. Methods Appl. Sci. **21**(8), 1685–1712 (2011)
15. R. Cockett, L.J. Heagy, E. Haber, Efficient 3D inversions using the Richards equation. Comput. Geosci. **116**, 91–102 (2018)
16. P. Vabishchevich, Numerical solution of the problem of the identification of the right-hand side of a parabolic equation. Russ. Math. (Iz. VUZ) **47**(1), 27–35 (2003)
17. S.I. Lyashko, D.A. Klyushin, V.V. Semenov, K.V. Schevchenko, Identification of point contamination source in ground water. Int. J. Ecol. Dev. **5**, 36–43 (2006)
18. A. Tymoshenko, D. Klyushin, S. Lyashko, Optimal control of point sources in Richards-Klute equation. Adv. Intel. Syst. Comput. **754**, 194–203 (2019)
19. E.A. Nikolaevskaya, A.N. Khimich, T.V. Chistyakova, Solution of linear algebraic equations by gauss method. Stud. Comput. Intell. **399**, 31–44 (2012)
20. D.F. Shulgin, S.N. Novoselskiy, Mathematical models and methods of calculation of moisture transfer in subsurface irrigation, *Mathematics and Problems of Water Industry* (Naukova Dumka, Kiev, 1986), pp. 73–89. (in Russian)
21. S.I. Lyashko, D.A. Klyushin, V.V. Onotskyi, N.I. Lyashko, Optimal control of drug delivery from microneedle systems. Cybern. Syst. Anal. **54**(3), 1–9 (2018)
22. S.I. Lyashko, D.A. Klyushin, D.A. Nomirovsky, V.V. Semenov, Identification of age-structured contamination sources in ground water, in *Optimal Control of Age-Structured Populations in Economy, Demography, and the Environment*, ed. by R. Boucekkline, et al. (Routledge, London, New York, 2013), pp. 277–292
23. S.I. Lyashko, D.A. Klyushin, L.I. Palienko, Simulation and generalized optimization in pseudohyperbolical systems. J. Autom. Inf. Sci. **32**(5), 108–117 (2000)
24. S.I. Lyashko, Numerical solution of pseudoparabolic equations. Cybern. Syst. Anal. **31**(5), 718–722 (1995)
25. S.I. Lyashko, Approximate solution of equations of pseudoparabolic type. Comput. Math. Math. Phys. **31**(12), 107–111 (1991)

Artificial Intelligent System for Grape Leaf Diseases Classification

Kamel K. Mohammed, Ashraf Darwish, and Aboul Ella Hassenian

Abstract In this paper, we built up an artificially intelligent technique for grape foliage disease detection and classification. The suggested method comprises four stages, including enhancement, segmentation, extraction of textural features, and classification. The stretch-based enhanced algorithm has been adapted for image enhancement. Then the method of grouping k-means is used for fragmentation. Textural features are extracted from the segmented grape foliage using texture features. Finally, two classifiers including multi Support Vector Machine and Bayesian Classifier are proposed to determine the type of grape foliage disease. The dataset consists of 400 grape foliage images in total divided into 100 Grape leaf Black rot, 100 Grape leaf Esca (Black Measles), 100 Grape Leaf blight (Isariopsis Leaf Spot) and 100 Grape leaf healthy. The data divided into 40 samples for the testing phase and 360 samples for the training phase. The experiment results evaluated by four parameters including accuracy, sensitivity, specificity, and precision. The proposed approach yielded the best performance with the highest average accuracy of 100%, the sensitivity of 100% and specificity of 100% by the testing dataset for the multi Support Vector Machine classifier and 99.5% for the training dataset.

Keywords Artificial intelligence · Grape leaf diseases · Classification

K. K. Mohammed (✉)
Center for Virus Research and Studies, al-Azhar University, Cairo, Egypt
e-mail: tawfickamel@gmail.com
URL: http://www.egyptscience.net

A. Darwish
Faculty of Science, Helwan University, Helwan, Egypt

A. E. Hassenian
Faculty of Computer and Artificial Intelligence, Cairo University, Cairo, Egypt

© The Editor(s) (if applicable) and The Author(s), under exclusive license to Springer Nature Switzerland AG 2021

A. E. Hassanien et al. (eds.), *Artificial Intelligence for Sustainable Development: Theory, Practice and Future Applications*, Studies in Computational Intelligence 912,
https://doi.org/10.1007/978-3-030-51920-9_2

1 Introduction

The worldwide economy relies heavily on the productivity of agriculture. The identity of plant illness performs a main position within the agricultural area. On the off hazard that sufficient plant care isn't always taken, it makes extreme plant effects and influences the sum or profitability of the relating thing. A dangerous place of plant leaves is the place on a leaf this is stricken by the disease, to reduce the first-rate of the plant. The automated disorder detection method is useful at the preliminary degree for detecting sickness. The present approach of detecting disease in plants is professional naked eye commentary. This requires a huge crew of experts and continuous monitoring of the plant, which for massive farms prices very high. Farmers in a few locations do not have good enough gadgets or even the idea of contracting experts, because of which consulting specialists even cost excessive and it is time-consuming too. In such conditions, the suggested method is useful for monitoring massive fields of plants. Automatically detecting illnesses utilizing simply looking on the signs and symptoms on leaves makes it less complicated and cost-powerful. This offers aid for system vision to present photo-primarily based automatic procedure manage, robotic steerage. The detection of plant disease by the visible way is hard as well as less correct. where as automated disease detection is used then it's going to give extra accurate effects, within less time and fewer efforts. Image segmentation can be done in numerous manners ranging from a simple threshold method to an advanced shade photo segmentation approach. This corresponds to something that the human eye can without problems separate and consider as an individual object. Recent traditional techniques are not able to recognize the objects in acceptable accuracy. For example, Authors in [1] have built up recognition and categorization of grape foliage illnesses utilizing Artificial Neural Networks (ANN). The framework comprises of leaf image as info and threshold extended to cover green pixels. An anisotropic dispersion utilized to eliminate noise. After that by utilizing K-means grouping grape foliage illness separation is done. Utilizing ANN, the unhealthy grape section identified. In [2] a correlation of the effect recognizes different types of color space in the disorder blot method. In [2] a relationship between the impact of various kinds of color space in the method of blot illness. All color methods (CIELAB, HSI, and YCbCr) looked at lastly A segment for CIELAB color model is utilized. At long last, by utilizing the Otsu technique on the color segment, the threshold can be determined. In [3] authors gave quick and exact determination and categorization of plant illness. In this technique, K-means grouping utilized for separation disorder blots on plant foliage, and ANN is utilized as a categorization utilizing some texture feature set. Above mentioned suffering from precisely describe grape leaf disease images with many feature extractions. Texture analysis approaches were commonly used to examine photographs of grape leaf disease because they provide information about the spatial arrangement of pixels in the image of the grape leaf disease the texture is one of the major grape leaf disease image characteristics for classification. Therefore, we extract and use 47 texture features for the analysis of grape leaf disease images.

2 Materials and Methods

2.1 K-Means Algorithm for Fragmentation

The K-means grouping is a method that divides the group data in the image into one cluster based on the similarity of features about them and each group data are dissimilarity. The grouping is completed by reducing the group data and the respective centroid group. Mathematically, due to a lot of specimens (s_1, s_2, …, s_n), where every specimen has an actual d-dimensional vector, k-means grouping divisions the m specimens in $k(\leq m)$ specimens $S = \{S_1, S_2, \ldots, S_n\}$ to reduce the number of squares within-cluster. The goal, then, is to find $\arg_s \min \sum_{i=1}^{k} \sum x \in S_i \|x - \mu_i\|^2$. Here μ_i is the mean of points in S_i [4].

2.2 Multiclass Support Vector Machine Classifier

It is a supervised learning classifier. The training section of the SVM technique is to find out the hyperplane with the largest margin which separates the more dimensional characteristic feature gap with fewer error in classification. The suggested technique needs multi-class categorization due to four groups of grape foliage illnesses is considered. There are 2 techniques for multi-class SVM:

- One-against-one: Several binary classifiers are combined [5].
- One-against-all: It takes all the data into account at the same time [5].

In the proposed system, the one-against-all method is utilized for multi-class categorization.

The one-against-all technique is likely the oldest used implementation for multi-class classification of SVM. It forms k SVM models, where k is the number of classes. The mth SVM is educated for all cases of positive labels in the mth class and all those for negative labels. Thus the l training data are given (x_1, y_1), …, (x_1, y_l), where $x_i \in \Re^n$, $i = 1, \ldots, l$, and $y_i \in \{1, \ldots, k\}$ is the category of x_i, the mth SVM explains the next issue:

$$\min_{W^m, b^m, \varepsilon^m} \; 1/2(W^m)^T W^m + C \sum_{i=1}^{l} \varepsilon_i^m$$

$$\begin{aligned} (W^m)^T \phi(x_i) + b^m &\geq 1 - \varepsilon_i^m, \quad \text{if } y_i = m \\ (W^m)^T \phi(x_i) + b^m &\geq 1 - \varepsilon_i^m, \quad \text{if } y_i \neq m \\ \varepsilon_i^m &\geq 0, i = 1, \ldots, l, \end{aligned} \tag{1}$$

The penalty parameter is where the training data x_i is represented into a more dimensional space by the function Ø and C.

Minimizing $1/2(w^m)^T w^m$ means optimizing $2/\| W^m \|$, the difference between two classes of data. If the data is not linearly separable, there is a penalty term $C \sum_i^l \varepsilon_i^m$ can be used to reduce the number of training errors. The core idea of SVM is to find a balance between the regularization term $1/2(w^m)^T w^m$ and the training errors. After solving (1), there are k decision functions:

$$\left(w^1\right)^T Ø(x_i) + b^1$$

$$\left(w^k\right)^T \dot{Ø}(x_i) + b^k$$

We assume that x is in the cluster with the highest decision function value:

$$\text{class of } x \ \arg \max_{m=1,\dots,k} \left(\left(w^m\right)^T Ø(x) + b^m \right). \tag{2}$$

The dual problem of Eq. (1) for the same number of variables as the number of data in (1) is solved. k, l-variable problems of quadratic programming are solved [5].

3 The Proposed Artificial Intelligent Based Grape Leaf Diseases

Figure 1 shows the architecture of the proposed plant leaf disease detection system.

3.1 Dataset Characteristic

The dataset taken from Kaggle-dataset [6] which contains the plant's disease images the dataset consisted of 400 grape foliage images. We have trained the proposed classifier using 360 images divided into 90 Grape Black rot, 90 Grape Esca (Black Measles), 90 Grape Leaf blight (Isariopsis Leaf Spot), and 90 Grape healthy as shown in Fig. 2. Additionally, we have tested our classifier using 40 images divided into 10 Grape Black rot, 10 Grape Esca (Black Measles), 10 Grape Leaf blight (Isariopsis Leaf Spot) and 10 Grape healthy.

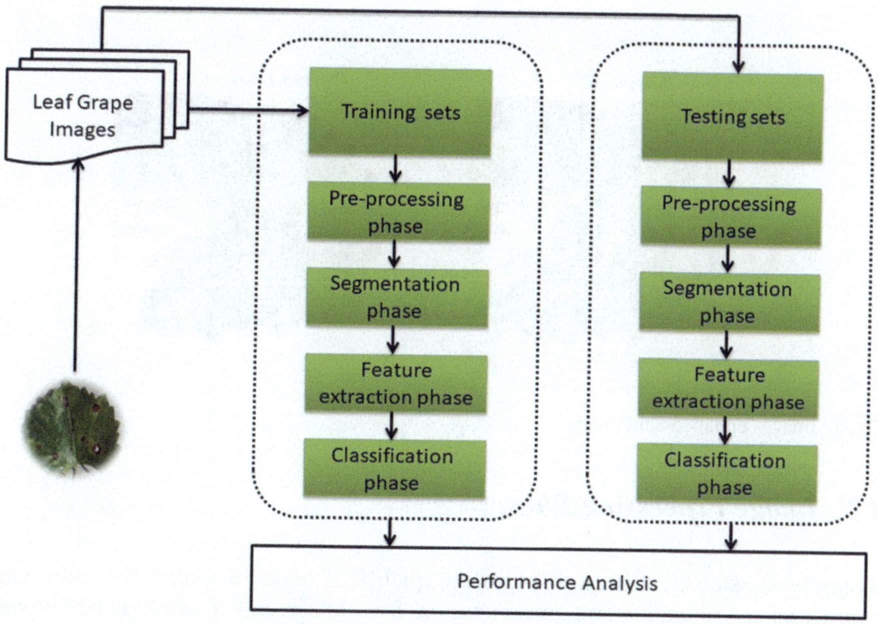

Fig. 1 The general architecture of the proposed leaf grape diagnosis system

Fig. 2 Images database **a** Grape black rot disease. **b** Grape Esca (black measles) leaf disease. **c** Grape leaf blight (Isariopsis leaf spot) leaf disease. **d** Healthy grape leaf

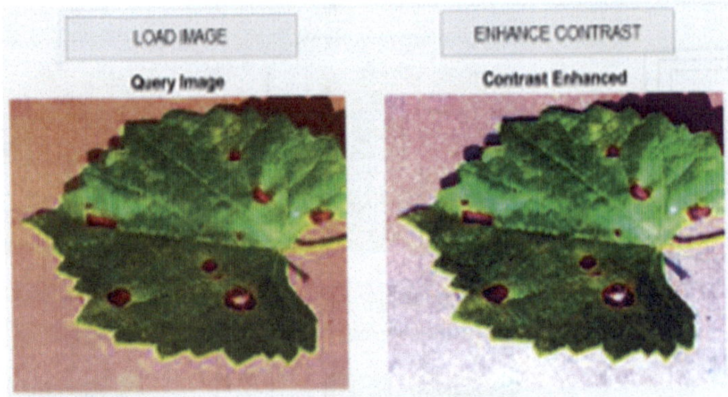

Fig. 3 Images enhancement results

3.2 *Image Processing Phase*

Image processing is utilized to boost the quality of image essential for additional treating and examination and determination. Leaf image enhancement is performed to increase the contrast of the image as shown in Fig. 3. The proposed approach is on the basis of gray level transformation that uses the intensity transformation of gray-scale images. We used imadjust function in Matlab and automatically adjust low and high parameters by using stretchlim function in MatLab.

3.3 *Image Segmentation Phase*

It implies a description of the picture in increasingly important and simpler to examinations way. In division a computerized picture is apportioned into various sections can characterize as super-pixels. A k-means clustering technique is adapted to cluster/divide the object on the basis of the feature of the leaf into k-number of groups. In this paper, $k = 3$ was utilized to two groups and is produced as output.

3.4 *Feature Extraction Phase*

Texture content counting is in the main approach for region description. After image segmentation, the statistical features extraction are 46 features [7].

3.5 *Classification Phase*

The supervised classifier is partitioned to the stage of the training and testing stage. The framework was trained during the training stage how to distinguish between Grape Black rot, Grape Esca (Black Measles), Grape Leaf blight (Isariopsis Leaf Spot), and Grape healthy is learned by utilizing known different grape leaf pictures. In the testing stage, the presentation of the framework is tested by entering a test picture to register the correctness level of the framework choice by utilizing obscure grape leaf pictures. The detection output of the classifiers was evaluated quantitatively by computing the sensitivity and specificity of the data. The Multi Support Vector Machines and Bayesian Classifier. Bayesian Classifier efficiency is accurately evaluated. The output produced by Bayesian Classifier is a disease name. Bayesian Classifier is a probabilistic classifier, which operates on the Bayes theorem principle. This needs conditional independence to reduce the difficulty of learning during classification modeling. To estimate the classifier parameters, the maximum likelihood calculation is used [8].

4 Results and Discussion

Every picture processing, segmentation, feature extraction, and MSVM categorization method in our proposed technique reproduced in MATLAB 2017b in an independent PC utilizing CPU Type Intel(R) Core (TM) i7-2620 M CPU @ 2.70 GHz Intel i7 3770 processor and 64-bit Windows 10 operating system. Total 400 image samples of grape images having grape black rot, grape Esca (Black Measles), grape Leaf blight (Isariopsis Leaf Spot), and Grape healthy and they were collected from plants diseases images "Kaggle-dataset". Among 400 samples, 360 samples are utilized for the training phase which composed of 90 samples of grape black rot, 90 samples of grape Esca (Black Measles), 90 samples of grape Leaf blight (Isariopsis Leaf Spot) and 90 samples grape healthy. Forty-six features are obtained from these samples after preprocessing and segmentation steps and a matrix of 360×46 features is created as outlined in Sect. 3 and those features matrix are input to Multiclass SVM for the training stage and Bayesian Classifier. The presented method consists of three main stages are grape foliage disease segmentation as shown in Fig. 4, feature extraction for the segmented grape foliage, and grape foliage disease classification. The accuracy test evaluates the performance of the classifier. Outcomes of the training data of the Bayesian Classifier show that overall accuracy of 95.28%, the sensitivity of 95.67%, specificity of 98.51% as shown in Fig. 5. The number of the input images loaded in the Bayesian Classifier was 360 samples and 343 samples were correctly classified and 17 samples were misclassified by this network. Results of the testing data of the Bayesian Classifier show that overall accuracy of 100%, the sensitivity of 100%, the specificity of 100%. The number of the input images loaded in the Bayesian Classifier were 40 samples that are utilized for testing phase which composed 10 Grape

Fig. 4 The input image and segmentation results

Black rot, 10 Grape Esca (Black Measles), 10 Grape Leaf blight (Isariopsis Leaf Spot) and 10 Grape healthy and 40 samples were correctly classified, and 0 samples was misclassified by this classifier as shown in Fig. 6. Multi-class SVM classifier trained utilized different kernel functions. Table 2 shows that using the polynomial kernel, the MSVM classifier can achieve overall maximum accuracy after training with 99.5%. Trained SVM classifier applied on four different test set of grape leaf image samples consisting of 90 samples of grape black rot, 90 samples of grape Esca (Black Measles), 90 samples of grape Leaf blight (Isariopsis Leaf Spot) and 90 samples grape healthy respectively. True positives, True negatives, False positives, and False negatives are defined and explained in [4]. Additionally, the performance

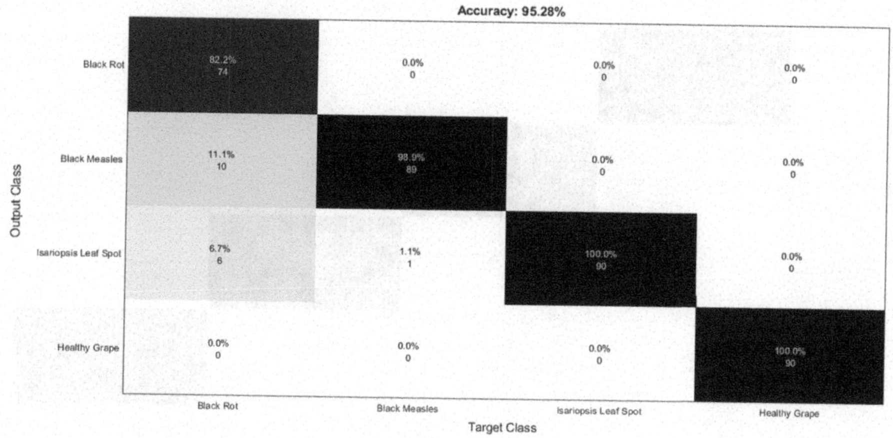

Fig. 5 Confusion matrix of training data set of bayesian classifier

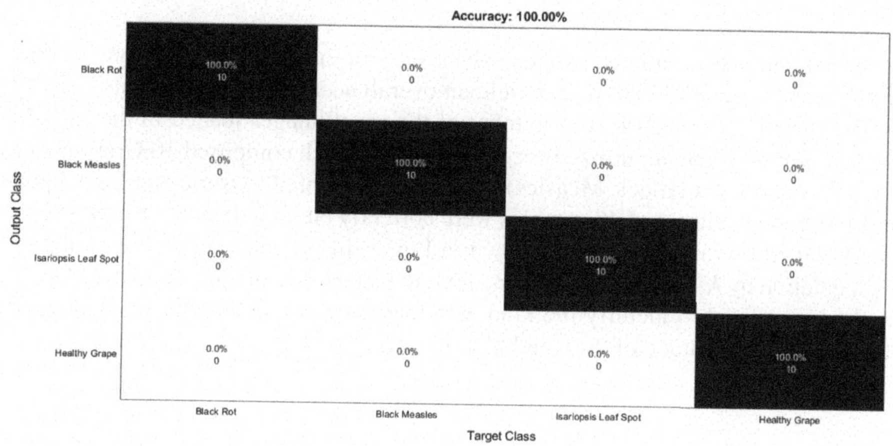

Fig. 6 Confusion matrix of testing data set of Bayesian Classifier

Table 2 Overall Performance evaluation of kernel functions utilized in training the multi-class SVM classifier for 4 different test sets of picture specimens

Kernel function	Accuracy for 300 images samples without 500 iterations	Accuracy for 300 images samples with 500 iterations
Linear	94%	98.2%
Quadratic	97.5%	98.2%
Polynomial	99.5%	98.2%
Rbf	96%	98.2%

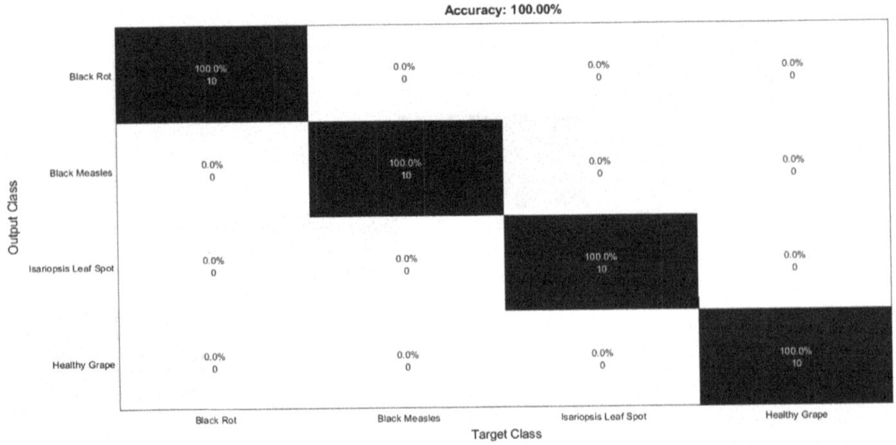

Fig. 7 Confusion matrix of testing data set of MSVM

of the MSVM was calculated by the analysis of a confusion matrix. Outcomes of the testing data of the SVM show that yield an overall accuracy of 100%, sensitivity of 100%, specificity of 100%. The number of the input images loaded in the MSVM were 40 samples that are utilized for testing phase which composed 10 Grape Black rot, 10 Grape Esca (Black Measles), 10 Grape Leaf blight (Isariopsis Leaf Spot) and 10 Grape healthy and 40 samples were correctly classified, and 0 samples were misclassified by this classifier as shown in Fig. 7. In [9] and [10] authors utilizing segmentation by *K*-means grouping and texture features are obtained and the MSVM method is utilized to identify the kind of foliage illness and classify the examined illness with an accuracy of 90% and 88.89% respectively.

5 Conclusions

In this paper, we have built up an intelligent that can computerize the classification of three unhealthy plant grape leaf diseases namely grape Esca (Black Measles), grape black rot, and grape foliage blight (Isariopsis Leaf Spot) and one healthy plant grape leaf. For the categorization stage, the multiclass SVM classifier is utilized which is much effective for multiclass classification. The 47 features extracted supported to design of a structure training data set. The proposed approach was varsities on four kinds of grape leaf diseases. The empirical outcomes demonstrate the proposed technique can perceive and classify grape plant diseases with high accuracy.

References

1. S.S. Sannakki, V.S. Rajpurohit, V.B. Nargund, P. Kulkarni, Diagnosis and classification of grape leaf diseases using neural networks, in *IEEE 4th ICCCNT* (2013)
2. P. Chaudhary, A.K. Chaudhari, A.N. Cheeran, S. Godara, Color transform based approach for disease spot. Int. J. Comput. Sci. Telecommun. **3**(6), 65–70 (2012)
3. H. Al-Hiary, S. Bani-Ahmad, M. Reyalat, M. Braik, Z. ALRahamneh, Fast and accurate detection and classification of plant diseases. IJCA **17**, (1), 31–38 (2011)
4. A. Dey, D. Bhoumik, K.N. Dey, Automatic multi-class classification of beetle pest using statistical feature extraction and support vector machine, in *Emerging Technologies in Data Mining and Information Security, IEMIS 2018*,vol. 2 (2019) pp. 533–544
5. C.-W. Hsu, C.-J. Lin, A comparison of methods for multi-class support vector machines. IEEE Trans. Neural Netw. **13**(2), 415–425 (2002)
6. L.M. Abou El-Maged, A. Darwish, A.E. Hassanien, Artificial intelligence-based plant's diseases classification, in *Proceedings of the International Conference on Artificial Intelligence and Computer Vision (AICV 2020)* (2020), pp. 3–15
7. K.K. Mohammed, H.M. Afify, F. Fouda, A.E. Hassanien, S. Bhattacharyya, S. Vaclav, Classification of human sperm head in microscopic images using twin support vector machine and neural network. Int. Conf. Innov. Comput. Commun. (2020)
8. R.O. Duda, P.E. Hart, D.G. Stork, *Pattern Classification* (Wiley, New York, USA, 2012)
9. N. Agrawal, J. Singhai, D.K. Agarwal, Grape leaf disease detection and classification using multi-class support vector machine, in *Proceedings of the Conference on Recent Innovations is Signal Processing and Embedded Systems (RISE-2017)* 27–29 Oct 2017
10. A.J. Ratnakumar, S. Balakrishnan, Machine learning-based grape leaf disease detection. Jour Adv Res. Dyn. Control. Syst. **10**(08) (2018)

Robust Deep Transfer Models for Fruit and Vegetable Classification: A Step Towards a Sustainable Dietary

Nour Eldeen M. Khalifa, Mohamed Hamed N. Taha, Mourad Raafat Mouhamed, and Aboul Ella Hassanien

Abstract Sustainable dietary plays an essential role in protecting the environment to be healthier. Moreover, it protects human life and health in its widest sense. Fruits and vegetables are basic components of sustainable dietary as it is considered one of the main sources of healthy food for humans. The classifications of fruits and vegetables are most helpful for dietary assessment and guidance which will reflect in increasing the awareness of sustainable dietary for consumers. In this chapter, a robust deep transfer model based on deep convolutional neural networks for fruits and vegetable classification is introduced. This presented model is considered a first step to build a useful mobile software application that will help in raising the awareness of sustainable dietary. Three deep transfer models were selected for experiments in this research and they are Alexnet, Squeeznet, and Googlenet. They were selected as they contained a small number of layers which will decrease the computational complexity. The dataset used in this research is FruitVeg-81 which contains 15,737 images. The number of extracted classes from the dataset is 96 class by expanding three layers of classifications from the original dataset. Augmentation technique (rotation) was adopted in this research to reduce the overfitting and increase the number of images to be 11 times larger than the original dataset. The experiment results show that the Googlenet achieves the highest testing accuracy with 99.82%.

N. E. M. Khalifa (✉) · M. H. N. Taha
Information Technology Department, Faculty of Computers and Artificial Intelligence, Cairo University, Giza, Egypt
e-mail: nourmahmoud@cu.edu.eg

M. H. N. Taha
e-mail: mnasrtaha@cu.edu.eg

M. R. Mouhamed
Faculty of Science, Helwan University, Cairo, Egypt
e-mail: mouradraafat@yahoo.com

N. E. M. Khalifa · M. H. N. Taha · M. R. Mouhamed · A. E. Hassanien
Scientific Research Group in Egypt (SRGE), Cairo, Egypt
e-mail: aboitcairo@cu.edu.eg
URL: http://www.egyptscience.net

© The Editor(s) (if applicable) and The Author(s), under exclusive license to Springer Nature Switzerland AG 2021

A. E. Hassanien et al. (eds.), *Artificial Intelligence for Sustainable Development: Theory, Practice and Future Applications*, Studies in Computational Intelligence 912,
https://doi.org/10.1007/978-3-030-51920-9_3

Moreover, it achieved the highest precision, recall, and F1 performance score if it is compared with other models. Finally, A comparison results were carried out at the end of the research with related work which used the same dataset FruitVeg-81. The presented work achieved a superior result than the related work in terms of testing accuracy.

Keywords Deep transfer models · Googlenet · Fruits · Vegetables · Classification · Sustainable dietary

1 Introduction

Food production and consumption usage and patterns are among the main sources of the burden on the environment. The term "Food" related to vegetables and fruits growing farms, animal farm production, and fishing farms. It is considered as a burden on the environments for its processing, storage, transport, and distribution up to waste disposal. So, there is a need to leave this burden off the environment to recover its health which will reflect human health and life.

The term sustainability means "meeting the needs of the present without compromising the ability of future generations to meet their own needs" according to Brundtland Report [1]. Merging the term of sustainability with Food production and consumption will produce a new term of sustainability food or sustainability dietary. There are related terms in the field of sustainability food (sustainability dietary) and they are illustrated in Table 1 [2].

Table 1 Terms related to sustainability food (sustainability dietary)

Term	Definition
Food systems [33]	Consist of all the essentials component that include (environment, people, processes, infrastructures, institutions etc.) and activities that relate to the production, processing, distribution, preparation and consumption of food and the outcomes of these activities, namely nutrition and health status, socioeconomic growth and equity, and environmental sustainability
Sustainable food systems [34]	A food system that guarantees food security and nutrition that the social, economic, and environmental foundation meant to generate food security and nutrition of future generations are not compromised
Diets [2]	A role of food systems, diets are the foods a person regularly consumes in daily bases including habits, routines and traditions around food
Sustainable dietary [35]	Diets that are protective and respectful of biodiversity and ecosystems, culturally acceptable, accessible, economically fair and affordable, nutritionally adequate, safe and healthy, while at the same time optimizing natural and human resources

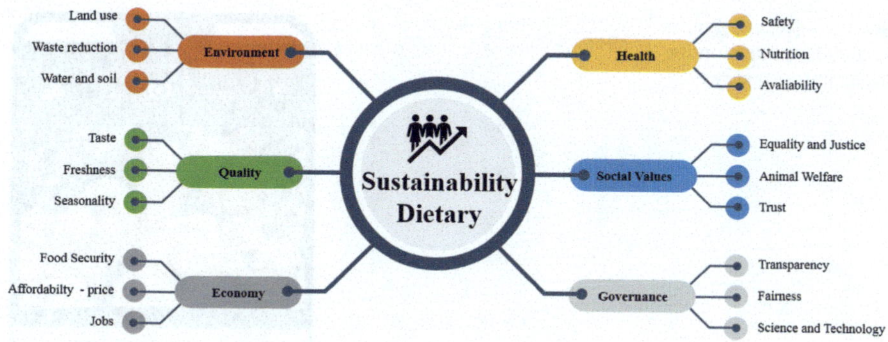

Fig. 1 Sustainability dietary effect on different domains

There are many advantages of sustainability food (sustainability dietary), below a shortlist of these advantages [3]:

- Reduces the negative impacts on the environment.
- provisions the health, human rights, and economic security of the people producing, preparing, and eating the food.
- reinforces connections within and between communities.
- guarantees humane treatment of livestock by reducing the demand for meat.
- conserves environmental and biological resources for future generations.

Figure 1 illustrates the expected potentials of using sustainability dietary on different domains and they are environment, health, quality, social values, economy, and governance.

As shown in Fig. 1, sustainability dietary will affect many domains, in the environment, it will help in reducing the land use for animal and fish farming by investing more in vegetables and fruits framing which is very effective for environmental health [4, 5]. In health, it helps in lowering the risk of weight gain, overweight, and obesity, which will lead to making the human-less vulnerable to diseases. Moreover, it helps to increase food safety, and make better nutrition for humans. In the food quality, better food will be planted, fresh food will be available with better taste. In social values, sustainability dietary will provide justice for workers and equality for the consumer to have better food for all social layers. It will grantee more humane treatment for animals and increase the trust between food consumers and producers. In the economy, it will provide food security and flatten the prices and create more jobs. Finally, in governance, sustainability dietary will depend on science and technology, which means more control, fairness, and transparency.

Technology and science proved its efficiency to provide various means of controls in many academic and industrials domains. In this chapter, a preliminary step towards empowering the use of science and technology in the field of sustainability dietary will be presented. The intended objective is to build a mobile application that will help in the detection of fruits and vegetables using a mobile camera. The application first will detect the fruit and vegetable type using deep learning models which is

Fig. 2 Concept design for the mobile application for sustainability dietary

the purpose of this chapter. Then after the identification process, it will display information about the detected fruit or vegetables that will help the consumer to mind his/her thinking of considering purchasing the fruit or vegetable or not. Figure 2 presents the concept of the mobile application as a final product of the proposed research model.

The consumer will use the previous mobile application in the market, then use the camera inside the application to recognize the fruit or the vegetable in front of him/her. The application will capture two images and send it to a computer server using cloud computing infrastructure. The deep learning model will detect the fruit or the vegetable, retrieve the required information of the detected fruit or vegetable and send back the information to the consumer mobile application and display the information as illustrated in Fig. 2. The information will include many items such as calories, carbs, fiber, protein, fat, available vitamins, folate, potassium magnesium, and average international price according to the current year. Figure 3 presents the steps of the proposed model for the mobile application.

In this chapter, only the part of the detection will be introduced in detail. The presented model can classify 96 class of fruits and vegetables depending on deep transfer learning which relies on deep learning methodology.

Deep Learning (DL) is a type of Artificial Intelligence (AI) concerned with methods inspired by the functions of people's brain [6]. For the time being, DL is quickly becoming an important method in image/video detection and diagnosis [6]. Convolutional Neural Network (ConvNet or CNN) is a mathematical type of DL architectures used originally to recognize and diagnose images. CNN's have

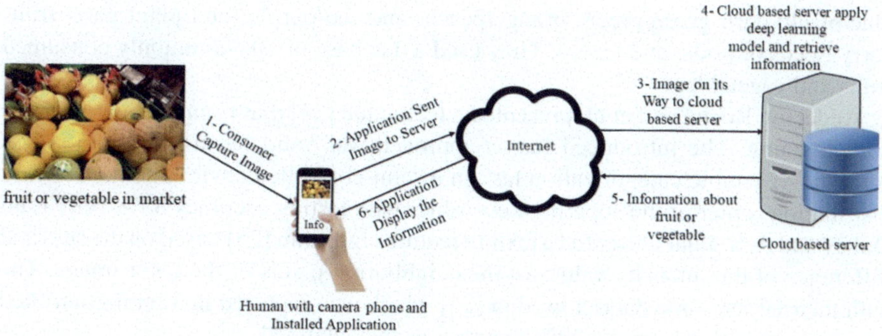

Fig. 3 The concept of the proposed model for sustainability dietary

masterful unusual success for medicinal image/video diagnoses and detection. In 2012, [7, 8] introduced how ConvNets can boost many image/vision databases such as Modified National Institute of Standards and Technology database (MNIST) [9], and big-scale ImageNet [10]. Deep Transfer Learning (DTL) is a CNN architecture that storing learning parameters gained while solving the DL problem and execute DTL to various DL problem. Many DTL models were introduced like VGG [11], Google CNN [12], residual neural network [13], Xception [14], Inception-V3 [15], and densely connected CNN [16]. DTL has been used in many domains such as medical X-rays diagnoses [17, 18], medical diabetic retinopathy detection [19], gender classification through iris patterns [20], and pests recognition [21] and achieved remarkable results in terms of testing accuracy.

The rest of the chapter is organized as follows. Section 2 presents a list of related works. Section 3 illustrates the data set characteristics, Sect. 4 discusses the proposed methodology, while Sect. 5 identifies the carried-out results and discussion. Finally, Sect. 6 provides conclusions and future directions for the proposed model.

2 Related Works

Consumption of fruits and vegetables is important for human health because these foods are primary sources of some essential nutrients and contain phytochemicals that may lower the risk of chronic disease [22]. Using computer algorithms and artificial intelligence techniques, the classification of fruits and vegetables automatically attracts the attention of many researchers during the last decade.

Jean A. T. Pennington and R. A. Fisher introduced a mathematical clustering algorithm [23] to group the foods into homogeneous clusters based on food component levels and the classification criteria. Most useful in categorizing were the botanic families rose, rue (citrus), amaryllis, goosefoot, and legume; color groupings blue/black, dark green/green, orange/peach, and red/purple; and plant parts fruit-berry, seeds or pods, and leaves. They used a database of 104 commonly consumed fruits and vegetables.

Anderson Rocha and et al. presented a technique [24] that is amenable to continuous learning. The introduced fusion approach was validated using a multi-class fruit-and-vegetable categorization task in a semi-controlled environment, such as a distribution center or the supermarket cashier with testing accuracy 85%. Shiv Ram Dubey and A. S. Jalal presented a texture feature algorithm [25] based on the sum and difference of the intensity values of the neighboring pixels of the color image. The authors used the same dataset used in [24] which was captured in a semi-controlled environment and achieved a 99% accuracy as they claimed.

Khurram Hameed et al. in [26] presented a comprehensive review of fruit and vegetable classification techniques with different machine learning techniques, for example, Support Vector Machine (SVM), K-Nearest Neighbour (KNN), Decision Trees, Artificial Neural Networks (ANN) and Convolutional Neural Networks (CNN) for fruit and vegetable classification in many real-life applications. The survey presents a critical comparison of different state-of-the-art computer vision methods proposed by researchers for classifying fruit and vegetable.

Georg Waltner and et al. in [27] introduced a personalized dietary self-management mobile vision-based assistance application using is FruitVeg-81 which they presented in their paper. The authors achieved a testing accuracy with 90.41%.

The mentioned works above used different datasets with different conditions for controlled or semi-controlled environments except for the research presented in [27]. The survey in [26] illustrated the researcher's work throughout the years in a comprehensive matter. The presented work in this paper used the same dataset introduced in [27] which was released in 2017 and comparative results will be illustrated in the results and discussion section.

3 Dataset Characteristics

The dataset used in this research is FruitVeg-81 [27]. It has been collected within the project MANGO (Mobile Augmented Reality for Nutrition Guidance and Food Awareness). It contains 15,737 images (all images resized to 512 * 512px). The dataset consists of fruit and vegetable items with hierarchical labels. It is structured as follows:

- The first level depicts the general sort of food item (apples, bananas, … etc.)
- The second level collects food cultivars with similar visual appearance (red apples, green apples, … etc.)
- The third level distinguishes between different cultivars (Golden Delicious, Granny Smith, … etc.) or packaging types (boxed, tray, … etc.).

This chapter adopts a combination of the original three levels of the original dataset which results in the increased number of classes. The original dataset consists of 81

Fig. 4 Examples of images for FruitVeg-81 with different conditions

classes on the first level only. We increased the classes to include the second and the third class which raises the number of classes to be 96 class.

Figure 4 represents a sample of images from the dataset. The dataset images were captured using different mobile devices such as Samsung Galaxy S3, Samsung Galaxy S5, HTC One, HTC Three and Motorola Moto G. Using different mobile devices poses new challenges in the dataset which includes the difference in the appearance, scale, illumination, number of objects and fine-grained differences.

4 Proposed Methodology

The proposed methodology relies on the deep transfer learning models. The selected models in this research are Alexnet [8], SqueezNet [13], and Googlenet [12] which consist of 16, 18, and 22 layers respectively as illustrated in Fig. 5. The previously stated pre-trained deep transfer CNN models had a quite few numbers of layers if it is compared to other large CNN models such as xception [14], densenet [16], and inceptionresnet [28] which consist of 71, 201 and 164 layers accordingly.

Choosing a less deep transfer deep learning models in the number of layers will reduce the computational complexity and thus decrease the time needed for the training, validation, and testing phase. Figure 5 illustrated the proposed deep transfer learning customization for Fruit and Vegetable classification used in this research.

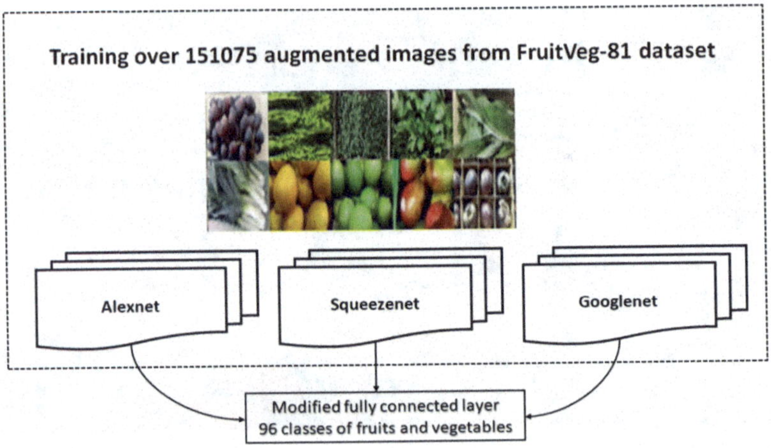

Fig. 5 Proposed methodology deep transfer learning customization for fruit and vegetable classification

4.1 Data Augmentation Techniques

The most common method to overcome overfitting is to increase the number of images used for training by applying label-preserving transformations [20]. Besides, data augmentation schemes are applied to the training set to make the resulting model more invariant for any kind of transformation. The adopted augmentation technique in this research is the rotation technique by 30, 60, 90, 120, 150, 180, 210, 240, 270, 300, 330 angles. The image transformation using the rotation technique is calculated using Eqs. (1) and (2).

$$x2 = \cos(\theta) * (x1 - x0) + \sin(\theta) * (y1 - y0) \tag{1}$$

$$y2 = -\sin(\theta) * (x1 - x0) + \cos(\theta) * (y1 - y0) \tag{2}$$

where the coordinates of a point $(x1, y1)$, when rotated by an angle θ around $(x0, y0)$, become $(x2, y2)$ in the augmented image. The adopted augmentation technique has raised the number of images of the dataset to be 11 times larger than the original dataset. The dataset raised to 173,107 images which are used for the training and the verification and the testing phases. This will lead to a significant improvement in CNN testing accuracy and make the proposed models more robust for any type of rotation. Figure 6 illustrates examples of different rotation angles for the images in the dataset.

Fig. 6 Examples of different rotation angles for the images in the dataset

5 Experimental Results

The proposed methodology was developed using a software package (MATLAB). The implementation was GPU specific. All experiments were performed on a computer with core i9 (2 GHz), 32 GB of RAM with Titan X GPU.

As stated in the proposed methodology section, we have selected three deep transfer model and they are Alexnet, Squeeznet, and Goolgenet with fine-tuning the last fully connected layers to classify 96 classes of fruits and vegetables. The dataset was divided into two parts with a dividing percentage of 80 and 20%. The 80% assigned to the training phase and the 20% percentages for the testing phase. The selected dividing percentage proved it is efficient in terms of training speed and testing accuracy in research [19, 21, 29].

This work was conducted first to build its deep neural networks implied in research works [30–32] to classify the 96 class but the testing accuracy was not acceptable. Therefore, the choice of pre-trained deep transfer models was the optimal solution to achieve a competitive testing accuracy with the other performance measure as it will be illustrated in the following subsections.

The first metric to be measured for the selected models is the confusion matrix. The confusion matrix illustrates the testing accuracy of every class and the overall testing accuracy for the model. As mentioned before the dataset consists of 96 class the confusion matrix will construct a matrix of 96 * 96 class. This matrix will be unfit to be displayed throughout the chapter. Figures 7 presents a visual representation of heatmaps for the confusion matrix for Alexnet, Squeeznet, and Goolgenet accordingly. The confusion matrix was calculated over the 34,621 images which represent 20% of the dataset total images after the augmentation process for the testing phase.

The previous figures consisted of the X-axis and Y-axis which presented the number for a class between 1 and 96. Every number maps to a class in the dataset.

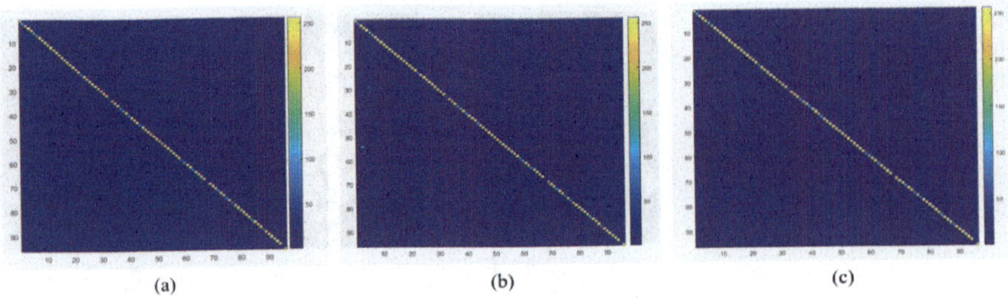

Fig. 7 Heatmap confusion matrix representation for **a** alexnet, **b** squeezenet, and **c** googlenet

Table 2 Testing accuracy for different deep transfer models

Model	Alexnet	Squeezenet	Googlenet
Testing accuracy (%)	99.63	98.74	99.82

The blue color presents the zero occurrences of misclassified class and the yellow color present 260 which reflect the largest occurrence of correctly classified class.

One of the measures to prove the efficiency of the model is the testing accuracy. The testing accuracy is calculated using the confusion matrix for every model and using Eq. (3). Table 2 presents the testing accuracy of the three selected models throughout this research. Table 2 illustrates that the Googlenet model achieves the best testing accuracy if it is compared with the other related model which includes Alexnet and Squeeznet.

Figure 8 illustrates the testing accuracy for different images from the dataset using Googlenet deep transfer model which achieves the best overall testing accuracy. The figure showed that the proposed model achieved 100% for testing accuracy in many classes such as honeydew, avocado, turnips, cabbage green, eggplant, apricot, mangosteen box, and peach tray.

To evaluate the performance of the proposed models, more performance matrices are needed to be investigated through this research. The most common performance measures in the field of deep learning are Precision, Recall, and F1 Score [4], and they are presented from Eqs. (4) to (6).

$$\text{Testing Accuracy} = \frac{(\text{TN} + \text{TP})}{(\text{TN} + \text{TP} + \text{FN} + \text{FP})} \tag{3}$$

$$\text{Precision} = \frac{\text{TP}}{(\text{TP} + \text{FP})} \tag{4}$$

$$\text{Recall} = \frac{\text{TP}}{(\text{TP} + \text{FN})} \tag{5}$$

$$\text{F1 Score} = 2 * \frac{\text{Precision} * \text{Recall}}{(\text{Precision} + \text{Recall})} \tag{6}$$

Fig. 8 Testing accuracy for samples of images in the dataset

where TP is the count of True Positive samples, TN is the count of True Negative samples, FP is the count of False Positive samples, and FN is the count of False Negative samples from a confusion matrix.

Table 3 presents the performance metrics for the different deep transfer models. The table illustrates that the Googlenet model achieved the highest percentage for precision, recall, and F1 score metrics with a percentage of 99.79, 99.80, and 99.79% accordingly.

Table 4 presents a comparative result with the related work in [27]. The presented work in [27] published the dataset which is used in this research. It is clearly shown that our proposed methodology using Googlenet and the adopted augmentation technique (rotation) led to a significant improvement in testing accuracy and super passed the testing accuracy presented in the related work.

Table 3 Performance metrics for the different deep transfer models

Metric/Model	Alexnet	Squeeznet	Googlenet
Precision (%)	99.63	99.04	99.79
Recall (%)	99.61	98.37	99.80
F1 score (%)	99.62	98.71	99.79

Table 4 The comparative result with related work

	Description	Testing accuracy (%)
Hameed et al. [27]	Shallow convolutional neural network	90.41
Proposed method	Googlenet + augmentation rotation technique	99.82

6 Conclusion and Future Works

Sustainable dietary plays an essential role in protecting the environment. Moreover, it protected human life and health in its broadest sense. Fruits and vegetables are basic components of sustainable dietary as it is considered one of the main sources of healthy food for humans. In this chapter, a deep transfer model based on a deep convolutional neural network for fruit and vegetable classification is introduced. Three deep transfer models were selected in this research and they are Alexnet, Squeeznet, and Googlenet. They were selected as they contained a small number of layers which will decrease computational complexity. The dataset used in this research is FruitVeg-81 which contains 15,737 images. The extract classed from the dataset according to merge three layers of classification is 96 class of fruits and vegetables. Augmentation technique (rotation) was adopted in this research to reduce the overfitting and increase the number of images to be 11 times larger than the original dataset. The experiment results show that the Googlenet achieves the highest testing accuracy with 99.82%. Moreover, it achieved the highest precision, recall, and F1 performance score if it is compared with other models. Finally, A comparison results were carried out at the end of the research with related work which used the same dataset FruitVeg-81. The presented work achieved a superior result than the related work in terms of testing accuracy. One of the potential future works is applying new architectures of deep neural networks such as Generative Adversarial Neural Networks. GAN will be used before the proposed models. It will help in generating new images from the trained images, which will reflect on the accuracy of the proposed models. Additionally, to expand the current work is to use large deep learning architecture such as Xception, DenseNet, and InceptionResNet to get better accuracies and implement the mobile application suggested for sustainable dietary.

Acknowledgements We sincerely thank the Austrian Research Promotion Agency (FFG) under the project Mobile Augmented Reality for Nutrition Guidance and Food Awareness (836488) for the dataset used in this research. We also gratefully acknowledge the support of NVIDIA Corporation, which donated the Titan X GPU used in this research.

References

1. B.R. Keeble, The brundtland report: 'our common future'. Med. War **4**(1), 17–25 (1988)
2. A.J.M. Timmermans, J. Ambuko, W. Belik, J. Huang, Food losses and waste in the context of sustainable food systems (2014)
3. T. Engel, Sustainable food purchasing guide. Yale Sustain. Food Proj. (2008)
4. C. Goutte, E. Gaussier, A probabilistic interpretation of precision, recall and F-score, with implication for evaluation, in *European Conference on Information Retrieval* (2005), pp. 345–359
5. A.A. Abd El-aziz, A. Darwish, D. Oliva, A.E. Hassanien, Machine learning for apple fruit diseases classification system, in *AICV 2020* (2020), pp. 16–25
6. D. Rong, L. Xie, Y. Ying, Computer vision detection of foreign objects in walnuts using deep learning. Comput. Electron. Agric. **162**, 1001–1010 (2019)
7. D. Ciregan, U. Meier, J. Schmidhuber, Multi-column deep neural networks for image classification, in *2012 IEEE Conference on Computer Vision and Pattern Recognition* (2012), pp. 3642–3649
8. A. Krizhevsky, I. Sutskever, G.E. Hinton, ImageNet classification with deep convolutional neural networks, in *ImageNet Classification with Deep Convolutional Neural Networks* (2012), pp. 1097–1105
9. Y. Lecun, L. Bottou, Y. Bengio, P. Haffner, Gradient-based learning applied to document recognition. Proc. IEEE **86**(11), 2278–2324 (1998)
10. J. Deng, W. Dong, R. Socher, L. Li, L. Kai, F.-F. Li, ImageNet: a large-scale hierarchical image database, in *2009 IEEE Conference on Computer Vision and Pattern Recognition* (2009), pp. 248–255
11. S. Liu, W. Deng, Very deep convolutional neural network based image classification using small training sample size, in *2015 3rd IAPR Asian Conference on Pattern Recognition (ACPR)* (2015), pp. 730–734
12. C. Szegedy et al., Going deeper with convolutions, in *Proceedings of the IEEE Computer Society Conference on Computer Vision and Pattern Recognition* (2015) 07–12 June, pp. 1–9
13. K. He, X. Zhang, S. Ren, J. Sun, Deep residual learning for image recognition, in *2016 IEEE Conference on Computer Vision and Pattern Recognition (CVPR)* (2016), pp. 770–778
14. F. Chollet, Xception: deep learning with depthwise separable convolutions, in *2017 IEEE Conference on Computer Vision and Pattern Recognition (CVPR)* (2017), pp. 1800–1807
15. C. Szegedy, V. Vanhoucke, S. Ioffe, J. Shlens, Z. Wojna, Rethinking the inception architecture for computer vision, in *Proceedings of the IEEE conference on computer vision and pattern recognition* (2016), pp. 2818–2826
16. G. Huang, Z. Liu, L.V.D. Maaten, K.Q. Weinberger, Densely connected convolutional networks, in *2017 IEEE Conference on Computer Vision and Pattern Recognition (CVPR)* (2017), pp. 2261–2269
17. M. Loey, F. Smarandache, N.E.M. Khalifa, Within the lack of chest COVID-19 X-ray dataset: a novel detection model based on GAN and deep transfer learning. Symmetry **12**, 651 (2020)
18. N.E.M. Khalifa, M.H.N. Taha, A.E. Hassanien, S. Elghamrawy, Detection of coronavirus (COVID-19) associated pneumonia based on generative adversarial networks and a fine-tuned deep transfer learning model using chest X-ray dataset. arXiv (2020), pp. 1–15
19. N. Khalifa, M. Loey, M. Taha, H. Mohamed, Deep transfer learning models for medical diabetic retinopathy detection. Acta Inform. Medica **27**(5), 327 (2019)
20. N. Khalifa, M. Taha, A. Hassanien, H. Mohamed, Deep iris: deep learning for gender classification through iris patterns. Acta Inform. Medica **27**(2), 96 (2019)
21. N.E.M. Khalifa, M. Loey, M.H.N. Taha, Insect pests recognition based on deep transfer learning models. J. Theor. Appl. Inf. Technol. **98**(1), 60–68 (2020)
22. Advisory Committee and others, Report of the dietary guidelines advisory committee dietary guidelines for Americans, 1995. Nutr. Rev. **53**, 376–385 (2009)
23. J.A.T. Pennington, R.A. Fisher, Classification of fruits and vegetables. J. Food Compos. Anal. **22**, S23–S31 (2009)

24. A. Rocha, D.C. Hauagge, J. Wainer, S. Goldenstein, Automatic fruit and vegetable classification from images. Comput. Electron. Agric. **70**(1), 96–104 (2010)
25. S.R. Dubey, A.S. Jalal, Robust approach for fruit and vegetable classification. Procedia Eng. **38**, 3449–3453 (2012)
26. K. Hameed, D. Chai, A. Rassau, A comprehensive review of fruit and vegetable classification techniques. Image Vis. Comput. **80**, 24–44 (2018)
27. G. Waltner et al., *Personalized Dietary Self-Management Using Mobile Vision-Based Assistance*, in Lecture Notes in Computer Science (including subseries Lecture Notes in Artificial Intelligence and Lecture Notes in Bioinformatics) (2017), pp. 385–393
28. C. Szegedy, S. Ioffe, V. Vanhoucke, A.A. Alemi, Inception-v4, inception-ResNet and the impact of residual connections on learning, in *31st AAAI Conference on Artificial Intelligence, AAAI 2017* (2017)
29. N.E.M. Khalifa, M.H.N. Taha, D. Ezzat Ali, A. Slowik, A.E. Hassanien, Artificial intelligence technique for gene expression by Tumor RNA-Seq data: a novel optimized deep learning approach. IEEE Access **8**, 22874–22883 (2020)
30. N.E. Khalifa, M. Hamed Taha, A.E. Hassanien, I. Selim, Deep galaxy V2: Robust deep convolutional neural networks for galaxy morphology classifications, in *2018 International Conference on Computing Sciences and Engineering, ICCSE 2018* (2018), pp. 1–6
31. N.E.M. Khalifa, M.H.N. Taha, A.E. Hassanien, A.A. Hemedan, Deep bacteria: robust deep learning data augmentation design for limited bacterial colony dataset. Int. J. Reason. Intell. Syst. **11**(3), 256–264 (2019)
32. N.E.M. Khalifa, M.H.N. Taha, A.E. Hassanien, Aquarium family fish species identification system using deep neural networks, in *International Conference on Advanced Intelligent Systems and Informatics* (2018), pp. 347–356
33. R. Valentini, J.L. Sievenpiper, M. Antonelli, K. Dembska, in *Achieving the Sustainable Development Goals Through Sustainable Food Systems* (Springer, Berlin, 2019)
34. P. Caron et al., Food systems for sustainable development: proposals for a profound four-part transformation. Agron. Sustain. Dev. **38**(4), 41 (2018)
35. A. Shepon, P.J.G. Henriksson, T. Wu, Conceptualizing a sustainable food system in an automated world: toward a 'eudaimonia' future. Front. Nutr. **5**, 104 (2018)

The Role of Artificial Neuron Networks in Intelligent Agriculture (Case Study: Greenhouse)

Abdelkader Hadidi, Djamel Saba, and Youcef Sahli

Abstract The cultivation under cover of fruits, vegetables, and floral species has developed from the traditional greenhouse to the agro-industrial greenhouse which is currently known for its modernity and its high level of automation (heating, misting system, air conditioning, control, regulation and command, supervision computer, etc.). New techniques have emerged, including the use of devices to control and regulate climatic variables in the greenhouse (temperature, humidity, CO_2 concentration, etc.). In addition, the use of artificial intelligence (AI) such as neural networks and/or fuzzy logic. Currently, the climate computer offers multiple services and makes it possible to solve problems relating to regulation, control, and commands. The main motivation in choosing an order by AI is to improve the performance of internal climate management, to move towards a control-command strategy to achieve a homogeneous calculation structure through a mathematical model of the process to be controlled, usable on the one hand for the synthesis of the controller and on the other hand by the simulation of the performances of the system. It is from this state, that begins this research work in this area include modelization an intelligent controller by the use of fuzzy logic.

Keywords Agriculture · Greenhouse · Artificial intelligence · Control · Fuzzy logic · Artificial neuron networks

A. Hadidi · D. Saba (✉) · Y. Sahli
Unité de Recherche en Energies Renouvelables en Milieu Saharien, URER-MS, Centre de Développement des Energies Renouvelables, CDER, 01000 Adrar, Algeria
e-mail: saba_djamel@yahoo.fr

A. Hadidi
e-mail: hadidiabdelkader@gmail.com

Y. Sahli
e-mail: sahli.sofc@gmail.com

© The Editor(s) (if applicable) and The Author(s), under exclusive license to Springer Nature Switzerland AG 2021
A. E. Hassanien et al. (eds.), *Artificial Intelligence for Sustainable Development: Theory, Practice and Future Applications*, Studies in Computational Intelligence 912,
https://doi.org/10.1007/978-3-030-51920-9_4

Abbreviations

AI	Artificial Intelligence
ANN	Artificial Neural Networks
CO_2	Carbon Dioxide
EAs	Evolution Algorithms
FAO-UN	Food and Agriculture Organization of the United Nations
FL	Fuzzy Logic
GA	Genetic Algorithms
H	Humidity
IT	Information Technology
LP	Linear Programming
MIMO	Multi-Input Multi-Output
NIAR	National Institute for Agronomic Research
PDF	Pseudo-Derivative Feedback
PE	Polyethylene
PID	Integral Controllers – Derivatives
PIP	Proportional-Integral-Plus
PVC	Polyvinyl Chloride
SISO	Single-Input, Single-Output
T	Temperature

1 Introduction

The agricultural sector will face enormous challenges to feed a world population which, according to the FAO-UN, should reach 9.6 billion people by 2050, technological progress has worked considerably in the development of agricultural greenhouses [1]. They are becoming very sophisticated (accessories and accompanying technical equipment, control computer). New climate control techniques have appeared, including the use of regulating devices, ranging from the classic to the application of AI, now known as neural networks and/or FL, etc. However, the air conditioning of modern greenhouses allows crops to be kept under shelter in conditions compatible with agronomic and economic objectives. Greenhouse operators opt for competitiveness. They must optimize their investments, the cost of which is becoming more and more expensive. The agricultural greenhouse can be profitable as long as its structure is improved. Well-chosen wall materials, depending on the nature and type of production, technical installations, and accompanying equipment must be judiciously defined. Numerous equipment and accessories have appeared to regulate and control state variables such as temperature, humidity, and CO_2 concentration. Currently, the climate computers in the greenhouses solve regulatory problems and ensure compliance with the climatic instructions required by the plants [2]. Now the climate computer is a dynamic production management tool, able to choose the most appropriate climate route [2]. According to Van Henten [3], the global approach to greenhouse systems is outlined as follows:

- **Physiological aspect**: this relatively complex and underdeveloped area requires total care and extensive scientific and experimental treatment. This allows us to characterize the behavior of the plant during its evolution, from growth to its final development; and to establish an operating model.
- **Technical aspect**: the greenhouse system is subject to a large number of data, decisions, and actions to be carried out on the plant's immediate climatic environment (temperature (T), humidity (H), CO_2 enrichment, misting, etc.). The complexity of managing this environment requires an analytical, digital, IT, and operational approach to the system.
- **Socio-economic aspect**: social evolution will be legitimized by a demanding and pressing demand for fresh products throughout the year; this state of affairs, leads all socio-economic operators, to be part of a scientific, technological, and cooking dynamic. This dynamic requires high professionalism.

New techniques have emerged, including the use of devices to control and regulate climatic variables in a greenhouse from the classic to the exploitation of AI, such as neural networks and/or FL [4, 5].

This document presents the techniques for monitoring and controlling the climatic management of agricultural greenhouses through the application of AI. These are especially ANN, FL, GA, control techniques, computing, and all the structures attached to them. These techniques are widely applied in the world of modern industry, in robotics, automation, and especially in the food industry. The agricultural greenhouse, to which plan to apply these techniques, challenges us to approach the system taking into account the constraints that can be encountered in a biophysical system, such as non-linearity, the fluctuation of state variables, the coupling between the different variables, the vagaries of the system over time, the variation of meteorological parameters, uncontrollable climatic disturbances, etc. All these handicaps lead us to consider the study and the development of an intelligent controller and the models of regulation control and command of the climatic environment of the internal atmosphere of greenhouses.

The objective of this document is to provide an information platform for the role of ANN in intelligent agriculture. Hence, the remainder of this paper is organized as follows. Section 2 presents the AI. Section 3 explains the agriculture and greenhouse. Section 4 explains the intelligent control systems. Section 5 details modern optimization techniques. Section 6 clarifies the fuzzy identification. Finally, Sect. 7 concludes the paper.

2 Overview of AI

Under the term AI, grouping all of the "theories and techniques used to produce machines capable of simulating intelligence" [6]. This practice allows Man to put a computer system on solving complex problems integrating logic. More commonly, when talking about AI, also mean machines imitating certain human features.

- **AI before 2000**: AI before 2000: the first traces of AI date back to 1950 in an article by Alan Turing entitled "Computing Machinery and Intelligence" in which the mathematician explores the problem of defining whether a machine is conscious or not [7]. This article will flow what is now called the Turing Test, which assesses the ability of a machine to hold a human conversation. Another probable origin in a publication by Warren Weaver with a memo on machine translation of languages which suggests that a machine could very well perform a task that falls under human intelligence. The formalization of AI as a true scientific field dates back to 1956 at a conference in the United States held at Dartmouth College. Subsequently, this field will reach prestigious universities such as Stanford, MIT, or even Edinburgh. By the mid-1960s, research around AI on American soil was primarily funded by the Department of Defense. At the same time, laboratories are opening up here and there around the world. Some experts predicted at the time that "machines will be able, within 20 years, to do the work that anyone can do". If the idea was visionary, even in 2018 AI has not yet taken on this importance in our lives. In 1974 came a period called "AI Winter". Many experts fail to complete their projects, and the British and American governments are cutting funding for academies. They prefer to support ideas that are more likely to lead to something concrete. In the 1980s, the success of expert systems made it possible to relaunch research projects on AI. An expert system was a computer capable of behaving like a (human) expert but in a very specific field. Thanks to this success, the AI market has reached a value of $1 billion, which motivates the various governments to once again financially support more academic projects. The exponential development of computer performance, in particular by following Moore's law, allowed between 1990 and 2000 to exploit AI on previously unusual grounds [7]. We find at this time data mining or medical diagnostics. It was not until 1997 that there was a real media release when the famous Deep Blue created by IBM defeated Garry Kasparov, world chess champion.
- **AI between 2000 and 2010**: in the early 2000s, AI became part of a large number of "science fiction" films presenting more or less realistic scenarios. The most significant of the new millennium being certainly Matrix, the first part of the saga released in theaters on June 23, 1999. Will follow A.I. by Steven Spielberg released in 2001, inspired by Stanley Kubrick, then I, Robot (2004) [8]. Metropolis (1927) Blade Runner (1982), Tron (1982), and Terminator (1984) had already paved the way but still didn't know enough about AI and its applications to imagine real scenarios. Between 2000 and 2010, the company experienced a real IT-boom. Not only did Moore's Law continue on its way, but so did Men. Personal computers are becoming more and more accessible, the Internet is being deployed, smartphones

are emerging… Connectivity and mobility are launching the Homo Numericus era. Until 2010, there are also questions about the ethics of integrating AI in many sectors. In 2007, South Korea unveiled a robot ethics charter to set limits and standards for users as well as manufacturers. In 2009, MIT launched a project bringing together leading AI scientists to reflect on the main lines of research in this area [8].

- **AI from 2010**: from the start of our decade, AI stood out thanks to the prowess of Watson from IBM. In 2011, this super-brain defeated the two biggest champions of Jeopardy. However, the 2010s marked a turning point in the media coverage of research. Moore's Law continues to guide advances in AI, but data processing reinforces all of this. Then, to perform a task, a system only needs rules. When it comes to thinking and delivering the fairest answer possible, this system has to learn. This is how researchers are developing new processes for machine learning and then deep learning [9]. These data-driven approaches quickly broke many records, prompting many other projects to follow this path. In addition, the development of technologies for AI makes it possible to launch very diverse projects and to no longer think of pure and hard calculation, but to integrate image processing. It is from this moment that some companies will take the lead. The problem with AI is no longer having the brains to develop systems, but having the data to process. That's why Google is quickly becoming a pioneer [10]. In 2012, the Mountain View firm had only a few usage projects, up from 2700 three years later [11]. Facebook opened the Facebook AI Research (FAIR) led by Castellanos [12]. Data management will allow AI to be applied to understand X-rays better than doctors, drive cars, translate, play complex video games, create music, see through a wall, imagine a game missing from a photograph,…The fields where AI performs are more than numerous and this raises many questions about the professional role of Man in the years to come [11]. The media position that AI now occupies hardly any longer places questions concerning this domain in the hands of researchers, but in public debate. This logically creates as much tension as excitement. Unfortunately, we are only at the beginning of the massive integration of these technologies. The decades to come still hold many surprises in store for us.

AI, which helps to make decisions, has already crept into cars, phones, computers, defense weapons, and transportation systems. But no one can yet predict how quickly it will develop, what tasks it will apply tomorrow and, how much,… Finally, artificial intelligence is integrated into most areas of life, such as transport, medicine, commerce, assistance for people with disabilities and other areas (Table 1).

3 Agriculture and Greenhouse

According to the FAO-UN [13], there will be two billion more mouths to feed by 2050, but the cultivable area can only increase by 4%. To feed humanity, therefore,

Table 1 Applications of artificial intelligence

Field	Description
Transport	• Mobility is a favorite field of AI • Some systems spot signs of fatigue on the driver's face • Take complete control of the vehicle, whether it's a passenger car or a semi-trailer • Become the "brain of the city", modeling the demand for public transport or adjusting traffic lights to optimize the flow of cars
Health	• Patient-doctor link, research, prevention, diagnosis, treatment • AI penetrates all segments of the medical industry • It is already better than the doctor to detect the cancerous nature of melanoma or to analyze a radio of the lungs. For example, IBM's Watson system provides diagnostic and prescribing assistance, particularly based on DNA sequencing of cancerous tumors. Other AIs play virtual shrinks
Commerce	• An online buyer provides treasures of data with which AI can create a tailor-made "customer experience". Because the past behavior of a consumer makes it possible to predict his needs • E-commerce thus multiplies chatbots and personalized virtual assistants • Some physical stores are testing facial recognition systems, tracking customer journey, and satisfaction
Personal assistant	• Having conquered tens of millions of homes connected to their designers' AI platforms, domestic voice assistants such as amazon echo or google home aspire to become our "digital valets" • They offer to manage home automation, to provide you with information, music, leisure programs, and to order errands, meals, and cars
Industry	• Many AI solutions are already deployed for the optimization of raw materials and that of stocks, predictive maintenance, or intelligent logistics in advanced factories • AI is one of the crucial ingredients of the so-called 4.0 industry, based on connected objects, digital solutions, cloud computing, robotics, and 3D printing
Environment	• AI could help us reduce our carbon footprint and facilitate sustainable production methods, for example: optimizing water filtration systems to regulating the energy consumption of smart buildings, or from promoting frugal agriculture as inputs to establishing short circuits or protecting biodiversity
Finance	• The comparative advantage of AI being to feed on millions of data to extract meaning, finance is one of its fields of play • Learning systems are already at work in the fields of customer relations, risk analysis, market forecasts, investment advice, and asset management. Some warn of the dangers of this automation, which in terms of trading has already worsened the crashes
Defense	• Already at work in corporate cybersecurity, AI is an "element of our national sovereignty", according to former defense minister Jean-Yves Le Drian • It can predict the risks of armed conflict, command a swarm of drones, and lead a fleet of fighter planes into combat • In terms of civil security, AI allows the police to deploy to prevent demonstrations

it is not so much a question of cultivating more than cultivating better. However, technology is already being used to increase yields: drones, thermal cameras, and other humidity sensors are already part of the daily lives of farmers around the world. With the explosion of data from these tools, the use of AI becomes essential to analyze them and help farmers make the right decisions.

The growing number of high-level meetings and conferences held annually around the world demonstrates the interest of researchers and scientists in the application of intelligent techniques in various fields; economic, industrial, automatic, biophysical, biological, etc. This new trend calls us to apply it in the field of climate management of agricultural greenhouses, for the purposes of quality production, productivity, profitability, food self-sufficiency, and why not in the case of excess production.

The greenhouse is considered to be a very confined environment where several components are exchanged between them. The main factors of the internal environment of the greenhouse are: temperature, light, humidity [14]. It is well known "greenhouse effect", that the ground and the plants located under shelters receiving the rays of the sun heat much more than in the open air: this is due to the suppression of the wind and the reduction of air convection, but also the physical properties of the greenhouse cover (transparent enough for solar radiation), but as an absorbent for infrared emitted by the soil placed at ordinary temperature, hence the "trapping" effect solar radiation. In summer a dangerous overheating is to be feared natural or forced ventilation (mechanical) is essential to cool the greenhouse. In winter, heating is generally required either in layers (heat of fermentation of manure or dead leaves), as well as other biotechnological procedures, either by other sources of energy (electricity, fuel, solar energy).

In addition, the temperature intervenes in a preponderant way in the growth and development of the vegetation. Then, the humidity increases in the greenhouses thanks to the transpiration of the plant, in the absence of wind and by evapotranspiration in the relatively closed enclosure. The concentrations of CO_2 and water vapor play a decisive role in the transpiration and photosynthesis of plants as well as in the development of fungal diseases. Solar radiation is also involved in photosynthesis.

A well-controlled control of the energy/mass balance of the climate, therefore, makes it possible to manage these parameters and improve the physiological functioning of plants. In this bibliographic study, we present the characteristics and climatic conditions of a greenhouse as well as the study of different foundations and essential tools for controlling the climatic parameters of the microclimate. The protected crop does not obey external random constraints and it is not affected by the problems encountered in field crops. Extreme unpredictable climatic conditions hamper production, while the modern greenhouse with its technical means meets the requirements for plant growth and development.

According to Duarte-Galvan et al. [15], some advantages of the agricultural greenhouse can be listed below:

- Satisfactory production and yield.
- Out-of-season production of fruits, vegetables and floral species.
- Significant reduction in plant pests thanks to air conditioning.

- Reduced exploitation of agricultural land.
- Quality and earliness of harvests. The modern agricultural greenhouse contributes greatly to the development and future strategy of the agricultural sector. It imposes itself with its new technologies and turns into a real industry.

Then, there are two types of greenhouses:

a. **Tunnel greenhouses and horticultural greenhouses**: the tunnel greenhouse consists of a series of juxtaposed elements each consisting of a steel tube frame and profiles assembled by bolts. The width is from (3 to 9) m. The plastic film is fixed by various clip systems which wedge the film against the profile or between two strips throughout the greenhouse. The conventional tunnel greenhouse is arched. There are also models with straight feet as for glass greenhouses, some of them are convertible for their covers [16].
b. **The chapel Greenhouses**: is the building unit of the greenhouse formed by two vertical sidewalls (or very slightly tilted) and a roof with two slopes, generally symmetrical, The chapel is characterized by its width [17], whose current dimensions are approximately between (3, 6, 9,12 and 16 m). When two consecutive chapels are not separated by an internal vertical wall, we speak of a multi-chapel greenhouse or twin chapels. The farm is the main supporting structure of the chapel, repeated at regular intervals. The module is a characteristic surface of the greenhouse obtained by making the product of the width of the chapel by the length of the truss; The gables are vertical walls forming the two ends of a chapel; The ridge is the line formed by the top of the chapel; The portico is a load-bearing structure, which existed in old greenhouses, it is supported by the farmhouse and by a beam joining the tops of the right feet.

The greenhouse is made up of two structures [18], a frame which constitutes the skeleton of the shelter, and a cover which makes the screen necessary for the creation of a microclimate specific to waterproofing.

a. **The frame**: this is the frame of the greenhouse, which must give rigidity to the entire structure and resist loads and the thrust of the winds. It can be made of concrete and wood (hard construction), galvanized steel, or anti-rust treated steel and aluminum. For shading, the dimensions of the arches, trusses, purlins, and all the elements constituting the height structure of the greenhouse must be as small as possible. The most suitable materials are steel and aluminum [19]. They have high resistance. Wood for equivalent resistance must have much larger dimensions and cause more shade. The use of aluminum or profiled steel has other advantages:

   - The use of standardized structural elements.
   - Easy assembly by the juxtaposition of elements.
   - Almost nonexistent deformation.
   - Reduced wear.
   - Installation in a reduced time.
   - Maintenance costs are minimal if not non-existent.

b. **Roofing materials**: their performance must be assessed on several levels in terms of their optical properties; by day: it is above all about presenting the best transmission to visible radiation useful for photosynthesis; at night: their emissivity in thermal infrared must be as low as possible, to limit the radiative losses. In other words, in terms of their thermal properties, their coefficients of conductivity and conduction losses must be as low as possible. They are therefore opaque infrared IR materials [20]. The coefficient of expansion of the material (wall) must be low, to avoid leakage problems and perpetuate the tightness of the system. Their lifespans and their weather resistance must be efficient. If the optical and mechanical properties of roofing materials in new condition are known, they are no longer known after a few months of use. The film undergoes alterations in the optical properties following photo-degradation and a weakening of the mechanical properties which are expressed in the form of tears, delamination, etc. In addition to soiling, the aging of the wall is declared and manifested by the yellowing of the cover.

In addition, several materials that are used to cover greenhouses such as glasses and plastics:

a. **Glass**: the point of view of light transmission, glass is the best material, especially special glasses. Its opacity to infrared radiation allows it to maintain and improve the greenhouse effect at best. Its weight and fragility mean that the use of these glass panels is cut in reduced dimensions to cover the walls, which consequently leads to a reinforcement of the frame which generates a little more shade. The structure must be very stable and requires a rigid foundation. The method of fitting the glass involves numerous joints producing imperfect caulking of the greenhouse. Even if the glass is almost unlimited in duration, it is still necessary to provide for a certain replacement rate following breakage. Since glass is not an insulating material, its use as a single covering leaves room for relative heat losses. The use of low-E glass allows savings of 20% with a reduction in brightness of around 10% while the use of double glass ("thermos" type) reduces heat loss by 40% [21].
b. **Plastic materials**: the use of plastic film has allowed a great development of greenhouses in recent decades. The most used material is PE polyethylene. It is robust, flexible, and mounts on a light structure. It is available in small thickness and very large width (12 m) [22]. Its transparency is high in the spectral ranges from 0.39 to 39 μm. It does not have the greenhouse effect ability except for special treatment or the presence of a film of condensed water on its inner face, infrared PE has the same capacities transparency than the previous one, but only allows a small proportion of long infrared to pass, thus equalizing the properties of the PVC film, its diffusing action eliminates direct shading on the ground caused by the structure. The easy installation of a polyethylene film and its low cost makes it the most used material for greenhouse coverings. The double-wall of blown polyethylene seems to be the most suitable, but its inability to retain the maximum of infrared radiation does not give it the greenhouse effect that glass has. However, installing a double wall leaving an insulating air space.

Reduces heat loss by around 40% compared to a single wall and considerably eliminates condensation inside compared to a single PE wall. The main weakness of polyethylene is its short lifespan due to aging problems and the appearance of mechanical breakdowns. In addition, the presence of dirt causes a decrease in light transmission.

4 Intelligent Control Systems (SISO and MIMO)

The principle of a basic feedback control system is to maintain the output of a process known as a "controlled variable" at the desired level called the reference or setpoint signal. In typical control applications, the process includes a single-input, single-output (SISO) or multi-input multi-output (MIMO) [23]. Depending on the types and requirements of the control system, different controllers are necessary to control such processes ranging from all or nothing regulators [23], to variations of proportional-integral controllers—derivatives (PID) [24], up to optimal adaptive controllers.

With the ever-increasing demands on control systems to adapt to other capacities and functionalities, these systems have become more and more complex. In addition, most of these systems are intrinsically non-linear. Temporal or dynamic variations are very complex and poorly understood, as is the case in most industrial process control systems [25]. However, recent advances in computer science, communication systems, and integrated control technologies have led experts in the field of engineering control to develop several new control strategies, architectures, and algorithms to meet the new requirements imposed on these systems and to cope with the complexities that arise. For this reason, the structures of the control system with a multi-controller function [26] have evolved considerably as the most dominant control architectures, where several controllers are used to controlling an entire system. Systems with multiple controllers have become standard in the control community. In addition, the subject of multi-controller systems has been developed on a variety of different research areas and mainly studied under various names and terms, such as distributed control [27], multiple-control [28], multi-agent control, cooperative control systems, collaborative control and distributed learning [29–32]. These architectures have been successfully used in different fields of application, such as industrial processes, power systems [33–35], telecommunications, robots [36] and automobiles, to name a few. Only a few areas of application. Thanks to a brief study of these architectures, it becomes clear that they have more advantages over a single control structure and that they have been proposed for two main purposes. First, to manage the complexity of these systems more effectively and, secondly, to achieve learning through collaboration between the different parts of a system [37, 38]. In the monitoring of complex process control systems, an architecture of a multi-agent system has been introduced in [29]. In the context of learning, various works have been reported. Researchers from [39] described a learning algorithm for small autonomous mobile robots, where two robots learn to avoid obstacles and the knowledge acquired is then transmitted from one to the other by intelligent communication.. An analytical study of a multi-agent environment has been carried out, where agents perform similar tasks and exchange information with each other. The results showed an improvement in performance and a faster learning rate for individual agents. Along with the aforementioned control architectures, intelligent control has emerged as one of the most dynamic fields in control engineering in recent decades. Intelligent control uses and develops algorithms and designs based on emulating intelligent behaviors of biological beings, such as how it performs a task or how it can find an optimal solution to a problem. These behaviors can include adapting to new situations, learning from experience, and cooperation in performing tasks. In general, intelligent control uses various techniques and tools to design intelligent controllers. The tools are commonly called soft computing and computational intelligence [32, 38], and the main, widely used examples includes: FL, ANN, and EAs.

4.1 Particular Aspects of Information Technology on Greenhouse Cultivation

In recent decades advances in IT have been applied to greenhouse cultivation, responding to the need for uniform production of plants throughout the year. Growing plants in a controlled environment is a very complicated process with many parameters that can directly or indirectly affect productivity. For these parameters to be controlled, all the physical phenomena of the greenhouse environment must be analyzed to calculate the energy and mass balances. Feedback control is only based on instantaneous measurements in real-time, but for optimal control and dynamic management of physical and biological [40] models are still being researched. Physical systems have been well defined and developed for a long time while biological systems are more complex and uncertain. Efforts in biophysical modeling have only recently reached a stage of practical use and have a long way to go to become a mature coupling of biophysics and technology. However, the societal requirements for respect for the environment and the quantitative and qualitative requirements of consumers, under the competition of world market prices, add new dimensions and constraints in the optimal management of a viable system. Integrated production management provides both the reason and the means to advance in this biophysical field (models concerning insect populations, diseases, production, etc.) and the IT implementations which will have to reach levels to become reliable and integrate as necessary inputs to the production process. Cultivation technologies (hydroponics, harvesting, robotics, plant factories, etc.) are becoming mature and less costly, because they gain wide acceptance, which stimulates knowledge needs as they move from the information age, from classical theoretical modeling to the era of knowledge and artificial intelligence. Efforts have been made based on modern communication technologies to provide the missing bridge connecting knowledge bases to emulation within intelligent command controllers [41].

4.2 Greenhouse Climate Control Techniques

Many studies have been carried out on greenhouse climate control. Among these studies, the PD-OF control structure to control the temperature of the greenhouse [42]. This diagram is a modification of the PDF algorithm. The PIP controller has also been used to control the ventilation speed in agricultural buildings to regulate its temperature [43]. Controlling the air temperature alone can only lead to poor greenhouse management. This is mainly due to the important role of relative humidity which acts on biological processes (such as perspiration and photosynthesis). This is the reason, why, that we pay in research, more attention to the coupling between the temperature of the indoor air of the greenhouse and the relative humidity. These variables were checked simultaneously using the PID-OF control structure, and later, the PI control structure. Although good results have been obtained using these conventional controllers, their strength has deteriorated under the effect of the operating conditions of the process. Smart control schemes are offered as an alternative option for controlling such complex, unreliable, and non-linear systems. So we can say that the basis for controlling the greenhouse environment consists of conventional control techniques such as the PID controller and artificial intelligence techniques such as neural networks and or FL, which we count, apply to climate control and regulation of the internal atmosphere of the greenhouse.

Plants are sensitive to light, carbon dioxide, water, temperature, relative humidity as well as to the movements of air which occur during aeration and the contribution of certain elements. (Supply of fertilizers, carbon dioxide enrichment, water supply, misting, etc.). These different factors act on the plant through:

- **Photosynthesis**: thanks to chlorophyllin assimilation, the plant absorbs carbon dioxide, rejects oxygen. This assimilation is only possible in the presence of light. Within certain limits, it becomes all the more active as the light is intense.
- **Breathing**: the plant absorbs oxygen and releases carbon dioxide. Breathing does not require light and continues both at night and during the day. It burns the reserves of the plant, while photosynthesis develops them.
- **Sweating**: the plant releases water vapor.

Despite these constraints, INRA offers temperature ranges to be respected depending on the stage of development of the plant classifies vegetable plants in four categories, according to their thermal requirements (Table 2) [44]:

- **Undemanding plants**: lettuces and celery.
- **The moderately demanding plants**: the tomato.
- **Demanding plants**: melon, chilli, eggplant, beans.
- **Very demanding plants**: cucumber.

Table 2 Needs of the vegetable species cultivated under shelters in the function of the development stage (INRA)

Vegetable species	Time between semi and start of harvest (days)	Flowering temperature (C°)		Relative humidity %	Critical temperature	
		Air	Ground		Air	Ground
Lettuce	110–120	04–06 (N) 08–10 (D)	08–10	60–70	−2	3
Tomato	110–120	15–10 N 22–28 D	16–20	60–65	+4	8
Cucumber	50–60	16–18 N 23–30 D	20–22	75–85	+6	12
Melon	115–125	16–18 N 25–30 J	18–20	50-60	+5	11
Chilli pepper	110–120	16–18 N 23–27 D	18–20	60–70	+5	10
Eggplant	110–120	16–18 N 23–27 D	18–20	60–70	+5	10
Bean	55–65	16–18 N 20–25 D		60–70	+4	08
Celery	110–120	16–18 N 20–25 D	12–20	60–70	−1	4

4.2.1 Classic Control

In classical control, the systems to be controlled are considered as input-output systems. The inputs are generally controlled disturbances, while the outputs are generally the variables to be controlled. In the greenhouse environment, the control inputs can be the amount of heating, the ventilation speed (opening windows, fan speed), the amount of additional lighting, the position of the screen, and the rate of CO_2 enrichment. Outdoor temperature and relative humidity, wind speed and direction, solar radiation, and CO_2 concentration are considered to be disturbances. The outputs are the interior temperature, the relative humidity, the CO_2 concentration and the light intensity at the level of the power plant.

The most widely used conventional control technique in greenhouse cultivation systems is feedback control. The regulator is often of the simple ON/OFF type or of the proportional-integral bypass (PID) type. A PID controller can manage set point changes, to compensate for load disturbances, and to cope with great uncertainty in the model. To improve the management and control of a greenhouse process, an adaptive PID control strategy (Fig. 1) can be applied to calculate the optimum control signals used for a function defined by cost/performance. Simpler versions of the PID regulator have also been used in monitoring greenhouse conditions.

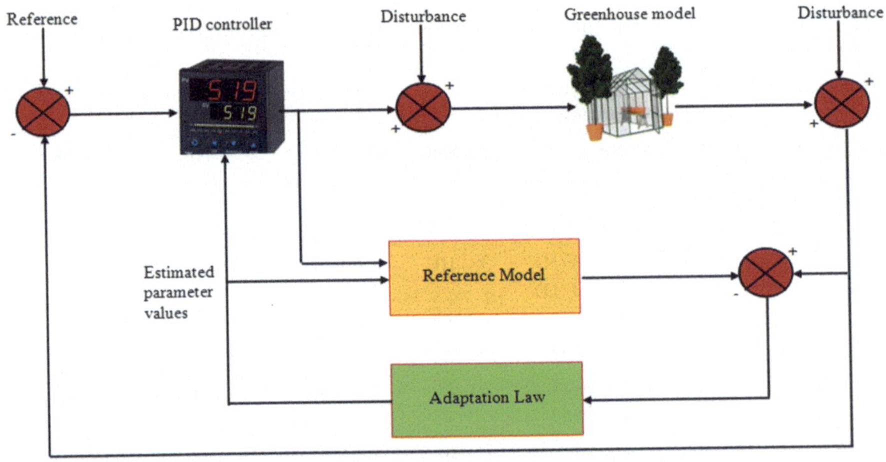

Fig. 1 Structure of the adaptive PID controller

4.2.2 Intelligent Control in Real-Time

In recent years, IT has played an important role in the development and materialization of control systems for greenhouse crops, In particular, the development of computer methodologies in the field of AI, which have been widely used to develop highly sophisticated intelligent systems for real-time control and management of surrounding installations, where conventional mathematical control approaches are difficult to apply [45]. ANN have been the most used tool for intelligent control of the greenhouse environment and hydroponics. Their main advantage is that they do not require an explicit evaluation of the transfer coefficients or any model formulation. They are based on the inherent data learning capacities of the process to be modeled. Initially, ANN was used in the modeling of the air environment of greenhouses, they are generally used as external environmental parameters of inputs (temperature, humidity, solar radiation, wind speed, etc.), control variables and state variables (instructions for cultivated plants). Simpler models for empty greenhouses that do not take into account plant conditions have also been successfully applied in temperature modeling. It should be noted here that the ANNs are generally a bad extrapolation, which means that they do not work satisfactorily under conditions different from those of the training data. In hydroponic systems, neural networks have been used to model with great precision the PH and electrical conductivity of the nutrient solution in deep culture systems as well as the rate of photosynthesis in cultivated plants. Also, ANNs have been used successfully in greenhouse environment control applications [46]. Very recently, their combination with GA in hydroponic modeling has been proven, and has given more success than the modeling of conventional neural networks [47].

GA is another AI technique that has been applied to the management and control of greenhouse crops. Their ability to find optimal solutions in large complex research spaces, as well as their innovative design capabilities inspired by the simulation of natural evolution, make them very powerful tools for design and optimization in several applications of engineering. They have been used as an optimization tool for the adjustment of greenhouse environment controllers also as methodologies for training agricultural models of neural networks as well as optimizers which determine optimal set values and as optimizers of other controllers (soft-computing-based) such as fuzzy controllers.

FL is an intelligent technique commonly used in advanced control and management of greenhouse cultivation systems. The complex processes and interactions of the greenhouse environment make this kind of flexible control by the application of FL, a powerful, efficient and successful tool in the precise control of the management of greenhouse systems or in combining with AGs and ANNs. It has been used to provide a larger scale between different sizes of production systems and loads in ventilation control and story heating and ventilation systems in greenhouses. It has also been used to provide real-time intelligent management decisions for controlling the greenhouse environment and hydroponics.

4.2.3 Adaptive Control

Adaptive controllers are essential in the area of greenhouse air conditioning, as greenhouses are continuously exposed to changing climatic conditions. For example, the dynamics of a greenhouse change with changes in the speed and direction of the outside air, The outside climate such as air temperature, humidity and CO_2 concentration, altitude of the greenhouse and the thermal effect on the growth of the plant inside the greenhouse. Therefore, the greenhouse moves between different operating points in the growing season and the controller is artificially aware of the operating conditions and adjusts to the new data. Research into adaptive control began in the early 1950s. An adaptive controller consists of two loops: a control loop and a parameter adjustment loop. The adaptive reference system model is an adaptive system in which the performance specifications are given by a reference model. In general, the model returns the desired response to a command signal. The parameters are changed based on the model error, which is the deviation of the plant's response from the desired response.

5 Modern Optimization Techniques

In recent years, several heuristic research techniques have been developed to solve combinatorial optimization problems. The word "heuristic" comes from the Greek word "heuriskein" which means "to discover or find" and which is also the origin of "Eureka", and resulting from the alleged exclamation of Archimedes [48]. However, three methods, which go beyond simple local search techniques and become particularly known as global optimization techniques and GA [49]. These methods all come at least in part from a study of the natural and physical processes which perform an optimization analogy. These methods are used to optimize an objective function with multiple variables [50]. The variable parameters are then changed logically or "intelligently" and presented to the objectivity function to determine whether or not this combination of variable parameters results in improvement.

5.1 Genetic Algorithms

The GA method is a global research technique based on an analogy with biology in which a group of solutions evolves through natural selection and survival of the fittest. The GA method represents each solution by a binary bit string or directly in its real value. Such a chain consists of substrings, each substring representing a different parameter. In GA terminology, bits are called "genes" and the whole chain is called a "chromosome". Several chromosomes representing the different solutions include a "population". This method is not based on the gradient; it uses an implicitly parallel sampling in the space of the solution. The population approach and multiple sampling mean that it is less subject to the trap in the local optima and those traditional optimization techniques explore a large space in the solution. The GA is powerful to reach an optimal or very close to the optimal solution.

The structure of GA is quite simple. GA begins with the random generation of the initial population chains, and the evaluation of each fitness chain. The algorithm proceeds by selecting, according to the strategy used, two "parental" solutions, exchanging portions of their strings and thus generating two "descending" solutions.

This process is called a "crossing". The process is repeated until the new population size is completed. The selection of a chromosome is generally based on its suitability for the suitability of other chromosomes in the population. In each generation, relatively the "good" chromosomes (solutions) are more likely to survive and produce offspring, and the "bad" chromosomes are doomed to die. To ensure an additional variety, the mutation operator acts with a small probability on the crossing for the random switching of one or more bits. Finally, the new population replaces the old (first). This procedure continues until a certain finalization condition is reached. A simple flowchart of a GA is presented in Fig. 2. There are several aspects in which GA differ from other research techniques:

- GAs optimizes the compromise between exploring new points in the research space and exploiting the information discovered so far.
- GAs has the property of implicit parallelism which means that the GAs effect is equivalent to an extensive search for hyper planes of the given space, without directly testing all the values of hyper planes.
- GAs is randomized algorithms, in the sense that they use operators whose results are governed by probability. The results of these operations are based on the value of a random number.

Fig. 2 GA flowchart

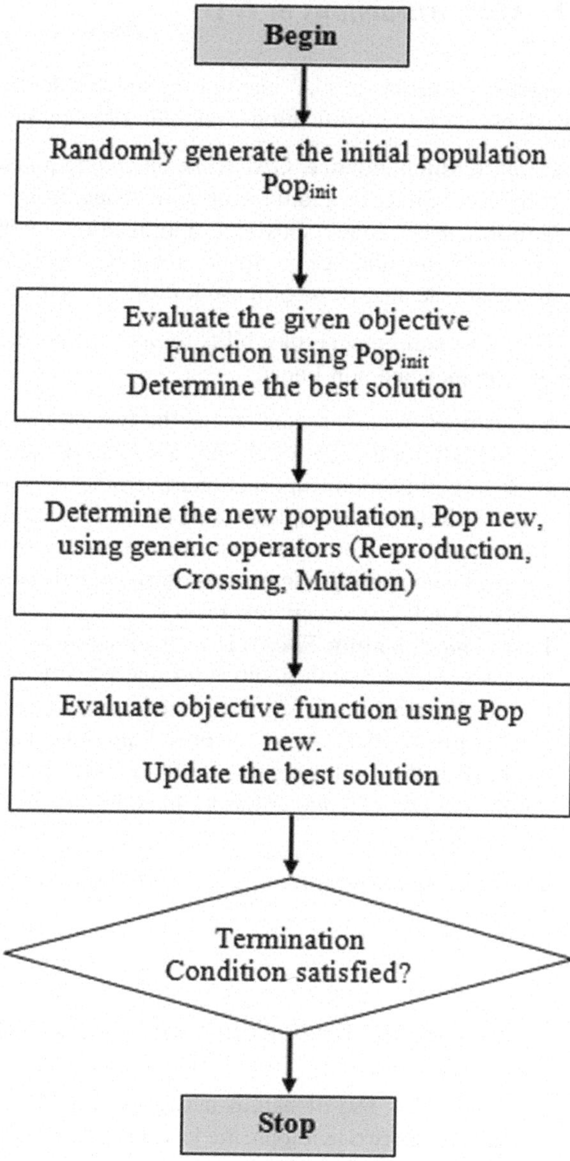

- GAs operates on several solutions simultaneously, gathering information from current research points to direct subsequent research. Their ability to maintain several solutions simultaneously makes AGs less sensitive to problems of the local optimum.

In recent years, interest in GAs has grown rapidly. Researchers are involved in various fields such as IT, engineering and operational research.

5.2 Main Attractions of GAs

The main attractions of GA listed in are independent of the domain, non-linearity, robustness, ease of modification, and multi objectivities.

- **Domain independence**: GAs work on coding a problem, so it's easy to write a general computer program to solve the many different optimization problems.
- **Non-linearity**: Many classical optimization methods depend on a hypothesis restricting the search space, for example linearity, continuity, convexity, differentiable, etc. None of these limitations is necessary for GAs.

The only requirement is the ability to calculate performance to some extent, which can be complex and non-linear.

- **Robustness**: As a consequence of the two properties listed above, the AGs are intrinsically robust. They can face a variety of types of problems; they not only work the highly nonlinear functions, but rather they process them in a very efficient way. In addition, the empirical data show that, although it is possible to refine an AG to better work on a given problem, it is nevertheless true that a wide range of GA parameter settings (selection criteria, population size, crossover and mutation rate, etc.) will give acceptable results.
- **Ease of modification**: Even relatively minor modifications of a particular problem can cause significant difficulties for many heuristic methods. On the other hand, it is easy to change a GA to model the initial variables of the problem.
- **Multi-objectivities**: One of the most important characteristics of GA; is that they can provide the multi-objectivity of the fitness function that can be formulated to optimize more than one criterion. In addition, GAs are very flexible in choosing an objective function.

These characteristics give GAs the ability to solve many complex problems in the real world.

5.3 Strong and Weak Points of FL and Neural Networks

The simultaneous use of neural networks and FL makes it possible to draw the advantages of the two methods: the learning capacities of the first and the readability and flexibility of the second. In order to summarize the contribution of the fuzzy neuron, groups together the advantages and disadvantages of FL and neural networks. Neuron-fuzzy systems are created to synthesize the advantages and overcome the disadvantages of neural networks and fuzzy systems. Learning algorithms can be used to determine the parameters of fuzzy systems. This amounts to creating or improving a fuzzy system automatically, using methods specific to neural networks. An important aspect is that the system always remains interpretable in terms of fuzzy rules since it is based on a fuzzy system.

6 Fuzzy Identification

In the context of control, identification refers to the determination of a model that is sufficient to allow the design of a climate controller for the system. The identification of time-invariant linear systems is carried out directly by conventional methods, so that fuzzy identification techniques do not follow such systems. Fuzzy techniques are useful for identifying non-linear systems, especially if the form of non-linearity is not known. The identifier takes measurements of the entrance and exit of the greenhouse and determines a model for it. Indeed, it is "enough" to determine the parameters of the controller (rules, the form of membership functions, the position of membership functions, etc.), to solve the non-linear optimization problem.

The general idea of ANFIS is to integrate in a single concept the advantages of the two domains: FL and Neural networks.

- **Fuzzy logic**: Introduction of prior knowledge to reduce the space of parameters to be optimized.
- **Neural networks**: Use of the "back propagation" optimization method. The first use is of course to obtain a model implantable in a computer of a nonlinear system. The interpretation of this model is still subject to caution.
  The synthesis of a regulator, fuzzy or not, obtained from a fuzzy identification, is one of the current themes of scientific research. For the moment, no clear methodology has emerged. Another application is quite common; it is the identification of more classic correctors. Indeed, the synthesis of fuzzy correctors is quite "fuzzy". Also, some authors recommend the identification of a classic corrector by a fuzzy system and extend it to new membership functions and new rules. The synthesis of a corrector follows the following algorithm:
  - Synthesis of a linear corrector around the operating point.
  - Fuzzy identification of this corrector.
  - Addition of rules and/or premises to extend it to the area of proper functioning. Another method giving good results consists in simulating the process to be commanded preceded by a crazy corrector, then in optimizing the parameters of this corrector or according to expected performances in closed loop.

7 Conclusion

If there is one field in full development at present, it is that of artificial intelligence. From face recognition, conversational assistants, to autonomous vehicles, and online shopping recommendation systems, these new technologies are invading our daily lives. Indeed, an artificial neuron network has already been explored for some time in Agriculture. First by the world of research then by that of research and development. At a time when the first commercial applications will arrive on the market, it seems important to us to be able to take an enlightened look at these technologies: understand what they are, what the applications, the limits are, and what are the questions that remain unanswered… This is what we wish to propose through this work.

As mentioned in the previous sections, we can say that the fuzzy controller has structures of different types. In addition, the components of a fuzzy controller have several parts, such as number; type; the position of the input and output membership functions; Entry and exit earnings; and the rules. These variations in the controller structure have significant effects on the performance of the fuzzy controller.

The problems of fuzzy controllers have been partially addressed by many researchers in the context of their applications. Due to the non-linearity, the inconsistency of the fuzzy controllers, difficulties arose when attempts were made to design a FL controller for general use.

Although valuable research has been carried out on the design of auto-tuning algorithms for fuzzy controllers, there is still a lack of study and empirical or analytical design covering the systematic auto-tuning method. In addition, most algorithms involve tuning multiple controller parameters that make the process of turning complex. In addition, the clear definition of physical parameters has been neglected, as is the case in the PID controller.

Indeed, adjustment efforts remain limited and local for a controller which retains the knowledge for future use and shares it with identical controllers with similar tasks.

The research work was started by a rich and interesting bibliographical study, which allowed us to discover this current field. A description of the types and models of agricultural greenhouses has been presented. Thermo hydric interactions, which occur within the greenhouse, have been approached. The biophysical and physiological state of plants through photosynthesis, respiration, and evapotranspiration were exposed while taking into account their influences on the immediate environment and the mode of air conditioning. Models of climate regulation and control have been discussed from the use of conventional devices to the use of artificial intelligence and/or FL. Knowledge models and IT techniques have been established following a well-defined approach and hierarchy for optimal climate management of greenhouse systems, while of course adopting Mamdani's method.

References

1. Z. Li, J. Wang, R. Higgs et al., Design of an intelligent management system for agricultural greenhouses based on the internet of things, in *Proceedings of the 2017 IEEE International Conference on Computational Science and Engineering and IEEE/IFIP International Conference on Embedded and Ubiquitous Computing, CSE and EUC* (2017)
2. D. Piscia, P. Muñoz, C. Panadès, J.I. Montero, A method of coupling CFD and energy balance simulations to study humidity control in unheated greenhouses. Comput. Electron. Agric. (2015). https://doi.org/10.1016/j.compag.2015.05.005
3. E.J. van Henten, Greenhouse climate management : an optimal control approach. Agric. Eng. Phys. PE&RC (1994)

4. R. Ben Ali, S. Bouadila, A. Mami, Development of a fuzzy logic controller applied to an agricultural greenhouse experimentally validated. Appl. Therm. Eng. **141**, 798–810 (2018). https://doi.org/10.1016/J.APPLTHERMALENG.2018.06.014
5. T. Morimoto, Y. Hashimoto, An intelligent control technique based on fuzzy controls, neural networks and genetic algorithms for greenhouse automation. IFAC Proc. **31**, 61–66 (1998). https://doi.org/10.1016/S1474-6670(17)42098-2
6. Y. Lu, Artificial intelligence: a survey on evolution, models, applications and future trends. J. Manag. Anal (2019)
7. J.C. van Dijk, P. Williams, The history of artificial intelligence. Expert. Syst. Audit. (1990)
8. A. Benko, C. Sik Lányi, History of artificial intelligence, in: *Encyclopedia of Information Science and Technology*, 2nd edn (2011)
9. B. van Ginneken, Fifty years of computer analysis in chest imaging: rule-based, machine learning, deep learning. Radiol. Phys. Technol. (2017)
10. J. Ring, in *We Were Yahoo! : From Internet Pioneer to the Trillion Dollar Loss of Google and Facebook*
11. M. Haenlein, A. Kaplan, A brief history of artificial intelligence: On the past, present, and future of artificial intelligence. Calif. Manage. Rev. (2019). https://doi.org/10.1177/0008125619864925
12. S. Castellanos, Facebook AI Chief Yann LeCun Says Machines Are Decades Away From Matching the Human Brain—CIO Journal. Wall Str. J. (2017)
13. A. Wennberg, Food and Agriculture Organization of the United Nations, in *Encyclopedia of Toxicology*, 3rd edn (2014)
14. N. Radojević, D. Kostadinović, H. Vlajković, E. Veg, Microclimate control in greenhouses. FME Trans. (2014). https://doi.org/10.5937/fmet1402167R
15. C. Duarte-Galvan, I. Torres-Pacheco, R.G. Guevara-Gonzalez et al., Review advantages and disadvantages of control theories applied in greenhouse climate control systems. Spanish J. Agric. Res. (2012). https://doi.org/10.5424/sjar/2012104-487-11
16. N. Choab, A. Allouhi, A. El Maakoul et al., Review on greenhouse microclimate and application: design parameters, thermal modeling and simulation, climate controlling technologies. Sol. Energy **191**, 109–137 (2019). https://doi.org/10.1016/j.solener.2019.08.042
17. E. Iddio, L. Wang, Y. Thomas et al., Energy efficient operation and modeling for greenhouses: a literature review. Renew. Sustain. Energy Rev. **117**, 109480 (2020). https://doi.org/10.1016/J.RSER.2019.109480
18. G. Ted, in *Greenhouse Management: A Guide to Greenhouse Technology and Operations*. (Apex Publishers, USA, 2019)
19. Structure used as greenhouse roof frame, greenhouse roof frame, greenhouse framework, greenhouse, and greenhouse framework building method (2004)
20. G.-W. Bruns, Experiences on damages to roofing-materials and greenhouse construction. Acta Hortic 127–132. https://doi.org/10.17660/ActaHortic.1985.170.14
21. Greenhouse construction (1910)
22. C. von Zabeltitz, Greenhouse Structures, *Integrated Greenhouse Systems for Mild Climates* (Springer, Berlin, 2011), pp. 59–135
23. S. Fang, C. Jie, I. Hideaki, in *MIMO Systems*. Lecture Notes in Control and Information Sciences (2017)
24. K.H. Ang, G. Chong, Y. Li, PID control system analysis, design, and technology. IEEE Trans. Control Syst. Technol. (2005). https://doi.org/10.1109/TCST.2005.847331
25. Industrial Process Automation Systems (2015)
26. O. Blial, M. Ben Mamoun, R. Benaini, An overview on SDN architectures with multiple controllers. J. Comput. Netw. Commun. (2016)
27. E. Camponogara, D. Jia, B.H. Krogh, S. Talukdar, Distributed model predictive control. IEEE Control Syst. (2002). https://doi.org/10.1109/37.980246
28. J.F. Cáceres, A.R. Kornblihtt, Alternative splicing: multiple control mechanisms and involvement in human disease. Trends Genet. **18**, 186–193 (2002). https://doi.org/10.1016/S0168-9525(01)02626-9

29. D. Saba, B. Berbaoui, H.E. Degha, F.Z. Laallam, A generic optimization solution for hybrid energy systems based on agent coordination. in eds. by A.E. Hassanien, K. Shaalan, T. Gaber, M.F. Tolba *Advances in Intelligent Systems and Computing* (Springer, Cham, Cairo, Egypte, 2018) pp. 527–536
30. D. Saba, H.E. Degha, B. Berbaoui et al., Contribution to the modeling and simulation of multiagent systems for energy saving in the habitat, in *Proceedings of the 2017 International Conference on Mathematics and Information Technology*, ICMIT (2017)
31. D. Saba, F.Z. Laallam, B. Berbaoui, F.H. Abanda, An energy management approach in hybrid energy system based on agent's coordination, in *Advances in Intelligent Systems and Computing*, 533rd edn., ed. by A. Hassanien, K. Shaalan, T. Gaber, A.T.M. Azar (Springer, Cham, Cairo, Egypte, 2017), pp. 299–309
32. D. Saba, F.Z. Laallam, A.E. Hadidi, B. Berbaoui, Contribution to the management of energy in the systems multi renewable sources with energy by the application of the multi agents systems "MAS". Energy Procedia **74**, 616–623 (2015). https://doi.org/10.1016/J.EGYPRO.2015.07.792
33. D. Saba, F.Z. Laallam, H.E. Degha et al., Design and development of an intelligent ontology-based solution for energy management in the home, in *Studies in Computational Intelligence*, 801st edn., ed. by A.E. Hassanien (Springer, Cham, Switzerland, 2019), pp. 135–167
34. D. Saba, R. Maouedj, B. Berbaoui, Contribution to the development of an energy management solution in a green smart home (EMSGSH), in *Proceedings of the 7th International Conference on Software Engineering and New Technologies—ICSENT 2018* (ACM Press, New York, NY, USA, 2018), pp. 1–7
35. D. Saba, H.E. Degha, B. Berbaoui, R. Maouedj, *Development of an Ontology Based Solution for Energy Saving Through a Smart Home in the City of Adrar in Algeria* (Springer, Cham, 2018), pp. 531–541
36. M. Pöller, S. Achilles, Aggregated wind park models for analyzing power system dynamics, in *4th International Workshop on Large-scale Integration of Wind Power and Transmission Networks for Offshore Wind Farms* (2003), pp. 1–10
37. D. Saba, F. Zohra Laallam, H. Belmili et al., Development of an ontology-based generic optimisation tool for the design of hybrid energy systems development of an ontology-based generic optimisation tool for the design of hybrid energy systems. Int. J. Comput. Appl. Technol. **55**, 232–243 (2017). https://doi.org/10.1504/IJCAT.2017.084773
38. D. Saba, F.Z. Laallam, A.E. Hadidi, B. Berbaoui, Optimization of a multi-source system with renewable energy based on ontology. Energy Procedia **74**, 608–615 (2015). https://doi.org/10.1016/J.EGYPRO.2015.07.787
39. V. Vanitha, P. Krishnan, R. Elakkiya, Collaborative optimization algorithm for learning path construction in E-learning. Comput. Electr. Eng. **77**, 325–338 (2019). https://doi.org/10.1016/J.COMPELECENG.2019.06.016
40. R.S. Epanchin-Niell, J.E. Wilen, Optimal spatial control of biological invasions. J. Environ. Econ. Manage. (2012). https://doi.org/10.1016/j.jeem.2011.10.003
41. M. Vassell, O. Apperson, P. Calyam et al., Intelligent dashboard for augmented reality based incident command response co-ordination, in *2016 13th IEEE Annual Consumer Communications and Networking Conference, CCNC 2016* (2016)
42. K. Lammari, F. Bounaama, B. Draoui, Interior climate control of Mimo green house model using PI and IP controllers. ARPN J. Eng. Appl. Sci. **12** (2017)
43. C.J. Taylor, P. Leigh, L. Price et al., Proportional-integral-plus (PIP) control of ventilation rate in agricultural buildings. Control Eng. Pract. (2004). https://doi.org/10.1016/S0967-0661(03)00060-1
44. M.-P. Raveneau, Effet des vitesses de dessiccation de la graine et des basses températures sur la germination du pois protéagineux
45. H.-J. Tantau, Greenhouse climate control using mathematical models. Acta Hortic 449–460 (1985). https://doi.org/10.17660/ActaHortic.1985.174.60
46. M. Trejo-Perea, G. Herrera-Ruiz, J. Rios-Moreno et al., Greenhouse energy consumption prediction using neural networks models. Int. J. Agric. Biol. (2009)

47. I. González Pérez, A. José, C. Godoy, Neural networks-based models for greenhouse climate control. J. Automática 1–5 (2018)
48. E.K. Burke, M. Hyde, G. Kendall et al., A classification of hyper-heuristic approaches (2010)
49. Genetic algorithms in search, optimization, and machine learning. Choice Rev. (1989). https://doi.org/10.5860/choice.27-0936
50. A. Konak, D.W. Coit, A.E. Smith, Multi-objective optimization using genetic algorithms: a tutorial. Reliab. Eng. Syst. Saf. (2006). https://doi.org/10.1016/j.ress.2005.11.018

Artificial Intelligence in Smart Health Care

Artificial Intelligence Based Multinational Corporate Model for EHR Interoperability on an E-Health Platform

Anjum Razzaque and Allam Hamdan

Abstract This study explores the improvement of efficiency in e-Health by standardizing access to electronic health records (EHRs). Without overlaid organizations, EHR will remain an uneven and fragmented network of lagging systems unable to achieve accuracy and consistency, thus efficiencies. A multinational corporation (MNC) model is proposed to reduce healthcare costs, and implement a coherent system where data, technology and training are uniformly upgraded to alleviate interoperability issues. The conclusion revealed from our review of literature suggests that EHR interoperability issues may be mitigated by creating common architectures that enable fragmented systems to interoperate under supra organizations. As a result, an Artificial Intelligence based model is proposed to facilitate the improvement of the efficacy of e-Health to standardize HER.

Keywords Artificial intelligence · E-Health · EHR · Multi-national corporation · Interoperability

1 Introduction

This study aims to reveal how a Multinational national Corporation (MNC) organizational model can be a private sector substitute to the UK-NHS government model, in places where a public sector model cannot be developed. The following discussion attempts to show why and how the MNC model can provide its own solutions to a viable and well-integrated EHR system. This chapter suggests that the quality of healthcare (HC) and efficiency of access to electronic health records (EHRs) can be improved if appropriate solutions can be found to the interoperability problem

---

A. Razzaque (✉) · A. Hamdan
Management Information Systems Department, Collage of Business and Finance, Ahlia University, Manama, Bahrain
e-mail: Anjum.razzaque@gmail.com

A. Hamdan
e-mail: allamh3@hotmail.com

© The Editor(s) (if applicable) and The Author(s), under exclusive license to Springer Nature Switzerland AG 2021

A. E. Hassanien et al. (eds.), *Artificial Intelligence for Sustainable Development: Theory, Practice and Future Applications*, Studies in Computational Intelligence 912,
https://doi.org/10.1007/978-3-030-51920-9_5

between institutions, organizations and computer systems. While healthcare (HC) is not primarily an area where efficiency is paramount, and instead effectiveness is valued, the efficiency of access to EHR is important due to the time sensitive nature of medical decision-making.

One reason for lack of efficiency is standardized access to EHRs. Without a supra organization, EHR systems will remain fragmented, giving rise to uneven networks of lagging and degraded systems, or subsystems unable to achieve accuracy, consistency and efficiencies in providing quality access to EHRs. At present such a supra organization exists in the form of the UK-NHS (serving 60 million people); some other countries have similar organizations. However, such public or government sector umbrellas are not acceptable in every country, like in the USA, with a population of 310 million people. The MNC organizational model could also be considered as an alternative form in the non-profit NGO sector, or as a for-profit MNC in regions like the GCC.

This chapter first proposes a basic technology related model highlighting the need to overcome interoperability issues, and then suggests an organizational model that would be the best suited solution for a coherent system where data, technology and training are continuously and uniformly updated and upgraded. Furthermore, this model is also underpinned by Artificial Intelligence so to facilitate EHRs interoperability on an e-Health platform. Also, the chapter defines EHRs and EPRs, describes the nature of the interoperability problem, delineates some of the barriers to implementing a viable EHR system, and finally, how they can potentially be overcome. For resolving the interoperability issues, the organizational model suggested is a particular type of MNC model related to Internalization Theory, one that can also achieve efficiencies, economies of scale, and reduce healthcare costs, which would otherwise be inherently higher due to fragmentation and issues of interoperability. In other words, the inference based on discussions of literature is that most interoperability issues of EHR maybe mitigated by creating one or few common infrastructures and architectures that enable fragmented systems to interact with each other under supra organizations. This chapter, however, does not propose a data or connectivity architecture, due to its complexity, although there is discussion of what in EPRs, EHRs, and clinical systems need to be connected.

2 The Common Goal to Reduce Margin of Error in the HC Sector

HC is a service industry where the margin of error must be extremely low compared to other services [24, 25]. In HC an error could be fatal, and cannot be reversed, much like an airline pilot error, even though there are redundant systems built into an aircraft. As such, while duplicate systems in HC may not be favored because of cost effectiveness issues, of paramount importance are accuracy, the development of information architectures within data flow highways to ensure quality of data, quality and similarity of technology, consistency of training and semantics among those entering the data, information, and decisions.

To achieve these goals, it is imperative that an organizational structure that looms large over 'subsidiary' units (or in this case hospitals and Health Care units) is needed to coordinate and achieve efficiencies in records entry and sharing in an accurate and timely manner. Without such a supra organization, the subsystems or '*subsidiary*' units will remain fragmented, segmented or fractured, giving rise to serious interoperability issues and delays in patient care. A lack of appropriate communication systems and information sharing among Health Care (HC) colleagues as well as with patients is also purported to be the main challenge, and is known to cause 1 in 5 medical errors as a result of access to incomplete information which therefore becomes a barrier to effective decision-making. Another consequence of fragmentation of computer systems is the duplication of procedures, which increases costs, as well as danger and pain to patients [17]. Literature suggests that Electronic Patient Records (EPRs) will reduce the high rate of medical errors. However, studies have shown that the attempt to use Electronic Health Record (EHR) systems slows down normal procedures since the interface is not user-friendly despite the fact that the interface was carefully designed with the preferences of professional users in mind, and who assessed it before mass use. Apparently, there is still no solution that provides for the utilization of more advanced information and communication technology (ICT) input/output systems [29].

It appears that EHRs and ICT related issues have not gone unnoticed. In the American context, Senator Hillary Rodham Clinton stated, "We have the most advanced medical system in the world, yet patient safety is compromised every day due to medical errors, duplication, and other inefficiencies. Harnessing the potential of information technology will help reduce errors and improve quality in our health system" It is also clear that chapter-based records are no longer in favor and need to be replaced with electronic medical records-EMRs [9].

3 Defining EPRs, EHRs and Clinical Systems

EPRs as defined by the UK National Health Services (UK-NHS) is a periodic record of patient care provided by an acute hospital. Such a hospital provides HC treatment for the patient during a specific period [17]. An EHR on the other hand, is a patient's life-long record composed of multiple EPRs [29]. An EHR is further defined by the International Standards Organization (ISO) as a bank of comprehensive and longitudinal (long term from cradle to grave) patient-centered health-related information, independent of EHR Information Systems (IS) that support efficiency and integration of HC quality of care by also providing plans, goals and evaluations of patient care [12]. Another description of EHR is that it is the assembly of distributed data of a patient [4]. As envisioned, before an EPR can become functional, a clinical record system needs to be put in place, which over time will advance into EPRs. Appropriate clinical systems will then link departments with a master patient index and the system will be integrated with an electronic clinical result reporting system. A clinical system, consisting of a secure computer system, is a prerequisite for developing an EHR and EPR that allows hospital computer systems to communicate with each other, and enables physicians to obtain patient data from different hospitals by directly accessing the inter-operating hospital IS. The entire system would be capable of integration with specialized clinical modules and document imaging systems to provide specialized support. In general, an EHR is activated when advanced multi-media and telemedicine are integrated with other communication applications [3].

4 Some Hurdles in an EHR System

Briefly, other than fragmented or mutually exclusive computer systems that cause inefficiencies, user related hurdles are also barriers to EHR systems in terms of [24]: (1) interfaces that need improvement to better report EHR, (2) how data is captured, (3) setup of rules and regulations to obtain patient feedback and consent when sharing their data on EHR, (4) technology issues to tackle EHR implementation due to huge data transfers, their privacy supervision and complexity of the system based upon available ICT infrastructures. (5) Data quality and its nationwide availability and acceptability by patients, physicians and nurses is a pre-requisite for EHR development. (6) Another noted hurdle in EHR use is that physicians are unable to do their jobs because of considerable data entry requirements to populate EHRs.

5 Overcoming Interoperability Issues

Interoperability of EHR occurs when medical record-based HC systems (centered on patients, doctors, nurses and other relevant HC departments) can talk with one another. Interoperability is seen when one application accepts data from another application and then performs satisfactory tasks [17]. In addition, interoperability means that the IS should support functions like: (1) physical access to patient information, (2) access among providers in different care settings, (3) access and order patient tests and medicines, (4) access to computerized decision support systems, (5) access wrapped in a secured electrical communication between patient and physician, (6) automated administrative processes e.g.: scheduling, (7) access for a patient to disease management tools, his/her patient record or health-based information resources, (8) access to automated insurance claims, and (9) access to database reports for patient safety and health service efficiency (Source: [9]).

To some extent interoperability problems have been alleviated by the ISO—International Organization for Standardization which has developed structure and function based EHR standards and EHR processing systems. The ISO has published 37 HC related standards dealing with compatibility and interoperability. These standards are divided into three sub-sections, namely: (1) EHR content, structure and context in which it is used and shared, (2) technical architecture-based specifications for new EHR standards to exchange EHR for developers refereed as open EHR, and (3) standards to achieve interoperability between HC applications and systems, along with and messaging and communicating criteria for developers to enable trustworthy information interchange [25]. The interoperability issues involve several hurdles, and a list would not be limited to legal issues, ICT protocols, EHR protocols, IT architectures, uniform technologies and training, consistency in upgrading technologies, accuracy of entry, and interpretation, etc. The organizational structure that deals with all the above issues does not necessarily have to be a government entity, but possibly a for-profit or non-profit organization that could be like the UK-NHS and established in countries like the US. The purpose and goals would be different in each geographical area, but the organizational model could be similar. In the US a nonprofit organization or for-profit corporation would be acceptable in the same way health insurance companies are – private health insurance companies have local or nationwide networks of medical care providers and facilities, but only the insurance company, the doctor or the medical facility, and the patient are the repositories of their own records and information. If a person changes the insurance company but not the doctor, then the doctor can use that EPR, but if the doctor is changed and another insurance company is involved, then there are several procedures to follow to obtain records, which most often are transmitted in the form of chapter records. This occurs primarily because of legal and privacy issues, and because there are no automated or computerized systems that 'talk' to each other for the data to flow seamlessly between two doctors or hospitals that belong to two different insurance companies.

The interoperability issue in the above situation can only be resolved by cooperation between two entities, however most likely than not, a third factor often becomes a hurdle, and that is the legal system under which the two entities operate. The complications arising from the legal issues involved could range from an unwillingness to share proprietary information, to opening oneself up to 'leaking' of information for use or misuse of the EPR by the legal system, the attorneys' of the patient, or of another entity. This chapter posits that such fragmentation exists not only due, among others, to the legal environment and the lack of connecting data architectures, but also due to an organizational structure, public or private, that can internalize all transactions – such a legal entity exists in the case of the UK-NHS, and similar organizations in other countries, but a supra organization in the NGO sector or the private sector should be considered as well.

6 Barriers in EHR Interoperability

The barriers to connecting and implementing EHR are: (1) adaptability of new systems and hence work procedure by doctors, (2) costs in terms of healthcare savings, government requirements and motivation, (3) connecting vendors who need to be pressurized to make interoperable systems, and (4) standards that need to be set to ensure communication. Legislation is also a concern and therefore the US government has established laws to regulate HC funding allocations, which if broken would incur punishment by jail and relevant charges. Furthermore, there are laws that prohibit interoperability. Due to lack of connecting technology to facilitate interoperability, hence reduce cost and improve healthcare quality, the US is far behind other countries in terms of technology to support interoperability, even though it is at the top for the per capita spending, along with second ranked Germany, and third ranked France [25]. Barriers also exist due to the lack of training, or implementation of the shared care concept. Shared care is an innovation in patient care, and is provided by an integration of care by multiple hospitals and clinicians and requires well structured and designed interoperable IT systems, which if not designed properly will reduce patient-care quality [12].

7 Characteristics and Improvements of the UK-NHS Model

As one of the more advanced HC systems, and HC organizations in the world, the UK-NHS has dealt with numerous issues, including those mentioned above. However, it is in the public sector, and has the advantages, as well as disadvantages of being a government organization. There is much to learn from this supra organization, but it is one that operates with the help of legislation and nearly unlimited funds at its disposal. However, not all countries think alike on this subject, and therefore while it is a model to replicate in the public sector, a private sector organization could still emulate and develop a coherent HC system, with or without the efficiencies obtained by the UK-NHS.

Among the many innovations it has been able to implement, the local and national NHS's IT systems still need to upgrade or replace existing IT systems to: (1) integrate them, (2) implementing new national systems, and (3) patch NHS CRS with related HC-based products and services, such as E-prescription, which improves patient care by reducing prescription errors, reducing data redundancy, staff time and cost. In addition, the right infrastructure (N3) can also provide the NHS with intelligent network services and high broadband connections to improve patient care procedures by accessing patient care records anytime and anywhere, hence saving HC costs by remotely providing patient care and saving time by speeding up patient care. (Source: [22]).

8 Summary of MNC/MNE Characteristics

The alternative organization model to the UK-NHS model proposed in this chapter has certain characteristics that do not necessarily deal with HC but have dealt with issues like interoperability in the context of cross-border environments of more than one country and jurisdiction. The multinational enterprise, or MNC, is defined as any company, that "owns, controls and manages income generating assets in more than one country" [11]. In the context of this chapter, the relevant factors of interest are control and asset management, or HC facility, in more than one jurisdiction. The following is a summary of a MNC's additional characteristics as stated in literature pertaining to the issues discussed in this chapter:

(1) MNC's are well known for their ability to transfer technology, stimulation of technology diffusion, and provision of worker training and management skill development [14]. In other words, as a HC provider it would be capable of introducing and implementing new ICT's and upgrade the skills of those involved.
(2) They are also able to plug gaps in technology between the foreign investor and the host economy [19].
(3) There is evidence of more intensive coaching for suppliers in terms of quality control, managerial efficiency, and marketing…. [23].
(4) [5] state that American MNEs stress formalization of structure and process while European MNEs place greater importance on socialization.
(5) Internalization theory explains the existence and functioning of the MNE/MNC [28]. It contributes to understanding the boundaries of the MNE, its interface with the external environment, and its internal organizational design. Williamson [32] asserted that due to missing markets and incomplete contracts that gives rise to opportunistic behavior by others, the firm replaces external contracts by direct ownership and internal hierarchies which facilitates greater transactional efficiencies.
(6) MNC—Internalization Theory has also been characterized as 'old' and 'new', but its relevance to this chapter is only in terms of firm structures and upgrades to its technology. The theory posits that since the transaction costs of doing business in other countries are high, an MNC can achieve both tangible and intangible savings and possibly efficiencies by carrying out all or as many activities as possible, within its own organizational structure. Establishing a form of control and accountability over its assets, both human and materiel, guards against leakage of processes and Intellectual Capital, and enables the MNC to achieve cost efficiencies via internal contractual accountability. The same activities if carried out through the open market (via different companies, suppliers, etc.), especially in more than one legal environment (like States, Counties and Cities in the USA) would open up the possibility of numerous and costly hesitations for smaller organizations, hospitals, due to compliance issues, as well as dependence on external players with their own agendas.
(7) Another characteristic of a modern MNE is its emergence as an eMNE, where the cyberspace is a global network of computers linked by high-speed data lines and wireless systems strengthening national and global governance…. Can an e-MNE be defined as a firm that has facilities in several countries and its management achieved via cyberspace? [33]. Most cyberspace MNCs have achieved economies of scale and are capable or proficient in reducing costs.

(8) Today the e-MNE can control and manage income generating assets in more than one country by the means of a network spread around the world and an electronic base located in a single building or place [33].

In examining the internalization theory, two parallels can be discerned; one with the circumstances and environment of the organization (as represented by the UK-NHS model), and the external market (in the form of system disparities evidenced in interoperability issues). The UK-NHS provides the umbrella for over 60 million people as an organization with its own forms of controls and accountabilities afforded to it by the legal authority of the UK government. The nature of interoperability issues in the UK are not so much in the realm of legal jurisdictions, but in technology, data architectures and human factors. However, jurisdictional problems do occur when one moves away from the UK's legal environment, and into the country environments of the USA, and other countries not at par with the UK or US legal environments.

9 Proposed Solutions—the UK-NHS Model or the MNC Organizational Model

An EHR system attempts to achieve what the financial, commercial, etc. industries have already done and succeeded in [9]. This raises an obvious question, why not follow instead of re-designing the wheel. The answer is that HC is complex and therefore requires a customizable model to cater to its and its patients' needs. In addition, authors state that a lack of information in EHR prevents clinicians from making sound decisions [18]. Therefore, much more needs to be done in terms of input and output coordination.

Given the above, this chapter proposes an organizational model and structure that best suits an environment where interoperability problems can be overcome when faced with two or more complex systems [26]. The NHS in UK is one such organizational model that attempts to overcome interoperability issues through its writ of legislation and the law, which it can also help enact because it is a government agency. However, despite a conducive legal and political environment, there are other interoperability issues, that remain due to technology, training and behavioral resistance. This chapter's proposed alternative solution to the NHS-model relies mainly on one organizational model developed by MNC over several years and is of relevance because they operate across several boundaries and legal systems. An examination of the literature on MNCs results in the finding that although they are private corporations operating under two or more complex countries, they have had to deal with many types of interoperability issues and consequently have been able to overcome the hurdles, partly due to their ability to solve issues via access and deployment of massive resources. Moreover, this MNC model can be deployed on the e-Health platforms facilitated by AI; as expressed in the next section.

10 E-Health and AI

Computing machines have changed the HC sector from various dimensions, e.g., Internet of Things (IoT) [16] with machine learning and AI as vital players [10]. The role of AI is expanding with its deployments in the HC sector, has been evidencing AI within the e-Health platform. AI is so attractive because of its readily available datasets and resources. AI is already serving in the HC sector: e.g., dermatology [1, 20] oncology [2, 13] radiology [6, 30] just to exemplify a few. Majority of AI and machine learning is appreciated as a support tool for knowledge-based medical decision-making during collaborative patient care [8, 15, 27, 31]. AI is currently applied on e-Health platforms where such platforms are integrated to transfer patient content, e.g., EHRs for in order to be acquired in multiple environments, e.g. within the environments of the patients' homes and also into a clinical warn room [21]. This is an innovative and a compete management information systems that forms a homecare AI based decision support system deployable on an e-Health platform.

11 Conclusions

The issues examined in this chapter point to solutions that are not insurmountable. The UK-NHS has proven that they can manage the HC of 60 million people, though with issues of interoperability still to overcome—countries that follow this public sector managed HC system can choose to adopt this model, if they have the political and economic will to do so. Those countries whose Constitutions, legal systems, political systems, or economic resources, among other reasons, are not conducive to implementing a UK-NHS, or similar model, could chose an alternative in the MNC model suggested above, regardless of whether it is designed as a non-profit NGO or a for-profit corporation.

Inside the government sector an organization has the help of the government and its legislators to pass laws that can enable the functioning of an HC, EHR, and EPR system where interoperability issues need only to be identified, and sooner or later can be overcome by fiat or the writ of the legislature. Outside of the government sector, the complex interoperability issues can also be overcome by the creation of an internal market under the umbrella of a NGO or a Corporation. This chapter has addressed the interoperability problem by suggesting a MNC organizational model was developed to overcome many interoperability issues between countries. The conclusion is that a MNC model, with its own internalized market to control, is well suited to overcome EHR interoperability issues, integrate the interrelated IS architectures, upgrade them across the board, and train the employees with some consistency. Regardless however, heterogeneity in HC software applications across EHR systems will likely remain a problem [7].

Another aspect to consider is that the MNC model has already dealt with software, privacy, jurisdictional and several other issues in the financial sector, while dealing with highly confidential financial information, and giving people worldwide access to their accounts. Thus while the issues and problems are not insurmountable, the HC sector is more complex because it involves not just the swipe of the card and recording of data, but considerable amounts of subjective interpretations and conclusions are made by HC providers of varied skills, and then passed on to other HC providers. Finally, it was pointed out that the difference between the UK-NHS model and the MNC model is that the former can operate by legislating laws, and the latter by signing contracts with people, and holding them accountable via the legal system. Finally, the concept of AI is introduced in this chapter so to emphasize its importance for its deployment within the e-Health platform, so to make globally facilitate the proposed MNC model.

References

1. H. Almubarak, R. Stanley, W. Stoecker, R. Moss, Fuzzy color clustering for melanoma diagnosis in. Information **8**(89) (2017)
2. A. Angulo, Gene selection for microarray cancer data classification by a novel rule-based algorithm. Information **9**, 6 (2018)
3. Avvon Health Authority 2000. *Electronic Patient Records Electronic Health Records*, Schofield, J, Bristol
4. A.R. Bakker, The need to know the history of the use of digital patient data, in particular the EHR. Int. J. Med. Inf. **76**, 438–441 (2007)
5. C.A. Bartlett, S. Ghoshal, *Managing across Borders: The Transnational Solution* (Harvard Business School Press, Boston, MA, 1989)
6. B. Baumann, Polarization sensitive optical coherence tomography: A review of technology and applications. Appl. Sci **7**, 474 (2017)
7. A. Begoyan, An overview of interoperability standards for electronic health records. *Society for Design and Process Science. 10th World Conference on Integrated Design and Process Technology; IDPT-2007*. Antalya, Turkey, June 3–8
8. K. Chung, R. Boutaba, S. Hariri, Knowledge based decision support system. Infor. Technol. Manag **17**, 1–3 (2016)
9. Commission on Systemic Interoperability, *Ending the Document Game* (Washington, U.S, Government Official Edition Notice, 2005)
10. R.C. Deo, Machine learning in medicine. Circulation **132**, 1920–1930 (2015)
11. J. Dunning, Multinational enterprises and the global economy, Addison-Wesley, Wokingham 1992, (pp. 3–4)
12. S. Garde, P. Knaup, E.J.S. Hovenga, S. Heard, Towards semantic interoperability for electronic health records: domain knowledge governance for open EHR archetypes. Methods Inform. Med. **11**(1), 74–82 (2006)
13. I. Guyon, J. Weston, S. Barnhill, V. Vapnik, Gene selection for cancer classification using support vector. Mach. Learn. **46**, 389–422 (2002)
14. A. Harrison, *The role o multinationals in economic development: the benefits of FDI*. Columbia J. World Bus. **29**(4), 6–11
15. D. Impedovo, G. Pirlo, Dynamic handwriting analysis for the assessment of neurodegenerative diseases. IEEE Rev. Biomed. Eng. **12**, 209–220 (2018)
16. S.M. Islam, D. Kwak, M.H. Kabir, M. Hossain, K. Kwak, The Internet of things for health care: IEEE. Access **3**, 678–708 (2015)
17. A. Jalal-Karim, W. Balachandran, The Influence of adopting detailed healthcare record on improving the quality of healthcare diagnosis and decision making processes. in *Multitopic Conference, 2008 IMIC, IEEE International*, 23–24 Dec 2008

18. A. Jalal-Karim, W. Balachandran, Interoperability standards: the most requested element for the electronic healthcare records significance. in *2nd International Conference–E-Medical Systems*, 29–31 Oct 2008, EMedisys 2008, IEEE, Tunisia
19. A. Kokko, Technology, market characteristics, and spillovers. J. Dev. Econ. **43**(2), 279–93 (1994)
20. Y. Li, L.S. Shen, lesion analysis towards melanoma detection using deep learning network. Sensors **18**, 556 (2018)
21. A. Massaro, V. Maritati, N. Savino, A. Galiano, D. Convertini, E. De Fonte, M. Di Muro, A study of a health resources management platform integrating neural networks and DSS telemedicine for homecare assistance. Information **9**, 176 (2018)
22. NHS National Program for Information Technology nd. Making Ithappen Information about the National Programme for IT. *NHS Inforamtoin Authority, UK*
23. W.P. Nunez, *Foreign Direct Investment and Industrial Development in Mexico* (OECD, Paris, 1990)
24. A. Razzaque, A. Jalal-Karim, The influence of knowledge management on EHR to improve the quality of health care services. in *European, Mediterranean and Middle Eastern Conference on Information Systems (EMCIS 2010)*. Abu-Dhabi, UAE (2010)
25. A. Razzaque, A. Jalal-Karim, Conceptual healthcare knowledge management model for adaptability and interoperability of EHR. in *European, Mediterranean and Middle Eastern Conference on Information Systems (EMCIS 2010)*. Abu-Dhabi, UAE (2010)
26. A. Razzaque, T. Eldabi, A. Jalal-Karim, An integrated framework to classify healthcare virtual communities. in *European, Mediterranean & Middle Eastern Conference on Information Systems 2012*. Munich, Germany (2012)
27. A. Razzaque, M. Mohamed, M. Birasnav, A new model for improving healthcare quality using web 3.0 decision making, in *Making it Real: Sustaining Knowledge Management Adapting for success in the Knowledge Based Economy* ed. by A. Green, L. Vandergriff, A. Green, L. Vandergriff, Academic Conferences and Publishing International Limited, Reading, UK. (pp. 375–368)
28. A.M. Rugman, *Inside the Multinationals: The Economics of Internal Markets*. Columbia University Press, New York. (1981) (Reissued by Palgrave Macmillan 2006)
29. O. Saigh, M. Triala, R.N. Link, Brief report: failure of an electronic medical record tool to improve pain assessment documentation. J. Gen. Int. Med. **11**(2), 185–188 (2007)
30. I. Sluimer, B. Ginneken, Computer analysis of computed tomography scans. IEEE Trans. Med. Imag **25**, 385–405 (2006)
31. D. Stacey, F. Légaré, K. Lewis, M. Barry, C. Bennett, K. Eden, M. Holmes-Rovner, Decision aids for people facing health treatment or screening decision. Cochrane Database Syst. Rev. **4**, CD001431
32. O.E. Williamson, *Markets and hierarchies, analysis and antitrust implications: a study in the economics of internal organizations* (Free Press, New York, 1975)
33. G. Zekos, Foreign direct investment in a digital economy. Eur. Bus. Rev. **17**(1), 52–68 (2005). Emerald Group Publishing Limited

Predicting COVID19 Spread in Saudi Arabia Using Artificial Intelligence Techniques—Proposing a Shift Towards a Sustainable Healthcare Approach

Anandhavalli Muniasamy, Roheet Bhatnagar, and Gauthaman Karunakaran

Abstract Medical data can be mined for effective decision making in spread of disease analysis. Globally, Coronavirus (COVID-19) has recently caused highly rated cause of mortality which is a serious threat as the number of coronavirus cases are increasing worldwide. Currently, the techniques of machine learning and predictive analytics has proven importance in data analysis. Predictive analytics techniques can give effective solutions for healthcare related problems and predict the significant information automatically using machine learning models to get knowledge about Covid-19 spread and its trends also. In a nutshell, this chapter aims to discuss upon the latest happenings in the technology front to tackle coronavirus and predict the spread of coronavirus in various cities of Saudi Arabia from purely a dataset perspective, outlines methodologies such as Naïve Bayes and Support vector machine approaches. Also, the chapter briefly covers the performance of the prediction models and provide the prediction results in order to better understand the confirmed, recovered and the mortality cases from COVID-19 infection in KSA regions. It also discusses and highlights the necessity for a Sustainable Healthcare Approach in tackling future pandemics and diseases.

Keywords Predictive analytics · Covid-19 · Machine learning · Naïve bayes (NB) · Support vector machine (SVM)

A. Muniasamy (✉)
College of Computer Science, King Khalid University, Abha, Saudi Arabia
e-mail: anandhavalli.dr@gmail.com

R. Bhatnagar
Department of CSE, Manipal University Jaipur, Jaipur, India
e-mail: roheet.bhatnagar@jaipur.manipal.edu

G. Karunakaran
Himalayan Pharmacy Institute, Sikkim University, Sikkim, India
e-mail: gauthamank@gmail.com

© The Editor(s) (if applicable) and The Author(s), under exclusive license to Springer Nature Switzerland AG 2021

A. E. Hassanien et al. (eds.), *Artificial Intelligence for Sustainable Development: Theory, Practice and Future Applications*, Studies in Computational Intelligence 912,
https://doi.org/10.1007/978-3-030-51920-9_6

1 Introduction

The outbreak of the new coronavirus (COVID-2019) to more countries enforce many challenges and questions that are of great value to global public-health research, and decision-making in medical analysis [1]. By May 1, 2020, a total of 3,175,207 cases had been confirmed infected, and 224,172 had died [2] and particularly in Saudi Arabia (KSA), a total of 24,104 had been confirmed infected and 167 deaths [2]. Also, early responses from the public, control actions within the infected area, timely prevention control the epidemic outbreak at its earliest stage, which increase the potential of preventing or controlling the later spread of the outbreak.

COVID-19, named as a family of Corona virus spread in the year 2019, can cause illnesses such as the fever, cough, common cold, shortness of breath, sore throat, headache etc. It has some similarity like severe acute respiratory syndrome (SARS) and Middle East respiratory syndrome (MERS) but has its own symptoms and named as SARS-CoV-2 also [3]. It was originated in China and the World Health Organization (WHO) announced the COVID-19 virus outbreak a pandemic on March 2020. World Health Organization generates COVID-19 case reports regularly. So, the identification and prevention of COVID-19 should reduce this growing death rate and also the timely data analytics may provide great value to public-health research and policy-making. The Saudi Ministry of Health provides a daily update on confirmed, death and recovered cases due to Covid-19 infections in Saudi Arabia.

As the COVID-19 spreads to KSA nowadays, the analysis of the information about this novel virus data is of great value to public-health research and policy-making as the confirmed cases with Covid-19 can lead to fatal problems. Machine learning techniques are the best to provide the useful approximation to the given data and have been widely applied in different applications. Machine learning techniques has proven importance in patient case diagnosis [4] to predict the total number of infected cases, confirmed cases, mortality count and recovered cases and have better understandings of it. The applications of predictive analytics, such as optimizing the cost of resources, the accuracy of disease diagnosis, and enhancement of patient care improves clinical outcomes [5]. In healthcare, the applications like predicting patient outcomes, ranking of hospitals, estimation of treatment effectiveness, and infection control [6] are based on the machine learning classification and prediction.

The chapter focuses on the prediction of COVID-19 case history using machine learning techniques such as Naïve Bayes, and Support vector machine (SVM) on COVID-19 dataset which is collected from the Saudi Ministry of health website [7], to gain knowledge and trends of Covid-19 spread in KSA. Following the introduction section, we highlight some of the related work in applications of machine learning techniques in healthcare. The methodology section covers the information about the dataset and its preprocessing steps, the concepts applied machine learning techniques. The results and analysis section report an analysis and findings of the machine learning classifiers and predicted results. Finally, the chapter concludes with recommendations for sustainable healthcare COVID 19 for Saudi Arabia and research directions with summary section.

2 Literature Review

This section covers the related applications of machine learning (ML) techniques in healthcare. The application of machine learning models in healthcare is a challenging task due to the complexity of the medical data. In [5, 8], the authors described the new challenges in the machine learning domain due to the emergence of healthcare digitization. The applications of various machine learning classifiers have great impact on the identification and the prediction of various leading death rate diseases globally. The application of ML techniques has great impact in diagnosis and outcome prediction of the medical field. So, it ensures the possibility for the identification of relapse or transition into another disease state which are high risk for medical emergencies.

In machine learning, classification comes under supervised learning approach in which the model classifies a new observation dependent on training data set collection of instances whose classification is known. The classification technique Naïve Bayes(NB), based on Bayes' Theorem, assumes that the appearance of a feature is irrelevant to the appearance of other features. It is mainly used to categorize text, including multidimensional training data sets. Some examples are famously document classification, span filtration, sentimental analysis, and using the NB algorithm, one can quickly create models and quickly predict models. To estimate the required parameters, a small amount of training data is required for NB.

Ferreira et al. [9] reported in their research that Naive Bayes classifier (NB), multilayer perceptron (MLP), and simple logistic regression are the best predictive models to improve the diagnosis of neonatal jaundice in newborns. [10] proposed a novel clarification on the classification performance of Naïve Bayes which explains the dependence distribution of all nodes in a class and the performance assessment has been highlighted. The comparison results of [6] showed that the performance of decision tree and Naive Bayes classifiers applied on the diagnosis and prognosis of breast cancer had comparable results. Bellaachia et al. [11] applied Naive Bayes (NB), back-propagated neural network (BPNN), and C4.5 decision tree classifiers to predict the survivability of breast cancer patients and their findings reported that the C4.5 model has best performance than NB and BPNN classifiers.

Afshar et al. [12] proposed prediction model for breast cancer patient's survival using Support Vector Machine (SVM), Bayes Net, and Chi-squared Automatic Interaction Detection. They compared these models in terms of accuracy, sensitivity, and specificity and concluded that SVM model showed the best performance in their research.

Sandhu et al. [13] proposed MERS-CoV prediction system based on Bayesian Belief Networks (BBN) with cloud concept for synthetic data of initial classification of patients and their model accuracy score is 83.1%. The stability and recovery from MERS-CoV infections model has been proposed by [14] using Naive Bayes classifier (NB) and J48 decision tree algorithm in order to better understand the stability and pointed that NB model has the best accuracy.

Gibbons et al. [15] proposed the models for identifying underestimation in the surveillance pyramid and compared multiplication factors resulting from those models. MFs show considerable between country and disease variations based on the surveillance pyramid and its relation to outbreak containment. Chowell et al. [3] provide a comparison of exposure patterns and transmission dynamics of large hospital clusters of MERS and SARS using branching process models rooted in transmission tree data and inferred the probability and characteristics of large outbreaks.

Support Vector Machine (SVM) is very popular prediction models among the ML community because of its high performance for accurate predictions in dataset categories or situations where the relationship between features and the outcome is non-linear. For the dataset with 'n' number of attributes, SVM maps each sample as a point or coordinates in a n-dimensional space for finding the class of the sample [16]. SVM finds a hyperplane to differentiate the two target classes for the sample classification. The classification process involves the mapping of the new sample into the n-dimensional space, based on which side of the hyperplane the new sample fall in. Burges [6] described SVM as the best tool to address bias-variance tradeoff, over-fitting, and capacity control to work within complex and noisy domains. However, the quality of training data [6] decides the accuracy of SVM classifier. Moreover, [17, 18, 19] concluded the scalability is the main issue in SVM. In addition, the results reported in [20, 17, 19] stated that the use of optimization techniques can reduce SVM's computational cost and increase its scalability.

The research works reviewed in this section reveal the important applications of classification, and prediction analysis using Naïve Bayes, and SVM classifiers. Our study focuses on the prediction model by standard machine learning techniques Naive Bayes and SVM for testing on COVID-19 datasets cases from KSA.

3 Experimental Methodology

Generally, conducting a machine learning analysis covers the following steps.

- Preparing the dataset.
- Model Preparation
  - Training the ML models.
  - Testing the ML models.
  - Evaluating the models using measures.

3.1 Dataset Description and Pre-processing

For the experiments, our dataset sample period is between March 2, 2020 to April 16, 2020. We considered these datasets from 12 regions of KSA namely Riyadh, Eastern Region, Makkah, Madina, Qassim, Najran, Asir, Jazan, Tabuk, Al baha, Northern Borders, Hail.

Table 1 Description of datasets

Date	Regions with cases counts	Class
Ranges from 2nd March to 16th April 2020	All 12 regions and their respective case counts	0—Reported case 1—Confirmed case 2—Recovered case 3—Mortality case

The dataset has 248 records (days) with 12 columns (regions) in which 62 records for case history, 62 records for confirmed cases, 62 records for mortality cases and 62 records for recovered cases for all the above mentioned 12 regions respectively. The dataset will most likely continue to change for different COVID-19 cases until the recovery of all infected cases. So, we have used the data for confirmed cases, mortality cases, recovered, and reported cases for all the analysis. Table 1 shows the description of the dataset structure.

The daily accumulative infection number of 2019-nCoV is collected from daily reports of the Ministry of Health [7, 21].

First, some exploratory analysis on the data was carried out along with and summarization of some statistics, plotting some trends in the existing data. Then we build the machine learning models and try to predict the count of cases in the upcoming days. The statistical analysis of all these four cases based on cumulative count on daily basin are shown in Figs. 1, 3, 5 and 7 and based on 12 regions of KSA in Figs. 2, 4, 6 and 8 respectively.

Figure 1 shows the ongoing COVID-19 pandemic cases reported and spread to Saudi Arabia from 2nd March to 16th April 2020 and the Ministry of Health confirmed the first case in the Saudi Arabia on March 2, 2020. As the reported cases gradually increased during this period, the government respond to control the cases effectively by closure of holy cities, temporary suspension of transports, curfews on limited timings in various cities.

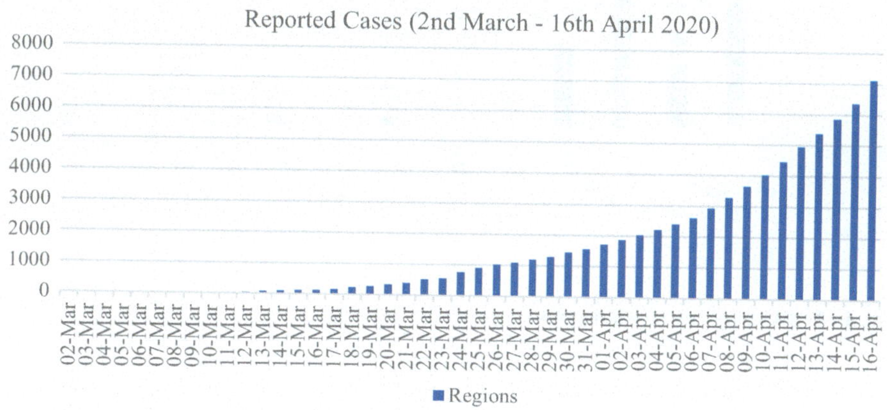

Fig. 1 Daily reported cases

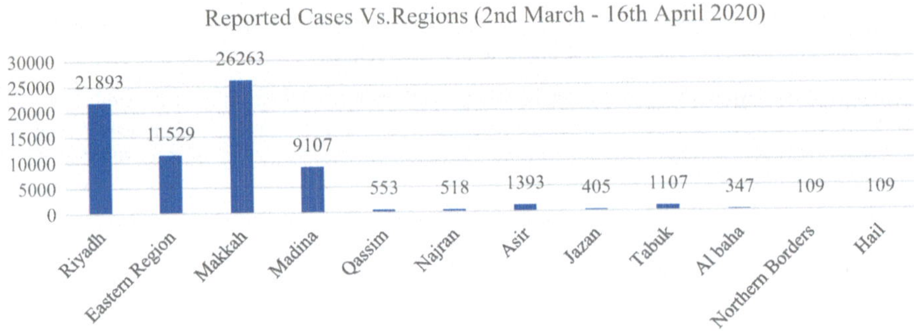

Fig. 2 Daily reported cases in 12 KSA regions

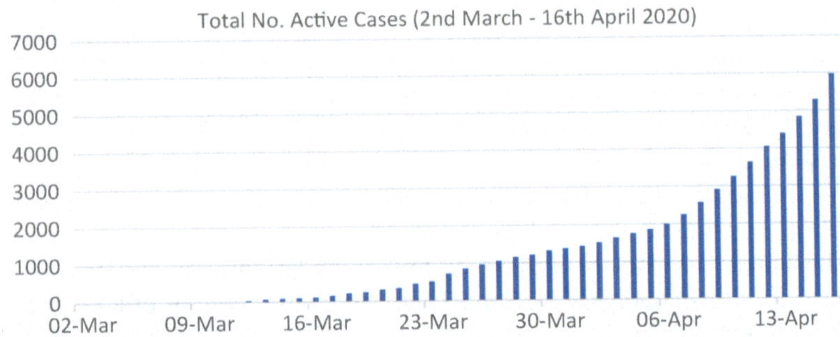

Fig. 3 Reported active cases

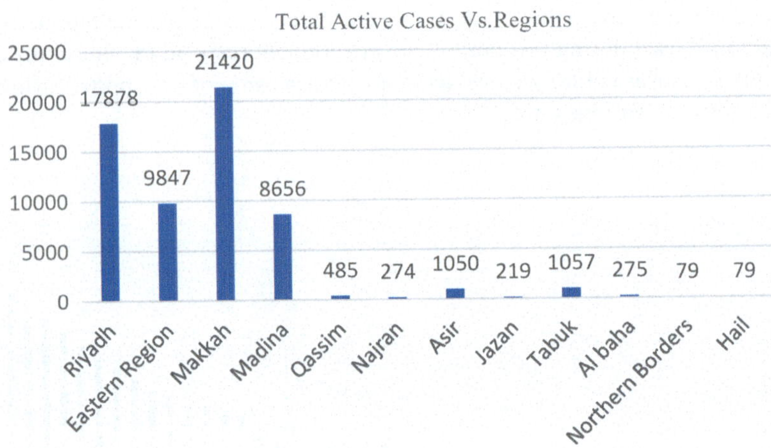

Fig. 4 Active cases in 12 KSA regions

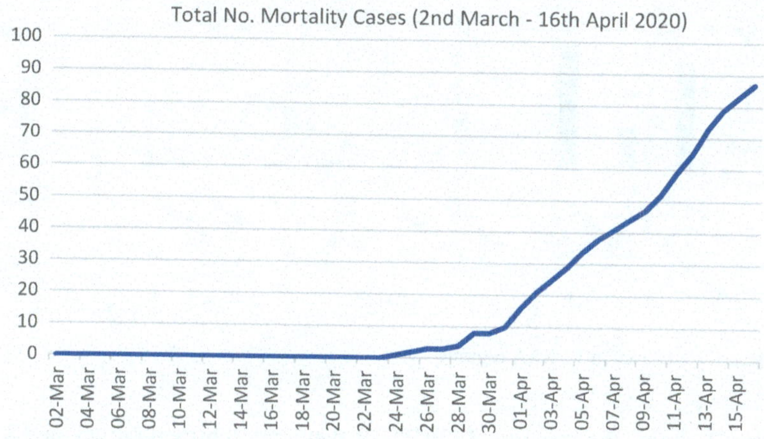

Fig. 5 Mortality cases

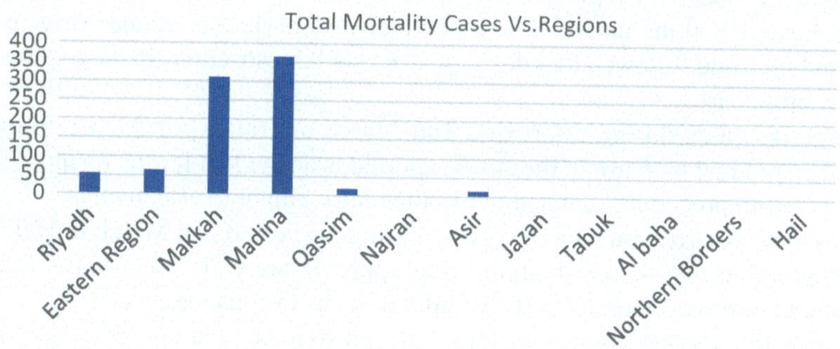

Fig. 6 Mortality cases in 12 KSA regions

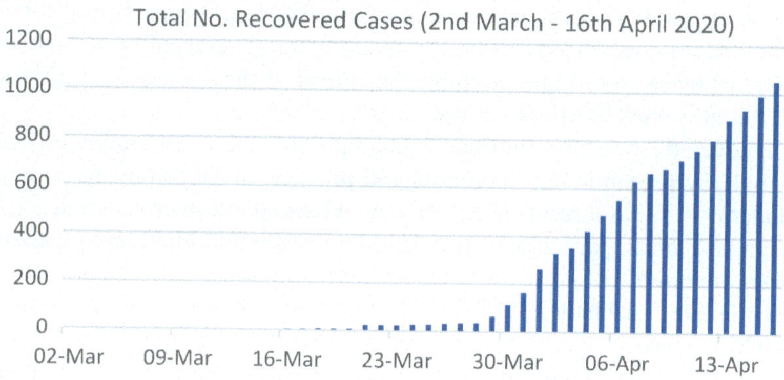

Fig. 7 Recovered cases

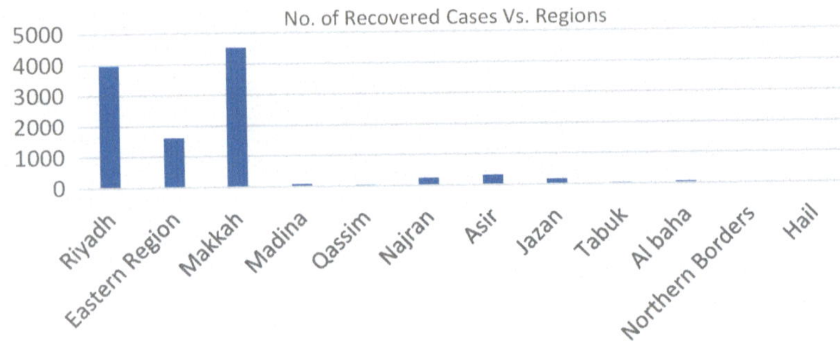

Fig. 8 Recovered cases in 12 KSA regions

Figure 2 shows the ongoing COVID-19 pandemic cases reported in 12 main regions of Saudi Arabia during the period 2nd March to 16th April 2020. Out of 12 regions, more cases were reported comparatively in four main cities namely Makkah, Riyadh, Eastern regions and Medina respectively. Authorities continue to urge people to stay at home and followed lockdown or strict social restrictions in the regions with more reported cases.

The active cases during the period 2nd March to 16th April 2020 is shown in Fig. 3. The gradual increase in the cases reported is an evident for the result of active medical testing procedures carried out in the entire kingdom effectively.

The active cases reported in 12 regions during the period 2nd March to 16th April 2020 is given in Fig. 4, which shows that approximately 20–80% of the reported cases were confirmed with COVID-19 infections in various regions.

Figure 5 reports that 2% of mortality rate approximately at the maximum during the period 2nd March to 16th April 2020.

There were more mortality cases in pilgrimage cities Makkah and Medina and the authorities reported that most of cases were suffering from chronic health conditions also. Saudi Arabia suspended entry and praying to the general public at the two Holy Mosques in Mecca and Medina to limit the spread of the coronavirus [22] on 20th March to control the COVID-19 cases.

The recovered cases given in Figs. 7 and 8 provided the information that nationalities abide by precautionary measures and practice social distancing to keep the virus under control, as a result of active testing carried out in crowded districts and other high-risk areas, particularly in cities like Riyadh and Makkah in which more cases were reported.

The complete case history of COVID-19 trends for the period of 2nd March - 16th April 2020 is given in Fig. 9. It is evident that mortality and recovered case rates are comparatively less than the reported and confirmed case rates. The mortality cases ratio for COVID-19 has been much lower than SARS of 2003 [23, 24] but the transmission has been significantly greater, with a significant total death toll [25].

Data preprocessing involves dividing the data into attributes and labels and dividing the data into training and testing sets. For data pre-processing, we split

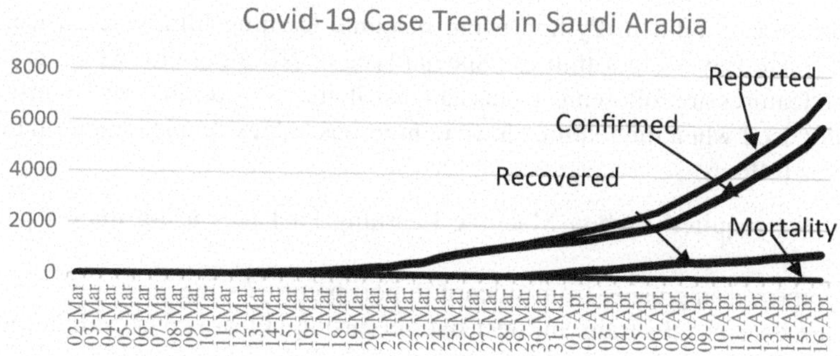

Fig. 9 Covid-19 case trend in Saudi Arabia

the dataset into two groups based on case categories. The first group consisted of recovery cases and mortality cases based on regions for predicting the recovery from Covid-19. Second group has the reported cases to be used to predict the stability of the infection based on the active cases. Columns are the same in this two dataset groups which are 12 KSA regions related to the number of Covid-19 cases i.e. Reported, Confirmed, Death and Recovered cases for the time period 2nd March–16th April 2020.

Before simulating the algorithms, the datasets are preprocessed to make them suitable for the classifier's implementation. First need to separate our training data by class.

3.2 *Building Models*

Classification is a widely used technique in health-care. Here, we build classification models to predict the frequency of Covid-19 infection cases. We applied two models namely Naive Bayes and SVM algorithms. The models are implemented in Python platform.

3.2.1 Naïve Bayes (NB)

Naive Bayes classifier is a classification algorithm for binary and multiclass classification problems using Bayes theorem and assumes that all the features are independent to each other. Bayes' theorem is based on conditional probability. The conditional probability calculates the probability that something will happen, given that something else has already happened.

Bayes' Theorem is stated as: P(class|data) = (P(data|class) * P(class))/P(data), where P(class|data) is the probability of class given the provided data.

NB classifier is built in Python using machine learning library scikit-learn. A Gaussian Naive Bayes algorithm is a special type of NB algorithm which assumes that all the features are following a gaussian distribution i.e., normal distribution. It's specifically used when the features have continuous values. Implementation details are given as follows:

- Import the required Python Machine Learning Packages using import pandas, numpy
- Data preprocessing using from sklearn import preprocessing
- Split the dataset into train and test datasets using sklearn.cross_validation and import train_test_split
- Model the Gaussian Navie Bayes classifier using sklearn.naive_bayes import GaussianNB
- Predict method of the GaussianNB class is used for making predictions.
- Calculate the accuracy score of the model using sklearn.metrics import accuracy_score

3.2.2 Support Vector Machine (SVM)

Support vector machine (SVM) classifier is a type of supervised machine learning classification algorithm. SVM differs from the other classification algorithms in the way that it chooses the decision boundary that maximizes the distance from the nearest data points of all the classes and finds the most optimal decision boundary. Implementation details simple linear SVM in Python are as follows:

- Import the required Python Machine Learning Packages using import pandas, numpy
- Data preprocessing using from sklearn import preprocessing
- Split the dataset into train and test datasets using sklearn.cross_validation and import train_test_split
- Model the SVC classifier with kernel type as linear using from sklearn.svm import SVC
- The predict method of the SVC class is used for making predictions.
- Calculate the accuracy score of the model using sklearn.metrics import accuracy_score and classification_report

After that, every dataset has been divided into, training and testing sets using the following ratios 80/20, 70/30, and 60/40, respectively. For the prediction models, the two models are applied to the original datasets and the performance of every classifier is analyzed using the metrics such as accuracy, precision and recall measures which are explained in the following section.

4 Model Evaluation Results and Analysis

We analyzed and evaluated NA and SVM machine learning classifiers using the performance metrics namely Classification accuracy, Precision, and Recall. The formulas for calculating these metrics are given in Table 2.

Performance measures, for the prediction of recover and mortality, namely classification accuracy percentage, Precision, Recall, of the models are presented in Table 3. The performance of SVM model is comparatively good in terms of classification accuracy, precision and recall values. The performance of NB model shows good results for the validation set with 70/30 for recovery-mortality dataset as shown in Table 3.

The performance of SVM classifier is good because all datasets have single-labels, which is the strongness of SVM for handling single-label data. SVM has better performance than NB with 2% classification accuracy.

In this work, two classification algorithms NB and SVM are used to produce highly accurate models for COVID-19 dataset. However, the performance of the these obtained models is little bit satisfactory for application in real pandemic of COVID-19 infection cases. We believe that there is a need to increase the size of the dataset in order to improve predictions because the main limitation lies in the size of the training dataset. In addition, more medical history of the patient information should be included in the future work.

Table 2 Description of metrics

Name of the metrics	Formula
Accuracy	$\frac{True\ Positives + True\ Negatives}{Poitives + Negatives}$
Precision	$\frac{True\ Positives}{True\ Positives + False\ Positives}$
Recall	$\frac{True\ Positives}{True\ Positives + False\ Negatives}$

Table 3 Predicted metrics

Method		Accuracy			Precision			Recall		
		80/20	70/30	60/40	80/20	70/30	60/40	80/20	70/30	60/40
Naïve bayes	Recovery—mortality	63	70.27	67.57	0.69	0.82	0.81	0.65	0.71	0.63
	Reported—confirmed	63.16	54.83	48.65	0.63	0.52	0.54	0.63	0.55	0.50
SVM	Recovery—mortality	0.79	0.64	0.62	0.80	0.67	0.70	0.79	0.64	0.62
	Reported—confirmed	0.70	0.63	0.61	0.76	0.69	0.67	0.73	0.61	0.60

5 Sustainable Healthcare Post COVID 19 for SA

Sustainability, as a concept has vastly benefited different sectors of business including energy, agriculture, forestry, construction and tourism. It is gaining popularity in the modern healthcare system which is predominant with contemporary pharmaceutical drugs & products [26]. But different instances have proved time and again, that the contemporary medication was not found to be an effective solution against various infectious and chronic diseases.

5.1 *Sustainable Healthcare*

Alliance for Natural Health, USA (ANH-USA) first defined Sustainable Health in 2006 as:

"A complex system of interacting approaches to the restoration, management and optimization of human health that has an ecological base, that is environmentally, economically and socially viable indefinitely, that functions harmoniously both with the human body and the non-human environment, and which does not result in unfair or disproportionate impacts on any significant contributory element of the healthcare system" [26].

Current COVID-19 pandemic, which has devastated the world and even the best healthcare systems have crippled under its pressure, points strongly in the direction of involving all kinds of healthcare systems to be bound with the principles of sustainability and demands a paradigm shift in healthcare approach by countries for the wellbeing of its citizens. Now the time has come where the countries must have to implement and practice Sustainable Healthcare for its citizens. Traditional Medicines and Alternative Medicines such as Homeopathy, Ayurveda, Yunani, Chinese medicine, Naturopathy etc. were always questionable for their scientific basis by the practitioners of Allopathy and/or the contemporary form of medication. But then the alternative form of medication has proved its effectiveness and efficiency time and again during challenging times and are practiced since many decades now.

There is a strong need to prepare/collect, use and analyse the data pertaining to the Traditional Form of medicines and its usefulness applying AI/ML techniques.

Following and subsequent section discusses some of the recommendations regarding the current pandemic, future directions towards a sustainable healthcare system in Saudi Arabia.

5.2 *Staff and Clinical Practice Sustainability During the Pandemic*

- Telehealth technology allows clinicians to monitor patients in-home and make treatment recommendations, A robust infrastructure for telemedicine is also required.
- COVID-19 has an overall lower case death rate than SARS or Middle East respiratory syndrome (MERS) [27], but the stress placed on healthcare systems globally is alarming [28]. The medical agency should have well improved plan for protecting healthcare workers from infection and exhaustion in this prolonged fight against COVID-19.
- Many healthcare offices have limited due to the lockdown timings, touch with policy makers as they design programs for government relief and support.

5.3 *Expand Hospital-at-Home During the COVID-19 Pandemic*

- Hospital-at-home program can reduce healthcare costs and mortality rates.
- During the pandemic, providing hospital care at home can reduce the risk of COVID-19 transmission, especially for vulnerable patients.
- For effective hospital-at-home, interdisciplinary social and behavioral health services are required.
- Primary care and hospital-at-home services should cover social and behavioral health requirements also.

5.4 *COVID-19 Pandemic and Sustainable Development Groups*

- Towards 2030, the World is expected to assure Peace and Prosperity for all People and the Planet through Partnerships (Governments-Private-NGOs-CSOs-Individuals) in the Social, Economic and Environmental Spheres. These 'COVID-19 Pandemic Benefits' should be optimized for 'Sustainable Development' by the nexus with the SDGs
- Medical council should continue the expansion of primary care and hospital-at-home services in remote areas as well. The patients and primary care teams should improve their services both during and after the pandemic.
- The technical guidance for strategic and operationally focused actions to support health service planners and health-care system managers in the Region to maintain the continuity and resourcing of priority services while mobilizing the health workforce to respond to the pandemic. This will help ensure that people continue to seek care when appropriate and adhere to public health advice.

5.5 Research Directions

The technology of machine learning can generate new opportunities for Sustainable Healthcare and the researchers can focus on the areas:

- Automated analysis and prediction of COVID-19 infection cases.
- Automated discovery of COVID-19 patient cases dynamically.
- Automation on existing consolidated portal to support future pandemics.
- Building a novel Pilot COVID-19 Data Warehouse for future reference.
- Improved techniques for capturing, preparing and storing data meticulously.
- Supportive platform for creating a community of medical practitioners for pandemic crisis.

6 Conclusion

Finding the hidden knowledge in the data is a challenging task in machine learning. This chapter focus on the classification and prediction by standard machine learning techniques (Naive Bayes and SVM) when tested on COVID-19 datasets cases from KSA. Covid-19 dataset was converted into patients' cases (reported and confirmed, recovered and death) classification problem, and the respective target prediction has been carried out. The performance of each model forecasts was assessed using classification accuracy, precision and recall. Our results demonstrate that Naive Bayes and SVM models can effectively classify and predict the cases of COVID-19 data and we discussed the sustainable healthcare of COVID 19 for Saudi Arabia.

This chapter also reports the applications of some of the well-known machine learning algorithms for the prediction of the frequency of COVID-19 disease. We found that SVM, and NB models can give relatively higher accuracy results.

The performance of the two models NB and SVM was evaluated and compared. In general, we found that the accuracy of the models is between 63% and 80%. In Future, the performance of the prediction models can be improved with the use of more COVID-19 datasets. The motivation of this chapter is to support the medical practitioners for choosing the appropriate machine learning classifiers for the analysis of various COVID-19 samples. For our future work on COVID-19 data, we plan to collect more data related to patients with COVID-19 cases directly from hospitals in KSA.

Together, we, as an organization, as a community, and as global citizens, can beat this disease, better prepare for the next pandemic, and ensure the safety and care of all our patients.

Acknowledgements We acknowledge MoH - KSA data repository for datasets.

References

1. V.J. Munster, M. Koopmans, N. van Doremalen, D. van Riel, E. de Wit, A novel coronavirus emerging in China—key questions for impact assessment. New England J. Med. (2020)
2. W.H. Organization, Novel coronavirus (2019-nCoV) situation reports, 2020
3. G. Chowell, F. Abdirizak, S. Lee et al., Transmission characteristics of MERS and SARS in the healthcare setting: a comparative study. BMC Med. **13**, 210 (2015). https://doi.org/10.1186/s12916-015-0450-0
4. B, Nithya, Study on predictive analytics practices in health care system. IJETTCS, **5** (2016)
5. D.R. Chowdhury, M. Chatterjee, R.K. Samanta, An artificial neural network model for neonatal disease diagnosis. Int. J. Artif. Intell. Expert Syst. (IJAE) **2**(3), (2011)
6. B. Venkatalakshmi, M. Shivsankar, Heart disease diagnosis using predictive data mining. Int. J. Innov. Res. Sci. Eng. Technol. **3**, 1873–1877 (2014)
7. Saudi Ministry of Health. https://covid19.moh.gov.sa/
8. K. Vanisree, J. Singaraju, Decision support system for congenital heart disease diagnosis based on signs and symptoms using neural networks. Int. J. Comput Appl. **19**(6), 0975–8887 (2011)
9. D. Ferreira, A. Oliveira, A. Freitas, Applying data mining techniques to improve diagnosis in neonatal jaundice. BMC Med. Inform. Dec. Mak **12**(143), (2012)
10. H. Zhang, The optimality of naïve bayes. Faculty of Computer Science at University of New Brunswick (2004)
11. A. Bellaachia, E. Guven, Predicting breast cancer survivabil-ity using data mining techniques, in *Ninth Workshop on Mining Scientific and Engineering Datasets in conjunctionwith the Sixth SIAM International Conference on Data Min-ing*, 2006
12. H.L. Afshar, M. Ahmadi, M. Roudbari, F. Sadoughi, Prediction of breast cancer survival through knowledge discovery in databases. Global J. Health Sci. **7**(4), 392 (2015)
13. R. Sandhu, S.K. Sood, G. Kaur, An intelligent system for pre-dicting and preventing MERS-CoV infection outbreak. J. Supecomputing 1–24 (2015)
14. I. Al-Turaiki, M. Alshahrani, T. Almutairi, Building predictive models for MERS-CoVinfections using data mining techniques. J. Infect. Public Health **9**, 744–748 (2016)
15. C.L. Gibbons, M.J. Mangen, D. Plass et al., Measuring underreporting and under-ascertainment in infectious disease datasets: a comparison of methods. BMC Public Health **14**, 147 (2014). https://doi.org/10.1186/1471-2458-14-147
16. C. Cortes, V. Vapnik, Support-vector networks, Mach. Learn. **20**(3), 273–297 (1995)
17. R. Burbidge, B. Buxton, An introduction to support vector machines for data mining. UCL: Computer Science Dept. (2001)
18. C. Burges, A tutorial on support vector machines for pattern recognition. Bell Laboratories and Lucent Technologies (1998)
19. R. Alizadehsani, J. Habibi, M.J. Hosseini, H. Mashayekhi, R. Boghrati, A. Ghandeharioun, B. Bahadorian, Z.A. Sani, A data mining approach for diagnosis of coronary artery disease. Comput. Methods Programs Biomed. **111**(1), 52–61 (2013)
20. I. Bardhan, J. Oh, Z. Zheng, K. Kirksey, Predictive analytics for readmission of patients with congestive heart failure. Inf. Syst. Res. **26**(1), 19–39 (2014)
21. Data Source. https://datasource.kapsarc.org/explore/dataset/saudi-arabia-coronavirus-disease-covid-19-situation-demographics, www.covid19.cdc.gov.sa
22. Entry and prayer in courtyards of the Two Holy mosques suspended. Saudigazette. 2020-03-20. Archived from the original on 2020-03-20. Retrieved 16 April 2020
23. Crunching the numbers for coronavirus. Imperial News. Archived from the original on 19 Mar 2020. Retrieved 16 Apr 2020
24. High consequence infectious diseases (HCID); Guidance and information about high consequence infectious diseases and their management in England. GOV.UK. Retrieved 16 Apr 2020
25. World Federation of Societies of Anaesthesiologists—Coronavirus. www.wfsahq.org. Archived from the original on 12 Mar 2020. Retrieved 16 Apr 2020

26. Sustainable Healthcare. https://anh-usa.org/position-papers/sustainable-healthcare/
27. J. Guarner, Three emerging coronaviruses in two decades. Am. J. Clin. Pathol. **153**, 420–421 (2020)
28. H. Legido-Quigley, N. Asgari, Y.Y. Teo, Are high-performing health systems resilient against the COVID-19 epidemic? Lancet **395**, 848–850 (2020)

Machine Learning and Deep Learning Applications

A Comprehensive Study of Deep Neural Networks for Unsupervised Deep Learning

Deepti Deshwal and Pardeep Sangwan

Abstract Deep learning methods aims at learning meaningful representations in the field of machine learning (ML). Unsupervised deep learning architectures has grown at a fast pace owing to their ability to learn intricate problems. Availability of large amount of labelled and unlabeled data with highly efficient computational resources makes deep learning models more practicable for different applications. Recently, deep neural networks (DNNs) have become an extremely effective and widespread research area in the field of machine learning. The significant aim of deep learning is to learn the primary structure of input data and also to investigate the nonlinear mapping between the inputs and outputs. The main emphasis of this chapter is on unsupervised deep learning. We first study difficulties with neural networks while training with backpropagation-algorithms. Later, different structures, namely, restricted Boltzmann machines (RBMs), Deep Belief Networks (DBNs), nonlinear autoencoders, deep Boltzmann machines are covered. Lastly, sustainable real applications in agricultural domain with deep learning are described.

Keywords Deep learning · Restricted boltzmann machines (RBMs) · Contrastive divergence · Deep belief network · Autoencoders

1 Introduction

For tasks such as pattern analysis, several layers in a deep learning system can be studied in an unsupervised way (Schmidhuber [1]. One layer at a time can be trained in a deep learning architecture, in which each layer is treated as an unsupervised restricted Boltzmann machine (RBM) [2]. The concept of unsupervised deep learning algorithms is significant because of the easy availability of unlabeled

D. Deshwal (✉) · P. Sangwan
Department of ECE, Maharaja Surajmal Institute of Technology, New Delhi, India
e-mail: deshwaldeepti@msit.in

P. Sangwan
e-mail: sangwanpardeep@msit.in

© The Editor(s) (if applicable) and The Author(s), under exclusive license to Springer Nature Switzerland AG 2021

A. E. Hassanien et al. (eds.), *Artificial Intelligence for Sustainable Development: Theory, Practice and Future Applications*, Studies in Computational Intelligence 912,
https://doi.org/10.1007/978-3-030-51920-9_7

data as compared to the labelled information [3]. A two-step process is used for applications with large volumes of unlabeled data. Firstly, pretraining of a DNN is performed in an unsupervised way. Later, a minor portion of the unlabeled data is manually labelled in the second step. The manually labelled data is further utilized for fine-tuning of supervised deep neural network. With the invention of several powerful learning methods and network architectures, the neural networks [4] was the most applied area in the field of machine learning in the late 1980s. These learning methods include multilayer perceptron (MLP) networks based on backpropagation algorithms and radial-based feature networks. Although neural networks [4] have given tremendous results in various domains, interest in this field of research later reduced. The concentration in research on machine learning shifted to other fields, such as kernel and Bayesian graphic approaches. Hinton introduced the concept of deep learning in the year 2006. Deep learning has since become a hot area in the field of machine learning, resulting in revival of research into neural networks [5]. Deep neural networks have produced incredible results in various regression as well as classification problems when properly trained. Deep learning is quite a forward-looking subject. Literature consists of different types of review articles on deep learning approaches covering all the aspects in this emerging area [6]. An older analysis is, and a strong introduction to deep learning is the doctoral theses [7]. Schmid Huber has given a short review listing more than 700 [1]. Work on deep learning is generally progressing very quickly, with the introduction of new concepts and approaches. In this chapter, Sect. 1 explains Feed forward neural network covering single and multilayer perceptron networks (MLP). Section 3 explains the concept of deep learning covering restricted Boltzmann machines (RBMs) as preliminary points for deep learning, and later move to other deep networks. The following unsupervised deep learning networks are explored in this chapter: restricted Boltzmann system, deep-belief networks [8], autoencoders. Section 4 covers applications of deep learning and lastly Sect. 5 covers the challenges and future scope.

2 Feedforward Neural Network

The primary and simplest form of artificial neural network (ANN) is the feedforward neural network [9]. This requires numerous neurons grouped in layers. Neurons from adjacent layers have interlinkages.

All of these relations have linked weights. Figure 1 provides an example of a feedforward neural network. Three types of nodes may form a feedforward neural network.

1. **Input Nodes**—These nodes deliver input to the network from the outside world and are called the "Data Layer" together. In any of the input nodes, no computation is done—they simply pass the information on to the hidden nodes.
2. **Hidden Nodes**—There is no direct connection between the Hidden Nodes and the outside world. This is the reason the name termed "hidden". Computations

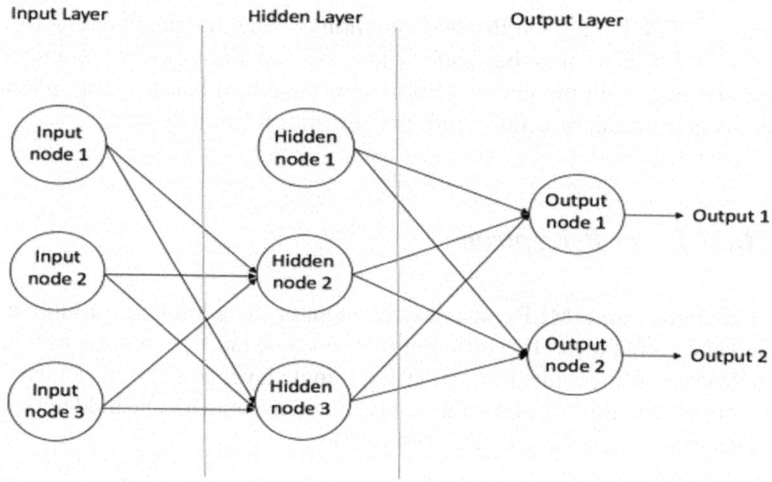

Fig. 1 Feed-forward neural network

are performed and information passes to the output nodes from the input nodes. Feedforward network consists of a single input and output layer, but the number of hidden Layers may vary.

3. **Output Nodes**—The output nodes are used for processing and transmitting information from the network to the outside world and are jointly referred to as the "Output Layer".

In a feedforward network, the information travels in a forward direction from the input to the hidden and finally to the output nodes. A feedforward network has no cycles or loops in comparison to the Recurrent Neural Networks where a cycle is produced due to the connections between the different nodes. Examples of feedforward networks are as follows:

1. **Single Layer Perceptron**—The simplest neural feedforward network with no hidden layer constitutes the single layer perceptron.
2. **Multi-Layer Perceptron**—One or more hidden layers have a multi-Layer perceptron. We'll just mention Multi-Layer perceptron below as they're more useful for practical applications today than Single Layer Perceptron.

2.1 Single Layer Perceptron

A single layer perceptron is the simplest form of a neural network used for the classification of patterns. Basically, it consists of a single neuron with adjustable synaptic weights and bias. It can be easily shown that a finite set of training samples can be classified correctly by a single-layer perceptron if and only if it is linearly separable (i.e. patterns with different type lie on opposite sides of a hyperplane).

Thus, for e.g. if we look at the Boolean functions (using the identification true = 1 and false = 0) it is clear that the "and" or the "or" functions can be computed by a single neuron (e.g. with the threshold activation function) but the "xor" (exclusive or) is not. A neuron can be trained with the perceptron learning rule.

2.2 *Multi-Layer Perceptron*

Multi-Layer Perceptron (MLP) includes one input layer, at least one or more hidden layers and one output layer. It is different from a single layer perceptron as it has the ability to learn non-linear functions whereas a single layer perceptron can only learn linear functions. Figure 2 displays a multilayer perceptron with one hidden layer and all the links have weights associated with them.

- **Input layer**: This layer consists of 3 nodes. The Bias node value is taken as 1. The other two nodes take X1 and X2 as external inputs. As conferred above, no computation is done in the input layer, so the node outputs 1, X1 and X2, in the input layer are fed in the Hidden Layer respectively.
- **Hidden Layer**: The Hidden layer also consists of 3 nodes. The Bias node with a value of 1 is assumed. The outputs (1, X1, X2) from the input layer and the weights associated with them decides the behaviour of the remaining 2 nodes in the hidden layer. Figure 2 represents the hidden nodes for performance measurement. Likewise, one can measure the output from another secret node. F denotes feature activation. Instead, the resultant outputs are further fed into the nodes of input layer.
- **Output Layer**: The Output layer consists of two nodes and the input is fed from the hidden layer. Similar computations are performed as shown for the hidden node. As a consequence of these computations the measured values (Y1 and Y2)

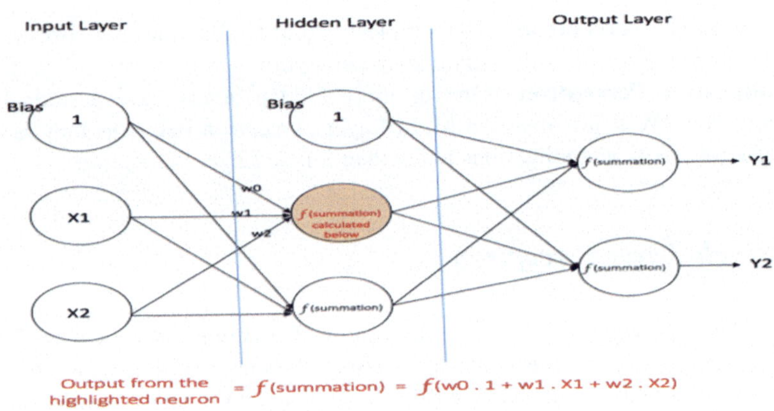

Fig. 2 Multi-layer perceptron having one hidden layer

serve as outputs of the Multi-Layer Perceptron. Figure 2 displays the input and output layer of an MLP network, the hidden layer of $L \geq 1$. The number of nodes in every layer will generally vary. The processing in the hidden layers of the multi-layer perception is generally nonlinear while the output layer processing may be linear or nonlinear. On the other hand, no computations occur in the input layer, only input components in each neuron are entered there.

The kth neuron operation in the lth hidden layer is defined by the equation below:

$$h_k^{[l]} = \emptyset\left(\sum_{j=1}^{m^{(l-1)}} w_{kj}^{[l]} h_j^{[l-1]} + b_k^{[l]}\right) \quad (1)$$

where $h_j^{[l-1]}$, $j = 1, \ldots, m^{[l-1]}$ are the $m^{[l-1]}$ input signals entering the kth neuron, and w_{kj}^{l}, $j = 1, \ldots, m^{[l-1]}$ are the input signals of the respective weights. In the lth layer the number of neurons is $m^{[l]}$. The input signals fed to the first hidden layer of the multi-layer perceptron are designated as $x_1, \ldots, x_p$. The weighted sum is added to the constant bias term b_k. The output vector y components are computed in the same way as the outputs of the lth hidden layer computation in Eq. (1). The function $\emptyset(t)$ represents the nonlinearity added to the weighted sum. Usually, it is preferred as hyperbolic tangent $\emptyset(t) = \tanh(at)$ where a is the logistic sigmoidal function

$$\emptyset(t) = 1/(1 + e^{-at}) \quad (2)$$

In case linear operation of a neuron is obtained then, $\emptyset(t) = at$[1; 2]. Though the computation inside a single neuron is generally easy but the result obtained is nonlinear. Such nonlinearities distributed in every neuron of each hidden layer and perhaps also in the output layer of the MLP network corresponds to high representational power, but then, make the mathematical analysis of the MLP network difficult. Besides that, it can lead to other problems such as local cost functional minima. Nonetheless, a multi-layer perceptron network with sufficient number of neurons in a single hidden layer can be used for performing any nonlinear mapping of input and outputs.

The extensive notations can quite complicate the learning algorithms of MLP networks. MLP networks are generally trained in a supervised way by N distinct training pairs $\{x_i, d_i\}$ where x_i denotes ith input vector and d_i is the desired output response. Later, vector x_i is entered into the MLP network, and the resultant y_i output is measured as a vector. The measure used for learning MLP network weights is usually the mean-square error.

$$E = E\{d_i - y_i^2\}, \text{ which is minimalized.} \quad (3)$$

The steepest descent learning rule in any layer for a weight w_{ji} is specified by

$$\Delta w_{ji} = -\mu \frac{\partial E}{\partial W_{ji}} \tag{4}$$

In reality, over 100–1000 training pairs replace the steepest descent by an instant gradient or a mini batch. For the neurons in the output layer, the necessary gradients are computed first by estimating their corresponding local errors. Later, the errors generated are further propagated in the backward direction to the former layer, and simultaneously, the weights of the neurons can be updated. Therefore, the name backpropagation for MLP networks is derived. The convergence usually requires numerous iterations and sweeps over the training data, particularly in the case of an instant stochastic gradient. Several ways of learning the backpropagation algorithm and alternatives for faster convergence have been introduced.

Generally, MLP networks are configured to have either one or two hidden layers due to its inefficacy to train additional hidden layers utilizing backpropagation algorithms based on steepest descent method. More hidden layers do not simply learn suitable features due to the fact that gradients decay exponentially w.r.t them. Learning algorithms utilizing only the steepest descent method have a disadvantage associated with them i.e. it leads to poor local optima, probably because of their inability to break the symmetry present in every hidden layer between many neurons.

3 Deep Learning

Nonetheless, designing a deep neural network with multiple hidden layers would be ideal. The intention is that the nearest layer to the data vectors has the ability to learn basic features, whereas the higher-level features can be learned from higher layers. For example, if we take the case of digital images the first hidden layer learns the low-level features such as edges and lines. Throughout higher-level layers, they are accompanied by structures, objects, etc. Human brains, specially the cortex, encompass deep neural biological networks that function in this manner. These are very effective in activities, such as different pattern recognition programs, which are difficult for computers.

Deep learning solves the different types of issues while applying backpropagation algorithms to multiple layer deep networks [10]. The prime idea is to understand the structure of the input data together with the nonlinear mappings between input and output vectors. This is achieved with the aid of unsupervised pretraining [11]. In practice, the creation of deep neural networks is accomplished by utilizing the chief building blocks such as RBMs or autoencoders in the hidden layers.

3.1 Restricted Boltzmann Machines (RBMs)

RBMs are a subset of neural networks implemented in 1980s [12]. These are based on statistical mechanics, and compared to most other neural network approaches [13], these use stochastic neurons. Simplified models of Boltzmann machines are RBMs as shown in Fig. 3. In RBMs, the relations in the original Boltzmann machines between the top hidden and among the bottom visible neurons are deleted. Only the neuronal connections in the visible layer remain with the hidden layer and the corresponding weights are grouped into matrix W. This interpretation makes RBM learning manageable compared to Boltzmann machines, where it rapidly becomes intractable due to various connections.

RBM is also termed as a generative model that has the ability to learn probability distribution over a certain set of inputs [14]. The term "restricted" refers to forbidden node's connection existing in the similar layer. RBMs are used to train different layers one at a time in large networks. RBM's training procedure involves changing the weights so that the probability of producing the training data is maximized. RBM comprises of 2 layers of neurons namely visible layer and hidden layer for vector v and vector h data. All the visible and hidden layer neurons are inter-connected to each other. There exists no intralayer connections between the visible and hidden neurons. Figure 3 illustrates the RBM construction, with m visible layers and n hidden layers. On the other hand, matrix W represents the corresponding weights between visible and hidden neurons. w_{ij} signifies weights amid ith visible and jth hidden neurons.

In RBM, the probability distributions of visible and hidden units over (v, h) are determined in the following manner:

$$p(v, h) = \frac{e^{-E(v,h)}}{\sum_{v,h} e^{-E(v,h)}} \tag{5}$$

where the denominator is a standardization constant (partition function) representing the number $e^{-E(v,h)}$ of total possibilities. $E(v, h)$ is the configuration energy (v, h) and is shown as follows:

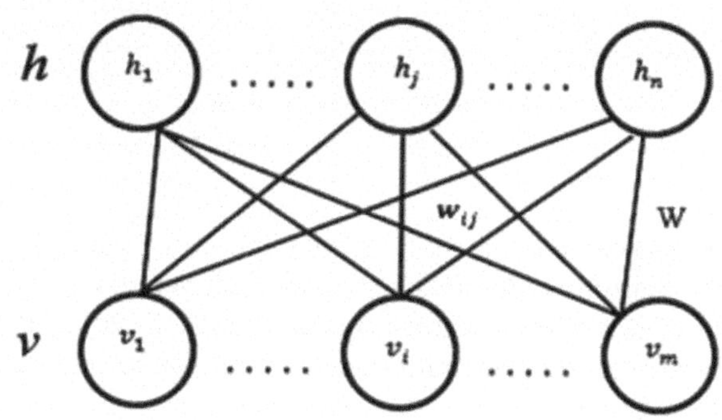

Fig. 3 Restricted boltzmann machine

$$E(v, h) = -\sum_{i} a_i v_i - \sum_{j} b_j h_j - \sum_{i} \sum_{j} v_i w_{ij} h_j \tag{6}$$

or in matrix notation

$$E(v, h; W, a, b) = -a^T v - b^T h - v^T W h \tag{7}$$

W reflects weights; b is the latent unit bias, and a is the obvious unit bias. The visible vector v states are correlated with the input data. On the other hand, hidden vector h depicts the internal neurons hidden characteristics. For an input data vector v, the conditional probability of is given as

$$p\left(h_j = 1/v\right) = \sigma(b_j + \sum_{i=1}^{m} w_{ij} v_i) \tag{8}$$

Where

$$\sigma = \frac{1}{1 + e^{-x}} \tag{9}$$

Equation 9 represents the sigmoid activation function. Data reconstruction is done utilizing hidden states. This is achieved by initiating the neurons in the visible layers with a conditional probability function given by

$$p(v_i = 1/h) = \sigma(a_i + \sum_{j=1}^{n} w_{ij} h_j) \tag{10}$$

3.1.1 Contrastive Divergence

RBMs are trained to improve the ability to reconstruct, thus maximizing the loglikelihood of training data for a given set of training parameters. The total likelihood of hidden vectors, for a visible input vector is derived as follows:

$$p(v) = \frac{\sum_h e^{-E(v,h)}}{\sum_{v,h} e^{-E(v,h)}} \tag{11}$$

It is possible to increase the likelihood of a training vector by changing the weights and biases in order to lower a particular vector energy and to increase the energy of all other vectors, sampled by that model. To adjust the weights and biases, the log probability derivative for the network parameters $\theta \varepsilon \{a_i, b_j, w_{ij}$ is calculated and is given by

$$\frac{\partial \log p(v)}{\partial \theta} = -\sum_{h} \underset{\text{Positive Phase}}{p\left(\frac{h}{v}\right)\partial \frac{E(v,h)}{\partial \theta}} + \sum_{v,h} \underset{\text{Negative phase}}{p(v,h)\partial \frac{E(v,h)}{\partial \theta}} \tag{12}$$

We need a strategy for sampling (h/v), and another strategy for sampling $p(v, h)$. The positive phase comprises of clamping the visible layer on the input data. Afterwards, sampling of $\boldsymbol{h}$ is done from $\boldsymbol{v}$, whereas in the negative phase sampling of $\boldsymbol{v}$ and $\boldsymbol{h}$ are to be sampled is performed from the base. First term calculation is usually simple, due to the fact that there exists no relation between the neurons of hidden and visible layers. Regrettably, it is hard to estimate the second term. Another possible strategy is to use the Alternating Gibbs Sampling (AGS) methodology.

Each AGS iteration updates all the hidden units using Eq. (8) in parallelly updating all the existing units utilizing the Eq. (10), and lastly again updating the hidden units using Eq. (8).

So, Eq. (12) is rephrased as

$$\frac{\partial \log p(v)}{\partial \theta} = \left\langle \partial \frac{E(v,h)}{\partial \theta} \right\rangle_0 + \left\langle \partial \frac{E(v,h)}{\partial \theta} \right\rangle_\infty \tag{13}$$

where $\langle \cdot \rangle_0 (p_0 = p(h/v) = p(h/x))$ and $\langle \cdot \rangle_0 (p_0 = p(h/v) = p(h/x))$ denotes the expectations described by the data and model under the distributions. The whole process is very time consuming, though, the convergence attained with this learning methodology is usually too sluggish. Solution to this problem adopted is the Contrastive Divergence (CD) method [15] where $\langle \cdot \rangle_\infty$ is substituted by $\langle \cdot \rangle_k$. The concept is essentially to adjust the neurons in visible layers utilizing a training sample. Thus, the hidden states can be inferred from the Eq. (8). Similarly, the visible states are deduced from hidden states using Eq. (10). That is similar to using $k = 1$ to run Gibbs sampling. This is shown in Fig. 4.

CD algorithm Convergence is guaranteed if the relationship which must be maintained by the Gibbs sampling step number and the learning rate is fulfilled in every step of the parameter updating. Consequently altering Eq. (13), the update rules are denoted as:

$$\Delta w_{ij} = \alpha\left(\langle v_i h_j \rangle_0 - \langle v_i h_j \rangle_1\right) \tag{14}$$

$$\Delta b_j = \alpha\left(\langle h_j \rangle_0 - \langle h_j \rangle_1\right) \tag{15}$$

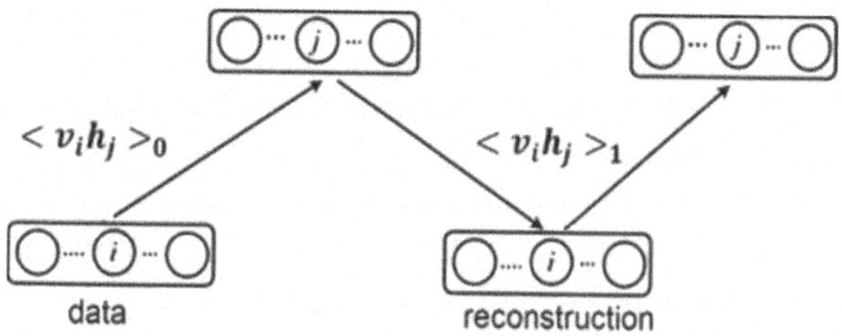

Fig. 4 Contrastive divergence training

$$\Delta a_i = \alpha(\langle v_i \rangle_0 - \langle v_i \rangle_1) \tag{16}$$

where α is the learning rate. The amendments are based on the difference between $\langle v_i h_j \rangle_0$ first value and $\langle v_i h_j \rangle_1$ last value. Weight modification w_{ij} depends only on device activations v_i and h_j.

The following steps constitutes the CD algorithm

1. A training sample x, $v^{(0)} \leftarrow x$ is considered.
2. Calculate hidden units $h^{(0)}$ binary states using Eq. (8)
3. Calculate the visible units $v^{(1)}$ reconstructed states using Eq. (10).
4. Calculate the hidden units' binary states utilizing the visible units reconstructed states obtained in step 3 using Eq. (8).
5. Update the neurons in the hidden and visible units as well as the weights utilizing Eqs. (14)–(16).

3.2 *Variants of Restricted Boltzmann Machine*

RBMs have many successful applications, including character recognition, labelling, subject modelling, dimension reduction, musical genre categorization, language identification feature learning, face recognition and video sequences. Scientists have suggested a number of variants of the RBMs. Such variations concentrate on various characteristics of the layout such as providing information of the relation between the hidden and visible units. Semi-restricted RBM is one such variation of RBM having adjacent connections amid visible units. Another variant is Temporal-Restricted Boltzmann Machines (TRBM) having guided connections amid visible and hidden units so as to transfer background information from previous states to current states. Also, TRBMs are used to model a complex sequence of time series, where the decision at each step involves past background knowledge. Recurrent Temporal-Restricted Boltzmann Machines (RTRBM) is one of the TRBM's extensions in which every individual RBM utilizes a hidden state attained from the preceding RBM. Such RBMs improves the output of predictions and are also used to identify significant patterns in the data. Another class of RTRBM is known by the name structured RTRBM (SRTRBM) and utilizes a graph for modelling the structure of the dependency. The conditional-restricted Boltzmann machine compute additions in both the visible and hidden units. In a fuzzy-restricted Boltzmann machine, fuzzy calculation is utilized to spread the relation, from constant to variable, between the hidden and the visible units. It substitutes standard RBM energy function by fuzzy energy function and fuzzy numbers by model parameters. Traditional RBMs utilizes various extensions and visible as well as hidden units in binary form such as the use of continuous units, SoftMax units and Poisson units.

Various RBM variants such as gated RBMs, spike-slab RBMs, mean-covariance RBMs, and factored three-way models were also used. A Robust-Restricted Boltzmann Machine (RoBM) has been used to eradicate the effect of noise in the input data, so as to attain an improved generalization by removing the effect of corrupted pixels. The RoBMs have given an impressive performance in the area of visual recognition in comparison to conventional algorithms by appropriately dealing with obstructions and noise. The consequence of the temperature parameter on RBM—Temperature-based Restricted Boltzmann Machine (TRBM) has also been considered. The temperature parameter in TRBMs is computed by setting the sigmoid function slope parameter to control the distribution of activity of the firing neurons.

3.2.1 Modeling Binary Data

The top layer in an RBM includes a set of stochastic binary functions h. That is, with a certain probability the status value of each neuron can be either 0 or 1. Stochastic visible binary variables x are present in the base layer. Joint Boltzmann distribution is denoted as follows

$$p(x, h) = \frac{1}{Z} \exp(-E(x, h)) \tag{17}$$

$E(x, h)$ represents the energy term denoted by

$$E(x, h) = -\sum_i b_i x_i - \sum_j b_j h_j - \sum_{i,j} x_i h_j W_i \tag{18}$$

Also, the normalization constant is as follows

$$Z = \sum_x \sum_h \exp(-E(x, h)) \tag{19}$$

The conditional Bernoulli distributions can be derived from the above equations:

$$p(h_j = 1|x) = \sigma\left(b_j + \sum_i W_{ij} x_i\right) \tag{20}$$

$\sigma(z)$ denotes logistic sigmoidal function

$$\sigma(z) = \frac{1}{1 + e^{-z}} \tag{21}$$

b_i, b_j are bias terms.
The visible vector x marginal distribution is

$$p(x) = \sum_h \frac{\exp(-E(x, h))}{\sum_{u,g} \exp(-E(u, g))} \tag{22}$$

The parameters performing gradient ascent in the log-likelihood turn into ($\langle\cdot\rangle$denotes expectation).

$$\Delta W_{ij} = \varepsilon\left(\langle x_i h_j\rangle_{\text{data}} - \langle x_i h_j\rangle_{\text{model}}\right) \tag{23}$$

In the distribution of data, x is derived from the input data set whereas h is derived from the model's conditional distribution $p(h/x, \theta)$. Both of these are taken from the model's joint distribution $p(x, h)$. One gets a similar but simpler equation, for the terms of bias. Computation of expectations is done using Gibbs sampling, where samples are produced from the probability distributions.

3.2.2 Modeling Real-Valued Data in RBMs

Restricted Boltzmann Machines can be generalized to exponential distributions within the family. For example, visible units with a Gaussian distribution may model digital images with real-valued pixels.

The hidden units determine the mean of this distribution in the following way:

$$p(x_i/h) = \frac{1}{\sqrt{2\pi}\sigma_i}\exp\left(-\frac{(x - b_i - \sum_j h_j w_{ij)}^2}{2\sigma_i^2}\right) \tag{24}$$

The marginal distribution over visible units x with an energy term is given by Eq. 25

$$E(x, h) = \sum_i \frac{(x_i - b_i)^2}{2\sigma_i^2} - \sum_j b_j h_j - \sum_{i,j} h_j w_{ij}\frac{x_i}{\sigma_i^2} \tag{25}$$

If for all the visible units i, variances are set to $\sigma_i^2 = 1$, same parameters are used as defined in Eq. (23).

3.3 *Deep Belief Network (DBN)*

DBNs utilizes RBM as major building blocks and comprises of an order of hidden stochastic variables, thereby also termed as probabilistic graphic models [16]. It also showed that DBNs are universal approximates. It has been applied to various issues namely handwritten digit identification, indexing of data, dimensionality reduction [3] and recognition of video and motion sequences. DBN is a subclass of DNNs comprising of several layers [17]. Each visible layer neurons represents input of the layer, whereas output of the layer is represented by hidden neurons. The preceding layer owns the visible neurons, for which such neurons are hidden. A DBN's distinctive feature is that there exist only symmetrical relations between the hidden and

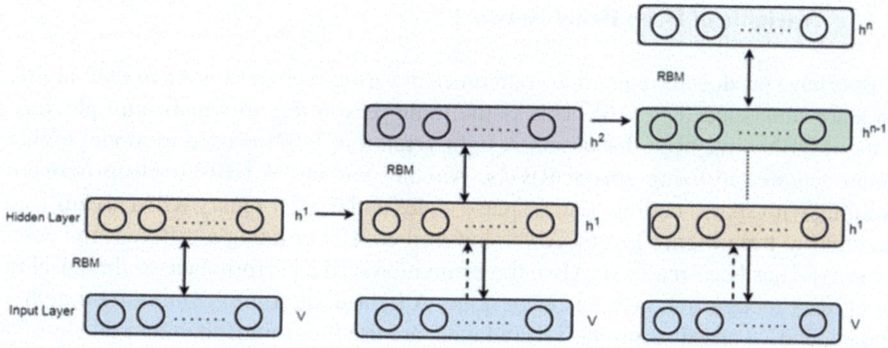

Fig. 5 A deep belief network

visible neurons. An example of DBN is shown in Fig. 5. DBNs have the capability just as the case with RBMs, to replicate, without control, the input data probability distribution. DBNs are better in terms of performance due to the fact that all the computations of probability distributions from the input data stream are performed in an unsupervised way, thereby, making them more robust than the shallow ones. Due to the fact that real-world data is frequently organized in hierarchical forms, DBN's stake profits from that. A lower layer learns features of low-level input, whereas higher layers learn features of high-level. DBNs are essentially trained in an unsupervised manner in contrast to RBMs that are trained in a supervised way. DBNs training is performed in two stages, one is the unsupervised pretraining phase, carried out in a bottom-up manner delivering the weights initialized in a better way as compared to the randomly initialized weights [11]. The next stage is the supervised fine tuning and is performed in order to change the entire network.

Due to the unsupervised training that is directed by the data, DBNs usually circumvent the difficulties of overfitting and underfitting. The parameters for every successive pair in the representational layers as shown in Fig. 5 are learned as RBM for unsupervised pretraining. In the first step the RBM at the bottom is trained utilizing the raw training data. After this the hidden activations of this RBM are utilized as inputs to the subsequent RBM so as to attain an encoded depiction of the training data.

Fundamentally, the hidden units existing in the previous RBM are fed as input to the subsequent RBM. Each RBM represents a DBN layer and the whole process is repeated for the chosen number of RBMS present in the network. Each RBM captures higher level relationships from the layers lying beneath. Stacking of the different RBMs in this way results in gradual discovery of functions Normally, a fine-tuning step is followed when the topmost RBM is equipped. This can be achieved either in a supervised way for classification and regression applications or in an unsupervised manner using gradient descent [18] on a log-likelihood approximation of the DBNs.

3.3.1 Variants of Deep Belief Network

DBNs have produced tremendous outcomes in various spheres owing to their ability to learn unlabeled data [16]. This is the main reason due to which multiple variants of DBN have been discussed. A light version of DBN is used to model higher order features utilizing sparse RBMs. Another variant of DBN to deep network training with sparse coding. Later, sparse codes and regular binary RBM are utilized as input to train higher layers. A version of the DBN utilizing a different top-level prototype has been realized. Also, the estimation of the performance of the DBN in 3D object recognition task has been done. A hybrid algorithm combining together the generative and discriminative gradients, are used to train Boltzmann third-order machine i.e. a top-level model. For increasing DBN's robustness to disparities such as obstruction and noise, a denoising and sparsification algorithm is proposed. To evade appalling forgetting during the course of unexpected changes in the input distribution, M-DBN is utilized as an unsupervised DBN in modular form to prevent disremembering of feature learning in continuous learning circumstances. M-DBNs comprises of multiple units, and the units that reconstructs a sample in the best way are only trained. Moreover, DBN practices batch-wise learning to fine-tune the learning rate of every module. M-DBN holds its efficiency even when there exist deviations in the input data stream distribution. This is different to monolithic DBNs which progressively overlook the earlier learned representations. Combinatorial DBN were used where one DBN extracts motion characteristics whereas the other DBN extract image characteristics. The output attained from both DBNs is used as an input to convolutional neural network for classification applications. Multi-resolution Deep Belief Network (MrDBN) learn features from multi-scale image representation. MrDBN includes creating the Laplacian Pyramid for individual picture, and then DBN training at each pyramid point is done separately. Next, both these DBNs are merged into a single network called MrDBN utilizing top-level RBM. DBN was also used in image classification through the use of the robust Convolutional Deep Belief Network (CDBN) and has also given good performance in various visual recognition tasks.

3.4 *Autoencoders (AEs)*

Autoencoders also known as auto associators are proficient in learning effective representations of the input signals without supervision means the training data is unlabeled [19]. Typically, such coding's have reduced dimensionality as compared to the input data, thereby making autoencoders an important application for dimensionality reduction [8]. These AEs also behave as influential feature detectors for pretraining deep neural networks without supervision [19]. Autoencoders serves as generative models and generates similar new data as the training data. For example, it is possible to train an auto encoder on facial pictures and then use it to generate new faces. A Multilayer Perceptron (MLP) can achieve dimensionality reduction and data compression in auto-association mode. One of the autoencoder's main tasks is to obtain a feature representation for reproducing high accuracy input data. The general process and representation of an auto encoder is presented in Figs. 6 and 7. The ultimate aim of the training is transformation of the input feature vectors into a coded vector with less dimensionality and to reconstruct the input data with least reconstruction error from the corresponding code. Essentially, the coding procedure includes learning functions from input signals. AE extracts useful features for each input vector, and filters unwanted information. The distinction between auto encoder and multi-layer perceptron is that MLP training is done to predict an output Y for a given input X whereas reconstruction of the input is performed using AE. All through the encoding, with the aid of weight matrix W, the AE transforms the input X to code h. Using the weight matrix W' reconstruction of $\tilde{X}$ is done all through the decoding process from h. Parameter optimization is practiced during the training of autoencoders to lessen the error between X and the $\tilde{X}$ reconstruction. Typically, if the internal layer dimension is less as compared to the input layer, autoencoder performs the process of reduction of dimensionality. On the opposite, if a bigger size hidden layer is considered, we will enter the realm of detection of features. In view of an unlabeled training set $\{x_1, x_2, \ldots\}$, where $x_i \in R^n$, the output vector $h_{(W,b)}$ is taken equal to the input vector x (i.e., $h_{(W,b)}(x_i) = x_i$). Reconstruction of the input vector is found with the help of autoencoders where the input vector is taken as the target vector. The basic auto encoder structure is depicted in Fig. 6. The first stage of an autoencoder consists of conversion of the input feature vectors to an internal representation, called the encoder. This phase is also called network recognition.

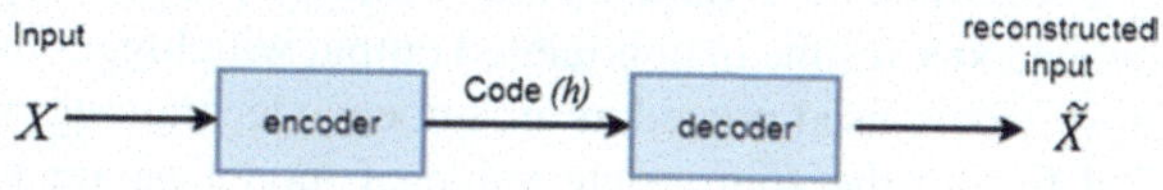

Fig. 6 The general process of an autoencoder

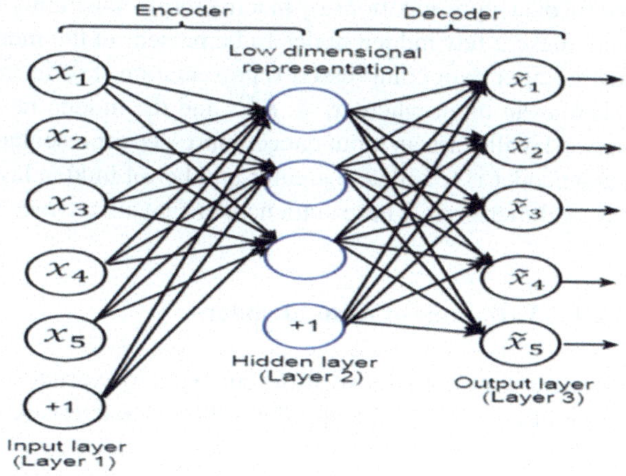

Fig. 7 Basic structure of an autoencoder

$$a^{(2)} = f\left(W^{(1)}x + b^{(1)}\right) \tag{26}$$

where $f(\cdot)$ denotes the encoder activation function. Next stage in an autoencoder consists of conversion of the internal representation into the target vector called the decoder.

$$h_{(W,b)}(x) = g\left(W^{(2)}a^{(2)} + b^{(2)}\right) \tag{27}$$

where $g(\cdot)$ denotes the decoder activation function. Minimizing a loss function L represents the learning process.

$$L(x, g(f(x))) \tag{28}$$

An autoencoder usually has similar architecture as MLP (Multi-layer Perceptron) apart from the fact that the output layer neurons must be equal to the neurons in the input layer. The below shown example represents an encoder-decoder network which is composed of only one hidden layer with three neurons in the encoder and a single output layer with five neurons in the decoder network. The autoencoder attempts to reconstruct the inputs and thus the outputs generated are also termed as reconstructions. The cost function encompasses a reconstruction loss to penalize the model with different reconstructions from the input. Sigmoid, identity (linear) function or hyperbolic tangent function are the most widely used activation functions for encoder and decoder. The encoder network encompasses a nonlinear activation function and a decoder network on the other hand encompasses a linear activation function when the input data are not limited to [0, 1] or [−1, 1]. This linear decoder autoencoder results in unbounded output with bigger than 1 value or less than 0. The most widely used procedure in autoencoder training is back propagation algorithm [20] to find the appropriate value of model parameters for reconstruction of the original input vector $\left\{W^{(1)}, b^{(1)}, W^{(2)}, b^{(2)}\right\}$. Autoencoders may be forced to learn useful data representations by placing those constraints upon them. These constraints may make a few hidden nodes to be present in the hidden unit, so that the network learns input data compressed representation. For instance, if a 30 by 30 image is taken as an input where $x_i \in R^{900}$ and the hidden neurons are taken to be 50, the network will learn an input data compressed representation. Another way to limit the autoencoder is to utilize a greater number of hidden layers as compared to the input vector dimensions. These autoencoders are also termed as regularized auto encoders.

3.4.1 Variations of Auto Encoders

Autoencoders consists of different types of variants. Table 1 lists some eminent variations and briefly recapitulates their characteristics and advantages.

Table 1 Autoencoder variants characteristics and advantages

Autoencoder	Characteristics	Advantages
Sparse autoencoders	Sparsity penalty is added in order to make the feature representation sparse	Performance of the network is improved thereby making the category more meaningful
Denoising autoencoders	Network is able to reconstruct the correct input from the corrupted data	Network is robust to noise
Contractive autoencoder	The reconstruction errors function is augmented with an analytic contractive penalty	Local directions of variation can be captured in a good way from the input data
Convolutional autoencoder	All locations in the input share weights	Allows to use 2D image structure
Zero bias autoencoder	Training of an autoencoder is done with without regularization and with the help of a suitable shrinkage function	High intrinsic dimensionality data attains better results

Denoising Autoencoders

The denoising autoencoder [21] differs from the autoencoder in one way. The input signal is initially corrupted partially in denoising autoencoder and later on it is fed to the network. The network training is done in a manner that the input data stream is restored from the moderately corrupted data. This criterion allows the AE to understand the primary structure of the input signals for adequately recreating the original input vector. to recreate the original input vector adequately [22]. Usually, autoencoders reduces the loss function L, which penalizes $g(f(x))$ because it is dissimilar to x.

$$L(x, g(f(x))) \tag{29}$$

A denoising autoencoder reduces the Loss function (L):

$$L\left(x, g\left(f\left(\hat{x}\right)\right)\right) \tag{30}$$

where $\hat{x}$ is a copy of x, corrupted with noise. Therefore, denoising autoencoders rather than just copying the data instead will reverse this noise corruption. Mechanism of the denoising auto-encoder is portrayed in Fig. 8. De-noising auto-encoder extracts the noise-free input data. In order to stochastically mitigate the adverse effects of noise corrupted input data, the essential statistical characteristics of input data can be taken into account. If it is possible to determine the form and level of corrupting noise, it is easier to implement the DAE.

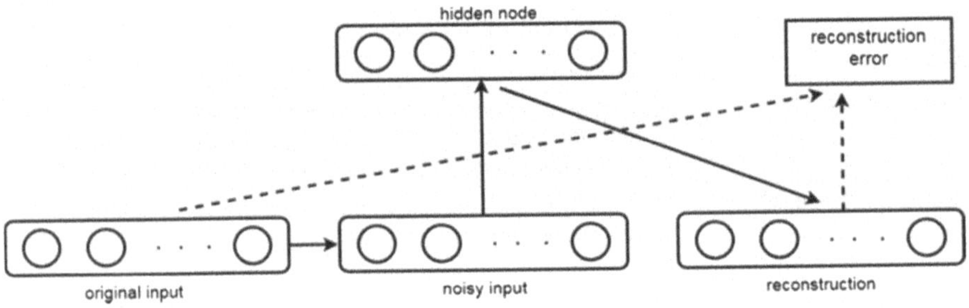

Fig. 8 Denoising autoencoder

Contractive Autoencoders (CAE)

CAE learns robust feature representations in a similar way as the denoising autoencoders [23]. In order to make the mapping reliable, a DAE adds noise to the training signals; and a CAE, on the other hand to realize robustness, during the reconstruction phase applies a contractive penalty to the cost function. The term "penalty" refers to precise function sensitivity to the input data. The implementation of a penalty word has been found to result in more robust applications that are resistant to minor changes in data. Also, the penalty addresses the trade-off between robustness and reconstruction accuracy. Contractive autoencoders [24] yield better results as compared to the other regularized autoencoders such as denoising autoencoders. A denoising autoencoder [21] with a very less amount of corruption noise is viewed as a form of CAE where both the encoder and the decoder are subject to the contractive penalty. CAEs serves as a good application in feature engineering due to the reason that only encoder part is utilized for feature extraction.

Deep Autoencoders (DAE)

DAE refers to auto associative networks having more than one hidden layer. Usually, DAE with a single layer cannot remove characteristics that are discriminatory and reflective of the unprocessed data. The concept of deep and stacked autoencoders was therefore put forward. The pictorial representation of deep stacked encoder is shown in Fig. 9. Addition of more layers assists the autoencoder in learning more complex codes. Though, care must be taken, not to specialize the auto encoder too much. An encoder basically specializes in learning the input mapping with an arbitrary number and the decoder performs the same function in a reverse way. No suitable general data representation can be acquired, though, such type of autoencoder can completely recreate the training data. Also, it is very improbable to generalize the training data efficiently into new occurrences. The stacked autoencoder architecture is usually proportioned with respect to the hidden central layer. It just seems like a sandwich, to put it simply. For example, a MNIST autoencoder may have 784 inputs, and a 300-neuron hidden layer, followed by a 150-neuron central

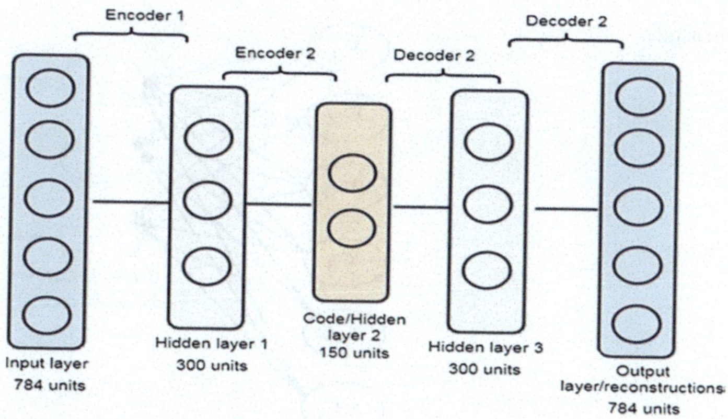

Fig. 9 Deep stacked autoencoder

hidden layer, a 300-neuron hidden layer, and lastly a 784-neuron output layer. Such stacked auto encoder is shown in Fig. 9. Except there are no labels, the stacked DAE can be realized in a similar way as a standard MLP. A series of autoencoder networks form the deep auto encoder network, stacked in a feature hierarchy one above the another. An autoencoder aims to reduce previous layer's reconstruction error. The stacked deep autoencoders training is usually done layer-wise utilizing greedy unsupervised learning followed by supervised fine tuning. This unsupervised pretraining is done to give a good initialization to the network weights until a supervised fine-tuning procedure is applied. In addition, unsupervised pretraining often results in improved models, as it depends primarily on unlabeled data. The consequent fine-tuning performed in a supervised manner includes altogether fine-tuning the weights learned with pretraining. The auto encoder is depicted in Fig. 7. In the first step training is performed with backpropagation algorithm utilizing gradient descent optimization [18] to acquire the features at the first level $h^{(1)(i)}$). Subsequently, the last layer of the decoder network is not utilized, whereas as in the encoder network the hidden layer having parameters $\{W^{(1)}, b^{(1)}\}$ is retained as depicted in Fig. 10. The second auto encoder is equipped with the characteristics attained from the first auto encoder as presented in Fig. 11. The first auto encoder parameters are kept unaffected while the second autoencoder is being trained. Therefore, the network training is done greedily in a layer by layer approach. For final supervised fine-tuning step, the weights obtained after the network training step are used as initial weights. This process is shown in Fig. 12. The first auto encoder is therefore trained on the *xi*-input data with backpropagation algorithm to attain the $h^{(1)(i)}$) features. The features attained from the first stage are used as inputs for subsequent autoencoder training. The second autoencoder is trained to generate another set of new representations $h^{(2)(i)}$ in a manner similar to the first auto encoder. Therefore, each autoencoder training is performed using the representations from the previous autoencoder. Only the currently trained autoencoder parameters are modified, while the preceding autoencoders parameters are kept unchanged. Lastly, an output layer is

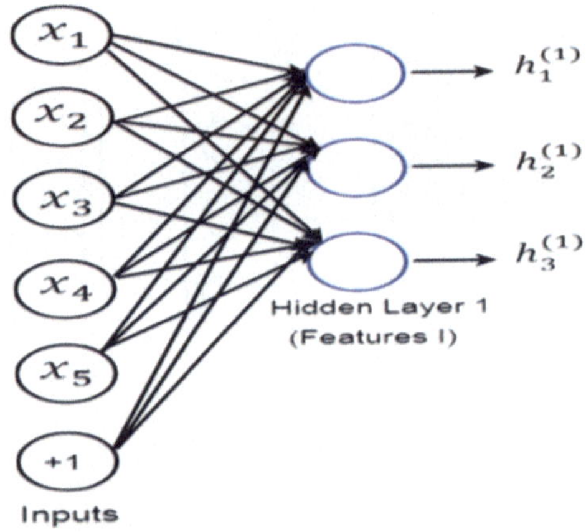

Fig. 10 Autoencoder training

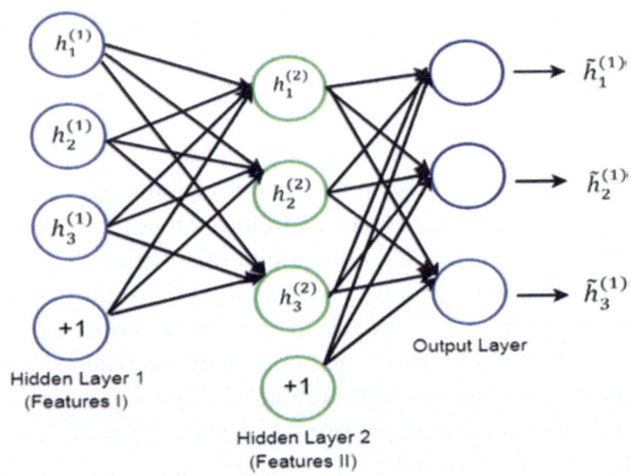

Fig. 11 Training second autoencoder

applied to this stack of qualified autoencoders, and a supervised learning algorithm (using labelled data) trains the entire network. At Fig. 12, two auto encoders are pre-trained, and a layer of output is then applied to form the final network.

Generative Adversarial Networks (GANs)

GAN models learn every data distribution and concentrating mainly on sampling from the learned distribution. They allow the creation of fairly realistic worlds which in any domain are indistinguishable to ours: audio, pictures, voice. A GAN consists of two prime components: Generator and Discriminative, which throughout the training process are in constant battle with each other.

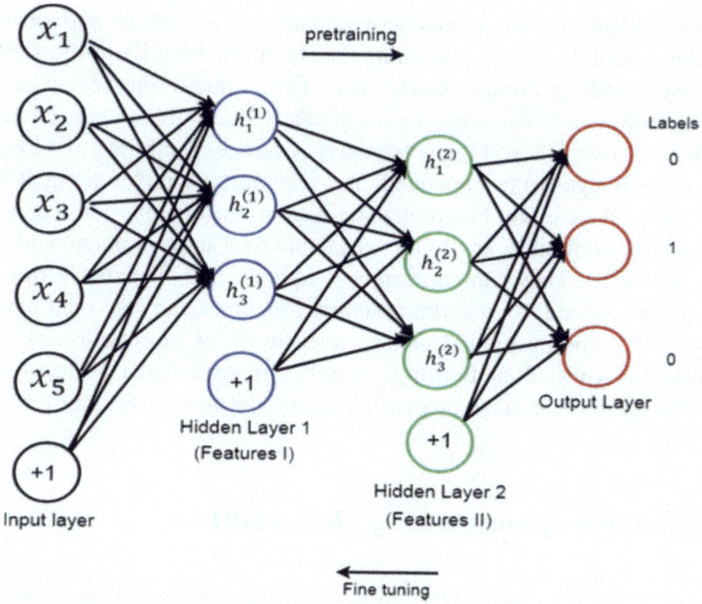

Fig. 12 Fine-tuning of the network

- Network generator—A generator $G(z)$ takes random noise as its input and attempts to produce a data sample.
- Discriminator network (or adversary)—the discriminator network $D(x)$ takes information from either the actual data or the data generated from the network and attempts to determine whether the input is real or created. It takes an input x from the actual pdata(x) distribution and then solves a question of binary classification giving output in the range from 0 to 1. The generator's task is basically to produce natural-looking images, and the discriminator's task is to determine whether the image is being generated or whether it is true.

4 Applications and Implications of Deep Learning

Deep Learning found applications in various domains such as computer vision, image processing, driving autonomous vehicles, natural language processing, and so on. A lot of data will be fed into the system in a supervised learning technique, so that the computer can determine whether the conclusion is correct or wrong due to the data labelling given.

There is no labelling in unsupervised machine learning and hence the algorithm has to find out for itself whether a certain decision was right or wrong due to the enormous amounts of data fed into the device. Then there is something called semi-supervised learning that works between supervised learning and unsupervised learning somewhere. Using deep learning Facebook can identify millions of images posted by users without human interference. This is due to the advent of machine learning that millions of images can be scanned at a time to find out the content of the image with greater accuracy. It then codes certain images according to the conditions imposed for the images to be separated. The machines by just looking at the given image and assign it a caption based on that image's constituents. This has been tried and tested and the computers are doing reasonably well at the moment, and with time it can only get better. They can generate a symphony, add elements with very good accuracy that are missing in a certain picture. Machines can also read handwriting so that they can come up with their own interpretations of conclusions and make sense of different types of handwritings. Another important area of Deep Learning's strength is the natural language processing sector and accent recognition.

4.1 Sustainable Applications of Deep Learning

Deep learning is a new, advanced technique for the processing of images and the analysis of data, with promising results and great potential. As deep learning has been implemented successfully in various domains, it has also recently entered the agricultural domain. Smart farming is critical in addressing the challenges of agribusiness in terms of efficiency, environmental impact, food security and sustainability. As the global population continues to grow, a significant increase in food production must be achieved, while at the same time maintaining availability and high nutritional quality throughout the world, protecting natural ecosystems by using sustainable farming methods. To address these problems, the dynamic, multivariate, ecosystems need to be better understood by constantly tracking, measuring, and analysing various physical aspects and phenomena. This includes analyzing large-scale agricultural data and using emerging information and communication technologies (ICT), both for short-scale crop/farm management and observation of large-scale ecosystems, improving existing management and decision/policy activities through context, situation and location. Large-scale observation is enabled by remote sensing using satellites, aircraft and unmanned aerial vehicles i.e. drones, offering wide-ranging snapshots of the agricultural environment. When applied to agriculture, it has many benefits, being a well-known, non-destructive method of collecting information on earth features while data can be collected systematically over broad geographic areas. A wide subset of data volume collected through remote sensing includes images. Images constitute a complete picture of the agricultural environments in many cases, and could address a variety of challenges. Imaging analysis is therefore an important area of research in the agricultural domain, and intelligent data analytics techniques are used in various agricultural applications for image identification/classification, anomaly detection, etc. DL in agriculture is a recent, modern and promising technique with increasing popularity, whilst DL's advances and applications in other fields indicate its great potential. Together with big data innovations and high-performance computing it has emerged to create new opportunities for unravelling, quantifying and understanding complex data processes in agricultural operating environments. Deep learning is the vanguard of the entire rising and harvesting process. It starts with a seed being planted in the soil—from soil preparation, seed planting, and water feed calculation—and ends with the aid of computer vision when robots pick up the harvest deciding ripeness. The various agricultural application areas of deep learning are as follows:

- **Species Breeding**

Species selection is a repetitive method of looking for different genes that inhibit the effectiveness of water and fertilizer usage, climate change adaptation, disease tolerance, fertilizer quality or a better taste. In particular, machine learning, deep learning algorithms, take decades of field data to evaluate crop production in different environments, and new features are being created in the process. They may create a probability model based on this data that will predict which genes would most likely contribute a beneficial trait to a plant.

- **Species Recognition**

While the conventional human approach to plant classification will be to match the color and shape of the leaves, deep learning will produce more precise and quicker results by analyzing the morphology of the leaf vein that carries more knowledge on the properties of the leaf.

- **Soil management**

Soil is a heterogeneous natural resource for agricultural specialists, with complex processes and vague mechanisms. The temperature alone can provide insight into the impact of climate change on the area yield. Deep learning algorithms study cycles of evaporation, soil moisture and temperature in order to understand ecosystem dynamics and the impingement in agriculture.

- **Water Management**

Agricultural water management impacts hydrological, climatic, and agronomic balance. To date, the most advanced DL-based applications are related to estimating daily, weekly, or monthly evapotranspiration allowing for more efficient use of irrigation systems and prediction of daily temperature at the dew point, which helps to recognize predicted weather patterns and estimate evapotranspiration and evaporation.

- **Yield Prediction**

Yield prediction is one of the most important and common topics in agriculture precision as it describes yield mapping and estimation, crop supply matching with demand, and crop management. State-of-the-art methods have gone well beyond simple prediction based on historical data, but integrate computer vision technologies to provide on-the-go data and detailed multidimensional analysis of crops, environment, and economic conditions to optimize yields for farmers and citizens.

- **Crop Quality**

Precise identification and classification of crop quality characteristics will improve the price of the commodity and reduce waste. Compared to the human experts, computers may use apparently irrelevant data and interconnections to discover and identify new qualities that play a role in the overall crop quality.

- **Weed Detection**

Besides pests, the most important threats to crop production are weeds. The greatest challenge in battling weeds is that they are hard to detect and dis-criminate from crops. Computer vision and DL algorithms can improve weed identification and discrimination at low cost and without environmental issues and side effects. These technologies will drive robots in the future which will destroy weeds, minimizing the need for herbicides.

While reading about the future is often interesting, the most significant part is the technology that paves the way for it. For example, agricultural deep learning is a collection of well-defined models that gather specific data and apply specific algorithms to achieve the expected results. Artificial and Deep Neural Networks (ANNs and DL), and Support Vector Machines (SVMs) are the most popular models in agriculture. While DL-driven farms are already evolving into artificial intelligence systems at the beginning of their journey. Currently, machine learning approaches resolve individual issues, but with further incorporation of automated data collection, data analysis, deep learning, and decision-making into an integrated framework, farming practices can be converted into so-called knowledge-based farming practices that could improve production rates and quality of goods.

5 Challenges and Future Scope

Despite the fact unsupervised learning systems have had a catalytic influence in revitalizing the attention in deep learning, additional research is required to develop different unsupervised algorithms based on deep learning. Generally, unsupervised algorithms are not good at separating the primary issues that account for how the learning data is spread in the hyperspace. Through developing unsupervised learning algorithms to disentangle the original issues that accounts for variations in hyperspace data, the information can be utilizes for efficient transfer learning and classification. We need to explore the advancement areas in the field of unsupervised learning by discovering new specifics of unlabeled data and mapping relationships between input and output. Taking advantage of the input output association is closely related to the conditional generative model development. Thus, generative networks provide a capable direction in the field of research. This advancement can return the spotlight of pattern recognition and machine learning in the near future for solving multiple tasks specifically in the agricultural domain making it a hot area for sustainable real applications.

References

1. J. Schmidhuber, Deep learning in neural networks: an overview. Neural Netw. 85–117 (2015)
2. B.Z. Leng, A 3D model recognition mechanism based on deep boltzmann machines. Neurocomputing **151**, 593–602 (2015)
3. G.E. Hinton, Reducing the dimensionality of data with neural networks. Science **313**(5786), 504–507 (2006)
4. S. Haykin, in *Neural Networks and Learning Machines,* 3rd edn (Pearson, Upper Saddle River, NJ, 2009), pp. 7458
5. Y.B. LeCun, Deep learning. Nature **521**(7553), 436–444 (2015)
6. Y. Bengio, Learning deep architectures for AI. Found. Trends® Mach. Learn. **2**(1), 1–127 (2009)
7. R. Salakhutdinov, Learning deep generative models. Doctoral thesis, MIT (2009). Available at http://www.mit.edu/_rsalakhu/papers/Russthesis.pdf
8. G.E. Hinton, A fast learning algorithm for deep belief nets. Neural Comput. **18**(7), 1527–1554 (2006)
9. N. Kermiche, Contrastive hebbian feedforward learning for neural networks. IEEE Trans. Neural Netw. Learn. Syst. (2019)
10. J.M. Wang, Deep learning for smart manufacturing: methods and applications. J. Manuf. Syst. **48**, 144–156 (2018)
11. D.B. Erhan, Why does unsupervised pre-training help deep learning? J. Mach. Learn. Res. **11**, 625–660 (2010)
12. X.M. Lü, Fuzzy removing redundancy restricted boltzmann machine: improving learning speed and classification accuracy. IEEE Trans. Fuzzy Syst. (2019)
13. A. Revathi, Emotion recognition from speech using perceptual filter and neural network, in *Neural Networks for Natural Language Processing* (IGI Global, 2020), pp. 78–91
14. R. Salakhutdinov, Learning deep generative models. Annu. Rev. Stat. Appl. **2**, 361–385 (2015)
15. E.M. Romero, Weighted contrastive divergence. Neural Netw. **114**, 147–156 (2019)
16. P.G. Safari, Feature classification by means of deep belief networks for speaker recognition, in *23rd European Signal Processing Conference (EUSIPCO)* (IEEE, 2015), pp. 2117–2121
17. Y.T. Huang, Feature fusion methods research based on deep belief networks for speech emotion recognition under noise condition. J. Ambient. Intell. Hum. Comput. **10**(5), 1787–1798 (2019)
18. Y.S. Bengio, Learning long-term dependencies with gradient descent is difficult. IEEE Trans. Neural Netw. **5**(2), 157–166 (1994)
19. D.P. Kingma, An introduction to variational autoencoders. Found. Trends® Mach. Learn. **12**(4), 307–392 (2019)
20. N.S. Rajput, Back propagation feed forward neural network approach for speech recognition. in *3rd International Conference on Reliability, Infocom Technologies and Optimization* (IEEE, 2014), pp. 1–6
21. P.L. Vincent, Stacked denoising autoencoders: Learning useful representations in a deep network with a local denoising criterion. J. Mach. Learn. Res. 3371–3408 (2010)
22. A.H. Hadjahmadi, Robust feature extraction and uncertainty estimation based on attractor dynamics in cyclic deep denoising autoencoders. Neural Comput. Appl. **31**(11), 7989–8002 (2019)

23. S.V. Rifai, Contractive auto-encoders: explicit invariance during feature extraction (2011)
24. E.Q. Wu Rotated sphere haar wavelet and deep contractive auto-encoder network with fuzzy gaussian SVM for pilot's pupil center detection. IEEE Trans. Cybern. (2019)

An Overview of Deep Learning Techniques for Biometric Systems

Soad M. Almabdy and Lamiaa A. Elrefaei

Abstract Deep learning is an evolutionary advancement in the field of machine learning. The technique has been adopted in several areas where the computer after processing volumes of data are expected to make intelligent decisions. An important field of application for deep learning is the area of biometrics wherein the patterns within the uniquely human traits are recognized. Recently, many systems and applications applied deep learning for biometric systems. The deep network is trained on the vast range of patterns, and once the network has learnt all the unique features from the data set, it can be used to recognize similar patterns. Biometric technology that is being widely used by security applications includes recognition based on face, fingerprint, iris, ear, palm-print, voice and gait. This paper provides an overview of some systems and applications that applied deep learning for biometric systems and classifying them according to biometrics modalities. Moreover, we are reviewing the existing system and performance indicators. After a detailed analysis of several existing approaches that combine biometric system with deep learning methods, we draw our conclusion.

Keywords Deep learning · Convolution neural network (CNN) · Machine learning · Neural networks · Biometrics

S. M. Almabdy (✉) · L. A. Elrefaei
Computer Science Department, Faculty of Computing and Information Technology, King Abdulaziz University, Jeddah, Saudi Arabia
e-mail: salmabdy@kau.edu.sa

L. A. Elrefaei
e-mail: laelrefaei@kau.edu.sa; lamia.alrefaai@feng.bu.edu.eg

L. A. Elrefaei
Electrical Engineering Department, Faculty of Engineering at Shoubra, Benha University, Cairo, Egypt

© The Editor(s) (if applicable) and The Author(s), under exclusive license to Springer Nature Switzerland AG 2021

A. E. Hassanien et al. (eds.), *Artificial Intelligence for Sustainable Development: Theory, Practice and Future Applications*, Studies in Computational Intelligence 912,
https://doi.org/10.1007/978-3-030-51920-9_8

1 Introduction

The machine learning has developments in the last few years. The important development is known as Deep learning (DL) technique. DL models are intelligent systems that simulator the workings of a human brain for manipulating complex data by considering real world scenarios then creating an intelligent decision. The structured of DL networks, known as hierarchical learning which is a methods of machine learning. Deep learning networks is applied for several recognition models, pattern recognition, processing of signal [1], computer vision [2], speech system [3, 4], language processing [5], audio system [6] etc. From the wide variety of deep learning architectures, Deep Neural Networks (DNNs) [7], Convolution Neural Networks (CNNs) [8], Recursive Neural Networks (RNNs) [9], and Deep Belief Networks (DBNs) [10], have been used for most of these systems. Among architectures, generally, CNNs have been effectively used in image, video, audio while RNNs have been used in processing sequential data such as text and speech [11, 12]. These systems assist in the experimental investigation of deep recurrent neural network recognition which is the perfect way for larger speech recognition. The main reasons for the success of deep learning are: the abilities of chip-based processing is improved, such as GPUs, computing hardware cost is significantly reduced, and the machine learning ML systems have an improvement [13].

1.1 Deep Learning

Machine learning (ML) refers to computer science field which enables computers to learn without being explicitly programmed. ML involves the usage of different techniques and development of algorithms to process vast amount of data and a number of rules to enable the user to access the results. It also refers to the development of fully automated machines governed simply by the development and running of algorithms and on a set of pre-defined rules. Algorithm developed for ML uses data and the set of pre-defined rules to execute and deliver optimum results. Depending on the nature of the learning "signal" or the "feedback" available to the system, Machine Learning can be broadly categorized into three categories [14, 15]:

- Supervised learning: An example of ideal input and desired output is fed into computer with the goal that it learns to map inputs into the desired outputs.
- Unsupervised learning: In this learning the computer is not fed with any structure to learn from and is thereby left to itself to understand its input. This learning is a goal itself where hidden patterns in data can be understood and can aid in future learning.
- Reinforcement learning: This involves a more interactive learning where the computer interacts to its dynamic environment in order to accomplish a certain goal such as playing a game with a user or driving a vehicle in a game. Feedback is provided to the system in terms of rewards and punishments.

Fig. 1 The framework of machine learning

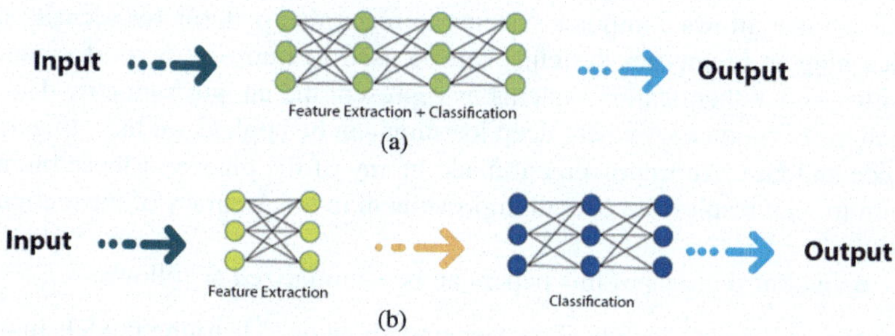

Fig. 2 The framework of deep learning

In the recent few years a method has been developed, which has given commendable results in many problems and has therefore affected the Computer Vision community. This method is known as, Deep Learning (DL) or more accurately Deep Neural Networks DNN.

The difference between traditional machine learning ML and deep learning DL algorithms is the feature engineering. Figure 1 showed the feature process in traditional ML [16], the process of feature extraction design to performs complex mathematics (complex design), wasn't very efficient. Then design model for classification to classify the extracted feature.

By contrast, in deep learning algorithms [17] as shown in Fig. 2 feature engineering is done automatically by implement classification and feature extraction in single stage as Fig. 2a, that means only one model is designed, or similar way to traditional machine learning as Fig. 2b. The feature engineering in DL algorithm is more accurate compared to traditional ML algorithm.

1.2 *Deep Learning for Biometric*

Recently, several methods of DL have been discussed and reviewed [13, 18–20]. DL techniques have been reported to show significant improvements in a range of applications, such as biometrics recognition and recognition of object. Deep learning techniques are being applied, in biometrics in different ways. It has been applied on biometrics modalities. Notably, there are apparent connections between the neural architectures of the brain and biometrics [21].

The use of biometric based authentication is constantly on the rise [22]. Biometric technology uses the unique biological properties to identify a person, that tend to remain consistent over one's lifetime e.g. face, iris, fingerprint, voice and gait. Unique data from these unique human traits are extracted, represented and matched to recognize or identify an individual. These biological properties allow humans identify several individuals depending on their behavioral and physical features as well as their correct use allows computer systems to recognize patterns for security tasks. Deep learning in biometrics systems, can be used to improve the performance of recognition and authentication systems by represent the unique biometric data. The typical areas of biometrics where deep learning can be applied are face, fingerprint, iris, voice and gait. An improvement made in any of the phases of these biometric applications, can result in an overall improvement in the accuracy of the recognition process.

The main contributions of this paper can be summarized as follows:

1. Reviews in details a technical background about deep learning models in neural networks such as Autoencoders AEs, Deep Belief Networks DBN, Recursive deep neural networks RNNs, and Convolution deep neural networks CNNs.
2. Gives a summary for the most common deep learning frameworks.
3. Reviews in details the deep learning techniques for biometric modalities based on biometrics characteristics.
4. State the main challenges for applying DL methods for biometric systems.
5. Summarizes the DL techniques for biometric modalities and show their model and performance of each application.

In the paper the applications of deep learning were categorized for biometrics identification systems according to biometrics type and modalities and present a review of these applications. Figure 3 showed the structure of the paper.

The rest of the paper is structured as follows: Sect. 2 provides a background about deep learning, Sect. 3 presents frameworks of deep learning, and Sect. 4 presents an overview of biometric system and present the deep learning techniques for biometrics modalities, reviews related work. In Sect. 5 the challenges. Finally, Sect. 6 states the discussion and conclusions.

2 Deep Learning in Neural Networks

A biological neural network (NN) comprises of a set of neurons which are associated with each other through axon terminals and the activation of neurons follows a linear path through these associating terminals. In a similar manner, in an artificial neural network, the associated artificial neurons perform activities based on connection weights and activation of their neighboring neuron. In such a system, a Neutral Network refers to a network which is enabled to use a number of networks such as recurrent network or feedforward, which may have one or two hidden layers. But if

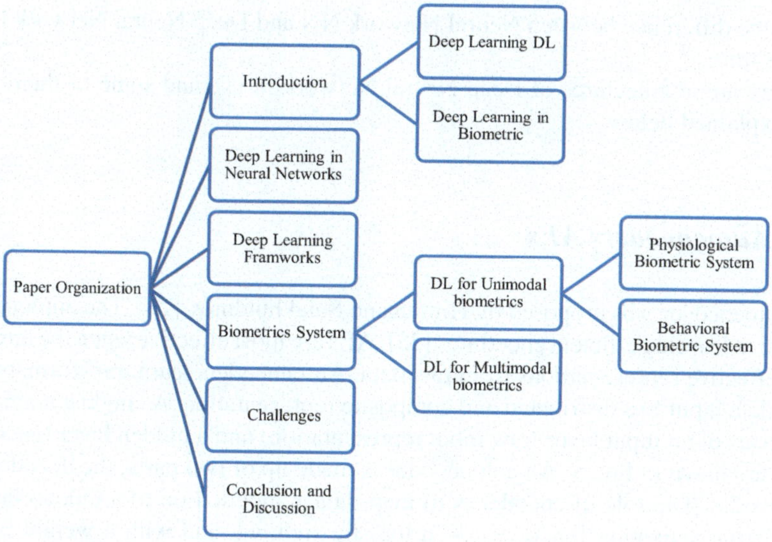

Fig. 3 Paper organization

the number of hidden layers becomes more than two, the network is known as Deep Neural Network DNN.

The architecture of deep network is consisting of hidden layers (typically 5–7) that is also termed as DNN [19]. The first deep architectures proposed are in research works [10, 23], which built for computer vision tasks. However, the process of training in DNN is implemented layer-wise by gradient descent. This layer-wise training enables DNN to learn the 'deep representations' that transform between the hidden layers. Usually, the training of layer-wise is unsupervised. Figure 4

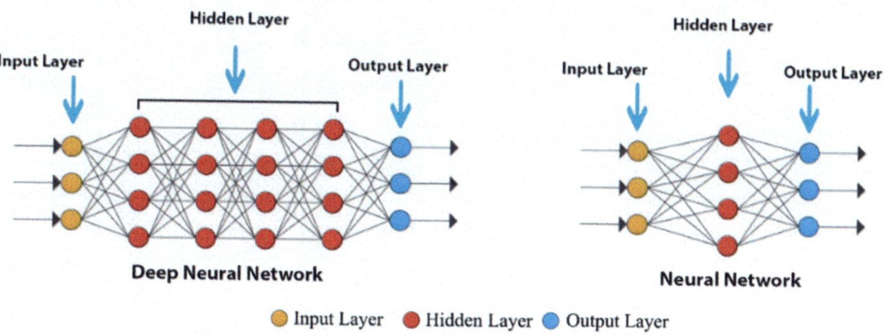

Fig. 4 The architecture of deep neural network DNN and neural network NN

shows the difference between Neural Network NN and Deep Neural Network DNN architecture.

There are architectures of Deep Neural Network in use and some of them have been explained below.

2.1 Autoencoders AEs

The autoencoder was proposed by Hinton and Salakhutdinov [24]. The autoencoder applied to learning efficient encodings [25]. AEs are most effective when the aim is to learn effective representations from raw data. Autoencoders learn transformation of a raw data input to a distributed and composite representation. A single autoencoder comprises of an input layer (raw input representation) and a hidden layer (encoding layer) as shown in Fig. 5, An autoencoder is made up of two parts; the decoder and the encoder. The role of encoder is to map data $\boldsymbol{x}$ inputted on to a hidden layer $\boldsymbol{h}$ by utilizing activation function; e.g. a logistic sigmoid, and with a weight matrix $\boldsymbol{w}$. The decoder later reconstructs it back to its original form. The encoder used transpose of weight matrix $\boldsymbol{W^T}$. Some autoencoders referred to as deep autoencoders are trained using back-propagation variants such as the method of conjugate gradient. The process of training and AE to be a deep AE can be broken down into two. The first step involves unsupervised learning where the AE learns the features then the second stage is where the network is fine-tuned by application of supervised learning.

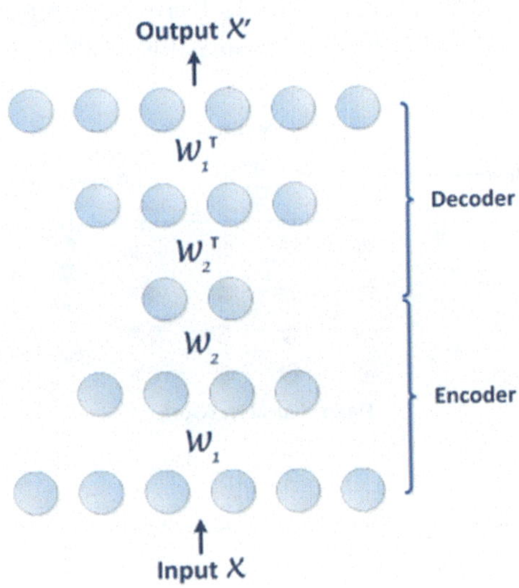

Fig. 5 Autoencoders architecture [24]

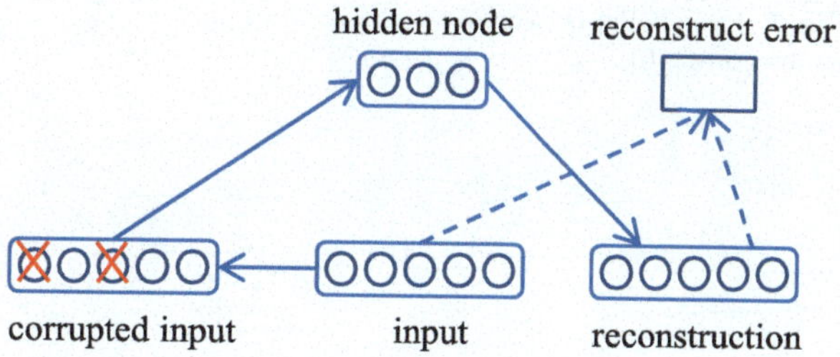

Fig. 6 Denoising autoencoder [30]

There are three variants of Autoencoders:

- **Sparse autoencoder**: The aim of the sparse AE is the extraction of sparse features from raw inputted data. This sparsity could either be extracted through the direct penalization of the output of the activation of hidden units [26, 27], or by penalization of the biases of hidden units [28, 29].
- **Denoising autoencoder (DAEs)**: Vincent et al. [30], proposed this model in order to increase how robust the model was by enabling it to recover the correct data inputted even if the data is corrupted, hence it could capture the input distribution structure. The DAE trains the AEs by adding noise into the training data intentionally in order to give the AEs the ability to recover data even when corrupted. The training process enables the DAE to gain the origin training data that is noise-free, hence suggesting increased process robustness in the DAE. DAE is shown in Fig. 6.
- **Contractive autoencoder (CAE)**: This technique was proposed by Rifai et al. [31]. Whereas the DAE trains the whole mapping by the injection of noise into the training set, the CAE adds analytic contractive penalty into the reconstruction error function hence increase the robustness of the system.

2.2 Deep Belief Networks DBN

DBNs presented by Hinton et al. [10]. These are similar to stacked Autoencoders and consist of stacks of simple learning methods known as Restricted Boltzmann Machines (RBM) [32]. RMB itself is a stack of two layers comprising a visible (input data) and a hidden layer $\boldsymbol{h}$ (enables learning of high correlation between data). All layers in a DBN interact with directed connections except for the top two, which form an undirected bipartite graph. Units belonging to the same layer (either visible or hidden) are not connected. The parameters of the DBN are weights $\boldsymbol{w}$ among the units of layers, and biases of the layer. Figure 7 showed an example Deep Belief Networks with 3 hidden layers. Every layer identifies correlations among the units of the layer beneath.

Fig. 7 Deep belief networks architecture [10]

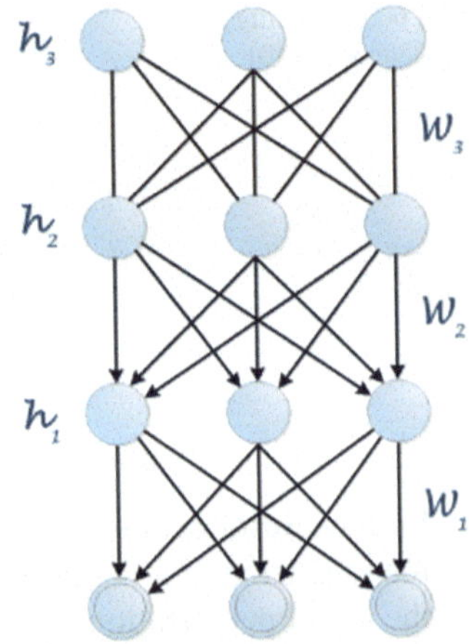

2.3 Recurrent Neural Networks RNN

RNN [33] is a powerful build that is applied in the modelling of sequential data such as text [34, 35] and sound [36]. An RNN usually has its parameters set by the use of three weight matrices and three bias vectors. The weight matrices used includes; hidden-to-hidden $\boldsymbol{W}_{hh}$, input-to-hidden, and $\boldsymbol{W}_{ih}$, hidden-to-output $\boldsymbol{W}_{ho,}$, whereas the bias vectors are the initial bias vector, the hidden vector and the output bias vector. With the input and the desired output given, a RNN is able to iteratively update its hidden state by the application of some nonlinearity over time; e.g. the sigmoid or the hyperbolic tangent, after which it can make a prediction of the output. To be specific, at each time-step **T**, the hidden network state is calculated as per three values: the first value is the input data, multiplied at this time-step by the input-to-hidden weight matrix. The second value is the hidden state of the preceding times-step which is multiplied by weight matrix of the hidden-to-hidden and the last value being the bias of the hidden later. In the same manner, each specific time-step's output of the network can be calculated by multiplying the sum of the output layer's bias and the time-step's state of the hidden layer with the hidden-to-output weight matrix [37]. This provides a connection between the input layer, the hidden layer, and the output layer as shown in Fig. 8. The matrices of weight of an RNN are distributed across the different time-steps because the same task is repeated in each step with only a change in the input data. This results to the RNN having less parameters when compared to a normal DNN.

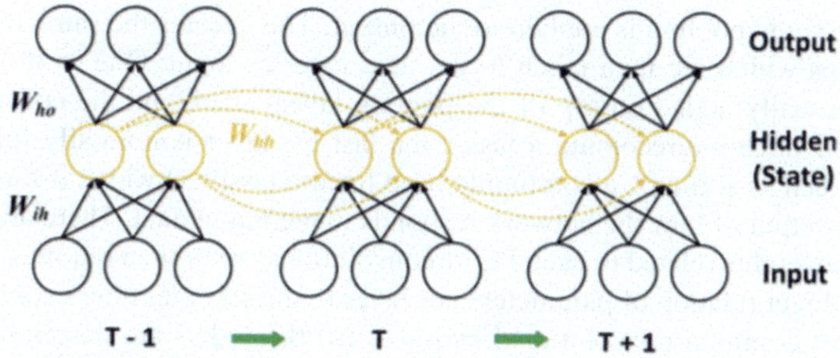

Fig. 8 RNN architecture [33]

2.4 *Convolutional Neural Networks CNNs*

Convolutional Neural Networks CNN is the most widely used Deep Neural Network in different problems of the Computer Vision, which is based in the *Multi-Layer Perceptron* architecture. CNN is a specialized form of neural network which comprises of a grid topology. It primarily consists of a number of filters which is applied at different locations of an organized input data in order to produce an output map. CNN was introduced by LeCun et al. [3] as a solution to the classification task created by Computer Vision. CNN simplified the tractability of training using simple methods of pooling, rectification and contrast normalization. The name "convolutional neural network" is derived from the term "Convolution" which is a special kind of linear operation used by the network [38]. Convolution networks have played a pivotal role in deep learning evolution and are a classic example of how information and insights from studying the brain can be applied to machine learning applications.

The architecture of CNN network shown in Fig. 9. It is normally made up of three main layers combined: the pooling-layers, the convolutional-layers, and the fully-connected-layers. To ensure that the input image and the output of the previous layer is convolved, several filters are used. The output values from such an operation are then taken through a nonlinear activation function (also known as nonlinearity)

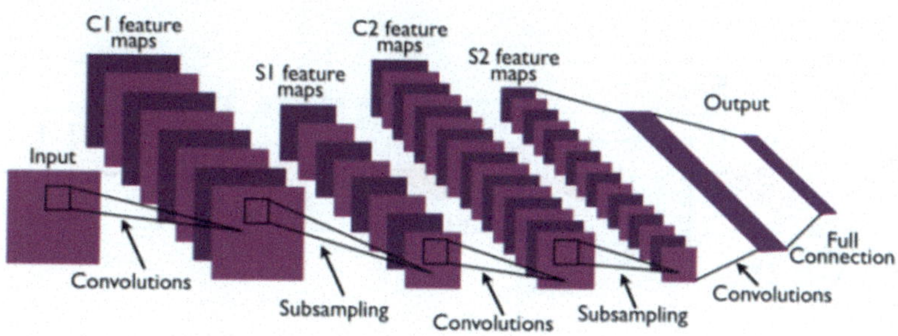

Fig. 9 A typical convolutional network architecture [8]

after which some pooling is applied to the output. This creates the same number of feature maps which are then taken to the next layer as input. One or more layers of FC are usually added on top of the pooling layers stack and the convolutional layer. In classification/recognition tasks, the last FC layer is normally linked to a classifier (such as softmax, a commonly used linear classifier) which is then able to provide an output of how the network responds to the initial data. There are specific parameters/weights related to each FC or convolutional layer than requires learning. There is a direct relation of parameters per layer to the size and filters applied [8].

The most common convolutional neural network models are describing as the following:

- **LeNet-5**: It is proposed by LeCun et al. [39]. It is consisting of seven convolutional layers. LetNet-5 was applied by a number of banks to recognize numbers of hand-written on cheques. The capability to process images with high resolution requires more convolutional layers. It is constrained by the availability of calculating resources. The Architecture of LetNet-5 shown in Fig. 10.
- **AlexNet**: The network proposed by Krizhevsky et al. [7], The network architecture is similar as LeNet network, but was deeper, includes more filters each layer, and stacked convolutional layers. AlexNet comprises of five convolutional layers, three max-pooling layers, and three fully-connected layers FC, as shown in Fig. 11. After inputting image with size (224 × 224), the network would frequently pool and convolve the activations, after that forward the output of feature vector to the FC layers. AlexNet was winning in the ILSVRC2012 competition [40].

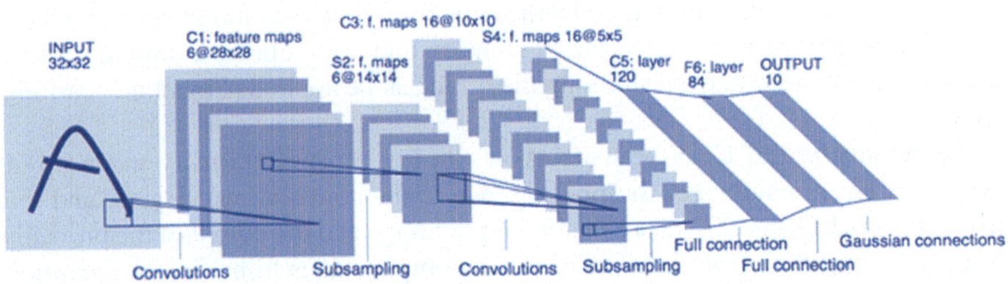

Fig. 10 Architecture of LetNet-5 [39]

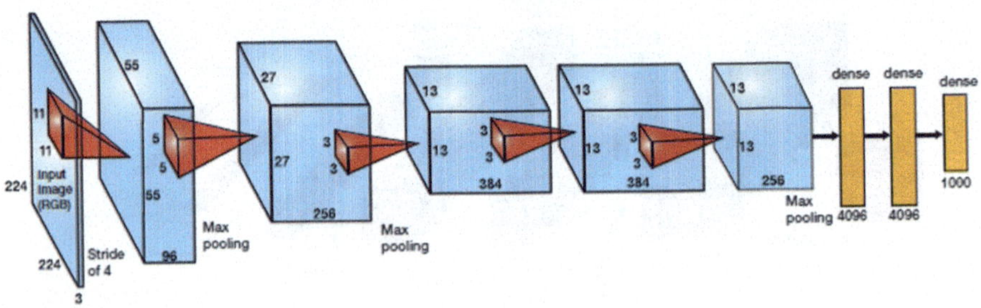

Fig. 11 Architecture of AlexNet [7]

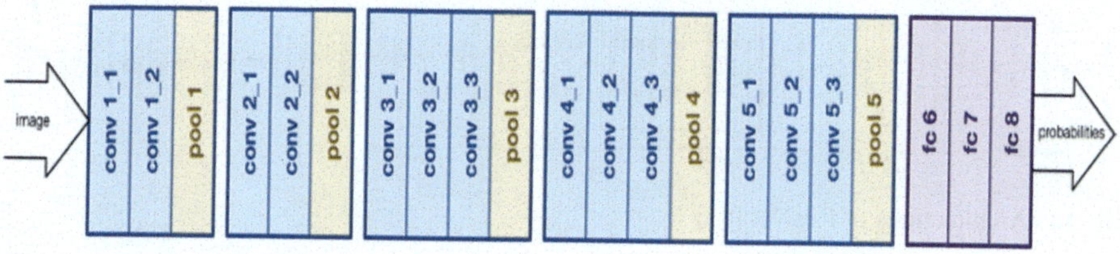

Fig. 12 Architecture of VGG [41]

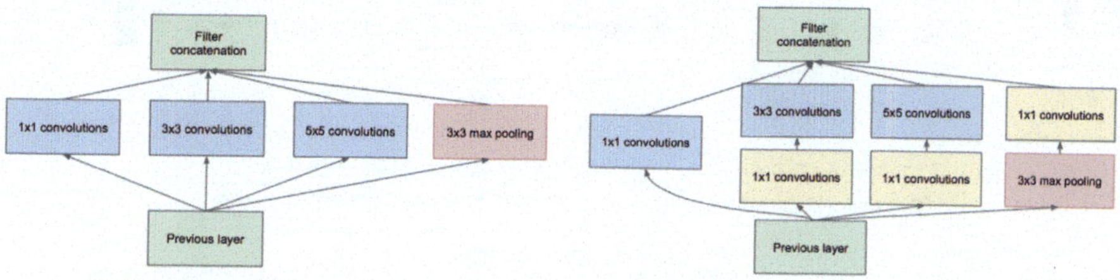

(a) Inception module, naïve version

(b) Inception module with dimension reductions

Fig. 13 Inception module [42]

- **VGG**: The approaches of VGG [41] increases the network' depth by increasing the convolutional layer and utilizing very little convolutional filters per layer. The VGG improves on the AlexNet by replacing large filters that are kennel-sized (with the first convolutional layer having 11 and the second one having 3) with multiple kernel-sized filter of size 3×3. Multiple stacked kernels of smaller sizes are advantageous over the large size kernel since the multiple layers that are non-linear increases the ability of the system to learn complex features at a lower cost by increase the network's depth. The Architecture of VGG shown in Fig. 12.
- **GoogLeNet**: The network also known as Inception (Fig. 13), which built with inception blocks. In the ILSVRC2014 competition [40], GoogLeNet has achieved leading performance. The architecture of the module is based on numerous very small convolutions, for the purpose of decrease the number of parameters. GoogLeNet architecture contained of 22 layers, and 40 million of parameters.
- **ResNet**: The ResNet architecture [43] was the winner architecture of ILSVRC2015 with 152 layers and consisted of so-called ResNet-blocks. The network designed by 3×3 convolutional layers. The residual block has two block of 3×3 convolutional layers with similar number of output channels. Also, has a ReLU activation function and batch normalization layer after each convolutional layers. The Details of different architecture of ResNet shown in Fig. 14.
- **DenseNet**: The Dense Convolutional Network (DenseNet) [44], is similar to ResNet network. It is built to solve the problem of vanishing gradient. In the network each layer takes input from preceding layer, then the features from the previous layer pass on to the subsequent layer. The Architecture of DenseNet shown in Fig. 15.

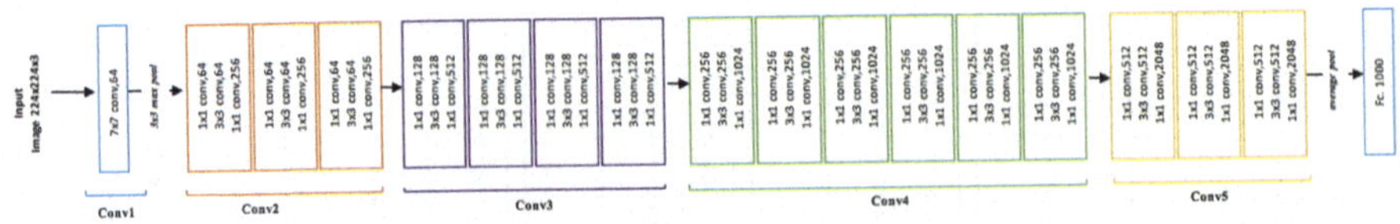

Fig. 14 Architectures of ResNet [81]

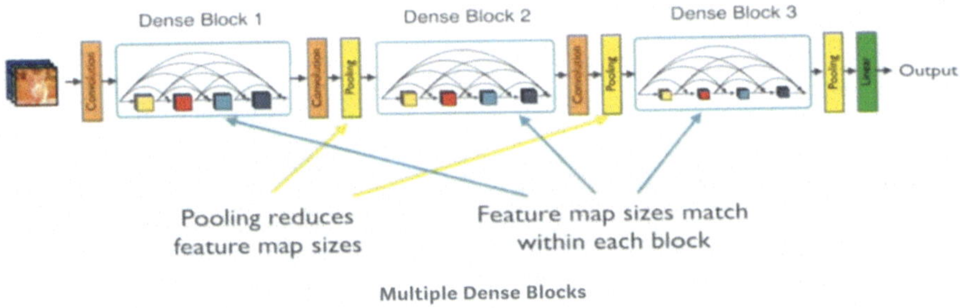

Fig. 15 Architectures of DenseNet [44]

- **Pyramidal Net**: The Deep Pyramidal Residual Networks proposed by Han et al. [45]. The main goal of the network improves the performance of image classification task by increases the feature map dimensions. The difference between Pyramid-Nets and other CNN architectures is the increase of channels dimension at all units. Whereas, in the other CNN models at the unit that execute down-sampling. there are two kinds of PyramidNets are Multiplicative PyramidNet and Additive PyramidNet. The Architecture of PyramidNet shown in Fig. 16.
- **ResNeXt**: The network proposed by Xie et al. [46]. ResNeXt designed for image classification task. Also, it is identified as Aggregated Residual Transformation Neural Network. It is being the winner architecture of ILSVRC 2016. ResNeXt consists of a stack of residual blocks that built by repeating a residual block with same topology to aggregates a set of transformations. The Architecture of ResNeXt shown in Fig. 17.

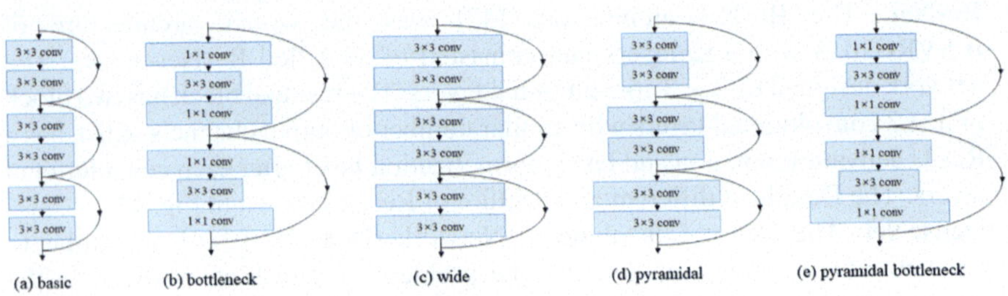

Fig. 16 Architectures of PyramidNet [45]

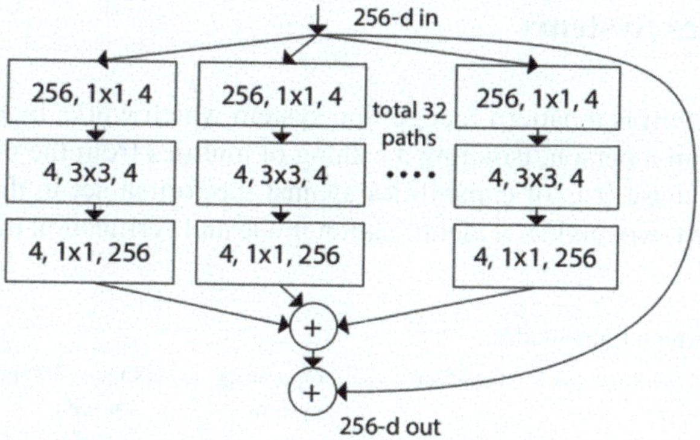

Fig. 17 Architectures of ResNeXt [46]

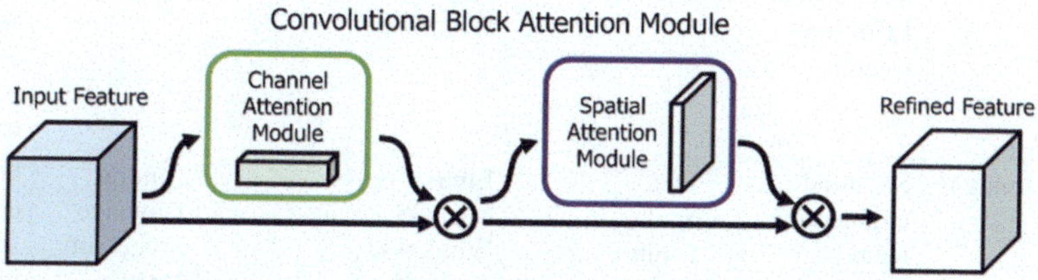

Fig. 18 Architectures of convolutional block attention module [47]

- **CBAM**: Convolutional Block Attention Module (CBAM) proposed by Woo et al. [47]. It is consisting of two Attention modules: spatial and channel. The refined feature obtained from the intermediate feature map through CBAM module by convolutional block attention of networks. In CBAM attention maps is infer sequentially by first passing feature-map through channel attention module then to spatial attention module, to obtain the refined feature. The Architecture of Convolutional Block Attention Module shown in Fig. 18.

3 Deep Learning Frameworks

It is possible to implement deep learning frameworks, although it could be time-consuming to start from scratch as they take time for optimization and maturity to occur. Fortunately, there are several open source frameworks that could be utilized more easily in the implementation and the deployment of deep learning algorithms.

These deep learning frameworks support languages like Python, C/C++, Matlab, and the Java language. An example of most common frameworks are Caffe, Keras, Torch, and Ternserflow. Table 1 provides a summarize information for the most popular frameworks with their developer, the support languages for the interface of frameworks, and the type of each framework with link for the website.

4 Biometrics Systems

A biometric system is a pattern recognition system which works by obtaining data of biometric from a person, isolating a catalog of abilities from the picked-up data, and comparing these lists of capabilities against the format set in the database. In biometric system two mode are identification mode and verification mode, as shown

Table 1 Deep learning frameworks

Framework	Developer(s)	Interface	Operating system	Open source yes: ✓ no: ✕	Type	Link
Caffe	Berkeley Vision And Learning Center	Python Matlab	Linux Mac OS Windows	✓	Library for DL	http://caffe.berkeleyvision.org/
Deeplearning4j	Skymind engineering team, Deeplearning4j community, originally Adam Gibson	Java Scala Clojure Python Kotlin	Linux Mac OS Windows Android	✓	Natural language processing Deep learning Machine vision Artificial intelligence	https://deeplearning4j.org/
Dlib	Davis King	C++	Cross-platform	✓	Library for machine learning	http://dlib.net/
Intel data analytics acceleration library	Intel	C++ Python Java	Linux Mac OS Windows	✓	Library framework	https://software.intel.com/en-us/blogs/daal
Intel math kernel library	Intel	C	Linux Mac OS Windows	✕	Library framework	https://software.intel.com/en-us/mkl
Keras	François Chollet	Python R	Linux Mac OS Windows	✓	Neural networks	https://keras.io/

(continued)

Table 1 (continued)

Framework	Developer(s)	Interface	Operating system	Open source yes: ✓ no: ✕	Type	Link
Microsoft cognitive toolkit	Microsoft Research	Python C++ Command line BrainScript	Linux Windows	✓	Library for ML and DL	https://www.microsoft.com/en-us/cognitive-toolkit/
Apache MXNet	Apache Software Foundation	Python Matlab C++ Go Scala R JavaScript Perl	Linux Mac OS Windows	✓	Library for ML and DL	https://mxnet.apache.org/
Neural designer	Artelnics	Graphical user interface	Linux MacOS X Windows	✕	Data mining ML Predictive analytics	https://www.neuraldesigner.com/
Tensorflow	Google Brainteam	Python (Keras) C C++ Java Go R	Linux MacOS Windows Android	✓	Library for ML	https://www.tensorflow.org/
Torch	Ronan, Koray, Clement, and Soumith	Lua LuaJIT C C++ OpenCL	Linux MacOS X Android	✓	Library for ML and DL	http://torch.ch/
Theano	University de Montréal	Python	Cross-platform	✓	Library for DL	http://www.deeplearning.net/software/theano/

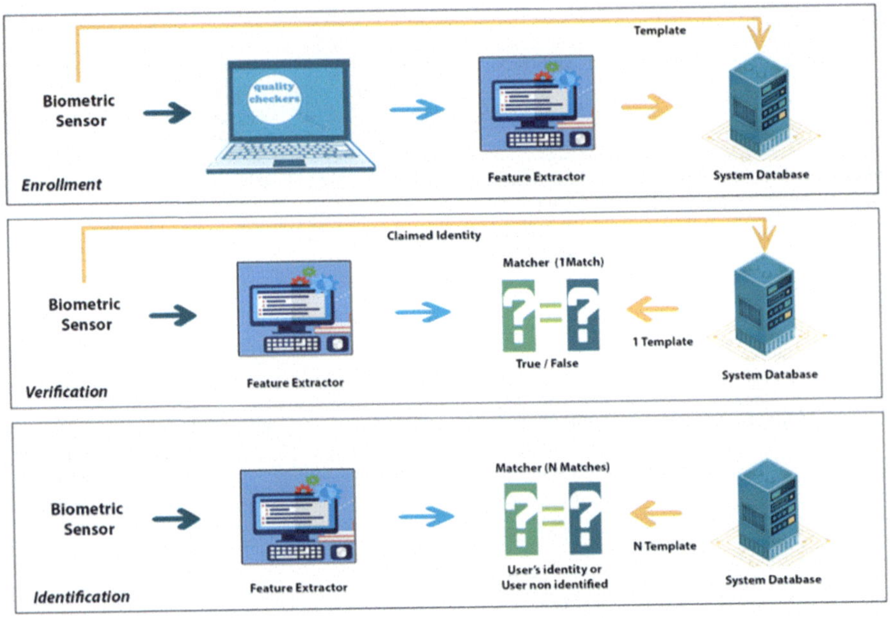

Fig. 19 Block diagrams of the main modules of a biometric system. Adopted from [48]

in Fig. 19. The verification mode involves the approval of a person's identity by the correlation of the captured biometric information with the biometric layout saved in the system database. The identification mode, on the other hand, involves the recognition of an individual by the framework, via looking through the layouts of all the clients in the database for a match [48].

A biometric structure as shown in Fig. 19 is made using four principle modules. First, the sensor module that captures the biometric information of a person. For example, a fingerprint sensor is one case of the sensor module, which captures the ridges and valley structures of the client's finger. Second, the feature extraction module, which involves the acquired biometric data being processed in order to derive a set of notable or discriminatory features. Third, the matcher module, which includes the examination of extracted features amid recognition against the saved formats to create matching scores. Finally, the system database module that is utilized by the biometric structure to save the biometric formats of the clients enlisted in the framework.

Biometric techniques are categorized based on the number of trials required for the identity of a person to be established, making for two categories [49]. There are Unimodal Biometric Techniques which make use of a single trait to identify a person. The other category of biometric techniques is the Multi-Biometric Techniques which utilizes multiple algorithms, traits, sensors or samples to identify a person.

In addition, the biometric techniques can be additionally classified into two based on the traits used to identify a person [17, 15]. Behavioral biometric system is those

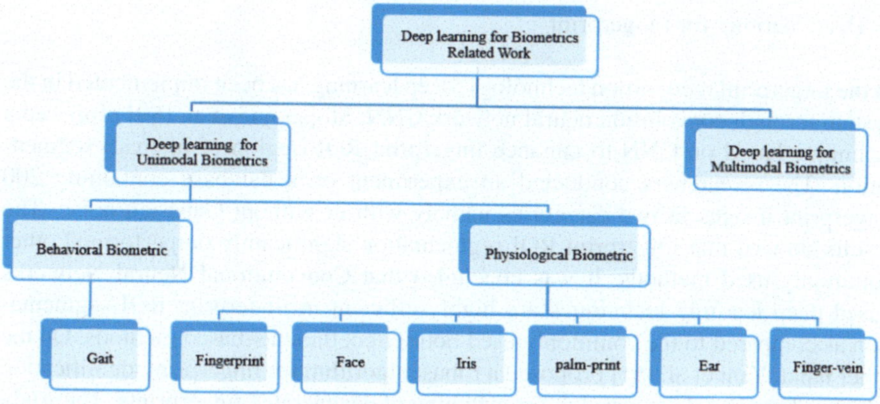

Fig. 20 Structure of deep learning for biometrics related work

which determine the identity of a person based on their behaviors as human beings, such as: gait, voice, keystroke, and handwritten signature. Whereas the physiological biometric system judges the person's identity by analyzing their physical characteristics, such as: face, fingerprint, ear, iris, and palm-print.

In this section we categorized the applications of deep learning for biometrics identification systems according to biometrics type and modalities and present a review of these applications as shown in Fig. 20.

4.1 Deep Learning for Unimodal Biometrics

Unimodal biometric identification systems are those which use a single biometric trait to identify and verify an individual. The greatest advantage of this single factor authentication is its simplicity. This makes unimodal biometric identification easier as there is not much need for user cooperation. It is also faster than the multi-biometric techniques.

In the following section we will surveying deep learning techniques with different modalities for biometric systems in two categories based on traits used for person identification physiological biometric and behavioral biometric.

4.1.1 Physiological Biometric

In this section surveyed the studies that applied on physiological biometric using deep learning techniques. And we found most of these studies applied for fingerprint, face, and iris modalities depend of that this section will categories as following:

- Deep learning for Fingerprint

In the fingerprint recognition technology, deep learning has been implemented in the system through convolution neural network CNN. Stojanović et al. [50] proposed a technique based on CNN to enhance fingerprint ROI (region of interest) segmentation. The researchers conducted an experiment on a database containing 200 fingerprint images in two categories namely with or without Gaussian noise. The results showed that fingerprint ROI segmentation significantly outperformed other commonly used methods. It was concluded that Convolutional Neural Networks based deep learning techniques are highly efficient in fingerprint ROI segmentation as compared to the commonly used Fourier coefficients-based methods. On the other hand, Yani et al. [51] proposed a robust algorithm for fingerprint identification which is based on deep learning for matching of degenerated fingerprints. The study employed an experimental study model involving the use of an algorithm for fingerprint recognition using CNN model. The results revealed that deep learning-based fingerprint recognition has a significantly higher robustness as compared to the traditional fingerprint identification techniques which primarily rely on matching of the feature points to identify similarities. The researchers concluded that deep learning can enhance the recognition of blurred or damaged fingerprints. Also, Jiang et al. [52] used a method of employing CNN in the direct extraction of minutiae from raw fingerprint images without preprocessing. The research involved a number of experiments using CNNs. The results showed that the use of deep learning technology significantly enhanced the effectiveness and accuracy of the extraction of minutiae. The researchers concluded that the approach performs significantly better than the conventional methods in terms of robustness and accuracy. In [53] they proposed a novel method for fingerprint based on FingerNet inspired by recent development of CNN. FingerNet has three major parts. The method is trained in the manner of pixels-to-pixels and end-to-end learning to enhance the output of the system. FingerNet evaluated on NIST SD27 dataset. Experimental results showed that the system improves the output and effectiveness.

Song et al. [54], proposed a novel aggregating model using CNNs. The method is composed of two modules: aggregation model and minutia descriptor, which are both learned by Deep CNN. the method is evaluated on five databases: NIST4, NIST14, FVC2000 DB2a, FVC2000 DB3a, and NIST4 natural database. the experiments result showed that the deep model improves the performance of the system.

The technologies of fingerprint classification is used to enhance identification in large fingerprint databases. In a research proposing a novel approach of using CNNs for the classification of large number of fingerprint captures, Peralta et al. [55] particularly used a number of experiments to test the efficiency and accuracy of the CNN based model. The findings revealed that the novel approach was able to yield significantly better penetration rate and accuracy than contemporary classifiers and algorithms such as FingerCode. Additionally, the networks tested also showed that the new deep learning method also resulted in improved runtime.

Wang et al. [56] focuses on the potential of deep learning technology of depth neural network in automatic fingerprint classification. The researchers used a quantitative research approach involving regression analysis (soft max regression) for the fuzzy classification of finger prints. The results showed that the Depth Neural Network algorithm had more than 99% accuracy in the finger print classification. It was concluded that deep networks can significantly enhance the accuracy of automatic fingerprint identification systems.

Wong and Lai [57], presented a CNN model for fingerprint recognition. The CNN model contains two networks single-task network and multi-task network. Single-task network is designed to reconstruct the fingerprint images in order to enhance the images. Multi-task network proposed to rebuild the image and orientation field simultaneously. The evaluation of the Multi-task CNN model conducted on IIIT-MOLF database. Experimental results showed that the model outperforms the state-of-the-art methods.

Since the incidents of spoofing biometric traits have increased, it has also been an application area for deep learning. According to Drahanský et al. [58], deep learning has a significant potential in the prevention of spoofing attacks particularly since the incidents of spoofing biometric traits in the past few years have increased. The rising cases of spoofing biometrics make it a potential area for the application of deep learning technology. The fingerprint proofing however complementarily invalidates the input images. The researchers provide inductive model of preparation of finger fakes (spoofs), summarization of skin disease and their influence as well as the proof detection methods. Nogueira et al. [59] proposed a system for software-based fingerprint liveness detection. The researchers used a mixed research methodology compare the effectiveness of the traditional learning methods with deep learning technology. The CNN system was evaluated in relation to the data sets used in the 2009, 2011 and 2013 liveness detection competition. The CNN based system detected almost 50,000 fake and real fingerprints images. For the authenticity of the experiment, four different CNN system models were compared; two CNN systems were fine-tuned and pre-trained on natural images with fingerprint images while the other two CNN systems had a classically modified binary pattern approach and random weights. In the findings, pre-trained CNNs yielded a state-of-the-art result which had no need for hyper-parameter selection or architecture and the overall achieves findings stood at an accuracy level of 97.1% as a correctly classified sample. Similarly, Kim et al. [60], proposed a system for fingerprint liveness detection using DBN. They used Restricted Boltzmann machine RBM with multiple layers, in order to identify the liveness and learn features from fake and live fingerprints. The detection method does not need an exact domain expertise regarding fake fingerprints or recognition model. The results demonstrate that the system achieved a high performance of detection for the liveness case on fingerprint detection model. Park et al. [61], proposed a CNN model for fake fingerprints detection. They considered the characteristic of the fingerprint' texture for fake fingerprint detection. The model evaluated on dataset include, LivDet2011, LivDet2013, and LivDet2015. The experiment results had an average detection error of 2.61%.

- Deep learning for Face

Deep learning has been at the center of the success experienced in the development of new image processing techniques and face recognition. For instance, CNNs are now being used in a wide range of applications to process images. Yu et al. [62], focused on the exploration of various methods for face recognition. They proposed the Biometric Quality Assessment (BQA) method as an applicable method in addressing this problem. The proposed method utilized light CNNs that made BQA robust and effective compared to the other methods. Their method evaluated on FLW, CASIA, and YouTube dataset. The results demonstrate that BQA method was effective.

CNNs have also been successfully use in the recognition of both the low and high features of an individual's thus making the method highly applicable, Jiang et al. [63] in their research proposed a multi-feature deep learning model, which can be used in gender recognition. They carried out experiments on the application of subsampling and DNN in the extraction of the face features of human beings. The results showed that higher accuracies were achieved using this method compared to the traditional methods.

Shailaja and Anuradha [64], proposed a model for face recognition based on the linear discriminant approach. The experiments were carried out to learn and analyze different samples in a face recognition based model. The authors concluded that the learning of face samples increased significant with the method. The performance of Linear Discriminant Regression Classification (LDRC) was also highly enhanced with the use of the method.

Sun et al. [65] an independent research conducted by Sun et al. sought to determine the application if hybrid deep learning in face recognition and verification. For the purposes of verification, the authors used CNNs based on RBM model. The results obtained showed that the approach improves the performance for the face verification.

In [66], the researchers used CNNs for the identification of new born infants within a given dataset. A class sample of approximately 210 infants was used for the study. At least 10 images were used for each infant. The results showed that the accuracy of identification does not related to the increasing of hidden layers. And they concluded that using large number of convolution layers also decrease the performance of the system. Also, Sharma et al. [67] proposed a method that uses generalized mean for faster convergence of feature sets and wavelet transform for deep learning to recognize faces from streaming video. The researcher employed a comparative study involving analysis of different methods. The proposed algorithm obtained frames by simply tracking the face images contained in the video. Feature verification and identity verification was then undertaken using a deep learning architecture. The algorithm was particularly tested on two of the popular databases namely YouTube and PaSC databases. The results showed that deep learning is effective in terms of identification accuracy when it comes to facial recognition.

As retouched images ruin the distinct features and lower the recognition accuracy, Bharati et al. [68] used a supervised Boltzmann machine deep learning algorithm to help identify the original and retouched images. The experimental approach involved identification of the original and retouched images. The research particularly demonstrated the impacts of digital alterations on the automatic face recognition performance. In addition, the research introduces a computer-based algorithm for classification and identification of face images as either retouched or original with a profound accuracy. The face recognition experiment herein shows that whenever a retouched image appears to match the original or unaltered image, then the identification experiment should be presumably disregarded; this is due to the matching accuracy drop by about 25%. However, when an image is retouched with a similar algorithm style, then the matching accuracy will mislead in comparison with the originally matching images. In order to undertake this research to its ultimate perfection, a novel supervised deep Boltzmann machine-based algorithm is used. In the proposed algorithm, there is a significant achievement in the supervised Boltzmann machine for detection retouching. The findings indicated that using deep learning algorithms significantly enhanced the reliability of biometric recognition and identification.

Many research efforts have been focused on how to enhance the recognition system accuracy, with ignoring for gathering samples with diverse variations. Especially, when there is only one image available for each person, Zhuo [69] in his research proposed a model based on neural a network that was capable of learning nonlinear mapping between images and components spaces. The researcher attempted separating components of pose against those of the persons through the use of DNN models. The results showed that the neural classifier produced better results when operating with virtual images compared to the training classifier working with frontal view images. Also, some studies purpose to reducing the computational cost and offering fast recognition system by addressing an intelligent recognition system for face that is can recognize face expression, pose invariant, occluded, and blurred faces by using efficient deep learning [70], the researcher presented a new approach, which is the fusion of higher-order novel neuron models with techniques of different complexities. In addition, different feature extraction algorithms were also used in the research thus, presenting classifiers of higher levels and improved complexities.

Illumination variation is an important cause that affect the performance of face recognition algorithms. For illumination variation issues, Guo et al. [71] proposed a system for face recognition, which applied for near-infrared and visible light image. Also, they designed an adaptive score fusion strategy that would be significant in the improvement of the performance of the use of infrared based CNN face recognition. Compared to the traditional methods, the designed method proved to be more robust in feature extraction. Specifically, it is highly robust in the variation of illumination. They evaluated the method on several datasets.

The research work in [72] proposes a face recognition approach referred to as WebFace. The method utilizes CNNs in learning the patterns applicable in face recognition. The research involved about 10,000 subjects and approximately 500,000 pictures contained in a database. In the study they train a much deeper CNN for face recognition. The architecture of WebFace contain 17 layers. It is comprised of 10 convolutional layers, 5 pooling layers, and 2 fully connected layers FC. WebFace proved to be quite effective in face recognition.

Despite the fact that CNNs have been commonly applied in face recognition since the year 1997 [73], continuous research has enabled the improvement of these methods. In DeepFace [74] Researchers have been able to develop an 8-layer deep face approach comprised of three conventional convolution layers, three connected layers, and two fully connected layers. It is however important to point out that DeepFace is trained on large databases that are comprised of about 4000 subjects and thousands of images.

DeepID [75], proposed by Y. Sun et al. the method operated through the training and building of CNN network fusion. In this method, each of the networks has a four convolution layers with 3 max-pooling layers, and 2 fully connected layers. The results obtained showed that the DeepID technique had an accuracy of 97.45% when implemented in a LFW dataset. Further improvements have been done of DeepID with the development of DeepID2 [76]. It used CNN for identification and verification. The method DeepID2+ [77], is more robust and overcomes some of the shortcomings of DeepID and DeepID2. DeepID2+ used a large set for training than DeepID and DeepID2.

The research by Lu et al. [78], proposed that use of the Deep Coupled ResNet (DCR) model in face recognition the method was comprised of a trunk network and two branch networks. The discriminative features on a face were extracted using the trunk network. The two branch networks transformed high resolution images to the targeted low resolution. Better results were achieved using the method compared to other traditional approaches.

Li et al. [79], proposed strategies using CNNs for face cropping and rotation by extracting only useful features from image. The proposed method evaluated on JAFFE and CK+ databases. The Experiments result achieved high recognition accuracies of 97.38 and 97.18%. Also, the results showed that the approach improve the recognition accuracy.

Ranjan et al. [80], proposed method called HyperFace. They used deep convolutional neural networks for face detection, landmarks localization, pose estimation, and gender recognition. HyperFace consist of two CNN architectures: HyperFace-ResNet and Fast-HyperFace based on AlexNet. They evaluated HyperFace on six datasets includes: AFLW, IBUG, AFLW, FDDB, Celeb A, PASCAL. The experiments results showed that HyperFace method achieves significantly better than many competitive algorithms.

Almabdy and Elrefaei [81], proposed face recognition system based on AlexNet and ResNet-50. The proposed model includes two approaches. The first approach using pre-trained CNN (AlexNet and ResNet-50) for feature extraction with support vector machine SVM. The second approach is transfer learning from AlexNet network for both feature extraction and classification. The system evaluated based on seven datasets include: ORL [82], GTAV [83], Georgia-Tech [84], FEI [85], LFW [86], F_LFW [87], and YTF [88]. The accuracy of approaches ranges of 94–100%.

Prasad et al. [89], proposed a face recognition system. The model built based on Lightened CNN and VGG-Face. They focused on face representation for some different conditions such as: illuminations, head poses, face occlusions, and alignment. The study conducted on AR dataset. The results of the recognition system of face images showed that the model is robust to several types of face representation like misalignment.

The researchers in [90], proposed a novel Hybrid Genetic Wolf Optimization that applied Convolution Neural Network for newborn baby face recognition. In the study the feature extraction process was performed by using four techniques then proposed a hybrid algorithm to combine these features as fusion between two algorithms are genetic algorithm and gray wolf optimization algorithm. CNN used for classification. The experiment evaluated on newborn baby face database. The accuracy of the proposed system is 98.10%.

In case of spoofing deep learning has a significant potential in the prevention of spoofing attacks, the authors in [91] proposed non-intrusive method detecting face spoofing attack from a video. Using DL technology to enhance computer vision. The researchers used a mixed study approach involving an experimental detection of spoofing attacks using a single frame from sequenced video frames as well as a survey of 1200 subjects who generated the short videos. The results suggested that the use of method achieved better results in the detection of face spoofing attack as compared to the conventional static algorithms results. The study concluded that deep learning is an effective technology which will significantly enhance the detection of spoofing attacks.

- Deep learning for Iris

In the iris technology, Nseaf et al. [92] suggest for iris recognition two DNN models from video data. These DNN models include the Bi-propagation and the Stacked Sparse Auto Encoders (SSAE). They first select clear and visible 10 images from each and every video to make a database. The second activity is the identification of localized iris region. This identification is made from the eye images using the Hough transformation mask process which is then complemented by the application of Daugman rubber sheet model and 1D Log-Gabor filter. The Log-Gabor filter feature extract and normalize the deep learning algorithm. In case the experiment becomes flowed, they should apply Bi-propagation and SSAE in a separate process for matching of the step. The results show the effective and efficient nature of the Bi-propagation in the training on both the video and SSAE. However, it is worth noting that both of these networks have achieved an irresistibly good and accurate results; the overall result in both algorithm networks are powerful and accurate for the iris matching step though can be entirely increased by segmentation steps enhancement. Considering iris segmentation using convolutional neural network CNNs, Arsalan et al. [93] proposed a scheme based on CNNs. They used visible light camera sensor for iris segmentation in noisy environments. The method evaluated on NICE-II and MICHE dataset. The results showed that the scheme outperformed the existing segmentation methods. CNN approach has been recently presented also in [94], their proposed method for iris identification based on CNN. The architecture of network consisted of 3 convolutional layers, and 3 fully-connected layers. In their experiment the results showed that improving the sensor model identification step can benefit the iris sensor interoperability.

Alaslani et al. [95], proposed a model for iris recognition system. The proposed of system is examined when extracting features from segmented image and normalized image. The proposed method evaluated on several datasets. The system achieved high accuracy. Also, in another study Alaslani et al. [96], they used transfer learning from VGG-16 network for extracting the features and classification. The iris recognition system evaluated on four datasets include, CASIA-Iris-thousand, iris databases CASIA-Iris-V1, CASIA-Iris-Interval and IITD. The proposed system achieved a very high accuracy rate.

Gangwar et al. [97] used two very deep CNN architectures for iris recognition, the first network built of five convolutional layers and two inception layers, and the second network built of eight convolutions. The researcher in this study found that the method more robustness for different kinds of error such as: rotation, segmentation and alignment.

Arora and Bhatia [98], presented a spoofing technique for iris detection. Deep CNN applied to detect print attacks in iris. The system trained to deal with three types of attacks. Deep networks used to feature extraction and classification. IIIT-WVU iris dataset was used to test iris recognition performance. The iris recognition techniques achieve higher performance in the detection of the attacks.

- Deep learning for other modalities

Recently, multispectral imaging technology has been used to make the biometric system more effective, for the purpose of increase the discriminating ability and the classification accuracy of the system Zhao et al. [99] in their study using the deep learning for a better performance. They presented a deep model for palm-print recognition implemented as a stack of RBMs at the bottom with a regression layer at the top, Deep Belief Network is efficient for feature learning with both supervised and unsupervised training.

The first approach for ear recognition using convolutional neural networks is proposed by Galdámez et al. [100], the approach used deep network to extracted features, which more robust than traditional systems, which used hand-crafted features. Almisreb et al. [101], investigated the transfer learning from AlexNet model in the domain of human recognition based for ear image. Also, in order to overcome the non-linear problem of the network, they added Rectified Linear Unit (ReLU). The result of the experiment achieved 100% validation accuracy.

Emeršič et al. [102] proposed pipeline consists of two models: RefineNet for ears detection and ResNet-152 for recognition of segmented ear regions. They conducted the experiments on AWE, and UERC dataset. The results of the presented pipeline are achieved 85.9% as recognition rate.

Ma et al. [103], proposed a technique for ear of winter wheat with segmentation. For the segmentation process they used Deep CNN. The evaluation of the method carried out on season 2018 dataset. Results showed that the method outperformed the state-of-the-art methods for ear of winter wheat segmentation at the flowering stage.

Liu et al. [104] presented a method for finger-vein recognition based on random projections and deep learning. They used secure biometric template scheme called FVR-DLRP. The results of the method showed that the identification accuracy provide better result in term of authentication.

Das et al. [105], proposed identification system for finger-vein based on CNN. The main goal of the system dealing with different image quality, to provides greatly accurate performance. They evaluated the system using four publicly datasets. The experiments result obtained identification accuracy greater than 95%.

Zhao et al. [106], proposed finger-vein recognition approach using lightweight CNN model to improve the robustness and the performance of the approach. The method used different loss function like triplet loss and softmax loss. Experiments were conducted on FV-USM and MMCBNU_6000 datasets. The approach achieved outstanding results and reduce the overfitting problem.

Al-johania and Elrefaei [107], proposed vein recognition system using Convolutional Neural Network. The systems include two approaches. The first approach using for extracting features three network includes; VGG16, VGG19, and AlexNet. And for classification task using two algorithm include; Support Vector machine (SVM) and Error-Correcting Output Codes (ECOC). The second approach applying transfer learning. The system achieved a very high accuracy rate.

4.1.2 Behavioral Biometric Systems

Some researches focused on gait identification systems more than a decade ago. In the gait technology. Most recently, Wu et al. [108] used the CNN for learning the distinct changes in the walking patterns and use these features to identify similarities in cross-view and cross-walking-condition scenarios. This method firstly evaluated the challenging datasets formally through the cross-view gait recognition. The results have shown to outperform the state-of-the-art technology for gait-based identification. A specialized deep CNN architecture for Gait recognition developed by Alotaibi and Mahmood [109], the model is less sensitive to several cases of the usual occlusions and variations that reduce the performance of gait recognition. The deep architecture be able to handle small data sets without using fine-tuning or augmentation techniques. The model evaluated on CASIA-B databases [110]. The results of the proposed model achieve competitive performance.

Baccouche et al. [111] carried out a study on the possible use of automated deep model learning technology in recognition and classification of human actions in the absence of any prior knowledge. The study employed a mixed study approach comprising of a literature review as well as a series of experiments involving the use of neural-based deep model in classifying human actions. The experiments particularly entailed using 3D CNNs to extract human behavioral features and resulted in an improved accuracy of the identification process. The results showed that proposed model outperforms existing models and were highly competitive results on KTH datasets [112] with 94.39% KTH1 dataset and KTH2 dataset 92.17%.

Sokolova and Konushin [113] combine neural feature extraction method for improve the representation. For the purpose of find the best heuristics, they compare various DNN architectures, feature learning and classification strategies. The study showed that available datasets are very small for training and not enough. Human gait recognition has been studied using machine learning algorithms includes; Hidden Markov Model (HMM) [114], SVM [115] and Artificial Neural Networks (ANN) [116].

The researchers in [117] present a DNN pipeline containing Deep Stacked Auto-Encoders (DSA) stacked below Softmax (SM) classifier for classification. The experiments conducted on CASIA Dataset [110]. The Deep Learning pipeline shows a significant improvement compared to other classifiers of state-of-the-art such as ANN and SVM, the accuracy of the recognition system is 92.30% with SVM, 95.26% with ANN, and 99.0% on Deep Stacked Auto-Encoders with Softmax.

Singh et al. [118], proposed a model for action recognition by make fusion of several VGG-Net. The dynamic images trained on VGG-Net for four views such as dynamic image, DMM-Top, DMM-Side, and DMM-Front. It is evaluated on UTD MHAD, MSR Daily Activity 3D, and CAD-60 datasets. The accuracy for the system in range of (94.80–96.38%).

Hasan and Mustafa [119], presented gait recognition approach using a RNN. The network used for features extraction to extract effective gait features. In the study they focused on human pose information. The approach evaluated on CASIA A and CASIA B datasets. Experimental depicts that the method improves the performance for gait recognition.

Tran and Choi [120], proposed an approach for gait using CNN to solve the problem of inertial gait data augmentation. The approach consists of two algorithms Stochastic Magnitude Perturbation (SMP) and Arbitrary Time Deformation (ATD). The approach was evaluated on OU-ISIR and CNU datasets. The experimental results showed that, using SMP algorithm and ATD algorithm improves the performance of the recognition rate.

4.2 Deep Learning for Multimodal Biometrics

Multimodal biometric systems combine more than two biometric technologies such as fingerprint recognition, facial detection, iris examining, voice recognition, hand geometry, etc. These applications take input data from biometric sensors for evaluating more than two different biometric characteristics [121]. A system fuse fingerprint and face characteristics for biometric identification is knows as a multimodal system. An example of multimodal systems would be a system which combines face recognition and iris recognition. This system accepts users to be verified using one of these modalities. In addition to enhancing the recognition accuracy, combining more than two modalities might be more suitable for different applications.

Menotti et al. [122] in their study investigated two deep learning presentation processes. The processes were comprised of learning from CNNs as well as iris spoof detection, face, and fingerprints in image detection. They proposed approaches focused on using back propagation to learning the weights of the networks in order to close the existing knowledge gap.

The research work in [123] proposed a multimodal recognition system for recognition of facial features based on facial video clips. The research methodology involved conducting various experiments on both constrained facial video dataset and unconstrained facial video datasets. The research methodology entailed the use of a deep network to extract distinct recognizing features of multiple biometric modalities (such as left/right ear, left/right face profile) from the facial video and give higher recognition accuracy. The results of the multimodal recognition system had a very high accuracy rate is 99.17% on constrained facial video dataset and 97.14% on unconstrained facial video datasets. Simón et al. [124] improved face recognition rate using deep CNN model. They proposed a multimodal facial recognition system. Also, applied a fusion of CNN features with various hand-craft features. The system evaluated on RGB-D-T database. Results of the experiments demonstrate that the combination between the classical feature and CNN feature enhance the performance of the system.

Meraoumia et al. [125] proposed PCANet, which a palm-print multimodal system based on deep learning technique, they applied matching score level fusion for each spectral band of palm-print. Also, in their study, features extracted by PCANet. PCANet evaluated on CASIA multispectral palm-print database. The results of the experiments showed that PCANet is very effective and improve the accuracy rate.

Neverova et al. [126] carried out a research on the possibility of using motion patterns to learn human identity. The researchers particularly used large scale study survey involving 1500 volunteer subjects who recorded their movements using smartphones for a number of months. A temporal DNN was then used in the interpretation of the human kinematics of the subjects and the results compared with those of other neural architectures. The results showed that human kinematics can reveal significant information on the identity of individuals.

Yudistira and Kurita [127], they proposed a multimodal model for action recognition based on CNN. The correlation network captured spatial and temporal streams over arbitrary times. The experiments conducted on the datasets of UCF-101 and HMDB-51. The multimodal method improved the video recognition accuracy.

Cherrat et al. [128], proposed a hybrid method. They were built fusion networks include three identification system. The proposed system combined three models: Random forest classifier, Softmax, and CNN. The hybrid method evaluated on SDUMLA-HMT database. And, the method achieved high performance. And they conclude that the pre-processing stage and dropout technique were effective to improve the rate of the recognition accuracy.

A comprehensive analysis of various deep learning DL based approaches for different biometrics modalities presented in Table 2. By evaluating the applications of deep learning in the fingerprint, face, iris, ear, palm-print, and gait biometric technology, a general observation is that deep learning neural networks such as Convolutional Neural Networks CNNs have shown high performance for application of biometrics identification and CNNs are an efficient artificial neural network method. Table 2 summarize the deep Learning techniques for biometric modalities and show their model and performance of each application.

Table 2 The deep learning techniques for biometric modalities

Ref.	Deep Learning Model	Deep learning used for	Dataset	Result
		Fingerprint Modality		
[50]	LeNet and AlexNet based on CNN contains 5 convolutional layers and 2 fully connected layers	Pre-processing	Db1 from FVC2002 (200 images)	-
[51]	CaffeNet based on CNN using 7 hidden layers	Classification	SF blind aligned fingerprint (20 images)	Recognition rate 94.73%
[52]	JudgeNet and LocateNet based on CNN with two-class classifier for JudgeNet and the nine-class classifier for LocateNet	Feature Extraction	SFinGe (1000 images) NIST-DB4 (1650 images)	Accuracy 94.59%
[53]	FingerNet based on CNN include encoding and two decoding.	Feature Extraction Classification/Matching	NIST-SD4(2000 images) NIST-SD14(27000 images) NIST-SD27(258 images)	-
[54]	DescriptorNet based on CNN, contain 10 convolutional layers, 1 max-pooling, and 2 fully-connected layers.	Feature Extraction Classification/Matching	FVC2000 DB2(800 images) FVC2000 DB3(800 images) NIST4(2000 images) NIST4 natural (2408 images) NIST14(2700 images)	Average penetration rate is: (1.02% to 0.06%)
[55]	CaffeNet based on CNN	Classification/Matching	SFinGe (1000 images) NIST-DB4 (1650 images)	Accuracy 99.60%
[56]	Stacked Sparse Autoencoders based on DNN used 3 hidden layers	Classification/Matching	NIST-DB4 (4000 images)	Accuracy 98.8%

(continued)

Table 2 (continued)

Ref.	Deep Learning Model	Deep learning used for	Dataset	Result
[57]	Multi-task CNN model consists of two networks: 1-Single-task network: 13 convolutional layers 2-OFFIENet: multi-task network 5 convolutional layers	Pre-processing Feature Extraction Classification/Matching	IIIT- MOLF FVC	-
[59]	(CNN-VGG, CNN-AlexNet, CNN-Random) based on CNN Local Binary Patterns (LBP)	Pre-processing	LivDet 2009,2011,2013 (50000 images)	Accuracy 97.1%
[60]	DBN With multiple layers of RBM	Feature Extraction	LivDet2013 (2000 live images) (2000 fake images)	Accuracy 97.10%
[61]	CNN model consist of 1×1 convolution layers, tanh nonlinear activation function, and gram layers.	Feature Extraction Classification/Matching	LivDet2011,2013,2015	Average detection error of 2.61%
		Face Modality		
[62]	A biometric quality assessment (BQA) contain Max Feature Map (MFM), and four Network in Network layer	Feature Extraction Classification/Matching	CASIA (494,414 images) FLW (13,233 images) YouTube,(2.15 videos)	Accuracy 99.01%
[63]	The joint features learning deep neural networks (JFLDNNs) based on CNN include two part: convolutional layers and max-pooling.	Feature Extraction Classification/Matching	FERET LFW-a CAS-PEAL Self-collected Internet face (13500 images)	Accuracy 89.63%

(continued)

Table 2 (continued)

Ref.	Deep Learning Model	Deep learning used for	Dataset	Result
[64]	Deep Learning Cumulative LDRC (DL-CLDRC)	Classification/Matching	YALE (165 faces) ORL(400 images)	Accuracy 92.8% YALE 87% ORL
[65]	hybrid convolutional network (ConvNet) and RBM model contain 4 convolutional layers and 1 max-pooling, and 2 fully-connected layers	Feature Extraction Classification/Matching	LFW CelebFaces (87,628 images)	Accuracy 97:08% CelebFaces 93.83% LFW
[124]	CNN-based include three convolutional layers and max-pooling.	Feature Extraction	RGB-D-T (45900 images)	EER 3.8 rotation 0.0 expression 0.4 illumination
[66]	DeepCNN include two convolution layers	Feature Extraction Classification/Matching	IIT(BHU) newborn database	Accuracy 91.03%
[123]	DNN based on Stacked Denoising Auto-encoder used DBN and Logistic regression layer	Feature Extraction Classification/Matching	FERET , J2, UND	Accuracy Ear = 95.04% Frontal Face = 97.52%; Profile Face= 93.39%; Fusion= 99.17%
[67]	Generalized mean Deep Learning Neural Network based on DNN	Feature Extraction Classification/Matching	PaSC and YouTube	Accuracy 71.8 %
[68]	Super- vised Restricted Boltzmann Machine (SRBM)	Feature Extraction Classification/Matching	ND-IIITD(4875images) Celebrity (330 images)	Accuracy 87% ND-IIITD 99% makeup
[69]	Nonlinear information processing model used two DNN with multi-layer autoencoder	Classification/Matching	BME (11 images) AUTFDB(960 images)	Recognition rate 82.72%

(continued)

Table 2 (continued)

Ref.	Deep Learning Model	Deep learning used for	Dataset	Result
[70]	DNN with different architectures of ANN ensemble	Feature Extraction Classification/Matching	ORL (400 images) Yale (165 images) Indian face(500 images)	Recognition rate 99.25%
[71]	DeepFace based on VGGNet	Feature Extraction Classification/Matching	LWF (13,000 images) YTF (3425video)	Accuracy 97.35%
[72]	Deep CNN model consist of 10 convolutional layer, 5 pooling layers and 1 fully-connected layers	Feature Extraction Classification/Matching	LFW (13,000 images) YouTube Faces (YTF) CASIA-WebFace (903,304 images)	Accuracy 97,73% LFW 92.24 % YTF
[75]	DeepID based on ConvNets model contain 4 convolutional layers, 1 max-pooling, and 1 fully-connected DeepID layer and a softmax layer	Feature Extraction Classification/Matching	CelebFaces (87, 628 images) LFW (13,233 images)	Accuracy 97.45%
[76]	DeepID2 consist of 4 convolutional layers 3 max-pooling layers, and a softmax layer	Feature Extraction Classification/Matching	LFW (13,233 images), CelebFaces	Accuracy 99.15%
[77]	DeepID2+ consist of 4 convolutional layers with 128 feature maps, first three layers followed by max-pooling and a 512-dimensional fully-connected layer	Feature Extraction Classification/Matching	LFW (13,233 images) YTF (3425 video)	Accuracy 99.47 % LFW 93.2 % YTF
[78]	Deep Coupled ResNet (DCR) model include 2 branch networks and 1 trunk network	Feature Extraction	LFW (13,233 images) SCface (1950 images)	Accuracy 98.7 % LFW 98.7 % SCface

(continued)

Table 2 (continued)

Ref.	Deep Learning Model	Deep learning used for	Dataset	Result
[79]	CNN model consist of 2 convolution layers, 3 max-pooling layers	Feature Extraction Classification/Matching	JAFFE (213 images) CK+(10,708 images)	Accuracy 97.38% CK+ 97.18% JAFFE
[80]	HyperFace based on CNN	Feature Extraction Classification/Matching	AFLW (25, 993 images) IBUG (135 images) AFLW (13,233 images) FDDB (2,845 images) CelebA (200,000 images) PASCAL(1335 images)	-
[81]	CNN model based on AlexNet and ResNet-50	Feature Extraction Classification/Matching	ORL (400 images) GTAV face (704 images) Georgia Tech face (700 images) FEI face (700 images) LFW (700 images) F_LFW (700 images) YTF (700 images)	Accuracy 94%-100%
[89]	CNN model based on Lightened CNN and VGG-Face	Pre-processing Feature Extraction Classification/Matching	AR face d (5000 images)	-
[90]	CNN model consists of convolutional layer, pooling layer, and fully connected layer.	Classification/ Matching	Newborn baby face dataset	Accuracy 98.10%
[91]	Specialized deep CNN model based on AOS-based schema, CNN consist of 6 layers	Feature Extraction Classification/Matching	Replay-Attack (1200 video)	Accuracy 17.37%
		Gait Modality		
[108]	CNN-based method with 3 different network architectures	Feature Extraction Classification/Matching	CASIA-B(124 subjects) OU-ISIR, USF	Accuracy 96.7 %

(continued)

Table 2 (continued)

Ref.	Deep Learning Model	Deep learning used for	Dataset	Result
[109]	Specialized deep CNN model consist of 4 convolutional layers and 4 pooling layers.	Feature Extraction Classification/Matching	CASIA-B (124 subjects)	Accuracy 98.3 %
[111]	Fully automated deep model based on CNN using 3D-ConvNets consists of 10 layers and RNN classifier	Feature Extraction Classification/Matching	KTH (25 subjects)	Accuracy 94.39% KTH1 92.17% KTH2
[113]	CNN based on VGG and CNN-M with Batch Normalization layer	Feature Extraction Classification/Matching	CASIA-B (124 subjects) TUM-GAID(305 subjects)	Accuracy 99,35% YUM 84,07%CASIA
[117]	Deep Stacked Auto-Encoders (DSA) based on Pipeline of DNN include a Softmax classifier and 2 Autoencoder Layers	Feature Extraction Classification/Matching	CASIA-B (9 different subject)	Accuracy 99.0%
[118]	CNN model Include 4 VGG-Net that consists of 5 convolutional layers, 3 pooling layers and 3 fully connected layers.	Feature Extraction Classification/Matching	UTD MHAD MSR Daily Activity 3D CAD-60	Accuracy (94.80% - 96.38%)
[119]	RNN model consists of 2 BiGRU layers, 2 batch normalization layer, and output softmax layer	Pre-processing Classification/Matching	CASIA A CASIA-B (124 subjects)	Recognition rate 99.41%
[120]	CNN model consists of 4 convolutional layers and 2 fully connected layers	Feature Extraction Classification/Matching	CNU) OU-ISIR	

(continued)

Table 2 (continued)

Ref.	Deep Learning Model	Deep learning used for	Dataset	Result
[126]	Dense Clockwork RNN (DCWRNN) used Long Short-Term Memory (LSTM)	Feature Extraction	Project Abacus (1,500 volunteers)	Accuracy 69.41%
[127]	CNN model based on two expert streams and one correlation stream structure of 3 layers of fully connected layers	Feature Extraction Classification/Matching	UCF101 (13 320 videos) HMDB51 (6766 videos)	Accuracy 94.4%
		Iris Modality		
[92]	Stacked Sparse Auto Encoders (SSAE) and Bi-propagation based on DNN	Classification/Matching	MBGC v1 NIR (290 video)	Recognition rate 95.51% SSAE 96.86% Bi-propagation
[93]	Two-stage CNN-based method used VGG-face, consist of 13 convolutional layers, 5 pooling layers, and 3 fully connected layers	Feature Extraction Classification/Matching	NICE-II (1000 images) MICHE	Segmentation Error is: 0.0082 NICE-II 0.00345 MICHE
[94]	AlexNet based on CNN consist of 3 convolutional layers followed by 3 fully connected layers	Feature Extraction Classification/Matching	ATVS-Fir (1600 images) CASIA-IrisV2, V4 (1200 images) IIIT-D CLI (6570 images) Notre Dame Iris Cosmetic Contact Lenses 2013 (2800 images)	Accuracy 98.09%
[95]	AlexNet based on CNN consist of 5 convolutional layers and 3 fully-connected layers	Feature Extraction Classification/Matching	IITD CASIA- Iris-V1 CASIA-Iris-thousand CASIA-Iris- V3 Interval	Accuracy (89% -100%)

(continued)

Table 2 (continued)

Ref.	Deep Learning Model	Deep learning used for	Dataset	Result
[96]	VGG-16 based on CNN consist of 5 convolutional layers and 5 pooling layers and 3 fully-connected layers	Pre-processing Classification/Matching	IIT Delhi Iris CASIA- Iris-V1 CASIA-Iris-Thousand CASIA-Iris-Interval	Accuracy (81.6% -100%)
[97]	DeepIrisNet, consist of two CNN: DeepIrisNet: 8 convolutional layers, 4 pooling layers. 2- DeepIrisNet-B: 5 convolutional layers, 2 inception layers, 2 pooling layers.	Feature Extraction Classification/Matching	ND-iris-0405 (64,980 images) ND-CrossSensor-Iris-2013	-
[98]	CNN model consists of 10 convolutional layers, 5 max pooling layers, and 2 fully connected layer	Feature Extraction Classification/Matching	IIIT- WVU iris	-
		Ear Modality		
[100]	CNN-based consist of alternating convolutional, max-pooling layers, and one or more linear layers.	Feature Extraction Classification/Matching	Bisite Videos Dataset and Avila's Police School (44 video)	Accuracy 98.03%
[101]	Transfer Learning from AlexNet CNN	Feature Extraction Classification/Matching	Ear Image Dataset (300 images)	Accuracy 100%
[102]	CNN model consists of two models: RefineNet and ResNet-152	Feature Extraction	AWE UERC	Accuracy 92.6%

(continued)

Table 2 (continued)

Ref.	Deep Learning Model	Deep learning used for	Dataset	Result
[103]	DCNN model consist of: 5 convolutional layer, 2 fully connected layer, 4 max-pooling,	Classification/Matching	Season 2018 (36 images)	F1 score 83.70%,
		Palm-print Modality		
[125]	PCANet based on DNN	Feature Extraction	CASIA multispectral palmprints (7200 images)	EER = 0.00%
[99]	DBN consist of two RBMs using layer-wise RBM and logistic regression	Feature Extraction Classification/Matching	Beijing Jiaotong University (800 images)	Recognition rate 0.89%
		Vein Modality		
[104]	FVR-DLRP based on DBN consist of two layers of RBM	Feature Extraction	FV_NET64 (960 images)	Recognition rate 96.9%
[105]	CNN model consists of five convolutional layers, three max-pooling, softmax layer, and one ReLU	Feature Extraction Classification/Matching	HKPU (3132 images) FV-USM SDUMLA (636 images) UTFVP(1440 images)	Accuracy (95.32%-98.33%)
[106]	CNN model consists of 3 convolutional layers, 3 max-pooling layers and 2 fully connected layers	Feature Extraction Classification/Matching	MMCBNU_6000 (6000 images) FV_USM	Accuracy 97.95%
[107]	Dorsal hand vein recognition system, based on CNN (AlexNet, VGG16 and VGG19)	Feature Extraction Classification/Matching	Dr. Badawi hand veins (500 images) BOSPPHORUS dorsal vein(1575 images)	Accuracy 100% Dr. Badawi 99.25 % BOSPPHORUS
		Iris, Face, and Fingerprint Modalities		

(continued)

Table 2 (continued)

Ref.	Deep Learning Model	Deep learning used for	Dataset	Result
[122]	Hyperopt-convnet for architecture optimization (AO) Cuda-convnet for filter optimization based on back-propagation algorithm	Feature Extraction Classification/Matching	LivDet2013 Replay-Attack, 3DMAD BioSec Warsaw MobBIOfake	-
		Face, Finger-vein, and Fingerprint Modalities		
[128]	CNN model consists of 3 CNN	Pre-processing Feature Extraction Classification/Matching	SDUMLA-HMT (41,340 images)	Accuracy 99.49%

where: **CNN** = Convolutional Neural Network, **DNN**=Deep Neural Network, **DBN**= Deep Belief Networks, **EER**= Equal Error Rate

5 Challenges

The challenges associated with biometrics system can be attributed to the following factors:

1. **Feature representation scheme**: the main challenges in biometrics is to extract features, for a given biometric trait by using the better method for representation. Deep learning can be implemented by hierarchical structure combined several processing layers, each of which extracts data from its input in the training process. The researchers in [63, 74, 76, 123] obtained learned features, from the internal representation of a CNN. They solved the problems for identify the best representation scheme and achieved an enhancement of their models.
2. **Biometric liveness detection**: in the case of spoofing detection methods for different modalities [59, 60, 91, 122], The researchers provide solutions to solve spoofing detection problem through techniques related to the texture patterns, modality, noise artifacts. The researchers found that the performance of such solutions vary significantly from dataset to dataset, and they proposed deep neural network techniques that automatically used deep representations schema to extract features from the dataset directly.
3. **Unconstrained cases**: data in datasets sometimes including many variations such as pose, expression, illumination, reflection from eye lashes, and occlusion. Which effect the biometric performance. The researchers in [71, 123, 124] applied DL techniques to improve system performance. They found that the procedure of a deep network extract robust recognizing features and give higher recognition accuracy.
4. **Noisy and distorted input**: biometric data collected in real-world applications are quite noisy and distorted due to noisy biometric sensors or other factors.

Stojanović et al. [50] and Arsalan et al. [93] applied deep technique based on CNN to enhance the performance with noisy data. The deep learning method efficient to enhance the system.

5. **Overfitting**: there is variance in the percentage of error occurred in training dataset and the percentage of error encountered in test dataset. It happens in complex models, such as having huge number of parameters relative to the observations number. The effectiveness of a system is judged by its ability to perform well on test dataset and not judged by its performance on dataset of training. To address this challenge, researchers in [94], has proposed techniques with transfer learning, to tackle the problem of limited training set availability and improve the system. Also, Song et al. [54] applied three different forms of data augmentation to overcome this problem.

6 Conclusion and Discussion

A comprehensive analysis presents in the paper of various deep learning DL based approaches for different biometrics modalities and discussed in detail the deep learning architectures that divided into four categories: Autoencoders, Convolutional Neural Network CNN, Deep Belief Networks DBN, and Recurrent Neural Networks RNN. By evaluating the applications of deep learning in the fingerprint, face, iris, ear, palm-print, and gait biometric technology, a general observation is that Convolutional Neural Network CNN have shown high performance for application of Biometrics identification and CNNs are an efficient artificial neural network method. Dynamic neural networks have been broadly used to provides solutions for recognition problems and applied to several systems. Currently, DNN has become a hot research area because of the benefits for companies. It should be pointed out that, so far, there have been a multitude of research results on the permanence analysis, stabilization and problems for various types of biometric systems and networks in the literature. The deep learning-based systems not only deliver better results, but they are also more robust than the traditional biometric systems. This is due to the high accuracy of CNNs in capturing and representing the data features.

References

1. L. Deng, D. Yu, Deep learning: methods and applications. Found. Trends® Signal Process. **7**(3–4), pp. 197–387 (2014)
2. D. Ciresan, U. Meier, J. Schmidhuber, Multi-column deep neural networks for image classification, in *Cvpr* (2012), pp. 3642–3649
3. Y. LeCun, Y. Bengio, G. Hinton, Deep learning. Nature **521**(7553), 436–444 (2015)
4. H. Al-Assam, H. Sellahewa, Deep Learning—the new kid in artificial intelligence news biometrics institute (2017). Online Available: http://www.biometricsinstitute.org/news.php/220/deep-learning-the-new-kid-in-artificial-intelligence?COLLCC=3945508322&. Accessed 06 Apr 2019

5. T. Mikolov, I. Sutskever, K. Chen, G.S. Corrado, J. Dean, Distributed representations of words and phrases and their compositionality. Adv. Neural Inf. Process. Syst. 3111–3119 (2013)
6. H. Lee, P. Pham, Y. Largman, A.Y. Ng, Unsupervised feature learning for audio classification using convolutional deep belief networks. Adv. Neural Inf. Process. Syst. **22**, 1096–1104 (2009)
7. A. Krizhevsky, I. Sutskever, G.E. Hinton, ImageNet classification with deep convolutional neural networks. Adv. Neural. Inf. Process. Syst. **25**, 1097–1105 (2012)
8. Y. LeCun, K. Kavukcuoglu, C. Farabet, Convolutional networks and applications in vision, in *ISCAS 2010–2010 IEEE International Symposium on Circuits and Systems. Nano-Bio Circuit Fabrics and Systems* (2010), pp. 253–256
9. Y. LeCun, Y. Bengio, G. Hinton, Deep learning. Nature **521**(7553), 436 (2015)
10. S.O. Y.-W.T. Geoffrey, E. Hinton, A fast learning algorithm for deep belief nets. Neural Comput. **18**(7), 1527–1554 (2006)
11. A. Graves, A.-R. Mohamed, G. Hinton, Speech recognition with deep recurrent neural networks, in *2013 IEEE International Conference on Acoustics, Speech and Signal Processing*, no. 6 (2013), pp. 6645–6649
12. D. Bahdanau, J. Chorowski, D. Serdyuk, P. Brakel, Y. Bengio, End-to-end attention-based large vocabulary speech recognition, in *2016 IEEE International Conference on Acoustics, Speech and Signal Processing (ICASSP)* (2016), pp. 4945–4949
13. L. Deng, A tutorial survey of architectures, algorithms, and applications for deep learning. APSIPA Trans. Signal Inf. Process. **3** (2014)
14. Y. Li, in *Deep Reinforcement Learning: An Overview*, pp. 1–85, (2017). Preprint at arXiv: 1701.07274
15. N. Ortiz, R.D. Hernández, R. Jimenez, Survey of biometric pattern recognition via machine learning techniques. Contemp. Eng. Sci. **11**(34), 1677–1694 (2018)
16. J. Riordon, D. Sovilj, S. Sanner, D. Sinton, E.W.K. Young, Deep Learning with microfluidics for biotechnology. Trends Biotechnol. 1–15 (2018)
17. K. Sundararajan, D.L. Woodard, Deep learning for biometrics : a survey. ACM Comput. Surv. **51**(3) (2018)
18. Y. Bengio, Deep learning of representations: looking forward. Stat. Lang. Speech Process. 1–37 (2013)
19. J. Schmidhuber, Deep learning in neural networks: an overview. Neural Netw. **61**, 85–117 (2015)
20. Y. Bengio, Learning deep architectures for AI. Found. Trends® Mach. Learn. **2**(1), 1–127 (2009)
21. K. Grill-Spector, K. Kay, K.S. Weiner, in *The Functional Neuroanatomy of Face Processing: Insights from Neuroimaging and Implications for Deep Learning Kalanit* (Springer, Berlin, 2017)
22. BCC Research, Adoption of Biometric Technologies in Private and Public Sectors Driving Global Markets, Reports BCC Research. *BCC Research* (2016). Online Available: http://www.marketwired.com/press-release/adoption-biometric-technologies-private-public-sectors-driving-global-markets-reports-2087162.htm. Accessed 06 Apr 2019
23. Y. Bengio, P. Lamblin, D. Popovici, H. Larochelle, Greedy layer-wise training of deep networks, in *Advances in Neural Information Processing Systems* (2007), pp. 153–160
24. G.E. Hinton, R.R. Salakhutdinov, Reducing the dimensionality of data with neural networks. Science **313**(5786), 504–507 (2006)
25. C. Liou, W. Cheng, J. Liou, D. Liou, Autoencoder for words. Neurocomputing **139**, 84–96 (2014)
26. Q.V. Le, J. Ngiam, A. Coates, A. Lahiri, B. Prochnow, A.Y. Ng, On optimization methods for deep learning, in *28th International Conference on Machine Learning* (2011), pp. 265–272
27. W.Y. Zou, A.Y. Ng, K. Yu, Unsupervised learning of visual invariance with temporal coherence, in *Neural Information Processing Systems Workshop on Deep Learning and Unsupervised Feature Learning*, vol. 3 (2011), pp. 1–9

28. M.A. Ranzato, C. Poultney, S. Chopra, Y.L. Cun, Efficient learning of sparse representations with an energy-based model, in *Proceedings of the NIPS* (2006)
29. H. Lee, C. Ekanadham, A.Y. Ng, Sparse deep belief net model for visual area V2, in *Proceedings of the NIPS* (2008)
30. P. Vincent, H. Larochelle, Y. Bengio, P.A. Manzagol, in Extracting and composing robust features with denoising autoencoders, in *Proceedings of the 25th International Conference on Machine Learning* (2008), pp. 1096–1103
31. S. Rifai, X. Muller, in *Contractive Auto-Encoders : Explicit Invariance During Feature Extraction*, pp. 833–840 (2011)
32. R. Salakhutdinov, G. Hinton, Deep boltzmann machines, in *Proceedings of the AISTATS* (2009)
33. B. Li et al., Large scale recurrent neural network on GPU, in *2014 International Joint Conference on Neural Networks (IJCNN)* (2014), pp. 4062–4069
34. N. Kalchbrenner, E. Grefenstette, P. Blunsom, A convolutional neural network for modelling sentences (2014) Preprint at arXiv:1404.2188
35. G. Sutskever, I. Martens, J. Hinton, Generating text with recurrent neural networks, in *Proceedings of the 28th International Conference on Machine Learning* (2011), pp. 1017–1024
36. G. Mesnil, X. He, L. Deng, Y. Bengio, Investigation of recurrent-neural-network architectures and learning methods for spoken language understanding, in *Interspeech* (2013), pp. 3771–3775
37. A. Ioannidou, E. Chatzilari, S. Nikolopoulos, I. Kompatsiaris, Deep learning advances in computer vision with 3D data. ACM Comput. Surv. **50**(2), 1–38 (2017)
38. I. Goodfellow, Y. Bengio, A. Courville, *Deep Learning* (MIT Press, Cambridge, MH, 2016)
39. Y. LeCun, L. Bottou, Y. Bengio, P. Haffner, Gradient-based learning applied to document recognition, in *Proceedings of the IEEE 86* (1998), pp. 2278–2324
40. O. Russakovsky et al., ImageNet large scale visual recognition challenge. Int. J. Comput. Vis. **115**(3), 211–252 (2015)
41. K. Simonyan, A. Zisserman, Very deep convolutional networks for large-scale image recognition (2014). Preprint at arXiv:1409.1556
42. C. Szegedy et al., Going deeper with convolutions, in *Proceedings of the CVPR* (2015)
43. K. He, X. Zhang, S. Ren, J. Sun, Deep residual learning for image recognition, in *2016 IEEE Conference on Computer Vision and Pattern Recognition* (2016), pp. 770–778
44. G. Huang, Z. Liu, L. Van Der Maaten, K.Q. Weinberger, Densely connected convolutional networks, in *Proceedings of the 30th IEEE Conference on Computer Vision and Pattern Recognition, CVPR 2017*, vol. 2017 (2017), pp. 2261–2269
45. D. Han, J. Kim, J. Kim, Deep pyramidal residual networks, in *CVPR 2017 IEEE Conference on Computer Vision and Pattern Recognition* (2017)
46. S. Xie, R. Girshick, P. Dollár, Z. Tu, K. He, Aggregated residual transformations for deep neural networks, in *Proceedings of the IEEE Conference on Computer Vision and Pattern Recognition* (2017), pp. 1492–1500
47. S. Woo, J. Park, J.Y. Lee, I.S. Kweon, CBAM: Convolutional block attention module, in *Proceedings of the European Conference on Computer Vision* (2018, pp. 3–19
48. A.K. Jain, A. Ross, S. Prabhakar, An introduction to biometric recognition. IEEE Trans. Circuits Syst. Video Technol. **14**(1), 4–20 (2004)
49. M.O. Oloyede, S. Member, G.P. Hancke, Unimodal and multimodal biometric sensing systems: a review. IEEE Access **4**, 7532–7555 (2016)
50. B. Stojanović, O. Marques, A. Neškovi, S. Puzovi, Fingerprint ROI segmentation based on deep learning, in *2016 24th Telecommunications Forum* (2016), pp. 5–8
51. W. Yani, W. Zhendong, Z. Jianwu, C. Hongli, A robust damaged fingerprint identification algorithm based on deep learning, in *2016 IEEE Advanced Information Management, Communicates, Electronic and Automation Control Conference (IMCEC)* (2016), pp. 1048–1052
52. L. Jiang, T. Zhao, C. Bai, A. Yong, M. Wu, A direct fingerprint minutiae extraction approach based on convolutional neural networks, in *International Joint Conference on Neural Networks* (2016), pp. 571–578

53. J. Li, J. Feng, C.-C.J. Kuo, Deep convolutional neural network for latent fingerprint enhancement. Signal Process. Image Commun. **60**, 52–63 (2018)
54. D. Song, Y. Tang, J. Feng, Aggregating minutia-centred deep convolutional features for fingerprint indexing. Pattern Recognit. (2018)
55. D. Peralta, I. Triguero, S. García, Y. Saeys, J.M. Benitez, F. Herrera, On the use of convolutional neural networks for robust classification of multiple fingerprint captures, pp. 1–22, (2017). Preprint at arXiv:1703.07270
56. R. Wang, C. Han, Y. Wu, T. Guo, Fingerprint classification based on depth neural network (2014). Preprint at arXiv:1409.5188
57. W.J. Wong, S.H. Lai, Multi-task CNN for restoring corrupted fingerprint images, Pattern Recognit. 107203 (2020)
58. M. Drahanský, O. Kanich, E. Březinová, Challenges for fingerprint recognition spoofing, skin diseases, and environmental effects, in *Handbook of Biometrics for Forensic Science*, (Springer, Berlin, 2017), pp. 63–83
59. R.F. Nogueira, R. de Alencar Lotufo, R.C. Machado, Fingerprint liveness detection using convolutional networks. IEEE Trans. Inf. Forensics Secur. **11**(6), 1206–1213 (2016)
60. S. Kim, B. Park, B.S. Song, S. Yang, Deep belief network based statistical feature learning for fingerprint liveness detection. Pattern Recognit. Lett. **77**, 58–65 (2016)
61. E. Park, X. Cui, W. Kim, H. Kim, End-to-end fingerprints liveness detection using convolutional networks with gram module, pp. 1–15 (2018). Preprint at arXiv:1803.07830
62. J. Yu, K. Sun, F. Gao, S. Zhu, Face biometric quality assessment via light CNN. Pattern Recognit. Lett. **0**, 1–8 (2017)
63. Y. Jiang, S. Li, P. Liu, Q. Dai, Multi-feature deep learning for face gender recognition, in *2014 IEEE 7th Joint International Information Technology and Artificial Intelligence Conference, ITAIC 2014* (2014), pp. 507–511
64. K. Shailaja, B. Anuradha, Effective face recognition using deep learning based linear discriminant classification, in *2016 IEEE International Conference on Computational* Intelligence *and Computing Research India* (2016), pp. 1–6
65. Y. Sun, X. Wang, X. Tang, Hybrid deep learning for computing face similarities. Int. Conf. Comput. Vis. **38**(10), 1997–2009 (2013)
66. R. Singh, H. Om, Newborn face recognition using deep convolutional neural network. Multimed. Tools Appl. **76**(18), 19005–19015 (2017)
67. P. Sharma, R.N. Yadav, K.V. Arya, Face recognition from video using generalized mean deep learning neural network, in *4th* 4th *International Symposium on Computational and Business Intelligence Face* (2016), pp. 195–199
68. A. Bharati, R. Singh, M. Vatsa, K.W. Bowyer, Detecting facial retouching using supervised deep learning. IEEE Trans. Inf. Forensics Secur. **11**(9), 1903–1913 (2016)
69. T. Zhuo, Face recognition from a single image per person using deep architecture neural networks. Cluster Comput. **19**(1), 73–77 (2016)
70. B.K. Tripathi, On the complex domain deep machine learning for face recognition. Appl. Intell. **47**(2), 382–396 (2017)
71. K. Guo, S. Wu, Y. Xu, Face recognition using both visible light image and near-infrared image and a deep network. CAAI Trans. Intell. Technol. **2**(1), 39–47 (2017)
72. D. Yi, Z. Lei, S. Liao, S.Z. Li, Learning face representation from scratch (2014). Preprint at arXiv:1411.7923
73. S. Lawrence, C.L. Giles, A.C. Tsoi, A.D. Back, Face recognition: a convolutional neural-network approach. IEEE Trans. Neural Netw. **8**(1), 98–113 (1997)
74. Y. Taigman, M. Yang, M. Ranzato, L. Wolf, DeepFace: closing the gap to human-level performance in face verification, in *Proceedings of the IEEE Conference on Computer Vision and Pattern Recognition* (2014), pp. 1701–1708
75. Y. Sun, X. Wang, X. Tang, Deep learning face representation from predicting 10,000 classes, in *Proceedings of the IEEE Conference on Computer Vision and Pattern Recognition* (2014), pp. 1891–1898

76. Y. Sun, Y. Chen, X. Wang, X. Tang, Deep learning face representation by joint identification-verification. Adv. Neural. Inf. Process. Syst. **27**, 1988–1996 (2014)
77. Y. Sun, X. Wang, X. Tang, Deeply learned face representations are sparse, selective, and robust, in *IEEE Conference on Computer Vision and Pattern Recognition* (2015), pp. 2892–2900
78. Z. Lu, X. Jiang, A.C. Kot, Deep coupled ResNet for low-resolution face recognition. IEEE Signal Process. Lett (2018)
79. K. Li, Y. Jin, M. Waqar, A. Ruize, H. Jiongwei, Facial expression recognition with convolutional neural networks via a new face cropping and rotation strategy. Vis. Comput. (2019)
80. R. Ranjan, V.M. Patel, S. Member, R. Chellappa, HyperFace : a deep multi-task learning framework for face detection, landmark localization, pose estimation, and gender recognition. IEEE Trans. Pattern Anal. Mach. Intell. **41**(1), 121–135 (2019)
81. S. Almabdy, L. Elrefaei, Deep convolutional neural network-based approaches for face recognition. Appl. Sci. **9**(20), 4397 (2019)
82. ORL face database. Online Available: http://www.uk.research.att.com/facedatabase.html. Accessed 06 Apr 2019
83. F. Tarres, A. Rama, GTAV face database (2011). Online Available: https://gtav.upc.edu/en/research-areas/face-database. Accessed 06 Apr 2019
84. A.V. Nefian, Georgia tech face database. Online Available: http://www.anefian.com/research/face_reco.htm. Accessed 06 Apr 2019
85. C.E. Thomaz, FEI face database (2012). Online Available: https://fei.edu.br/~cet/facedatabase.html. Accessed 06 Apr 2019
86. G.B. Huang, M. Ramesh, T. Berg, E. Learned-Miller, Labeled faces in the wild: a database for studying face recognition in unconstrained environments (2007)
87. Frontalized faces in the wild (2016). Online Available: https://www.micc.unifi.it/resources/datasets/frontalized-faces-in-the-wild/. Accessed 06 Apr 2019
88. L. Wolf, T. Hassner, I. Maoz, Face recognition in unconstrained videos with matched background similarity. in *2011 IEEE Conference on Computer Vision and Pattern Recognition* (2011), pp. 529–534
89. P.S. Prasad, R. Pathak, V.K. Gunjan, H.V.R. Rao, Deep learning based representation for face recognition, in *ICCCE 2019* (Singapore, Springer, 2019), pp. 419–424
90. R.B. TA Raj, A novel hybrid genetic wolf optimization for newborn baby face recognition, Paid. J. 1–9 (2020)
91. A. Alotaibi, A. Mahmood, Enhancing computer vision to detect face spoofing attack utilizing a single frame from a replay video attack using deep learning, in *Proceedings of the 2016 International Conference on Optoelectronics and Image Processing-ICOIP 2016*, (2016), pp. 1–5
92. A. Nseaf, A. Jaafar, K.N. Jassim, A. Nsaif, M. Oudelha, Deep neural networks for iris recognition system based on video: stacked sparse auto encoders (SSAE) and bi-propagation neural. J. Theor. Appl. Inf. Technol. **93**(2), 487–499 (2016)
93. M. Arsalan et al., Deep learning-based iris segmentation for iris recognition in visible light environment, Symmetry (Basel) **9**(11) (2017)
94. F. Marra, G. Poggi, C. Sansone, L. Verdoliva, A deep learning approach for iris sensor model identification. Pattern Recognit. Lett. **0**, 1–8 (2017)
95. M.G. Alaslani, L.A. Elrefaei, Convolutional neural network based feature extraction for iris. Int. J. Comput. Sci. Inf. Technol. **10**(2), 65–78 (2018)
96. M.G. Alaslani, L.A. Elrefaei, Transfer lerning with convolutional neural networks for iris recognition. Int. J. Artif. Intell. Appl. **10**(5), 47–64 (2019)
97. A.J. Abhishek Gangwar, DeepIrisNet: deep iris representation with applications in iris recognition and cross-sensor iris recognition, in *2016 IEEE International Conference on Image Processing* (2016), pp. 2301–2305
98. S. Arora, M.P.S. Bhatia, Presentation attack detection for iris recognition using deep learning. Int. J. Syst. Assur. Eng. Manage. 1–7 (2020)

99. D. Zhao, X. Pan, X. Luo, X. Gao, Palmprint recognition based on deep learning, in *6th International Conference on Wireless, Mobile and Multi-Media (ICWMMN 2015)* (2015)
100. P.L. Galdámez, W. Raveane, A. González Arrieta, A brief review of the ear recognition process using deep neural networks. J. Appl. Log. **24**, 62–70 (2017)
101. A.A. Almisreb, N. Jamil, N.M. Din, Utilizing AlexNet deep transfer learning for ear recognition, in *Proceedings of the 2018 4th International Conference on Information Retrieval and Knowledge Management Diving into Data science CAMP 2018* (2018), pp. 8–12
102. Ž. Emeršič, J. Križaj, V. Štruc, P. Peer, Deep ear recognition pipeline. Recent Adv. Comput. Vis. Theor. Appl. 333–362 (2019)
103. J. Ma et al., Segmenting ears of winter wheat at flowering stage using digital images and deep learning. Comput. Electron. Agric. **168**, 105159 (2020)
104. Y. Liu, J. Ling, Z. Liu, J. Shen, C. Gao, Finger vein secure biometric template generation based on deep learning. Soft Comput. (2017)
105. R. Das, E. Piciucco, E. Maiorana, P. Campisi, Convolutional neural network for finger-vein-based biometric identification. IEEE Trans. Inf. Forensics Secur. **14**(2), 360–373 (2018)
106. D. Zhao, H. Ma, Z. Yang, J. Li, W. Tian, Finger vein recognition based on lightweight CNN combining center loss and dynamic regularization. Infrared Phys. Technol. 103221 (2020)
107. N.A. Al-johania, L.A. Elrefaei, Dorsal hand vein recognition by convolutional neural networks: feature learning and transfer learning approaches. Int. J. Intell. Eng. Syst. **12**(3), 178–191 (2019)
108. Z. Wu, Y. Huang, L. Wang, X. Wang, T. Tan, A comprehensive study on cross-view gait based human identification with deep CNNs. IEEE Trans. Pattern Anal. Mach. Intell. **39**(2), 209–226 (2017)
109. M. Alotaibi, A. Mahmood, Improved gait recognition based on specialized deep convolutional neural network, Comput. Vis. Image Underst. 1–8 (2017)
110. Center for biometrics and security research, CASIA Gait Database. Online Available: http://www.cbsr.ia.ac.cn. Accessed 06 Apr 2019
111. M. Baccouche, F. Mamalet, C. Wolf, C. Garcia, A. Baskurt, in *Sequential Deep Learning for Human Action Recognition* (Springer, Berlin, 2011), pp. 29–39
112. C. Schuldt, I. Laptev, B. Caputo, Recognizing human actions: a local SVM approach, in *Proceedings of the 17th International Conference on Pattern Recognition,* vol. 3 (2004), pp. 32–36
113. A. Sokolova, A. Konushin, Gait recognition based on convolutional neural networks. ISPRS Int. Arch. Photogramm. Remote Sens. Spat. Inf. Sci. **42**, 207–212 (2017)
114. J.M. Baker, L. Deng, J. Glass, S. Khudanpur, C.H. Lee, N. Morgan, D. O'Shaughnessy, Developments and directions in speech recognition and understanding, Part 1 [DSP Education]. IEEE Signal Process. Mag. **26**(3), 75–80 (2009)
115. C. Chang, C. Lin, LIBSVM: a library for support vector machines. ACM Trans. Intell. Syst. Technol. **2**, 1–39 (2013)
116. M. Kubat, Artificial neural networks, in *An Introduction to Machine Learning* (Springer, Berlin, 2015), pp. 91–111
117. D. Das, A. Chakrabarty, Human gait recognition using deep neural networks, pp. 5–10 (2016)
118. R. Singh, R. Khurana, A.K.S. Kushwaha, R. Srivastava, Combining CNN streams of dynamic image and depth data for action recognition. Multimed. Syst. 1–10 (2020)
119. M.M. Hasan, H.A. Mustafa, Multi-level feature fusion for robust pose-based gait recognition using RNN. Int. J. Comput. Sci. Inf. Secur. **18**(1), 20–31 (2020)
120. L. Tran, D. Choi, Data augmentation for inertial sensor-based gait deep neural network. IEEE Access **8**, 12364–12378 (2020)
121. K. Delac, M. Grgic, A survey of biometric recognition methods, in *Proceedings of the Elmar-2004. 46th International Symposium on Electronics in Marine 2004* (2004). pp. 184–193
122. D. Menotti et al., Deep representations for iris, face, and fingerprint spoofing detection. IEEE Trans. Inf. Forensics Secur. **10**(4), 864–879 (2015)
123. S. Maity, M. Abdel-Mottaleb, S.S. Asfour, Multimodal biometrics recognition from facial video via deep learning. Int. J. **8**(1), 81–90 (2017)

124. M. Simón et al., Improved RGB-D-T based face recognition. IET Biom. 297–304 (2016)
125. A. Meraoumia, L. Laimeche, H. Bendjenna, S. Chitroub, Do we have to trust the deep learning methods for palmprints identification? in *Proceedings of the Mediterranean Conference on Pattern Recognition and Artificial Intelligence 2016* (2016), pp. 85–91
126. N. Neverova et al., Learning human identity from motion patterns. IEEE Access **4**, 1810–1820 (2016)
127. N. Yudistira, T. Kurita, Correlation net: spatiotemporal multimodal deep learning for action recognition. Signal Process. Image Commun. **82**, 115731 (2020)
128. E.M. Cherrat, R. Alaoui, H. Bouzahir, Convolutional neural networks approach for multimodal biometric identification system using the fusion of fingerprint, finger-vein and face images, Peer J. Comput. Sci. **6**, e248 (2020)

Convolution of Images Using Deep Neural Networks in the Recognition of Footage Objects

Varlamova Lyudmila Petrovna

Abstract In the problems of image recognition, various approaches used when the image is noisy and there is a small sample of observations. The paper discusses nonparametric recognition methods and methods based on deep neural networks. This type of neural network allows you to collapse images, to perform downsampling as many times as necessary. Moreover, the image recognition speed is quite high, and the data dimension is reduced by using convolutional layers. One of the most important elements of the application of convolutional neural networks is training. The article gives the results of work on the application of convolutional neural networks. The work was carried out in several stages. In the first stage was carried out the modeling of the convolutional neural network and was developed its architecture. In the second stage, the neural network was trained. The third phase produced Python software. The software health check and video processing speed were then performed.

Keywords Nonparametric methods · Small sampling · Image recognition · Convolutional neural networks · Training algorithm

1 Introduction

In the problems of image recognition in the conditions of small samples of observations and with uneven distribution, the presence of interference or noise, they face problems of eliminating interference and maximum recognition of image objects. The main indicators are the stability of recognition algorithms in the presence of applicative interference (image noise due to object shadowing, the presence of affected areas), when there is a small sample of observations and a priori data are not available, when applying the Laplace principle is difficult or impossible. For example, in [1–3], the problems of applying approaches to overcoming the problem of a small number of

V. L. Petrovna (✉)
Department of Multimedia Technologies, Tashkent University of Information Technologies, Amir Temur Str. 108A, Tashkent 100083, Uzbekistan
e-mail: vlp@bk.ru; dimirel@gmail.com

© The Editor(s) (if applicable) and The Author(s), under exclusive license to Springer Nature Switzerland AG 2021

A. E. Hassanien et al. (eds.), *Artificial Intelligence for Sustainable Development: Theory, Practice and Future Applications*, Studies in Computational Intelligence 912,
https://doi.org/10.1007/978-3-030-51920-9_9

samples were considered using the methods of reducing dimensionality and adaptive nonparametric identification algorithms, and discriminant analysis methods [2, 3].

The problems should be divided into tasks with severe restrictions for small samples, but with the presence of a sufficient number of reference images and the problem of classifying images with small samples, but with a large dimension and the smallest number of reference images.

The purpose of this work is to compare the use of nonparametric methods with convolutional neural networks in image recognition problems in the conditions of small observation samples.

2 Statement of the Problem

The using of contours is common in the tasks of combining a pair of images or an image with a vector model (for example, a location map or a detail drawing), the description of the shape of objects or areas along their contours, for example, using mathematical morphology methods [4], to solve the problem of stereoscopic vision. Contour or spatial representations also serve as the basis for constructing structural descriptions of images.

When comparing contour methods such as the Otsu binarization method; the border detectors Roberts, Prewitt, Sobel, Kenny Laplace, Kirsch, Robinson, then in terms of processing speed they have differences. Thus, among these methods of constructing a gradient field of a halftone image, the Kenny operator algorithm performs a segmentation procedure in a record short time of 0.010 s, and segmentation based on the Otsu binarization method is carried out in 0.139 s. On average, segmentation using the Roberts, Sobel, Laplace, Prewitt, Kenny operators is 0.0158 s. There are differences in the results of the segmentation carried out. Figure 1 shows the results of construction of gradient field of halftone image using specified operators.

Let the mathematical model of the source image be a two-dimensional discrete sequence $\overline{N}$, $j = \overline{1, M}$ of the form:

$$X_{i,j} = S_{i,j} + \eta_{i,j}, i = \overline{1, N}, j = \overline{1, M} \quad (1)$$

where $S_{i,j}$ is a useful two-dimensional component (the original undistorted image); $\eta_{i,j}$ is the additive noise component; N is the number of rows; M is the number of columns of a two-dimensional image array. Image $X_{i,j}$ of height N with width M in pixels.

The task of constructing a segmented image allows you to remove part of the noise by averaging or smoothing the histogram, and obtaining brightness values at the boundaries of image objects. In this case, the method of randomly choosing brightness values using a 3×3 window was used.

All contour methods discussed above are spatial and can be described by expressions of the form [4]

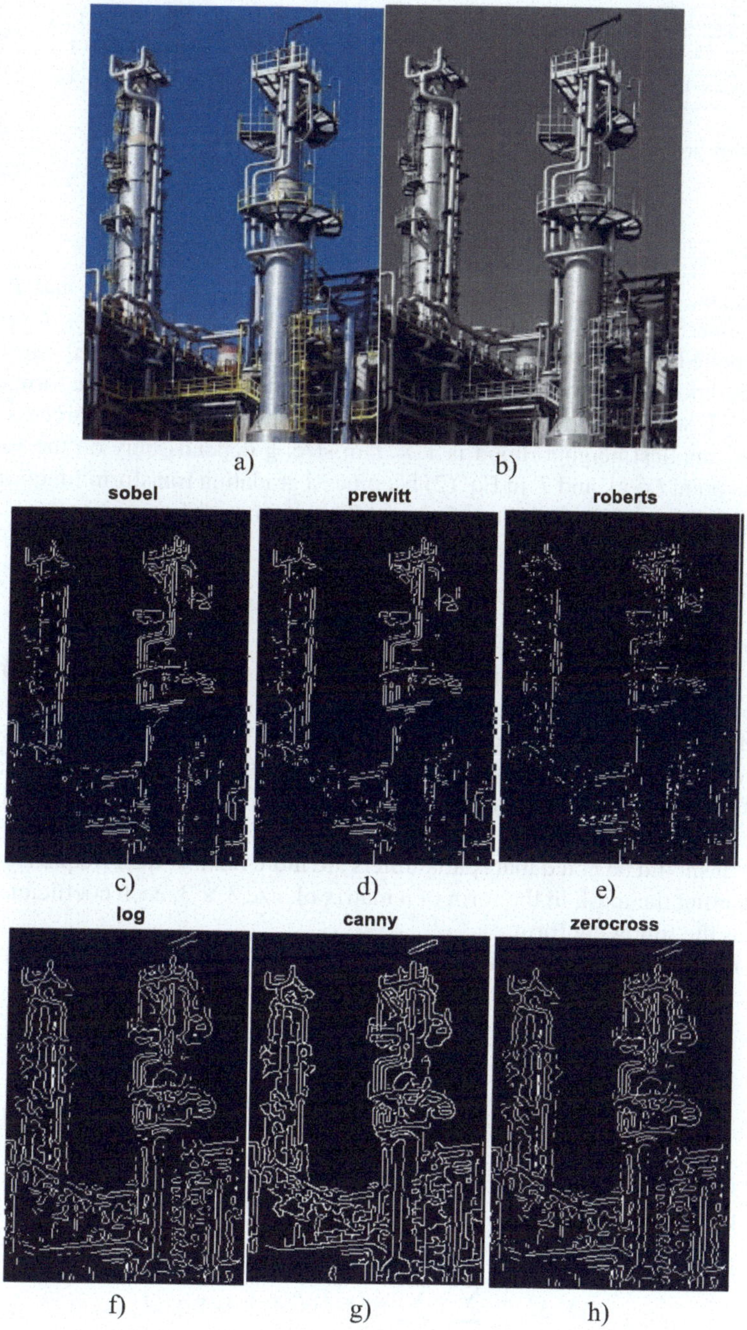

Fig. 1 Image processing using operators. **a** and **b** Source images: color and gray scale; **c** Sobel; **d** Prewitt; **e** Roberts; **f** Laplacian-Gaussian; **g** Kenny; **h** Robinson

f(x-1, y-1)	f(x-1,y)	f(x-1, y+1)
f(x, y-1)	f(x,y)	f(x, y+1)
f(x+1, y-1)	f(x+1, y)	f(x+1, y+1)

$$\begin{matrix} w(-1,-1) & w(-1,0) & w(-1,1) \\ w(0-1) & w(0,0) & w(0,1) \\ w(1,-1) & w(1,0) & w(1,1) \end{matrix}$$

Fig. 2 Image area elements

$$I_m(x, y) = T[I_{i,j}(x, y)], \tag{2}$$

where $I_{i,j}$ *(x, y)* is the input image, I_m *(x, y)* is the processed image, and T is the operator over $I_{i,j}$ defined in some neighborhood of the point *(x, y)*. The T operator can be applied to a single image and is used to calculate the average brightness over a neighborhood of a point (pixel). The neighborhood of elements in the form of 3 × 3, or 4 × 4, was called the core or window, and the spatial filtering process itself. Since the smallest neighborhood is 1 × 1 in size, g depends only on the value of $I_{i,j}$ at the point (x, y), and T in Eq. (2) becomes a gradation transform function, also called a brightness transform function or a display function having view

$$s = T(r), \tag{3}$$

where r and s are variables that denote respectively the brightness values of the images $I_{i,j}$ *(x, y)* and I_m *(x, y)* at each point (x, y). Based on (2) and (3), the images shown in Fig. 1a, b were processed.

In the process of spatial image processing, after the stage of brightness transformations, the selection of contours, as a rule, the filtering process follows. This implies the execution of operations on each element or pixel. The spatial filtering scheme can be represented as moving a mask or window across each image element (Fig. 2). It should be noted that spatial filters are more flexible than frequency filters.

Presenting the mask in the form of a matrix of size 3 × 3, each coefficient of the mask has the following form:

The response $g(x, y)$ at each point in the image is the sum of the products

$$\begin{aligned} g(x, y) &= w(s, t) \times f(x, y) \\ g(x, y) &= w(-1, -1)f(-1, -1) + w(-1, 0)f(x-1, y) + w(-1, 1)f(x-1, y+1) \\ &\quad + w(0, -1)f(x, y-1) + w(0, 0)f(x, y) + w(0, 1)f(x, y+1) \\ &\quad + w(1, -1)f(x+1, y-1) + w(1, 0)f(x+1, y) + w(1, 1)f(x+1, y+1) \end{aligned} \tag{4}$$

The image the $M \times N$ size, a mask $m \times n$ or 3 × 3, filtration taking into account (4) has an appearance

$$g(x, y) = \sum_{s=-a}^{a} \sum_{t=-b}^{b} w(s, t)f(x+s, y+t), \tag{5}$$

w(s,t)—filter mask coefficients, $a = \frac{m-1}{2}, b = \frac{n-1}{2}$.

The work carried out image processing and obtained values $g(x, y)$ at each point of the image (Fig. 1b) with size Matrix 274 × 522. A disadvantage of this technique is the occurrence of undesirable effects, i.e. incomplete image processing when the edges of the image remain untreated due to the nonlinear combination of mask weights. Adding zero elements at the edge of the image results in bands.

Carrying out the equalization of the histogram is similar to averaging the values of the elements along the vicinity of the mask-covered filter, the so-called sliding window method. It consists in determination of the size of a mask of the $m \times n$ filter for which the arithmetic average value of each pixel is calculated [5]

$$\bar{g}(x, y) = \frac{1}{m \cdot n} \cdot \sum_{s=-a}^{a} \sum_{t=-b}^{b} g(x, y) \tag{6}$$

If we look at the filtering result in the frequency domain, the set of weights is a two-dimensional impulse response. Such filter will be the FIR-filter with final pulse characteristic (finite impulse response) if area $\bar{g}(x, y)$ of course and pulse characteristic has final length. Otherwise, the impulse response has an infinite length and the IIR-filter is an infinite impulse response filter. However, in this work such filters will not be considered [5].

The correlation is calculated by window filtering, but if to rotate the filter 180°, the image is convolved [4, 6].

In the case where the image is very noisy, there is a small sample and the application of the above methods does not produce results for its processing, consider the application of nonparametric methods.

3 Image Processing by Non-Parametric Methods

It is necessary to distinguish objects in the presence of noise. Consider the selection using the Parzen window method. Since the distribution of objects in the image is highly uneven using the Parzen method, we will use a variable-width window with a decisive rule [1]

$$\begin{gathered} a\left(x,\ X^{l}, k,\ K\right) = \arg\max_{y \in y} \lambda_y \sum_{i=1}^{l} [y_i \equiv y] K\left(\tfrac{\rho(x, x_i)}{h}\right), \\ h = \rho\left(x, x^{(k+1)}\right), \end{gathered} \tag{7}$$

$X_{i,j}, i = \overline{1, N}$, with evaluation of species density

$$p_{y,h}(x) = \frac{1}{l_y V(h)} \sum_{i=1}^{l} [y_i \equiv y] K\left(\frac{\rho(x, x_i)}{h}\right), \tag{8}$$

$K(\theta)$ is an arbitrary even function of the kernel or window of width h, does not increase and positive on the interval [0,1] with weight

$$\begin{aligned} w(i, x) &= K\left(\frac{\rho(x, x_i)}{h}\right), \\ K(\theta) &= \tfrac{1}{2}[|\theta| < 1]. \end{aligned} \tag{9}$$

By small sample size, the matrix of two-dimensional distribution parameters becomes singular, and for small window widths this method reduces to the k-nearest neighbor's method [1], which has its own characteristics such as dependence on the selected step and instability to errors. In this case, it becomes necessary to impose conditions on the distribution density, the function $p_{y,h}(x)$ and the width of the window. Accordingly, the amount of data [2] in the image set is growing. However, a similar problem can be solved by methods of reducing the dimension or by methods of discriminant analysis [2]. Moreover, to reduce the volume of the data set, an external image used database. Then the task of constructing a classifier [1, 6, 7] is greatly simplified and the problem with a minimum sample and the least number of standards is reduced to a problem with a minimum sample, which is solved by nonparametric methods [6, 8].

The distribution density calculated using window (nuclear) functions is described by expression (7) of the form

$$\hat{p}(x)\frac{1}{nh}\sum_{i=1}^{n} K\left(\frac{x - x_i}{h}\right), \tag{10}$$

where n—sample size; K is the nuclear (window) function; h is the window width; x is a random sample; x_i is the ith implementation of a random variable.

In the multidimensional case, the density estimate, taking

$$\hat{p}(x) = \frac{1}{n}\sum_{i=1}^{n}\prod_{j=1}^{m}\frac{1}{h^j}K\left(\frac{x^j - x_i^j}{h^j}\right), \tag{11}$$

where m—space size, *kernel*—a function used to restore the distribution density, a continuous bounded function with a unit integral

$$\begin{aligned} &\textstyle\int K(y)dy = 1, \\ &\textstyle\int y\, K(y)dy = 0, \\ &\textstyle\int y^i K(y)dy = k_i(K) < \infty. \end{aligned} \tag{12}$$

Function (12) with properties $K(y) \geq 0$, is $K(y) = K(-y)$. The kernel is a non-negative bounded symmetric real function whose integral is equal to unity; the statistical moments must be infinite. The order ν of function (12) is equal to the order of the first moment, which is not equal to zero. If $k_1(K) = 0$ and $k_2(K) > 0$, then K is a second-order kernel ($\nu = 2$).

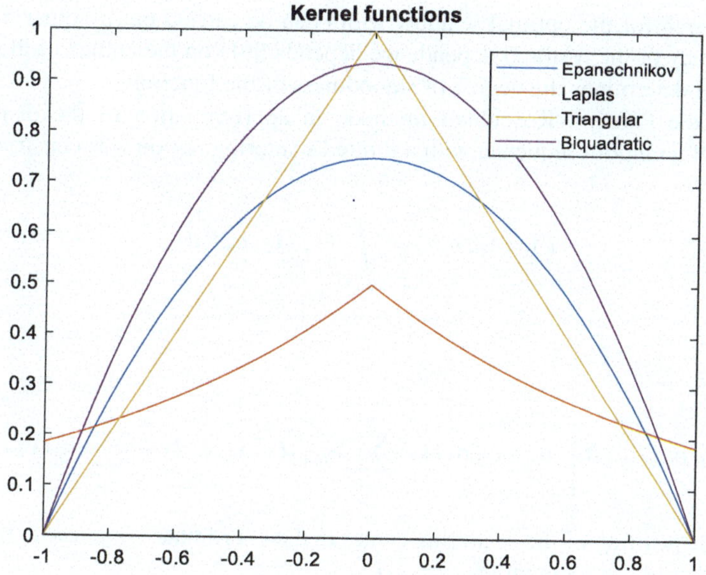

Fig. 3 Examples of various kernel functions

Known nuclear functions of the second order (Fig. 3):

- Epanechnikov kernel $K(y) = \frac{3}{4}(1 - y^2)$;
- Gauss kernel $K(y) = \frac{1}{\sqrt{2\pi}}e^{-0.5y^2}$;
- Laplace kernel $K(y) = \frac{1}{2}e^{-|y|}$;
- Uniform kernel $K(y) = \frac{1}{2}, |y| \leq 1$;
- Triangular kernel $K(y) = 1 - |y|, |y| \leq 1$
- Biquadratic kernel $k(y) = \frac{3(1-y^2)}{4}, |y| \leq 1$.

The optimal values of the nuclear function and parameter h are found from the condition that the functional $J = \int \ln K(y) \cdot K(y)dy$ reaches the maximum value.

Or in other words: to restore the empirical distribution density using the Parzen-Rosenblatt window, the unknown parameter is the window width h in expression (10). Therefore, to determine the empirical density, it is necessary to solve the problem of finding the window width, so as to find the optimal h_{opt}. Finding the optimal window size is made from the condition

$$\left(\frac{R(K)}{k_2^4(K)S_D\big(a\big(h_{\text{opt}}\big)\big)}\right)^{1/5} n^{1/5} - h_{\text{opt}} = 0. \tag{13}$$

The Parzen—Rosenblatt method allows one to construct an approximation of the distribution function of any finite random sequence, which, provided the parameter h is optimized, turns out to be quite smooth [7, 8].

The search for the optimal window width can be carried out by other methods. The accuracy of the restored dependence depends little on the choice of the kernel. The kernel determines the degree of smoothness of the function.

Using the Parzen—Rosenblatt method, an approximation of the distribution function of a random sequence with a limited scattering region was constructed

$$F(x; x_0, \sigma, l) = \int_{x_{\min}}^{x} f(\xi; x_0, \sigma, l) d\xi, \tag{14}$$

where

$$f_{\lim}(x; x_0, \sigma, l) = K\left[\phi(x; x_0, \sigma, l) + \sum_{n=0}^{\infty} \phi_{2n+1}^{\pm}(x; x_0, \sigma, l) + \sum_{n=1}^{\infty} \phi_{2n}^{\pm}(x; x_0, \sigma, l)\right],$$

x_0—the position of the scattering center in the coordinate system with the origin in the center of the segment $[x_{\min}, x_{\max}]$,
σ—standard deviation (SD) of a random function in the absence of restrictions,
$l = x_{\max} - x_{\min}$—span scatter,
K—normalization coefficient [9],
$x_{2n+1}^{\pm}, x_{2n}^{\pm}$ determined by the formulas:

$$x_{2n}^{\pm} = \pm 4nl + x_0, \; x_{2n+1}^{\pm} = \pm(4n+2)l - x_0,$$

When analyzing the quality of the approximation of the distribution function of a random sequence by the Parzen-Rosenblatt method and the *k*-nearest neighbors (*k*-NN) method in the scattering region $[-5, 5]$, with the standard deviation of the random variable $\sigma = [1, 3, 5\text{–}7, 10]$, the results were obtained shown in Tables 1 and 2. Graphs of convergence are shown in Figs. 4 and 5.

The use of the approximation of the distribution function of a random variable by the methods of Parzen-Rosenblatt and *k*-NN method with different nuclear functions showed the relative proximity of the approximating function and the true distribution

Table 1 The error of the estimation of the distribution function by the Parzen–Rosenblatt method

SD	Diapazon				
	−5	−3	0	3	5
1	0.001432	0.00078	0.0001389	0.00079	0.001428
3	0.000227	0.00023	0.00008821	0.000398	0.000553
5	0.000279	0.00022	0.0001638	0.00018	0.000201
7	0.0002	0.000181	0.0001298	0.000125	0.000152
10	0.000143	0.000138	0.0001379	0.000161	0.000147

Table 2 Error of estimation of distribution function by k-nearest neighbor's method

SD	Diapazon				
	−5	−3	0	3	5
1	0.0006412	0.00004523	0.00001498	0.00004267	0.0001125
3	0.00007934	0.00005424	0.00002274	0.00005917	0.00009254
5	0.00002315	0.00002884	0.00004868	0.00005254	0.00004132
7	0.00003157	0.00005793	0.00002141	0.00006232	0.00006778
10	0.00005682	0.0006147	0.00002798	0.00001356	0.00001067

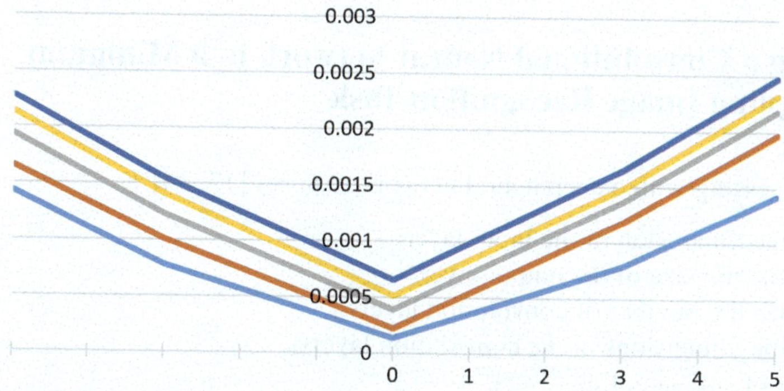

Fig. 4 Restoration of the distribution density function by the Parzen–Rosenblatt method (standard deviation-SD)

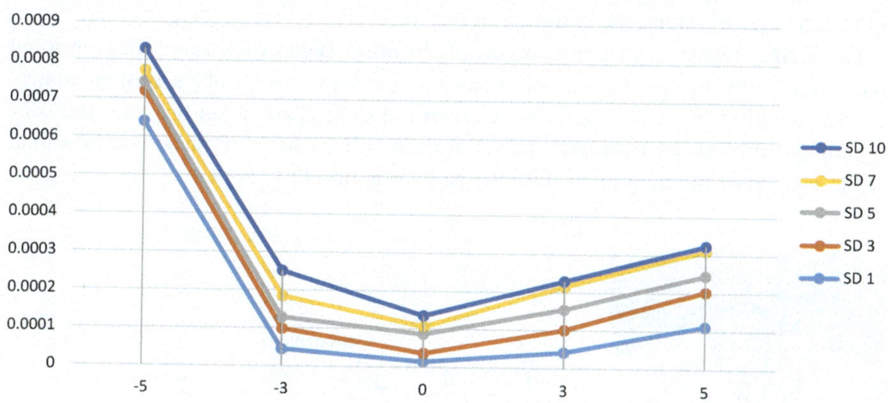

Fig. 5 Recovery of the density function by the method of k-NN

function. In the literature [10–12], there are a number of works with analytical data on the issue of comparing the Parzen–Rosenblatt method with imaginary sources, histograms.

If we have large data sets, an effective mechanism is required to search for neighboring points closest to the query point, since it takes too much time to execute the method in which the distance to each point is calculated. The proposed methods to improve the efficiency of this stage were based on preliminary processing of training data. The whole problem is that the methods of maximum likelihood, *k*-nearest neighbors or minimum distance, do not scale well enough with an increase in the number of dimensions of space. Convolutional Neural Networks are an alternative approach to solving such problems. For training a convolutional neural network, for example, databases of photographs of individuals available on the Internet can be used [13].

4 Using a Convolutional Neural Network in a Minimum Sampling Image Recognition Task

Image processing with a deep neural network requires [13, 14]:

- define the dimension of the input layer;
- determine the size of the output layer.
- computes the number of convolution layers.
- define the dimensions of the convolution layers.
- number of sub-sampling layers;
- define the dimensions of the subsampling layers.

The construction of the classifier with the help of a deep neural network, a neural network of direct propagation, begins with the first screwing layer. The architecture of the screw neural network is shown in Fig. 6.

The CNN shown in Fig. 6 consists of different types of layers: convolutional layers, subsampling layers. The convolution operation uses only a limited matrix of small weights (Fig. 6), which moves over the entire processed layer (at the very beginning—directly on the input image), forming after each shift an activation signal for the neuron of the next layer with the same position [13, 14].

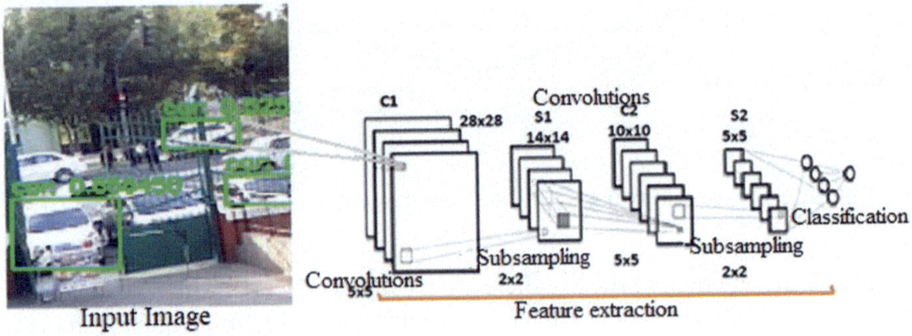

Fig. 6 Convolutional neural network architecture

A limited small scale matrix (Fig. 2) called a kernel is used to perform the convolution operation. The kernel moves along the entire processed layer (at the very beginning—directly on the input image), after each shift an activation signal is generated for the neuron of the next layer with the same position [10, 11].

The convolutional neural network architecture includes a cascade of convolution layers and sub-sampling layers (stacked convolutional and pooling layers), usually followed by several fully connected layers (FL), allowing local perception to be produced, layer weights to be separated at each step, and data to be filtered. When moving deep into the network, the filters (matrices w) work with a large perception field, which means that they are able to process information from a larger area of the original image, i.e. they are better adapted to the processing of a larger area of pixel space. The output layer of the convolutional network represents a feature map: each element of the output layer is obtained by applying the convolution operation between the input layer and the final sub band (receptive field) with the application of a certain filter (core) and the subsequent action of a non-linear activation function.

Pixel values are stored in a two-dimensional grid, that is, in an array of numbers (Fig. 6) that is processed by the kernel and the value written to the next layer [12, 13].

Each CNN layer converts the input set of weights into an output activation volume of neurons. Note that the system does not store redundant information, but stores the weight index instead of the weight itself. The direct passage in the convolution layer takes place in exactly the same way as in the full-knit layer—from the input layer to the output layer. At the same time, it is necessary to take into account that the weights of neurons are common [10, 14].

Let the image be given in the form of the matrix X and W—the matrix of weights, called the convolution kernel with the central element-anchor.

The first layer is an inlet layer. It receives a three-dimensional array that specifies the parameters of the incoming image

$$F = m \times n \times 3,$$

where F is the dimension of the input data array, $m \times n$ is the size of the image in pixels, "3" is the dimension of the array encoding the color in RGB format. The input image is "collapsed" using the matrix W (Fig. 7) In layer C_1, and a feature map is formed.

The convolution operation is determined by the expression

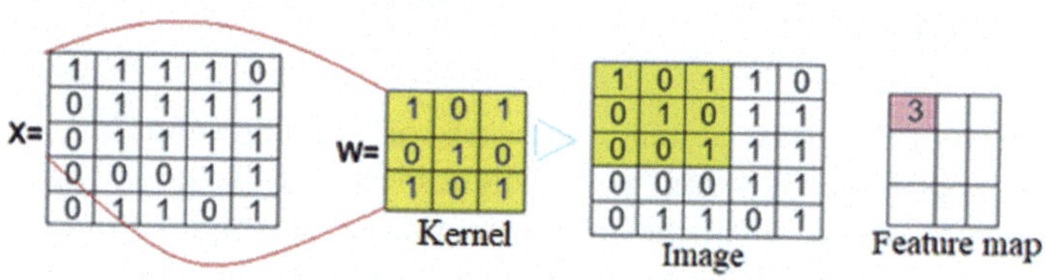

Fig. 7 Image convolution algorithm

$$y_{i,j} = \sum_{s=1}^{K}\sum_{t=1}^{K}(w_{s,t}, x_{(i-1)+s,(j-1)+t}), \quad (15)$$

where $w_{s,t}$ is the value of the convolution kernel element at the position (s,t), $y_{i,j}$ is the pixel value of the output image, $x_{((i-1)+s,(j-1)+t)}$ is the pixel value of the original image, K is the size of the convolution kernel.

After the first layer, we get a 28 × 28 × 1 matrix—an activation function or a feature map, that is, 784 values. Then, the matrix obtained in layer C_1 passes the operation of subsampling (pooling) using a window of size—$k \times k$. At the stage of subsampling, the signal has the form:

$$y_{i,j} = \max\left(x_{(ik+s,jk+t)}\right),$$

where $y_{(i,j)}$ is the pixel value of the output image, $\left(x_{(ik+s,jk+t)}\right)$ is the pixel value of the initial image and so on to an output layer.

The pooling layer resembles the convolution layer in its structure. In it, as in the convolution layer, each neuron of the map is connected to a rectangular area on the previous one.

Neurons have a nonlinear activation function—a logistical or hyperbolic tangent. Only, unlike the convolution layer, the regions of neighboring neurons do not overlap. In the convolution layer, each neuron of the region has its own connection having a weight.

In the pooling layer, each neuron averages the outputs of the neurons of the region to which it is attached. It turns out that each card has only two adjustable weights: multiplicative (weight averaging neurons) and additive (threshold). The pooling layers perform a downsampling operation for a feature map (often by calculating a maximum within a certain finite area).

Parameters of CNN (the weight of communications the convolutional and full-coherent layers of network) as a rule are adjusted by application of a method of the return distribution of a mistake (backpropagation, BP) realized by means of classical gradient descent (stochastic gradient descent) [14–18]. Alternating layers of convolution and subsampling (pooling) are performed to ensure extraction of signs at sufficiently small number of trained parameters.

5 Deep Learning

The application of the artificial neural network training algorithm involves solving the problem of optimization search in the weight space. Stochastic and batch learning modes are distinguished. In stochastic learning mode, examples from the learning sample are provided to the neural network input one after the other.

After each example, the network weights are updated. In the packet training mode, a whole set of training examples is supplied to the input of the neural network, after which the weights of the network are updated. A network weight error accumulates within the set for subsequent updating.

The classic error measurement criterion is the sum of the mean square errors

$$E_n^p = \frac{1}{2}E_{rr}^2 = \frac{1}{2}\sum_{j=1}^{M}(x_j - d_j)^2 \rightarrow \text{min}, \tag{16}$$

where M is number of output layer neurons, j is number of output neuron, x_j is real value of neuron output signal, d_j is the expected value. To reduce the quadratic error, the neural network will be trained by gradient descent, calculating the frequency derivative of the E_n^p with respect to each weight. We get the following ratio:

$$\frac{\partial E_n^p}{\partial w_i} = (x_j - d_j) \times \frac{\partial(x_j - d_j)}{\partial w_i} = (x_j - d_j) \times \frac{\partial}{\partial w_i} g\left(Y - g\left(\sum_{j=0}^{n} w_j x_j\right)\right)$$
$$= -(x_j - d_j) \times g'(\text{in}) \times x_j, \tag{17}$$

$$\frac{\partial E_n^p}{\partial w_i} = x_{n-1}^j \cdot \frac{\partial E_n^p}{\partial y_n^i} \tag{18}$$

$$\frac{\partial E_n^p}{\partial y_n^i} = g'(x_n^j) \cdot \frac{\partial E_n^p}{\partial x_n^i} \tag{19}$$

where g'—derivative activation function

$$\frac{\partial E_n^p}{\partial y_n^i} = x_n^i - d_n^i, \tag{20}$$

x_{n-1}^j is the output of the jth neuron of the $(n-1)$th layer, y_n^i is the scalar product of all the outputs of the neurons of the $(n-1)$th layer neurons and the corresponding weighting coefficients.

The gradient descent algorithm provides error propagation to the next layer and

$$\frac{\partial E_{n-1}^p}{\partial x_{n-1}^i} = \sum_i w_n^{ik} \cdot \frac{\partial E_n^p}{\partial y_n^i},$$

if we need to reduce E_n^p, then the weight is updated as follows

$$w_i \leftarrow w_i + \alpha \times (x_n^j - d_n^j) \times g'(\text{in}) \times x_i, \tag{21}$$

where α is training speed.

If the error $E_{rr} = \left(x_n^j - d_n^j\right)$ is positive, then the network output is too small and therefore the weights increase with positive input data and decrease with negative input data. With a negative error, the opposite happens. This error obtained in calculating the gradient can be considered as noise, which affects the correction of weights and can be useful in training.

Mathematically, the gradient is a partial derivative of the loss over each assimilable parameter, and one parameter update is formulated as follows [19, 20]:

$$w_i := w_i - \alpha * \frac{\partial L}{\partial w_n}, \tag{22}$$

Where, L is the loss Function.

The gradient of the loss function with respect to parameters is calculated using a subset of the learning dataset (Fig. 8) called the mini-package applied to parameter updates.

This method is called mini-packet gradient descent, also often called stochastic gradient descent (SGD), and the size of the mini-lot is also a hyperparameter [18–20].

Stochastic learning has some advantages over batch learning:

- in most cases, much faster than batch;
- can be used to track changes;
- often leads to better recognizers.

If the training sample size 500 consists of 10 identical sets of 50 examples, the average gradient across a thousand examples would produce the same result as a gradient calculation based on fifty examples. Thus, batch learning calculates the same value 10 times before updating the weights of the neural network. Stochastic

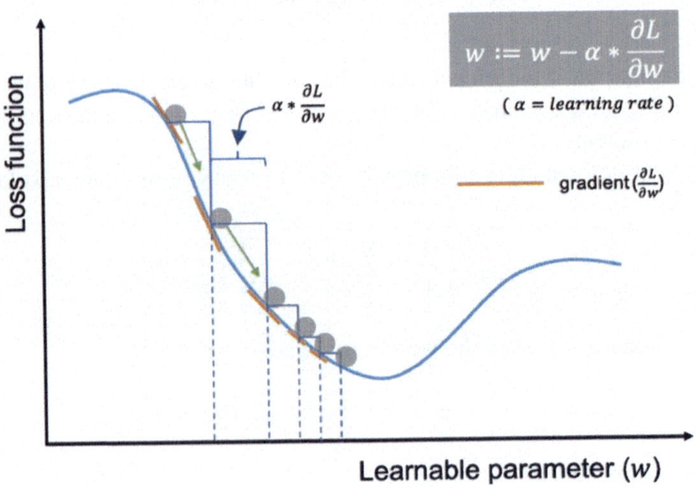

Fig. 8 Finding the loss function gradient to a calculated parameter (weight) [37]

learning, by contrast, will present an entire era as 10 iterations (eras) on a learning set of length 50. Typically, examples are rarely found more than once in a learning sample, but clusters of very similar examples may be found [21].

Nonlinear networks often have many local minima of different depths. The task of training is to hit the network in one of the lows. Batch training will lead to a minimum, in the vicinity of which weights are originally located. By stochastic learning, noise appearing when the weights are corrected causes the network to jump from one local minimum to another, possibly deeper [21].

Let's now consider the advantages of the batch mode of learning over stochastic [21, 22]:

- Convergence conditions are well studied;
- A large number of training acceleration techniques work only with batch mode;
- Theoretical analysis of the dynamics of changes in weights and convergence rate is simpler.

These benefits arise from the same noise factor that is present in stochastic learning. Such noises are removed by various methods. Despite some advantages of the batch mode, the stochastic method of training is used much more often, especially in those tasks when the training sample is large [23].

The learning process can be divided into several stages: training, verification and a set of tests (Fig. 9).

Learning data involves the use of a learning model with or without a teacher. To verify the correct model selection, performance monitoring is carried out, then the hyperparameter is set up and the model is finally selected. And to check the correct network settings, testing is carried out with an assessment of the final performance [24].

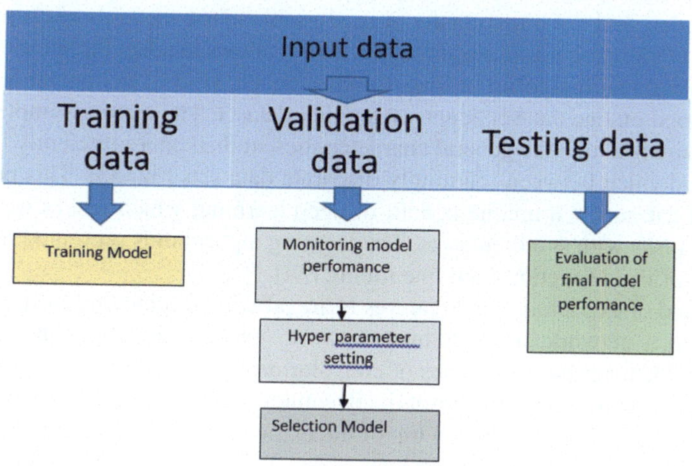

Fig. 9 The training process

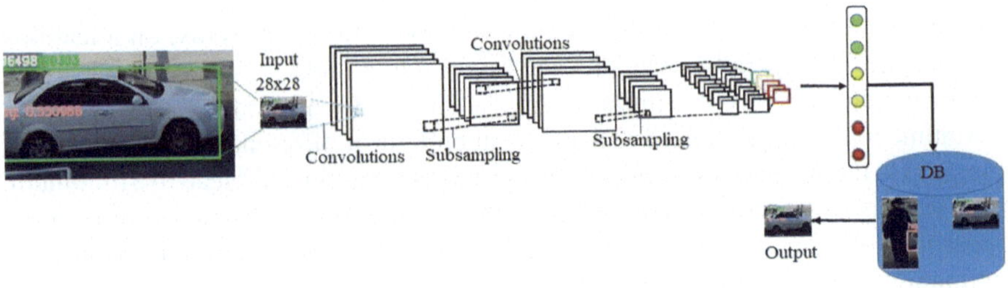

Fig. 10 The recognition process

The process of recognition and extraction of signs, the formation of a database of objects is shown in Fig. 10.

In the training process a convolutional neural network, when forming a database of objects, steps of calculating a gradient based on the use of a subset of a training set of data are sequentially passed. A network trained on a larger data set generalizes better [25–28]. In the event of noise, the use of filtering methods is shown depending on the type of noise [28].

When creating software for recognizing and classifying objects, we used the teaching method with a teacher, i.e. first, linear regression was calculated (16–19), then by the least squares method, the loss function (18) [27, 29–31].

6 Presence of Small Observations Samples

In the case when the task of training a model for a smaller data set is considered: increasing data and teaching transfer, it is advisable to use methods of deep learning [18, 29]. Let's stop on transfer training because it allows to adapt the selected model, for example, on an ImageNet or *lasagne* data sets [9, 31–33]. In transfer learning, a model trained on one dataset adapts to another dataset. The main assumption about transfer training is that the general characteristics studied on a sufficiently large data set can be divided between seemingly disparate data sets [32, 33]. This portability of studied datasets is a unique benefit of deep learning, which makes itself useful in various tasks with small datasets. The learning algorithm is presented in Fig. 11: extraction of fixed functions and fine tuning [34].

A method of extracting fixed features is the process of removing fully connected layers from a network, a pre-trained network, while maintaining the remaining network, which consists of a series of convolutional and combining layers, called a convolutional base, as an extractor of fixed features. In this case, the machine learning classifier adds random weights on top of the extractor of fixed functions in ordinary fully connected convolutional neural networks. As a result, training is limited to the added classifier for a given dataset.

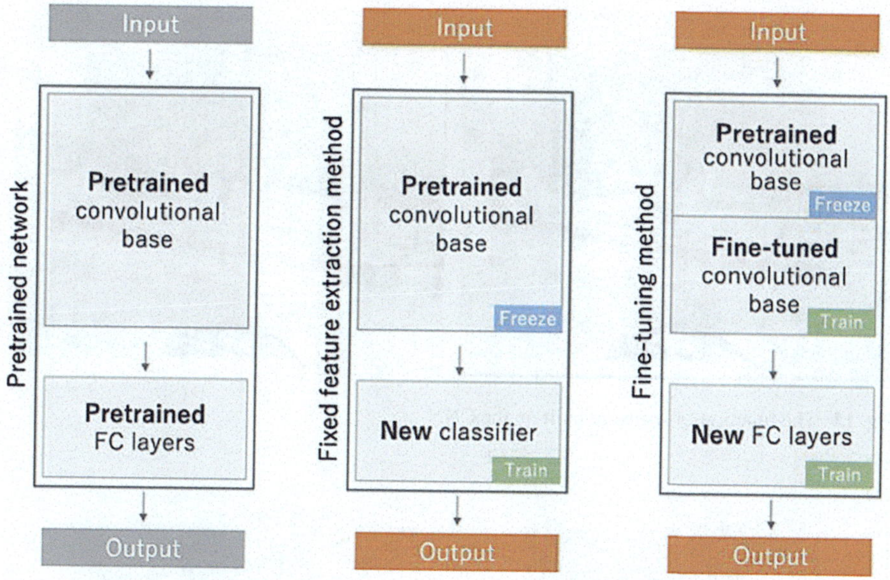

Fig. 11 Transfer training

The fine tuning method is not only to replace fully connected layers of a pre-prepared model with a new set of fully connected layers to retrain a given dataset, but also to fine tune all or part of the cores in a pre-trained convolutional basis using reverse propagation (Figs. 6, 7 and 10).

All layers of the convolutional layer can be fine-tuned as an alternative, and some earlier layers can be fixed by fine-tuning the remaining deeper layers [18, 35, 36].

7 Example of Application of a Convolutional Neural Network

In the work, the CNN was chosen with one input layer, two convolutional and two layers of subsampling. Dimension of an entrance layer $1 \times 28 \times 28$, the first convolutional layer $32 \times 24 \times 24$, the first layer of subsample $32 \times 12 \times 12$. The given layers consist of 10 feature cards. The second convolution layer has a dimension of 10×10, the subsampling layer is -5×5. The network structure is shown in Fig. 9.

By training, varying, and testing the selected network, the optimal number of epochs (iterations) was determined. As a result, the loss function L amounted to 14–15, and the recognition rate at various objects ranged from 56 to 97. The database of program objects includes about 80 objects, including people, animals, plants, automobile transport, etc. (Figures 12 and 13).

Fig. 12 The function of software built on the CNN

Fig. 13 The function of software built on the CNN

At the CNN output, the probabilities of matching the object in the database are obtained. A frame is selected with maximum probability and is taken as the final one at the moment. The number of errors was 3–12%. In each frame, the recognized object is highlighted by a rectangular frame, above which the coincidence or recognition coefficient is indicated (Figs. 12 and 13).

8 Conclusion

In this article proposed using of nonparametric analysis algorithms in comparison with convolutional neural network algorithms, which showed good results and significantly smaller data size. Regardless of the number of objects that appear in the frame, they were all recognized and the class specified.

However, despite the results obtained, the recognition coefficient for some classes was low. Therefore, need to pay attention to the process of normalizing the input data for training and verification images.

References

1. R. Duda, P. Hart, Pattern recognition and scene analysis. in Translation from English.ed. by G.G. Vajeshtejnv, A.M. Vaskovski, V.L. Stefanyuk (MIR Publishing House, Moskow, 1976), p. 509
2. V.V. Mokeev, S.V. Tomilov, On the solution of the problem of small sample size when using linear discriminant analysis in face recognition problems. Bus. Inform. **1**(23), 37–43 (2013)
3. A.V. Lapko, S.V. Chencov, V.A. Lapko, Non-parametric patterns of pattern recognition in small samples. Autometry (6), 105–113 (1999)
4. Authorized translation from the English language edition, entitled DIGITAL IMAGE PROCESSING: International Version, 3rd edn. ed by C. Gonzalez Rafael, E. Woods Richard, published by (Pearson Education, Prentice Hall, 2008). Copyright ©2008 by Pearson Education, Inc ISBN: 0132345633
5. V.V. Voronin, V.I. Marchuk-Shakhts, Methods and algorithms of image recovery in conditions of incomplete a priori information: monograph. VPO "JURGUES," (2010), p. 89
6. E. Parzen, On estimation of a probability density function and mode. Annal. Math. Statistics. **33**, 1065–1076 (1962)
7. L. Bertinetto, J.F. Henriques, J. Valmadre, P. Torr, A. Vedaldi, Learning feed-forward one-shot learners, in *Advances in Neural Information Processing Systems 29: Annual Conference on Neural Information Processing Systems 2016* (2016), pp. 523–531
8. L. Varlamova Lyudmila, Non-parametric classification methods in image recognition. J. Xi'an Univ. Arch. Technol. **XI**(XII), pp. 1494–1498 (2019). https://doi.org/20.19001.JAT.2020.XI.I12.20.1891
9. O. Russakovsky, J. Deng, H. Su, J. Krause, S. Satheesh, S. Ma, Z. Huang, A. Karpathy, A. Khosla, M. Bernstein, A.C. Berg, F.F. Li, Imagenet large scale visual recognition challenge. Int. J. Comput. Vis. **115**(3), 211–252. https://doi.org/10.1007/s11263-015-0816-y
10. V. Katkovnik, Nonparametric density estimation with adaptive varying window size, in *Signal Processing Laboratory* (Tampere University of Technology, 2000). http://www2.mdanderson.org/app/ilya/Publications/europtoparzen.pdf
11. A.J. Izenman, Recent developments in nonparametric density estimation. J. Am. Statistical Assoc. **86**, pp 205–224 (1991)
12. B. Jeon, D. Landgrebe, Fast parzen density estimation using clustering-based branch and bound. IEEE Trans. Pattern Anal. Mach. Intell. **16**(9), 950–954 (1994)
13. V. Lempitsky, Convolutional neural network. Available at: https://postnauka.ru/video/66872
14. K. Fukushima, Neocognitron: a self-organizing neural network model for a mechanism of pattern recognition unaffected by shift in position. Biol. Cybern. **36**, 193–202 (1980). https://doi.org/10.1007/BF00344251
15. D.H. Hubel, T.N. Wiesel, Receptive fields and functional architecture of monkey striate cortex. J. Physiol. **195**, 215–243 (1968). https://doi.org/10.1113/jphysiol.1968.sp008455
16. S. Russell, P. Norvig, in Artificial Intelligence: A Modern Approach, 2nd edn. (Williams Publishing House, Chicago, 2006), 1408p
17. N. Qian, On the momentum term in gradient descent learning algorithms. Neural Netw. **12**, 145–151 (1999). https://doi.org/10.1016/S0893-6080(98)00116-6
18. D.P. Kingma, J. Ba, Adam: a method for stochastic optimization (2014). Available online at: https://arxiv.org/pdf/1412.6980.pdf
19. S. Ruder, An overview of gradient descent optimization algorithms (2016). Available online at: https://arxiv.org/abs/1609.04747

20. Y. Bengio, Y. LeCun, D. Henderson, Globally trained handwritten word recognizer using spatial representation, space displacement neural networks and hidden Markov models, in *Advances in Neural Information Processing Systems*, vol. 6 (Morgan Kaufmann, San Mateo CA, 1994)
21. K. Clark, B. Vendt, K. Smith et al., The cancer imaging archive (TCIA): maintaining and operating a public information repository. J. Digit. Imaging **26**, 1045–1057 (2013). https://doi.org/10.1007/s10278-013-9622-7
22. X. Wang, Y. Peng, L. Lu, Z. Lu, M. Bagheri, R.M. Summers, ChestX-ray8: hospital-scale chest X-ray database and benchmarks on weakly-supervised classification and localization of common thorax diseases. in *Proceedings of the 2017 IEEE Conference on Computer Vision and Pattern Recognition (CVPR)* (2017), pp. 3462–3471. https://doi.org/10.1109/cvpr.2017.369
23. Y. LeCun, Y. Bengio, G. Hinton, Deep learning. Nature **521**, 436–444 (2015)
24. A. Marakhimov, K. Khudaybergenov, Convergence analysis of feedforward neural networks with backpropagation. Bull. Natl. Univ. Uzb.: Math. Nat. Sci. **2**(2), Article 1 (2019). Available at: https://www.uzjournals.edu.uz/mns_nuu/vol2/iss2/1
25. C. Tomasi, R. Manduchi, Bilateral filtering for gray and color images, in *Proceedings of the IEEE International Conference on Computer Vision* (1998), pp. 839–846
26. K. Overton, T. Weymouth, A noise reducing preprocessing algorithm, in *Proceedings of the IEEE Computer Science Conf. Pattern Recognition and Image Processing* (Chicago, IL, 1979), pp. 498–507
27. C. Chui, G. Chen, in *Kalman Filtering with Real-Time Applications*, 5th edn. (Springer, Berlin, 2017), p. 245
28. A.R. Marakhimov, L.P. Varlamova, Block form of kalman filter in processing images with low resolution. Chem. Technology. Control. Manag. (3), 57–72 (2019)
29. J. Brownlee, A gentle introduction to transfer learning for deep learning (2017). Available at: https://machinelearningmastery.com/transfer-learning-for-deep-learning/
30. D.H. Lee, Pseudo-label: the simple and efficient semi-supervised learning method for deep neural networks, in *Proceedings of the ICML 2013 Workshop: Challenges in Representation Learning* (2013). Available online at: https://www.researchgate.net/publication/280581078_Pseudo-Label_The_Simple_and_Efficient_Semi-Supervised_Learning_Method_for_Deep_Neural_Networks
31. https://pythonhosted.org/nolearn/lasagne.html
32. K. He, X. Zhang, S. Ren, J. Sun, Deep residual learning for image recognition, in *Proceedings of the 2016 IEEE Conference on Computer Vision and Pattern Recognition (CVPR)* (2016). https://doi.org/10.1109/CVPR.2016.90
33. C. Szegedy, W. Liu, Y. Jia et al., Going deeper with convolutions, in *Proceedings of the 2015 IEEE Conference on Computer Vision and Pattern Recognition (CVPR)* (2015). https://doi.org/10.1109/CVPR.2015.7298594
34. G. Huang, Z. Liu, L. van der Maaten, K.Q. Weinberger, Densely connected convolutional networks, in *Proceedings of the 2017 IEEE Conference on Computer Vision and Pattern Recognition (CVPR)* (2017). https://doi.org/10.1109/CVPR.2017.243
35. M.D. Zeiler, R. Fergus, Visualizing and understanding convolutional networks. In: *Proceedings of Computer Vision – ECCV 2014*, vol. 8689, pp. 818–833 (2014)
36. J. Yosinski, J. Clune, Y. Bengio, H. Lipson, How transferable are features in deep neural networks? arXiv (2014). Available online at: https://scholar.google.com/citations?user=gxL1qj8AAAAJ&hl=ru
37. R. Yamashita, M. Nishio, R.K. Do, K. Togashi, Convolutional neural networks: an overview and application in radiology. Insights Imaging **9**(4): 611–629 (2018). Published online 2018 Jun 22. https://doi.org/10.1007/s13244-018-0639-9. Available online at: https://www.ncbi.nlm.nih.gov/pmc/articles/PMC6108980/
38. A. Marakhimov, K. Khudaybergenov, "Neuro-fuzzy identification of nonlinear dependencies". Bull. Natl. Univ. Uzb.: Math. Nat. Sci. **1**(3), Article 1 (2018). Available at: https://www.uzjournals.edu.uz/mns_nuu/vol1/iss3/1

39. L.P. Varlamova, K.N. Salakhova, R.S. Tillakhodzhaeva, Neural network approach in the task of data processing. Young Sci. **202** (Part 1), 99–101 (2018)
40. A.R. Marakhimov, K.K. Khudaybergenov, A fuzzy MLP approach for identification of nonlinear systems. Contemporary Mathematics. Fundam. Dir. **65**(1), 44–53 (2019)

A Machine Learning-Based Framework for Efficient LTE Downlink Throughput

Nihal H. Mohammed, Heba Nashaat, Salah M. Abdel-Mageid, and Rawia Y. Rizk

Abstract Mobile Network Operator (MNO) provides Quality of Services (QoS) for different traffic types. It requires configuration and adaptation of networks, which is time-consuming due to the growing numbers of mobile users and nodes. The objective of this chapter is to investigate and predict traffic patterns in order to reduce the manual work of the MNO. Machine learning (ML) algorithms have used as necessary tools to analyze traffic and improve network efficiency. In this chapter, a ML-based framework is used to analyze and predict traffic flow for real 4G/LTE-A mobile networks. In the proposed framework, a clustering model is used to identify the cells which have the same traffic patterns and analyze each cluster's performance, and then, a traffic predicting algorithm is proposed to enhance the cluster performance based on downlink (DL) throughput in the cells or on edge. The experimental results can be used to balance the traffic load and optimize resource utilization under the channel conditions.

Keywords 4G/LTE-A · KPIs · Machine learning · Traffic load balance · Traffic predication

N. H. Mohammed · H. Nashaat · R. Y. Rizk (✉)
Electrical Engineering Department, Port Said University, Port Said 42523, Egypt
e-mail: r.rizk@eng.psu.edu.eg

N. H. Mohammed
e-mail: nihalhossny@eng.psu.edu.eg

H. Nashaat
e-mail: hebanashaat@eng.psu.edu.eg

S. M. Abdel-Mageid
Computer Engineering Department, Collage of Computer Science and Engineering, Taibah University, Medina, Saudi Arabia
e-mail: sabdelmageid@taibahu.edu.sa

© The Editor(s) (if applicable) and The Author(s), under exclusive license to Springer Nature Switzerland AG 2021

A. E. Hassanien et al. (eds.), *Artificial Intelligence for Sustainable Development: Theory, Practice and Future Applications*, Studies in Computational Intelligence 912,
https://doi.org/10.1007/978-3-030-51920-9_10

1 Introduction

Improving network performance and Quality of Services (QoS) satisfaction are the two greatest challenges in 4G/LTE- networks. Key Performance Indicators (KPIs) are used to observe and enhance network performance. The KPIs indicate service quality and accomplish resource utilization. KPI could be based upon network statistics, user drive testing, or a combination of both. The KPIs are considered as an indication of the performance of the crowning periods. With unacceptable network performance, it is highly desirable to search for throughput enhancing techniques, particularly on the downlink traffic of wireless systems [1–4].

Recently, Machine learning (ML) is used to analyze and optimize the performance in 4G and 5G wireless systems. Some studies show that it is possible to deploy ML algorithms in cellular networks effectively. Evaluation of the gains of a data-driven approach with real large-scale network datasets is studied in [5]. In [6], a comprehensive strategy of using big data and ML algorithms to cluster and forecast traffic behaviors of 5G cells is presented. This strategy uses a traffic forecasting model for each cluster using various ML algorithms. The Self-Optimized Network (SON) functions configuration is updated in [7] such that the SON functions contribute better toward achieving the KPI target. The evaluation is done on a real data set, which shows that the overall network performance is improved by including SON management. Also, realistic KPIs are used to study the impact of several SON function combinations on network performance; eight distinct cell classes have been considered enabling a more detailed view of the network performance [8]. Moreover, ML is used to predict traffic flow for many real-world applications. This prediction can be considered as a helpful method in improving network performance [9].

In this chapter, a real 4G mobile network data set is collected hourly for three weeks in a heavy traffic compound in Egypt to analyze user QoS limitations. This limitation may be corresponding to system resources and traffic load. A ML-based framework is introduced to analyze the Real 4G/LTE-A mobile network, cluster, predict, and enhance the DownLink (DL) throughput of a considerable number of cells. It uses visualization, dimension reduction, and clustering algorithms to improve user DL throughput in the cell or on edge and balances the load. Then, an approach of using ML algorithms to effectively cluster and predict hourly traffic of the considered Real 4G/LTE-A mobile network is used. The mobile network has three bands (L900, L1800, and L2100). Spectrum efficiency is collected hourly for these sites to analyze user QoS limitations.

The rest of this chapter is organized as follows: Sect. 2 describes KPIs types and their usage. The ML algorithms used in the proposed framework are introduced in Sect. 3. Section 4 presents the ML-based framework for efficient DL throughput. Experimental results and discussion are introduced in Sect. 5. Finally, Sect. 6 presents the main conclusion and future work.

2 4G/LTE Network KPIs

The main purpose of Radio Access Network (RAN) is to check the performance of the network. Post-processing usually checks, monitors, and optimizes KPIs values and counters to enhance the QoS or to get better usage of network resources [10, 11].

KPIs are categorized to radio network KPIs (from 1 to 6) and service KPIs (7 and 8) [12]:

1. Accessibility KPI measurements assist the network operator with information about whether the services requested by a user can be accessed with specified levels of tolerance in some given operating conditions.
2. Retainability KPIs measure the capacity of systems to endure consistent reuse and perform its intended functions. Call drop and call setup measure this category.
3. Mobility KPIs are used to measure the performance of a network that can manage the movement of users and keep the attachment with a network such as a handover. The measurements include both intra and inter radio access technology (RAT) and frequency success rate (SR) handover (HO).
4. Availability KPIs measure the percentage of time that a cell is available. A cell is available when the eNB can provide radio bearer services.
5. Utilization KPIs are used to measure the utilization of network and distribution of resources according to demands. It consists of uplink (UL) resource block (RB) utilization rate and downlink (DL) RB utilization rate.
6. Traffic KPIs are used to measure the traffic volumes on LTE RAN. Traffic KPIs are categorized based on the type of traffic: radio bearers, downlink traffic volume, and uplink traffic volume.
7. Integrity KPIs are used to measure the benefits introduced by networks to its user. This indicates the impact of eNBs on the service quality provided to the user, such as what is the throughput for cell and user and latency, which users are served.
8. Latency KPIs measure the amount of service latency for the user or the amount of latency to access a service.

In our research, three types of KPIs are analyzed to notice Cell Edge User (CEU) throughput and its relation with traffic load among bands. These are Integrity KPIs, utilization KPIs, and traffic KPIs.

3 ML Algorithms Used in the Framework

In this section, ML algorithms used in the proposed framework are described. There are three ML algorithms: Dimension reduction, K-means clustering, and Linear regression with polynomial features.

3.1 Dimension Reduction Algorithm

Principle component analysis (PCA) helps us to identify patterns in data based on the correlation between features. PCA aims to find the directions of maximum variance in high-dimensional data and projects it onto a new subspace with equal or fewer dimensions than the original one. It maps the data to a different variance-based arranged coordinate system. The points in the new coordinate system are arranged in descending. That transformation is done as an orthogonal linear mapping by analyzing the eigenvectors and eigenvalues. Eigenvectors of a dataset are computed then gather them in a projection matrix. Each of these eigenvectors is associated with an eigenvalue, which can be interpreted as the magnitude of the corresponding eigenvector. When the eigenvalues have a larger magnitude than others, the dataset is reduced to a smaller dimensional by dropping the less valuable data. Therefore, a d-dimensional dataset is reduced by projecting it onto an m-dimensional subspace (where d < n) to increase the computational efficiency [13].

3.2 K-Means Clustering Algorithm

Clustering is implemented in a way to configure the cells into groups. The K-means clustering algorithm is used for unlabeled datasets for more visualization and clarification. The K-means clustering algorithm is widely used because of its simplicity and fast convergence. However, the K-value of clustering needs to be given in advance, and the choice of K-value directly affects the convergence result. The initial centroid of each class is determined by using the distance as the metric. The elbow method is used to determine the number of clusters. It is assumed that $U = \{u_1, u_2, u_3, \ldots, u_n\}$ is the set of cells and $V = \{v_1, v_2, \ldots, v_k\}$ is the set of centers. To cluster the cells into K clusters, K number of centroids is taken initially at random places. The path-loss from each cell is calculated to all centroids. Then, each cell is assigned to the cluster whose path-loss from cluster center is the minimum of all cluster centers. The new cluster centroid is recalculate using the next formula [14]:

$$v_i = \frac{1}{c_i} \sum_{j=1}^{c_i} u_j \tag{1}$$

where, c_i represents the number of cells in ith cluster.

3.3 Linear Regression Algorithm with Polynomial Features

Linear Regression is a ML algorithm based on supervised learning. It performs a target prediction value based on independent variables. It is mostly used for finding out the relationship between variables and forecasting. Different regression models differ based on the kind of connection between the dependent and independent variables [15]. The number of independent variables being used have to be considered. Linear regression performs the task to predict a dependent variable value (y) based on a given independent variable (x). So, this regression technique finds out a linear relationship between x (input) and y (output). Hence, the name is Linear Regression. Hypothesis function for Linear Regression is:

$$y = \theta_1 + \theta_1 x \tag{2}$$

In the standard linear regression case, a model for three degree-dimensional data which is called polynomial features using linear regression have to be used as in the case of our framework:

$$y(\theta, x) = \theta_1 + \theta_1 x_1 + \theta_2 x_2 + \theta_3 x_3 \tag{3}$$

Some parameters can be used to evaluate the success degree of the prediction process [6, 15]. The mean absolute error (E_{ma}) can be formulated as:

$$E_{ma} = \sum_{i=1}^{\pi} \| P_{iIa} - r_{iIa} \| \tag{4}$$

where P_{iIa} is the predicted value and r_{iIa} is the real value and n is the total number of test points. The root mean square error (E_{rms}) can be calculated as:

$$E_{ma} = \text{sqrt}\left(\frac{1}{n} \sum_{i=1}^{\pi} (P_{iIa} - r_{iIa})^2 \right) \tag{5}$$

The coefficient of determination (R^2) which shows how well the regression model fits the data. It's better to reach the value of one and the correlation coefficient (R) also better to reach the value of one. It can be formulated as:

$$R^2 = 1 - \sum_{i=1}^{n} (P_{iIa} - r_{iIa})^2) / \sum_{i=1}^{n} (r_{iIa} - \frac{1}{n} \times \sum_{i=1}^{n} r_{iIa})^2) \tag{6}$$

4 A ML-Based Framework for Efficient LTE Downlink Throughput

Figure 1 shows the ML-based framework for efficient downlink throughput. The proposed structure comprises three phases. These phases investigate the network

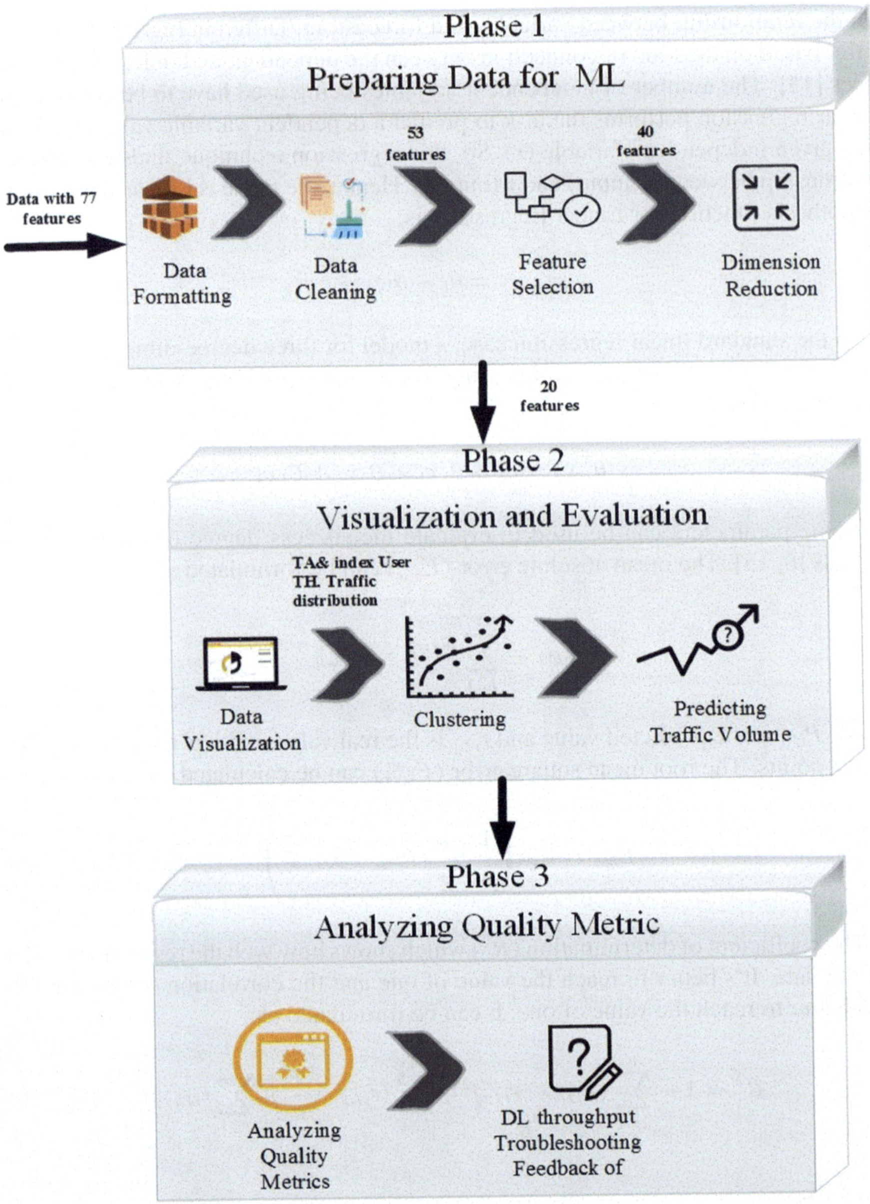

Fig. 1 The main phases of the ML-based framework for efficient LTE downlink throughput

performance retrieving management information and managing the performance of networks. These phases are described as follows.

4.1 Phase 1: Preparing Data for ML

A data set used in behavior evaluation is based on monitoring of logs generated by 104 eNBs (312 cells). The selected area is "Tagamoaa Elawal" in Egypt, which is a massive traffic area. It has more than 4,743,470 user elements (UEs) per day. The base stations in the dataset belong to a 4G LTE-A, 2 × 2 Multiple Input Multiple Output (MIMO) deployment with three bands of the three frequencies that exist in Egypt applied in each cell: 2100, 1800, and 900 MHz with 10, 10, and 5 MHz Bandwidth (BW); respectively assigned to each band. It represents the most advanced cellular technology commercially in Egypt deployed on a large scale. Figure 2 is a screen shoot of a part from Data log file. This data is collected hourly for three weeks as a 104 Megabyte log file with more than 77 features and 259,224-time rows.

There are four steps to prepare data for ML algorithms: Formatting, data cleaning, features selection, and dimension reduction.

4.1.1 Formatting

ML algorithms can acquire their knowledge by extracting patterns from raw data. This capability allows them to perform tasks that are not complicated to humans, but require a more subject and intuitive knowledge and, therefore, are not easily described using a set of logical rules. Log files collected from the network optimizer should be entered into the machine in excel or CSV file format.

Time	hours	eNodeB I	freq	Cell Name	Traffic DL volun	Cell DL Avg. TH.	User DL Avg.TH.	Avg. suitable selection MSC	Avg. PRB utiliza	CQI 0-4 percent	CQI 5-9
9/17/19	12:00:00 AM	95	L1200	1	2738.7715	16.5842	13.3816	16.7104	37.0003	6.5843	
9/17/19	12:00:00 AM	95	L1800	30	1006.0637	7.1678	6.6892	19.5284	34.3149	3.6509	
9/17/19	12:00:00 AM	15	L900	60	238.6368	6.0362	5.5831	13.2766	10.8028	36.8611	
9/17/19	12:00:00 AM	15	L1200	1	529.2536	19.6093	25.0732	16.2551	7.601	6.073	
9/17/19	12:00:00 AM	46	L900	62	383.3166	5.882	5.3922	15.3544	17.1316	5.0202	
9/17/19	12:00:00 AM	53	L1200	2	478.993	11.551	14.9269	17.1675	9.7956	2.5278	
9/17/19	12:00:00 AM	53	L900	8	461.0029	7.9433	6.9815	15.5843	25.7549	3.5771	
9/17/19	12:00:00 AM	80	L1200	3	1409.1239	19.2761	20.9471	17.4641	22.6624	7.8298	
9/17/19	12:00:00 AM	80	L900	61	16.0372	8.6492	8.2305	13.9763	6.4711	14.8651	
9/17/19	12:00:00 AM	88	L1200	1	2306.0453	13.5024	9.7957	14.5295	39.168	10.54	
9/17/19	12:00:00 AM	67	L900	62	16.1724	6.0967	7.7537	8.5198	4.818	5.9505	
9/17/19	12:00:00 AM	101	L900	62	976.3247	9.8299	8.3084	17.8025	40.8092	9.327	
9/17/19	12:00:00 AM	101	L1200	2	2715.8083	17.2417	14.3936	18.1271	37.9691	10.9465	
9/17/19	12:00:00 AM	37	L900	61	16.9944	3.1745	3.2556	8.6478	5.923	12.8049	
9/17/19	12:00:00 AM	59	L900	60	116.7249	7.3081	5.6572	17.3114	8.5678	10.1791	
9/17/19	12:00:00 AM	60	L1200	1	2120.6457	14.4293	14.9702	19.2648	29.4839	10.3319	
9/17/19	12:00:00 AM	60	L1800	31	19.916	5.1894	5.2635	16.4081	5.5739	3.9534	
9/17/19	12:00:00 AM	77	L900	61	0.1015	1.072	1.0739	11.7984	4.1346	21.0459	
9/17/19	12:00:00 AM	77	L1200	3	1033.635	13.8086	12.4995	15.0793	21.5168	24.2357	
9/17/19	12:00:00 AM	95	L900	60	542.0848	6.9984	6.8608	17.8654	24.7177	3.2184	
9/17/19	12:00:00 AM	95	L1800	33	363.4042	10.6522	10.4959	17.6709	15.0457	7.2844	
9/17/19	12:00:00 AM	95	L1200	2	479.1888	10.5589	12.2267	15.1364	13.753	20.5621	
9/17/19	12:00:00 AM	98	L900	61	860.6515	8.4121	7.3184	16.7137	40.5508	4.1177	

Fig. 2 Data log file screen shoot

4.1.2 Data Cleaning

Pandas data frame [16] provides a tool to read data from a wide variety of sources. Either Jupiter notebook or Goggle Collab is used for that step. Data cleaning and preparation is a critical step in any ML process. Cleaning data is to remove any null or zero value and its corresponding time row using python codes to avoid any mistake during ML algorithms later. After the cleaning step in our framework, data is reduced to 53 features and 222,534-time lines.

4.1.3 Features Selection

This step aims to select and exclude features. Measured features after data cleaning are summarized in Table 1. It considers the necessary parameters for the 4G/LTE-A network, such as DL traffic volume, average throughput distributed for a specific cell, average throughput for users, maximum and average number of UEs in a particular cell, and network utilization. Utilization physical resource block (PRB) can be considered as PRB percentage, which represents the percentage of resources distribution of each band according to demands and available frequencies BW. The scheduler should take into account the demand BW and load of traffic when assigning to the band. Therefore the scheduler doesn't allocate PRBs to users who are already satisfied with their current allocation. Moreover, these resources are allocated to other users who need them according to band load and the available BW.

The Channel Quality Indicators (CQIs) have features number from 6 to 8. It represents the percentage of users in three categories of CQI; lowest, good and best, as in Table 1. The features with numbers from 13 to 19 represent the indexes with Timing Advance (TA). It can be considered as an indication of the coverage of each cell. The TA is located on each index, which is a negative offset. This offset is necessary to ensure that the downlink and uplink sub frames are synchronized at the eNB [17]. The used Modulation and Coding Scheme (MCS) (numbered in Table 1 from 21 to 52) is also taken into account. MCS depends on radio link quality and defines how many useful bits can be transmitted per Resource Element (RE). UE can use the MCS index (IMCS) from 0–31 to determine the modulation order (Qm), and each IMCS is mapped to transport block size (TBS) index to assess the number of physical resource blocks. In LTE, there are the following modulations supported: QPSK, 16QAM, 64QAM, and 256QAM, and to indicate if the most proper MCS level is chosen to use, an average MCS (feature number 4 in Table 1) is used. It takes the range from 1 to 30. It represents a lousy choice for MCS when it is fewer than eight, from 10 to 20 it is good and excellent MCS when it is above 20. Both MCS and CQI are used as an indication of radio condition [18].

By applying the sklearn's feature selection module [19] to the data set of 4G/LTE-A network, all features haven't zero difference, and there are no features with the same value in all columns. Therefore no features are removed when sklearn's feature selection module is used. The output of correlation code in python is applied on these 53 features. The closest the value to 1 is the highest correlation between two features,

Table 1 Used features after cleaning

No.	Feature name	Description	No.	Feature name	Description
*0	Traffic DL volume	Measure DL traffic volume on LTE radio access NW	*27	MCS.6	No. of users have Modulation (QPSK) and index TBS (6)
*1	Cell Name	Name of the cell	*28	MCS.7	No. of users have Modulation (QPSK) and index TBS (7)
*2	Cell DL Avg. TH.	Average throughput distributed for specific cell	*29	MCS.8	No. of users have Modulation (QPSK) and index TBS (8)
*3	User DL Avg.TH.	Average throughput for users on specific cell	*30	MCS.9	No. of users have Modulation (QPSK) and index TBS (9)
*4	Avg. suitable selection MSC	An indication for Efficient Selection MSC on specific cell	*31	MCS.10	No. of users have Modulation (16QAM) and index TBS (9)
*5	Avg. PRB utilization	Measure the system capability to meet the traffic demand	*32	MCS.11	No. of users have Modulation (16QAM) and index TBS (10)
*6	CQI 0-4 percentage	Percentage of users have Channel quality indicator QPSK (lowest)	*33	MCS.12	No. of users have Modulation (16QAM) and index TBS (11)
*7	CQI 5-9 percentage	Percentage of users have channel quality indicator 16QAM (good)	*34	MCS.13	No. of users have Modulation (16QAM) and index TBS (12)
*8	CQI 10-15 percentage	Percentage of users have Channel quality indicator level 64QAM (best)	*35	MCS.14	No. of users have Modulation (16QAM) and index TBS (13)
*9	CEU cell DL Avg. TH.	Avg. predicted DL throughput Cell edge user for specific cell	*36	MCS.15	No. of users have Modulation (16QAM) and index TBS(14)
*10	CEU user DL Avg. TH.	Avg. throughput for users on an edge	*37	MCS.16	No. of users have Modulation (16QAM) and index TBS (15)
*11	Avg. UE No.	Avg. No. of UE in a specific cell	*38	MCS.17	No. of users have Modulation (64QAM) and index TBS (15)

(continued)

Table 1 (continued)

No.	Feature name	Description	No.	Feature name	Description
*12	Max UE No.	Max. No. of UE in a specific cell	*39	MCS.18	No. of users have Modulation (64QAM) and index TBS (16)
*13	TA & Index0	eNB coverage 39 m and TA is 0.5 m	*40	MCS.19	No. of users have Modulation (64QAM) and index TBS (17)
*14	TA &Index1	eNB coverage 195 m and TA is 2.5 m	41	MCS.20	No. of users have Modulation (64QAM) and index TBS (18)
*15	TA & Index2	eNB coverage 429 m and TA is 5.5 m	42	MCS.21	No. of users have Modulation (64QAM) and index TBS (19)
*16	TA& Index3	eNB coverage 819 m and TA is 10.5 m	43	MCS.22	No. of users have Modulation (64QAM) and index TBS (19)
*17	TA &Index4	eNB coverage 1521 m and TA is 19.5 m	44	MCS.23	No. of users have Modulation (64QAM) and index TBS (20)
*18	TA & Index5	eNB coverage 2769 m and TA is 35.5 m	45	MCS.24	No. of users have Modulation (64QAM) and index TBS (21)
*19	TA &Index6	eNB coverage 5109 m and TA is 65.5 m	46	MCS.25	No. of users have Modulation (64QAM) and index TBS (22)
*20	L.PRB.TM2	Capacity monitoring by PRB	47	MCS.26	No. of users have Modulation (64QAM) and index TBS (23)
*21	MCS.0	No. of users have Modulation (QPSK) and index TBS (0)	48	MCS.27	No. of users have Modulation (64QAM) and index TBS (24)

(continued)

Table 1 (continued)

No.	Feature name	Description	No.	Feature name	Description
*22	MCS.1	No. of users have Modulation (QPSK) and index TBS (1)	49	MCS.28	No. of users have Modulation (64QAM) and index TBS (25)
*23	MCS.2	No. of users have Modulation (QPSK) and index TBS (2)	50	MCS.29	No. of users have Modulation (QPSK) and index TBS reserved
*24	MCS.3	No. of users have Modulation (QPSK) and index TBS (3)	51	MCS.30	No. of users have Modulation (16QAM) and index TBS reserved
*25	MCS.4	No. of users have Modulation (QPSK) and index TBS (4)	52	MCS.31	No. of users have Modulation (64QAM) and index TBS reserved
*26	MCS.5	No. of users have Modulation (QPSK) and index TBS (5)			

as in Fig. 3. It is clear from the figure that, a lot of features are highly correlated and redundant. Univariate feature selection works by selecting the best features based on univariate statistical tests [20]. Sklearn's SelectKBest [20] is used to choose some features to keep. This method uses statistical analysis to select features having the highest correlation to the target (our target here is user DL throughput in the cell and on edge), it is the top 40 features (denoted by * in Table 1).

4.1.4 Dimension Reduction

Many features are highly correlated (redundant) where it could be eliminated. Threrfore, dimensionality reduction transforms features to a lower dimension. PCA is a the reduction technique used to projects the data into a lower-dimensional space. Features are reduced by PCA to the first 20 features in Table 1 where they are less and medium correlated as in Fig. 4.

4.2 Phase 2: Data Visualization and Evaluation

This phase comprises three main stages. First, the data is visualized in order to provide an accessible way to see and understand trends, outliers, and patterns in it.

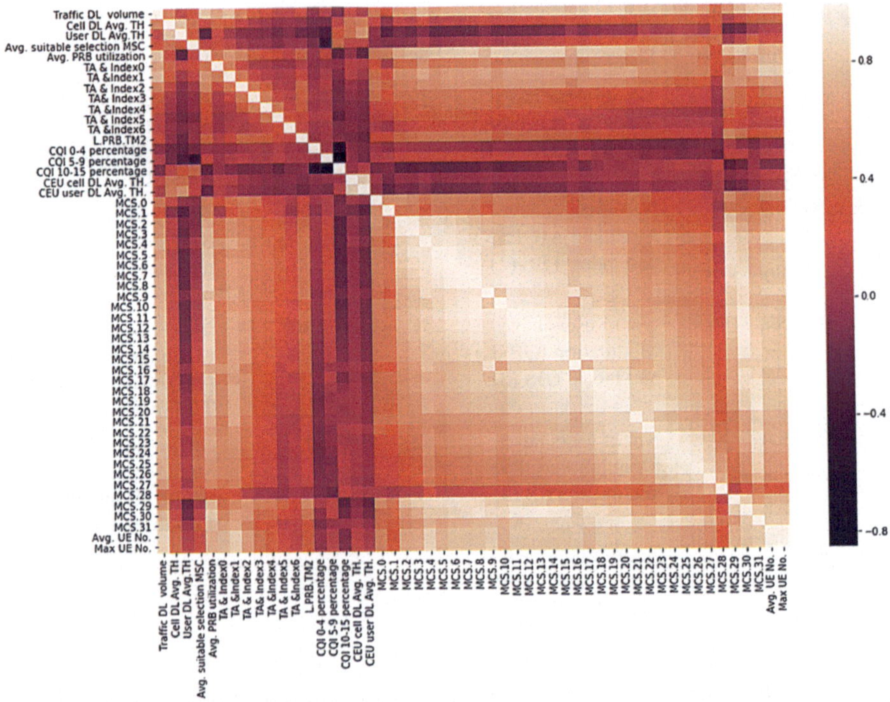

Fig. 3 Data features correlations

Then, ML-based traffic clustering and prediction algorithms are used to predict the traffic conditions for an upcoming traffic.

4.2.1 Visualization

Distribution of traffic, User DL throughput, and Indexes and TA are plotted to understand data characterization:

1. Distribution of traffic in three bands: Table 2 shows the traffic density of three bands in Megabyte (MB). The L2100 band has a huge traffic density, and most traffic is congested in its cells. Therefore, load balancing must be applied to transfer load from overloaded cells to the neighboring cells with free resources for more balanced load distribution to maintain appropriate end-user experience and performance.
2. A Scatter plot in Fig. 5 is used to represent the distribution between DL throughput, traffic volume and PRB utilization. An increase in usage of PRB and traffic causes a decrease in DL throughput for UEs. Also, average DL throughput for CEUs is plotted with average UEs number. It is found that, the increase of the number of UEs may lead to a decrease in CEU's throughput and vice versa with

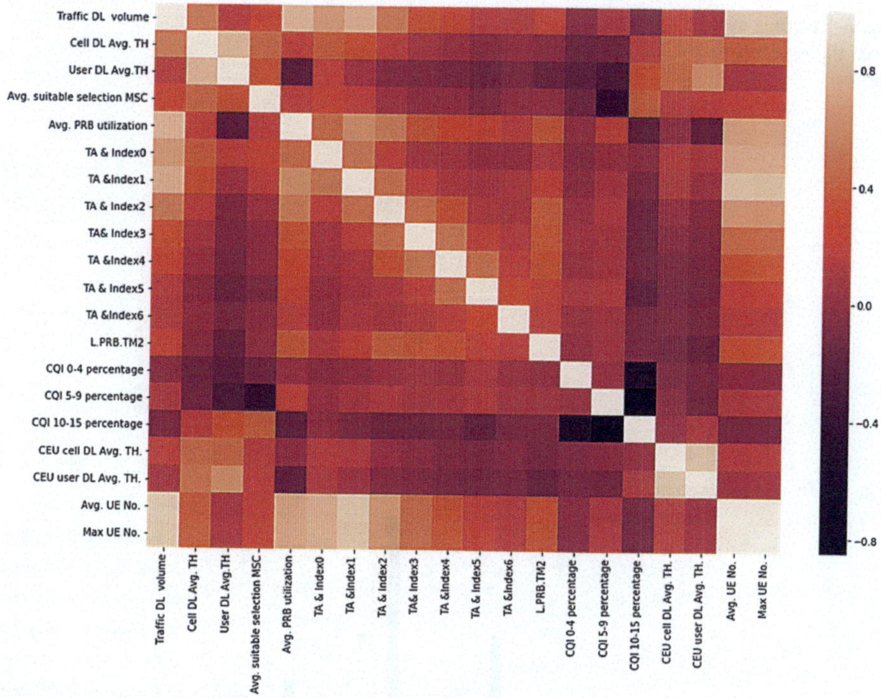

Fig. 4 Reduced features correlations

Table 2 Traffic volume distribution in three bands

	L900	L1800	L2100
Avg. DL traffic volume (MB)	297.177527	278.716868	1215.516581

Fig. 5 User DL TH according to traffic and utilization

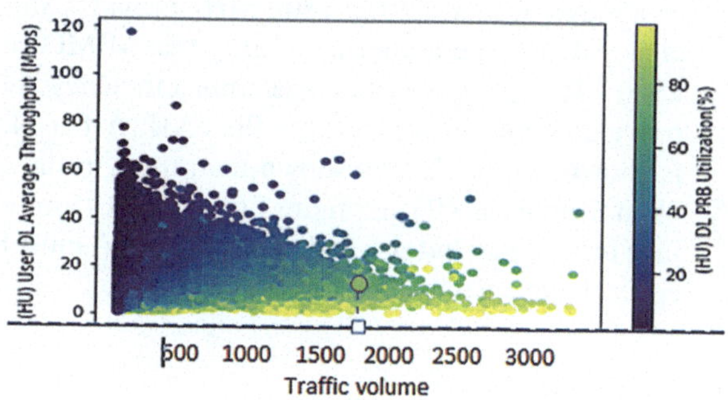

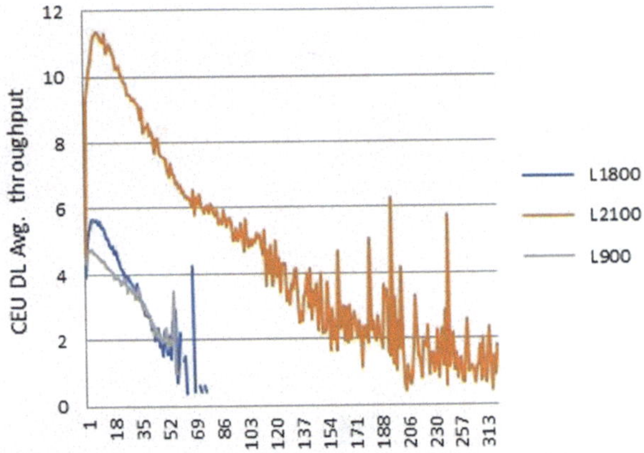

Fig. 6 Average user DL throughput versus max UEs number

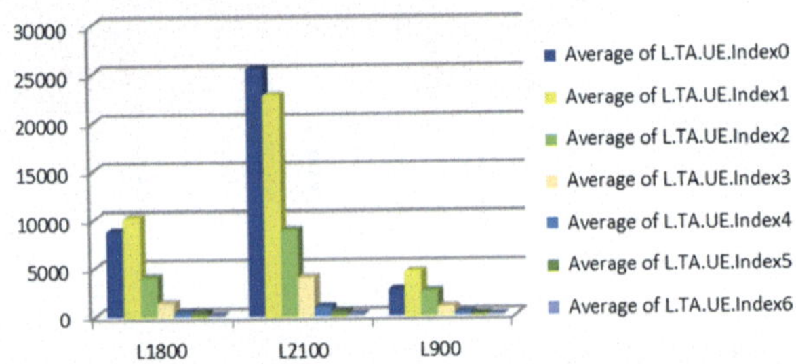

Fig. 7 TA and Indexes in three bands

the polynomial distribution. With an increasing number of UEs, DL throughputs decrease to reach zero during three bands, as in Fig. 6.

3. TA and index: There are significant differences between LTE bands in terms of performance. The 900 MHz band offers superior indoor penetration and rural coverage, while the 1800 MHz provides slightly improved spectrum efficiency due to the higher possibility that MIMO channels are available. Finally, 2100 MHz assigns better spectrum efficiency than 1800, and 900 MHz and provides better coverage near the eNB. A bar blot for the three band's index is shown in Fig. 7. It is shown that, most traffic comes from Index0 (distance 39 m from the eNB) and Index1 (distance 195 m from the eNB). However, other indexes such as Index4, Index5, and Index6 must be used with 1800 and 900 to cover the users on edge.

4.2.2 Clustering

For more visualization and clarification, the k-means clustering algorithm is used for unlabeled data. The K-means clustering algorithm is widely used because of its simplicity and fast convergence. However, the K-value of clustering needs to be given in advance, and the choice of K-value directly affects the convergence result. The initial centroid of each class is determined by using the distance as the metric.

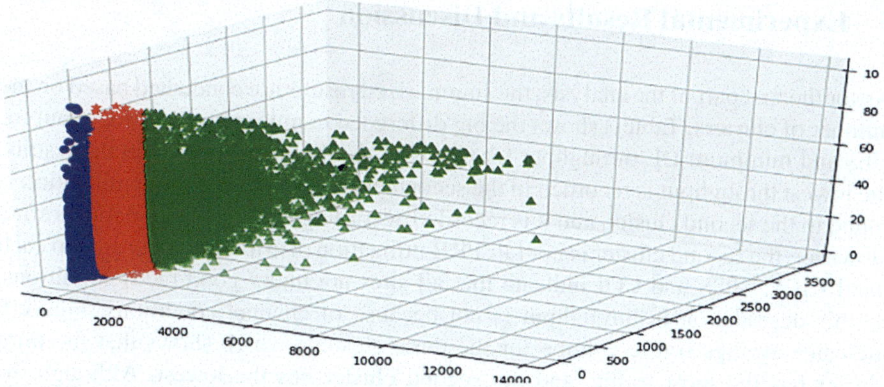

Fig. 8 Real data clustering

The elbow method is used to determine the number of clusters. Implementing the elbow method in our framework indicates that the number of clusters should be three clusters [21]. A Scatter plot in three dimensions verified the number of the clusters, as in Fig. 8.

4.2.3 Predicting Traffic Load

Traffic predicting plays a vital role in improving network performance. It can provide a behavior of future traffic of the cells in the same cluster. The traffic predicting models could be used to achieve the desired balanced throughput either in the cell or on edge or between bands in the same cluster. It could be an unsuitable traffic load and resource utilization distribution for different bands. For example, the L2100 and L1800 bands may have the most PRBs utilization percentage compared to L900. Also, this can cause degradation in DL throughput for UEs, especially during peak hours, when it has the lowest traffic volume and lowest PRB utilization. ML linear regression algorithm is used with the polynomial feature of third-degree for predicting process.

4.3 Phase 3: Analyzing Quality Metric

This phase is responsible for discovering and analyzing network performance. Twenty features that are the output of Phase 2 are used to find out the overall system performance decline. Therefore, all network problems such as throughput troubleshooting for UEs in the cell or on edge (which we focus on), traffic load balance, and PRB utilization distribution could be discovered during this phase. The analysis considers the overall DL throughput, the traffic volume, number of UEs, and network efficiency during peak hours.

5 Experimental Results and Discussion

As for the first part of the analysis, the summarized results are conducted based on the number of clusters. Table 3 shows the big difference in minimum DL throughput for UEs and minimum DL throughput for CEUs in the three clusters. As in the results, the lowest throughput is recorded in the second cluster. Also, minimum utilization is found in the second cluster, and it is recorded according to the most moderate traffic. However, the second cluster is not fair PRB utilization distribution according to each band's BW. MCS and CQI indicate that all sites are under good radio conditions, so this degradation in throughput is not because of channel conditions. Figure 9 indicates average traffic volume for the three clusters, which shows that the third cluster has the most traffic, and the second cluster has the lowest. Although the traffic volume in the three clusters are large varying, there is no much dissimilarity in average DL throughput, as shown in Fig. 10. Figure 11 shows that, the second

Table 3 Network performance for three clusters

Features		First Cluster	Second Cluster	Third Cluster
Avg. Traffic volume in Mbps	L900	1310.0152	189.68	3204.7704
	L1800	1256.146	194.039	3079.34776
	L2100	1490.93	396.053	3603.91808
Avg. UEs DL throughput in Mbps	L900	7.383044	7.99	7.76
	L1800	7.16169	7.81	9.44
	L2100	16.0265	18.55	12.60
Min. UEs DL throughput in Mbps	L900	1.2974	0.0549	1.6552
	L1800	0.4471	0.1297	2.4382
	L2100	2.9597	0.7952	0.7111
Min. CEU user DL throughput in Mbps	L900	0.0164	0.0022	0.043
	L1800	0.0174	0.0012	0.5553
	L2100	0.0462	0.0141	0.1055
Max UEs no. in each cluster	L900	62	230	97
	L1800	103	78	53
	L2100	169	150	340
PRB Utilization	L900	41.6%	12.7%	70%
	L1800	41.6%	12.8%	62.7%
	L2100	23.7%	8.8%	47.9%
Min DL user throughput during peak hours		Low (0.5–4 Mpbs)	Very low (0.2–3.8 Mpbs)	Reasonable (1–5 Mpbs)
Min. CEU DL throughput during peak hours		Low (0.5–1 Mpbs)	Very low (0.0039–0.15 Mpbs)	Low (0.5–0.3 Mpbs)

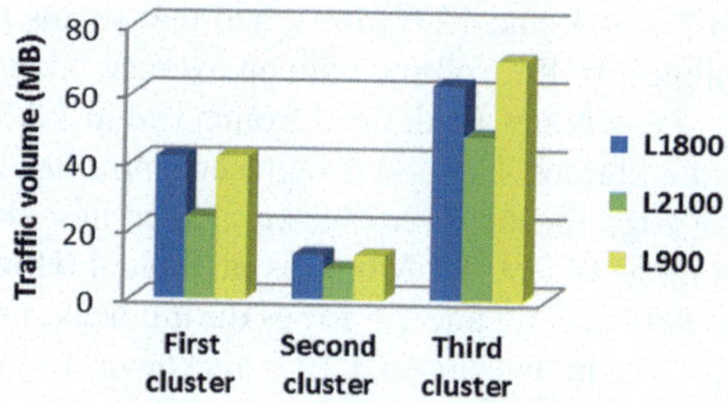

Fig. 9 Avg. Traffic volume

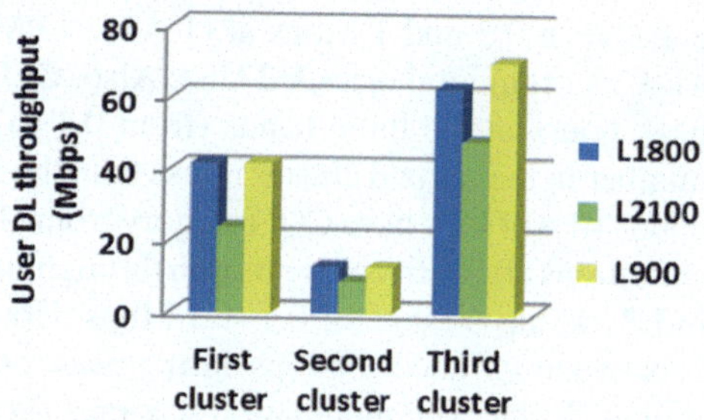

Fig. 10 Avg. DL user throughput

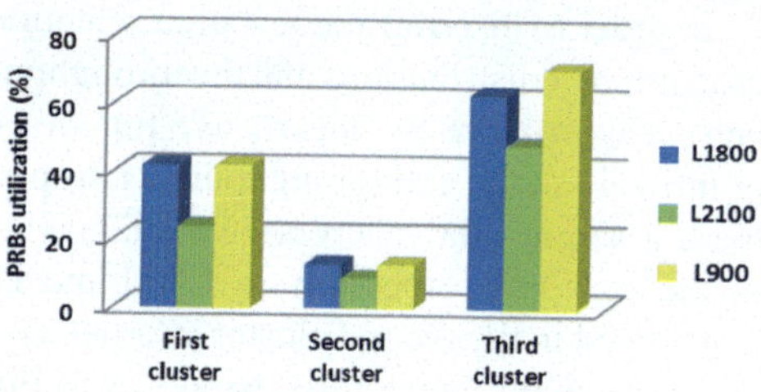

Fig. 11 Average DL PRB utilization for three clusters

cluster has the lowest traffic volume and lowest PRB utilization. DL throughput is supposed to inversely proportional to the average number of active UEs. However, the number of active UEs may be not the instinctive KPI for characterizing network load. A more common meaning of network load is the fraction of utilized PRBs. Therefore, load and user throughput are strongly related to PRB.

In order to evaluate the performance of the clusters, they are analyzed in Table 3. The number of rows associated to the first cluster is 165,645 for 103 eNB, Average DL UEs throughput is 7.3, 7.1, and 16 Mbps for L900, L1800, and L2100, respectively, and that seems to be suitable average throughput for the cells with medium average traffic volume (between 1.25 and 1.5 GB). In the second cluster, the number of output rows is a 10,953-time row for 99 eNBs. Average DL UEs throughput is 7.9, 7.8, and 18.5 Mbps for three bands, respectively. It considers low average throughput for the cells with the range 200–400 MB average traffic volume. Similar in the third cluster, the number of output rows is 43,047 rows for 100 eNB. Average DL user throughput is 7.7, 9.4, and 12.6 Mbps, and that seems to be good average throughput with the highest traffic volume with an average 3.25–3.6 GB.

Peak hours are defined from 5 PM to 1 AM according to maximum traffic volume time. Tables 4, 5 and 6 represent min throughput during these hours in the cell or on edge for the three clusters. In the first cluster, min DL throughput in L2100 has a range of 2.9–4.1 Mbps, as in Table 4. However, min DL user throughput in L900 is between 1.2 and 2.5 Mbps during peak, and that's not very bad for medium traffic volume in this cluster. CEUs also have very low DL throughput during peak hours in the three bands (from 0.9 to 0.1 Mbps). In the second cluster, max numbers of UEs are recorded at 7 PM as in Table 5. On the other hand, min DL throughput in L1800 is between 0.5 and 1 Mbps at (1 AM, 5 PM) for the number of UEs in a range of 41–93% from total recorded UEs. Also, CEUs have very low DL throughput during peak hours in the three bands (from 0.1 to 0.003 Mbps). The modulation scheme number in the second cluster is less than the first cluster. It is between 15 and 16, and about 40% of UEs have CQI category from 10 to 15, which represent acceptable radio conditions. Table 6 represents min throughput during peak hours. Min DL throughput in L2100 has a range of 0.7–1.5 Mbps. However, min DL user throughput in L900 is between 1.7 and 3.7 Mbps during peak, and that is suitable for high traffic volume in this cluster. The modulation schemes used during peak hours have about 50% of users during peak in this cluster which is the best CQI categories from 10 to 15.

In order to discover the resource selection behavior, it is important to analyze the utilization of distributions and the throughput. Utilization distribution and throughput behave approximately linearly as a function of radio utilization. For example, at 50% of utilization, the cell throughput has dropped to half. In comparison, for 75% radio load, a single user receives 25% of maximum throughput, which is not achieved in the real data, especially in L900 and L1800 bands. For example, one eNB is considered in the second cluster in order to study the effect of resource utilization on user throughput in the three bands, as in Fig. 12. It is found that, the relationship is not inverse linear proportion as it supposed to be in L900 and L1800, and it is much better in L2100. This situation could be considered as throughput troubleshooting for UEs in the cell or on edge and could be enhanced by balancing the traffic load. Therefore, the prediction of traffic load for the future period based on real traffic can improve the overall network performance. Figures 13 and 14 demonstrate that our proposed framework can obtain accurate traffic predictions in the second cluster as a case study.

To evaluate the success of the prediction process, the scatter plot is plotted between the original traffic load and the predicted traffic load to present a straight line, as in Fig. 15. It could be considered as an indication of choosing the right model. In addition, the parameters used to asses the success degree of the prediction process as in Eqs. (4–6) are calculated as $R^2 = 0.97$, $R = 0.98$, $E_{ma} = 79.78509343424172$, and $E_{rms} = 138.473$ where all have adequate values.

Table 4 Performance parameters during peak hours in the first cluster

Peak hours		12:00 AM	01:00 AM	05:00 PM	06:00 PM	07:00 PM	08:00 PM	09:00 PM	10:00 M	11:00 PM
Min DL throughput for UEs	L900	1.6703	2.4228	2.1976	2.3301	2.0757	1.6093	1.2974	1.4363	1.7741
	L1800	0.9651	1.4878	0.6872	0.7971	1.0529	0.6731	0.8365	0.727	0.7894
	L2100	3.0826	3.7448	3.4231	2.9597	3.4764	3.7711	3.6582	3.1529	4.1291
Min DL throughput. for CEUs	L900	0.0438	0.917	0.2775	0.1289	0.0164	0.0938	0.3494	0.1819	0.0762
	L1800	0.8225	0.4531	0.4262	0.5496	0.8258	0.4433	0.0174	0.4892	0.56
	L2100	0.0462	0.1406	0.3467	0.0762	0.1009	0.2902	0.0832	0.1803	0.1563
Max UEs number	L900	52	46	48	53	54	58	52	53	54
	L1800	44	50	72	60	103	59	62	54	52
	L2100	103	74	123	111	117	169	109	85	132
Average MCS	L900	17.8	17.8	17.8	18	17.9	17.9	17.7	17.7	17.9
	L1800	17.45	17.62	17.52	17.49	17.44	17.09	17.17	17.2	17.27
	L2100	17.43	17.62	17.4	17.37	17.3	17.176	17.11	17.15	17.243
Average CQI % of UEs 10–15	L900	43.7	44.1	47.4	48.08	47.6	47.3	45.56	44.8	44.1
	L1800	41.03	42.33	43.29	43.2	41.42	39.746	40.17	39.41	39.17
	L2100	50.05	48.81	50.11	50.56	50.81	51.1	51.28	51.51	51.38

Table 5 Performance parameters during peak hours in the second cluster

Peak hours		12:00 AM	01:00 AM	05:00 PM	06:00 PM	07:00 PM	08:00 PM	09:00 PM	10:00 PM	11:00 PM
Min DL throughput for UEs	L900	0.156	0.3787	0.416	0.5506	1.0832	0.8198	0.8327	0.8962	0.6756
	L1800	0.2306	0.1297	0.5426	0.3564	0.1982	0.5107	0.6064	0.5864	0.2293
	L2100	0.7952	2.9959	3.3255	3.8762	2.9897	1.8825	3.5418	1.9299	2.1229
Min DL throughput. for CEUs	L900	0.0121	0.0553	0.0885	0.0781	0.0039	0.0564	0.0393	0.0134	0.0438
	L1800	0.0278	0.0137	0.0078	0.0323	0.0078	0.0169	0.0689	0.0469	0.0443
	L2100	0.0248	0.0684	0.0652	0.1415	0.1563	0.1875	0.1523	0.2578	0.0547
Max UEs number	L900	36	46	43	228	230	34	34	32	44
	L1800	29	32	73	53	68	39	37	55	57
	L2100	70	48	99	84	71	82	76	66	60
Average MCS	L900	16.06	15.9	16.08	16.12	16.14	16.06	16.02	16.05	16.00
	L1800	16.09	16.11	16.05	15.99	15.97	15.93	15.93	15.862	15.927
	L2100	16.25	16.4	16.38	16.33	16.24	16.2	16.19	16.15	16.19
Average CQI % of UEs 10–15	L900	41.36	41.6	44.61	44.38	43.96	43	42.1	41.65	40.38
	L1800	39.44	40.95	42.52	42.27	42.074	40.347	39.97	39.65	39.3
	L2100	41.75	43.505	42.705	42.292	41.64	41.239	40.384	40.12	40.07

Table 6 Performance parameters during peak hours in the third cluster

Peak hours		12:00 AM	01:00 AM	05:00 PM	06:00 PM	07:00 PM	08:00 PM	09:00 PM	10:00 PM	11:00 PM
Min DL throughput for UEs	L900	1.9496	3.7219	2.3672	2.9945	1.7936	3.053	2.18	1.6552	1.8979
	L1800	5.0324	5.7957	4.0274	4.1089	3.5665	2.7681	2.4382	3.4332	5.5034
	L2100	0.9988	0.7863	1.2821	1.4058	1.4459	1.0811	0.8372	0.7111	1.0979
Min DL throughput. for CEUs	L900	0.5187	0.761	0.6571	1.1383	0.043	0.2413	0.4297	0.4033	0.8237
	L1800	1.5699	3.1615	2.09	1.8597	0.8588	1.3375	1.5791	1.496	2.5773
	L2100	0.3657	0.3791	0.4524	0.1717	0.4407	0.2401	0.2188	0.218	0.166
Max UEs number	L900	52	97	50	46	44	56	53	58	57
	L1800	44	35	48	45	46	53	47	44	43
	L2100	241	218	229	274	303	324	340	333	324
Average MCS	L900	19.4	20.01	19.22	19.35	19.55	19.27	18.81	19.02	18.61
	L1800	19.528	17.147	19.68	21.1	18.7	19.53	18.75	19.2	19.64
	L2100	17.77	17.96	17.55	17.62	17.56	17.56	17.51	17.49	17.52
Average CQI % of UEs 10–15	L900	48.56	48.73	49.91	43.02	46.26	46.47	48.27	46.22	44.75
	L1800	48	37.15	44.31	59.4	47.4	42.918	43.35	44.27	42.68
	L2100	39.417	41.527	39.315	39.076	38.422	38.2	37.49	37.31	37.59

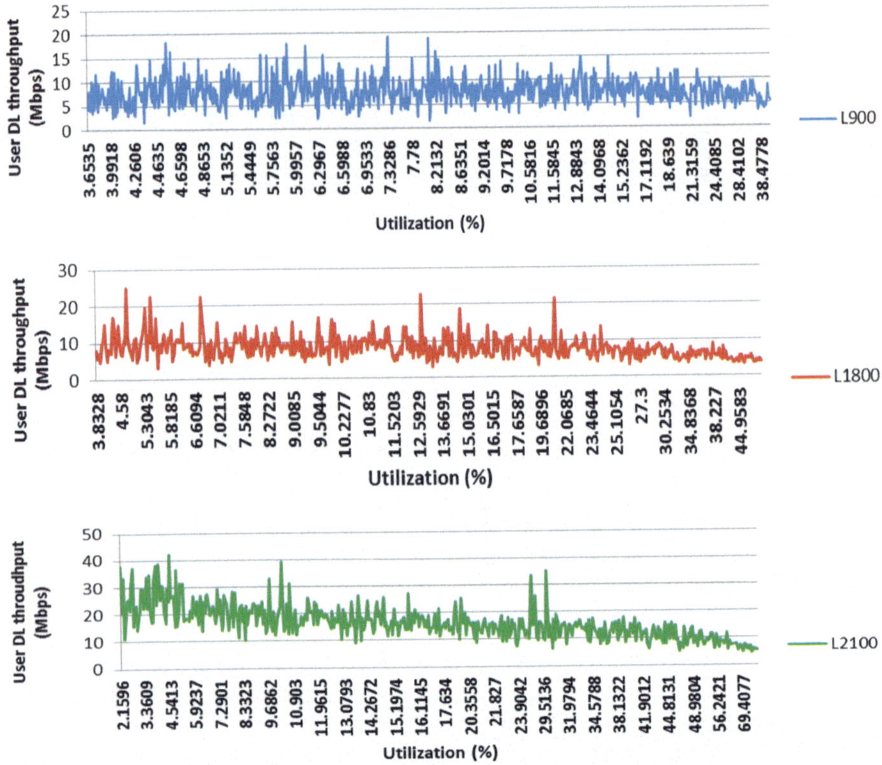

Fig. 12 Resource utilization % versus DL user throughput for one eNB in the second cluster

6 Conclusion

In this chapter, real mobile network problems are studied using real data LTE-A heavy traffic. A ML-based framework is proposed to analyze the traffic. Analyzing data set with 312 cells with 20 radio KPI features discovered that there are a number of problems. Timing advance and index indicate that all cell bands cover users near the site regardless of far users. Therefore, this is one of the reasons for bad DL throughput for CEU,s and the 1800 and 900 bands should cover users on the edge. PRB utilization is not distributed well. L2100 had the lowest utilization even though it has the largest BW (10 MHz), and also it has the largest traffic volume in all clusters. The second cluster has the lowest min DL throughput at beak hours. Moreover, all UEs (100% of max UEs) take this min throughput in this cluster, although CQI and MCS are good. In the second cluster, CEU has very bad throughput during the peak in all bands. Low demand throughput is due to lousy load distribution among three bands in each site and inadequate resource utilization where network parameters should be optimized to give users better QoS and to enhance coverage of each band. Therefore, an appropriate regression algorithm is proposed to record enhancement

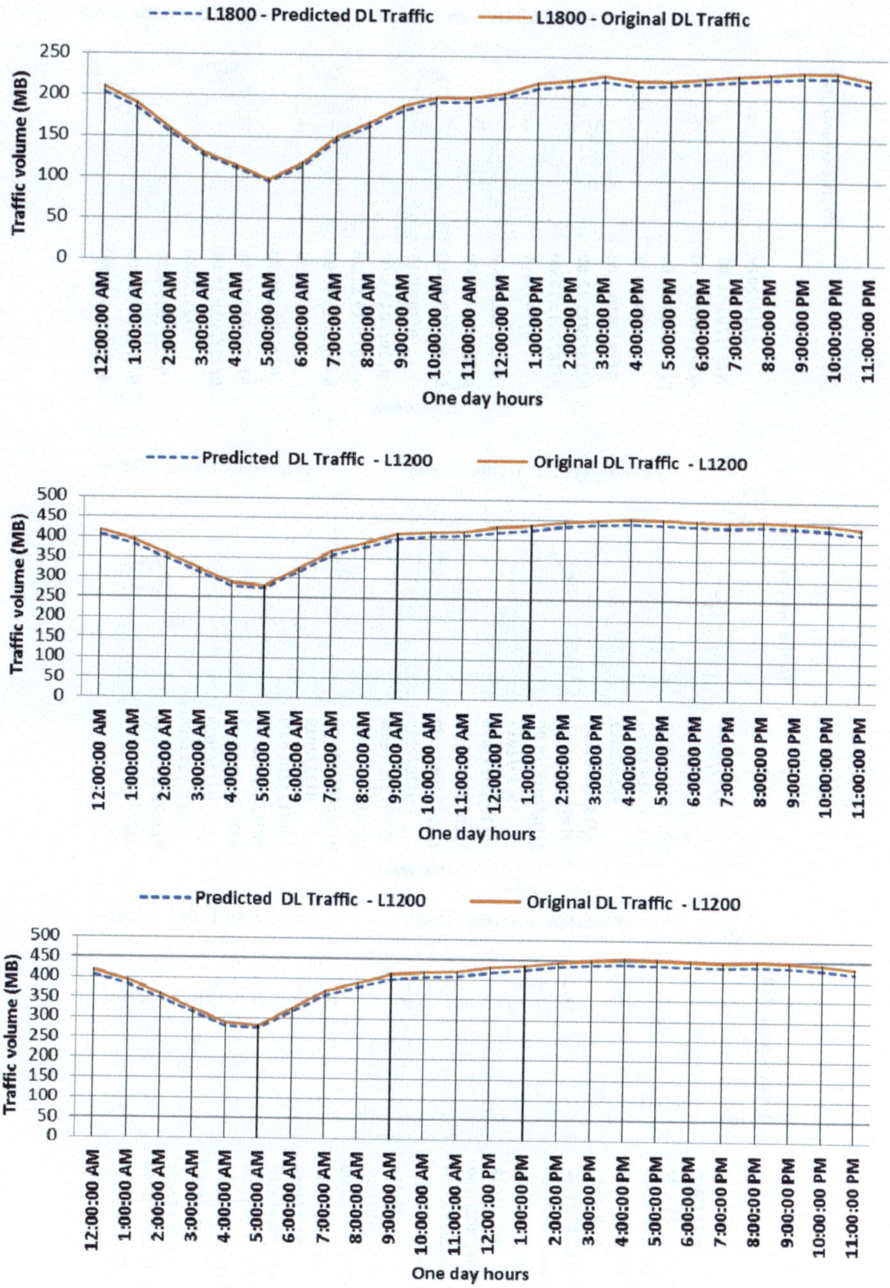

Fig. 13 Hourly original and predicted traffic volume for cells in the second cluster

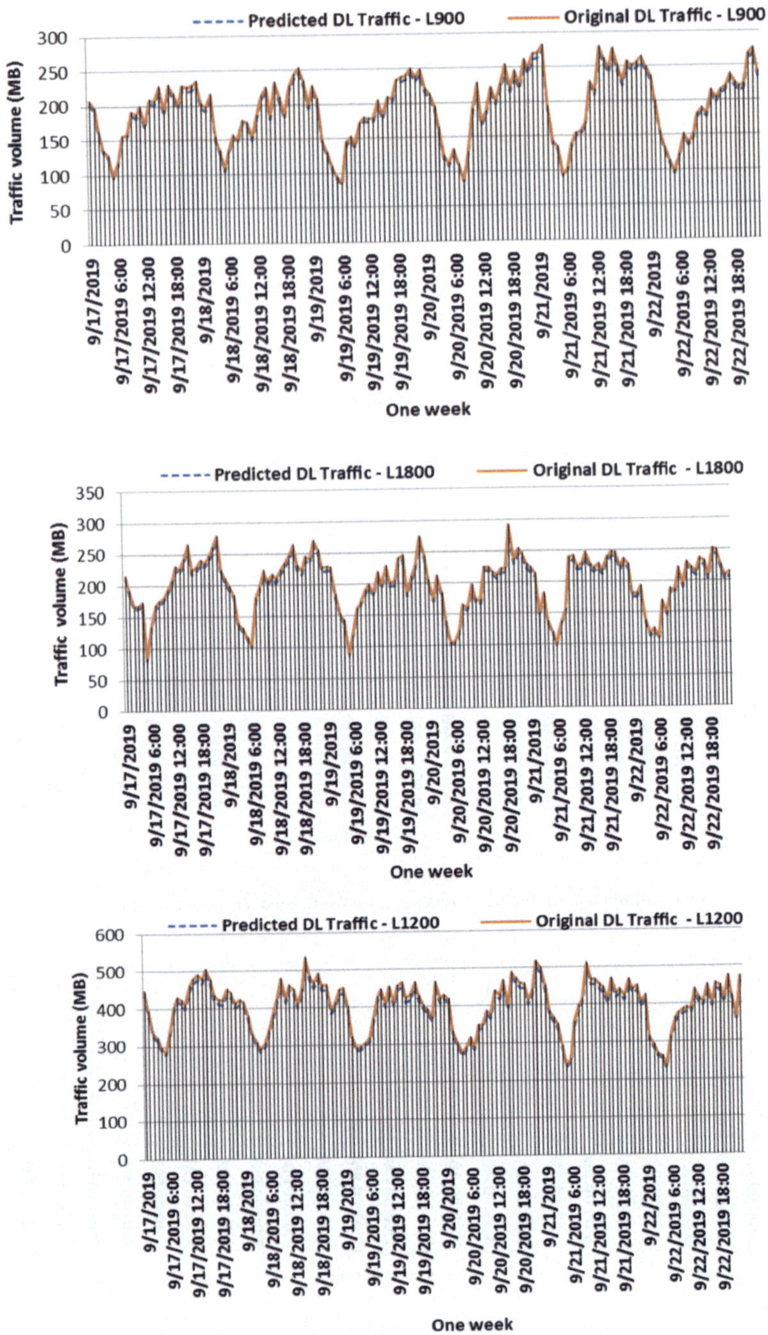

Fig. 14 Weekly original and predicted traffic volume for cells in the second cluster

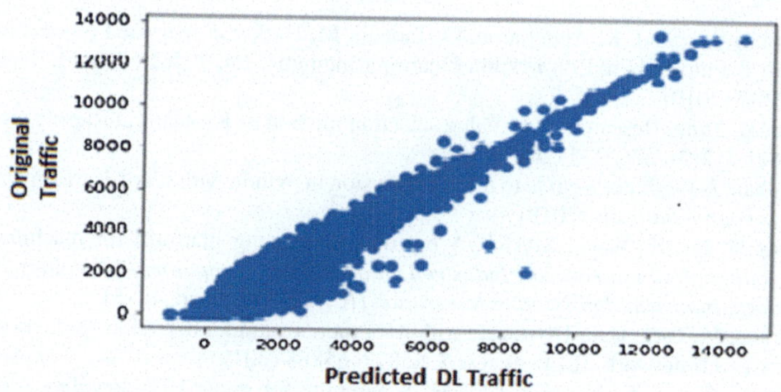

Fig. 15 Linear regression of predicted traffic and original traffic for cells in the second cluster

on spectrum efficiency. Trying to optimize network parameters using ML to enhance DL throughput, especially for CEU is the future work.

References

1. H. Nashaat, R. Rizk, Handover management based on location based services in F-HMIPv6 networks. KSII Trans. Internet Inf. Syst. (TIIS) **15**(4), 192–209 (2018)
2. R. Rizk, H. Nashaat, Smart prediction for seamless mobility in F-HMIPv6 based on location based services. China Commun. Springer **15**(4), 192–209 (2018)
3. H. Nashaat, QoS-aware cross layer handover scheme for high-speed vehicles. KSII Trans. Internet Inf. Syst. (TIIS) **12**(1), 135–158 (2018)
4. S. Kukliński, L. Tomaszewski, Key performance indicators for 5G network slicing, in *2019 IEEE Conference on Network Softwarization (NetSoft)*, Paris, France
5. J. Salo, B. EduardoZacarías, Analysis of LTE radio load and user throughput, in *Zacharias International Journal of Computer Networks and Communications (IJCNC)*, vol. 9, no. 6 (2017)
6. Luong,V., Do, S., Bao, S., Paul, L., Li-Ping, T.: Applying Big data, machine learning, and SDN/NFV to 5G traffic clustering, forecasting, and management, in *IEEE International Conference 2018 on Network Softwarization and workshops (Netsoft)*, Montreal, Canada (2018)
7. C. Lars, S. Shmelz, Adaptive SON management using KPI measurements, in *IEEE IFIP Conference on Network Operations and Management Symposium ((NOMS)*, Istanbul, Turkey (2016)
8. H. Soren, S. Michael, K. Thomas, Impact of SON function combinations on the KPI behavior in realistic mobile network scenarios, in *IEEE Wireless Communication and Network Conference Workshops (WCNCW)*, Barcelona, Spain (2018)
9. M. Kibria, K. Nguyen, G. Villardi, O. Zhao, K. Ishizu, F. Kojima, Big data analytics, machine learning and artificial intelligence in next-generation wireless networks. IEEE Access **6**, 32328–32338 (2018)
10. S. Abo Hashish, R. Rizk, F. Zaki, Joint energy and spectral efficient power allocation for long term evolution-advanced. Comput. Electr. Eng. Elsevier **72**, 828–845 (2018)
11. H. Nashaat, O. Refaat, F. Zaki, E. Shaalan, Dragonfly-based joint delay/ energy LTE downlink scheduling algorithm. IEEE Acess **8**, 35392–35402 (2020)
12. K. Ralf, Key performance indicators and measurements for LTE radio network optimization, in *LTE Signaling, Troubleshooting and Performance Measurement* (2015), pp. 267–336

13. H. Lahdhiri, M. Said, K. Abdellafou, O. Taouali, M. Harkat, Supervised process monitoring and fault diagnosis based on machine learning methods. Int. J. Adv. Manuf. Technol. **102**, 2321–2337 (2019)
14. C. Yuan, H. Yang, Research on K-Value selection method of K-means clustering algorithm. J. Multi. Sci. J. **2**(2), 226–235 (2019)
15. N. Chauhan, A Beginner's guide to linear regression in python with scikit-learn, in *A free online statistics course tutorials* (2019)
16. K. Wang, J. Fu, K. Wang, SPARK-A big data processing platform for machine learning, in *International Conference on Industrial Informatics—Computing Technology, Intelligent Technology, Industrial Information Integration (ICIICII)* 2016, pp. 48–51
17. J. Bejarano, M. Toril, Data-Driven algorithm for indoor/outdoor detection based on connection traces in a LTE network. IEEE Access **7**, 65877–65888 (2019)
18. M. Salman, C. Ng, K. Noordin, CQI-MCS mapping for green LTE downlink transmission. Proc. Asia-Pacific Adv. Netw. **36**, 74–82 (2013)
19. K. Yong seog, M. Filippo, W. Nick, **Feature** selection in data mining, in *Data Mining Opportunities and challenge* (2003), pp. 80–105
20. A Little Book of Python for Multivariate Analysis. https://python-for-multivariate-analysis.readthedocs.io/a_little_book_of_python_for_multivariate_analysis.html
21. T. Theyazn, M. Joshi, Integration of time series models with soft clustering to enhance network traffic forecasting, in *Conference: 2016 Second International Conference on Research in Computational Intelligence and Communication Networks (ICRCICN)*, Kolkata, India (2016)

Artificial Intelligence and Blockchain for Transparency in Governance

Mohammed AlShamsi, Said A. Salloum, Muhammad Alshurideh, and Sherief Abdallah

Abstract One of the influential research fields is the use of Artificial Intelligence and Blockchain for transparency in governance. The standard mechanisms utilized in governance are required to be transformed in respect of assorted parameters such as availability of data to users further as information asymmetries between the users should be minimized. And we did an in-depth analysis of the use of AI and Blockchain technologies for governance transparency. We've considered three qualitative approaches for evaluating the research within the proposed area, i.e., conceptual modeling, analysis based work, and implementation based work. We presented an in-depth overview of two research papers for each methodological approach. In terms of using AI and Blockchain technology for governance transparency, we have preferred conceptual modeling to support the prevalent work under the proposed research model.

Keywords Artificial intelligence · Blockchain · Transparency · Governance · Qualitative methods

M. AlShamsi · S. A. Salloum (✉) · S. Abdallah
Faculty of Engineering and IT, The British University in Dubai, Dubai, UAE
e-mail: ssalloum@sharjah.ac.ae

M. Alshurideh
University of Sharjah, Sharjah, UAE

Faculty of Business, University of Jordan, Amman, Jordan

S. A. Salloum
Machine Learning and NLP Research Group, Department of Computer Science, University of Sharjah, Sharjah, UAE

© The Editor(s) (if applicable) and The Author(s), under exclusive license to Springer Nature Switzerland AG 2021

A. E. Hassanien et al. (eds.), *Artificial Intelligence for Sustainable Development: Theory, Practice and Future Applications*, Studies in Computational Intelligence 912,
https://doi.org/10.1007/978-3-030-51920-9_11

1 Introduction

Artificial Intelligence and Blockchain technologies dramatically change the life way of citizens, and most of the routines of daily life are being influenced by such technologies [1–9]. Such technologies provide various reliable services, which are being trusted by all the users [10–15]. Therefore, it is an effective way to use artificial intelligence and Blockchain technologies for transparency in governance. Since the traditional mechanisms utilized in governance is require to be transform in respect of assorted parameters such availability of information to users as well as information asymmetries between the users should be minimized [16–20]. Similarly, the speed of transfer of data and the security of sensitive information are being improved. Furthermore, by applying check and balance mechanism on each and every aspect of governance through Blockchain technology, there should not be any room for corruption at all [21]. Blockchain recently has emerged as a technology that can enable or allow to do things that seemed impossible in that past like allowing to record assets, allocation of value and most importantly to register, monitor the footprint of electronic transactions without any central repository, i.e., decentralized, thus providing transparency, integrity, and traceability of information and data on a consensus-based approach where trusted and parties can validate and verify the information, eliminating the need of a central authority. Blockchain can take transparency and governance too much higher level as it can eliminate false transactions because of the distributed-ledger system capable of certifying records and transactions—or "blocks"—without the use of a central database and in a manner that cannot be erased, changed or altered. This offers the knowledge this handles an unparalleled level of integrity, confidentiality, and reliability, the risks associated with having one single point of failure [22–24].

On the other hand, AI being deployed for facial recognition and various decision making across applications and multiple sectors, the concern is with the transparency and responsibility to ensure that advanced or AI-powered algorithms are comprehensively verified and tested from time to. AI is one of the most rapidly changing and advancing technology that can bring a great amount of value in our today and future, but it needs to be fully controlled by producing transparency and establishing vibrant rules, strategies, and procedures when it comes to implement, create and utilize the AI applications. It would be of great importance to make sure that AI-powered algorithms function as planned and is optimized to capture the value for which they have been deployed [25]. The aim of this study is to analyze the qualitative method applied for the use of Artificial Intelligence and Blockchain for transparency in governance. To do this, three qualitative methods have been analyzed. Such qualitative approaches are analytical research, work based on analysis, and work based on implementation.

The objectives of this paper are to review a recently published research work on the proposed topic, to identify the qualitative research methods in each article, to select a qualitative research method for proposed work, and to compare and justify the preferred research methodology.

2 Literature Review of Research Paper

This section will review the most important research papers that use AI and Blockchain technologies for Transparency in Governance. We divided it into three categories including Conceptual Framework, Review based work, and Implementation based work.

2.1 Conceptual Framework

Blockchain-based sharing services Authors in [26] suggests a three-dimensional conceptual framework, i.e., (a) human, (b) technology, (c) organization to detect a feature for Smart Cities from a common economic perspective. The authors use the proposed framework to examine the impact of Blockchain on building smart cities, and to understand the development of smart cities from a shared services angle, the authors propose a literature-based smart city theoretical framework. According to previous studies on the theoretical classification of smart cities in technology, humans and organizations characterize the relations most often utilized to describe smart cities.

The city is human-based, technology-oriented, and can have service relationships. The use of technology ICT is based on changing lives and working in the city in relevant ways, the smart city organization focuses on government support and governance policies and includes a number of elements such as smart society, smart government, organized and open management, networking and partnerships. A keen government does maximum to manage the consequences of the social and economic system and also connects all the stakeholders like citizens, communities, and businesses. From the Sharing Service angle, the rule requires that users who share services be protected from fraud, inadequate service providers, and liability; there are six types of service between technology, human, and organization. In a Blockchain-based approach, staying confident is a major feature of the relationship among people [26].

The economist has described the Blockchain as a trust machine, indicating that it cares about trustworthy issues between the individuals. The economic system, "based on Blockchain technology, operates without people and therefore allows a trust-free transaction" [27]. Factually, the business of trust is very small, often with a trusted third party, which is classy. Blockchain technology provides a viable alternative to the removal of mediators, thus reducing operating costs and increasing the efficiency of exchanging services. With the help of Blockchain technology, one of the most basic business exchanges in the world can be re-imagined. For that reason, Reliability-sharing services have opened the door to new forms of digital interaction.

Finding the meaning of smart in the Smart City label from the perspective of a shared economy can assist us in understanding the needs of smart cities and how accepting new technologies. In this article, the authors explore how the functionality of Blockchain technology will contribute to the development of smart cities through shared services, based on a conceptual framework. Although the authors provide a goof framework in respect of Blockchain technology, however, the work is only limited to the sharing of services, whereas, rest of the parameters regarding governance are not being considered in the proposed work.

Governance on the Drug Supply Chain via Gcoin Blockchain Authors in [28] suggested governance on Blockchain technology as an innovative service platform for the reason that is managing the drug supply chain. In this regard, the authors suggest Gcoin Blockchain technology as the basic drug data flow for the creation of a transparent and independent drug transaction data.

Description, Pros and Cons From a development perspective of post-market, "drug life cycle, basic research, non-clinical trials, clinical trials, manufacturing, production and distribution/sales" (Taiwan Food and Drug Administration 2016 Annual Report). Good inspection and control are needed to accomplish good exercise at every step of the life cycle. One of the worst scenarios can be fake drugs when surveying the drug supply chain if it is incomplete or complicated. The World Health Organization (WHO) will describe the counterfeit drugs, which include "Products purposely and fraudulently produced and mislabeled about identity and source to make it appear to be a genuine product" [29]. The economic damage due to counterfeit drugs is difficult to quantify, and the world has no reliable basic statistics, such as the count of counterfeit drugs. "However, in recent years, public opinion and analysts have widely accepted the argument that 10% of medicinal products can be developed around the world" (WHO Drug Information 2006). Several recommendations to avoid counterfeit pharmaceutical items include: improving supply chain management, improving secondary drug market controls, and improving the technology used for tracking and tracking counterfeiting drugs. Keeping in view the suggestions referred above, the most straightforward and comprehensive approach is to improve drug supply chain management, from drug procurement, distribution and production to drug use, every step of the drug's supply plays a vital role in drug safety. The suggested platform Gcoin Blockchain is although an innovative technology that provides the Drug supply chain governance model transitioning from regulatory to net surveillance. However, the proposed work is limited only to the single factor, which is the drug supply chain, and it can be updated to overall transparency in governance [28].

2.2 *Review Based Work*

Blockchain-Powered Internet of Things, E-Governance and E-Democracy Authors in [30] investigates associated answers to all current issues: a technology innovation called Blockchain. Nekmudo first launched the Blockchain as a testimony to BitCoin, the first widespread digital cash deployed close to the old American gold standard currency [31]. Blockchain BitCoin will keep track of all consumers' deals and provide an illogical way of processing such transactions. Different from traditional financial facilities that employ a bank to validate every transaction, Blockchain does not require a central authority. Several Blockchain participants volunteered to verify each transaction, which would make Opex very rare. To guarantee performance, the volunteers are rewarded for their valid work, and sometimes because of their wrongdoing will be fined. In this way, Blockchain members rely only on trusted distributed instead of centralized options such as banks. Whereas on the other side, if a partner wants to interfere with past matters, he has to convince all rest of the users to work the same, which has proved to be a difficult task. Due to its Blockchain has emerged as an ideal technology for permanently storing documents including contracts, diplomas, and certificates at low cost and high level of security.

Description, Pros and Cons The authors investigated a new technology called Blockchain. The authors first define the Blockchain mechanism that enables non-existent parties to do the transaction in a decentralized manner with each other. The authors use specific examples to illustrate Blockchain wide use in the Internet of Things, E-governance, and e-democracy. Three aspects of Blockchain are the subject with such examples:

1. **Decentralization**: Blockchain reduces the cost of connected devices without the central option and prevents single-point interruption to the network.
2. **Automation**: The apps are smarter with self-service, and so on, with the use of smart contracts. Also, repetitive work can be done automatically for the government, raising operating costs and allowing the government to more effectively provide services.
3. **Security**: Blockchain is distributed to decrease the harmful effects of slashing a consumer. Consequently, every Blockchain process is transparent and registered with every user. Without such extensive monitoring, it is impossible to reveal absolute wrong behavior.

This work concentrated on the basic factors related to AI and Blockchain technology (BT) such as decentralization, automation, and security; however, the factors of data management and data flow were still intact [30].

Blockchain Technologies for Open Innovation (OI) Authors in [32] provide more information on theoretical grounds so that a brief overview of the prior research can be provided, and potential areas for future research can be highlighted. In addition, the authors aim to create a common understanding of the Blockchain technology theory in the field of open innovation. BT is still considered an advancement in the field of OI analysis and has not yet become part of mainstream OI research. It also supports the general scenario, which has focused primarily on Blockchain as a hidden economic system, e.g., BitCoin. The authors often consider the amount of literature in the field as a significant factor when determining the maturity of the concepts. The authors conclude that the BitCoin definition has been tested with 24,500 findings on Google 3 Scholar, with 17,500 results similar to Blockchain. This study aims to provide a new perspective on Blockchain technology by analyzing existing BT research and integrating it with other OI principles, such as Bluetooth, ecosystems, inventions, and technical features [32].

The survey is based on OI for SMEs (small and medium-sized enterprises) driven by the online platform and potential application of BT. This article analyzes and synthesizes the main results in the field of current and future BT to OI applications in key non-OI areas for smart cities, in particular, the digitization of collaborative work. And, through smart competitive openness, new sources of financing, open data, and even the creation of bots on the Blockchain, cooperation agreements are smooth across companies' limits. The main idea is that BT seems to be connected to creativity, with naturally distributed OI networks and online collaboration. There are massive benefits of using BT to improve the OI network, and most of them are still at a critical juncture to explore exciting new phases of technology development in both BT and OI. The proposed work is limited only to the extent of enterprise-level and requires to modify in respect to using AI and Blockchain technology for transparency in governance.

2.3 Implementation Based Work

Blockchain 3.0 Smart Contracts in E-Government 3.0 Applications Authors in [33] observes BC (Blockchain) 3.0 and SC (smart contracts) features and characteristics expected to take part in EG (E-Governance) 3.0 applications. They present designated practices for incorporating BC 3.0 and SC both into the application of information communication technology (ICT) Web 3.0 e-government resolutions and solutions.

Blockchain technology known as Blockchain 1.0 for the automation of intermediary financial payments, and this technology was recognized as Blockchain 2.0, followed by the Ethereum Project, in support of Smart Contracts (SC), which was different from BC 1.0. Certain ventures of BC 2.0 development include the HL Fabric, Sawtooth, Aroha, and R3 Carda hyper-leaders. SCs are code guidelines written to operate on the Blockchain and to provide enforcement and security systems so that the parties can agree on certain cases and steps to meet the requirements. Such features of the SCs have reshaped the supply chain mechanism to provide additional electronic measures such as asset tracking and, at the same time, the required functionality for non-supply chain business transactions. Blockchain is widely used in sectors such as government, health care, education, charities, properties, insurance, and banking. Blockchain 3.0 is the name of this growing field of BC-supported applications as solutions are not limited to funding and asset transfer. With the advent of Blockchain 3.0 technology, "BC systems were better, more effective, more scalable, more interoperable, and better user interface based on the directed acyclic graph (DAG) data structures" [33]. In these areas, government use cases are of particular interest because of their implications for the adoption of BC infrastructure. These may include issues of national governments, such as non-action and resistance by regulators, or external issues such as digital transformation laws and disadvantaged persons and officials' personal data [34]. BC's decentralized features provide zero time-to-date, ensure data and non-refundable tampering, enforce security with confidentiality to build trust between partners, and data Blockchain uses integrity, authentication, and the consensus algorithm for private and allowed scalability. The proposed work was implemented and outperformed with regard to two scenarios, i.e., health data and energy data. However, no specific E-Government application has been considered by the authors.

Towards Enabling Trusted Artificial Intelligence via Blockchain Authors in [35] depicts how Blockchain innovation can be utilized to handle different parts of trust in the simulated intelligence preparing process, including information and models' review provenance, information security, and reasonableness. Blockchain data can include: model formation history, a hash key information that can be used to guide models, model originality, and possibly the contribution of different participants in model creation. Models would then be able to be moved to various gatherings inside a similar association or different substances, and every association can peruse the chronicled account of the model from the Blockchain and decide to revise the model with extra information. For instance, in order to meet the lack of diversity, previous training set organizations can also read the results of the model's replication of pre-training from a Blockchain conducted by another community to see if the training was done earlier. Some of the data also affected the performance negatively [35].

In order to support model recovery in these different scenarios, we need to collaborate with Provenance on a model relevant to the AI model. While some work needs to be done on map design for AI models, we need to regularize the AI model definition in order to make it usable in cases where the models can be used. Providing information and data for AI models that can be interpreted as model-associated metadata and can be provided by any company distributing AI models such as AI Marketplace or cloud-based AI models. The following information is included in the Provenance data of the AI model [36].

- Information of the training data used to build an AI model.
- Description of the model pipeline used for the testing, as many models are created from a pipeline containing basic training.
- Description of the procedure used to train the AI model.
- Definition of any updates or changes that might have been made to the AI model
- Description of any experiments that were performed using the AI model and the test results.

The prudence information and data requires to be saved so that the descriptions can be retained in such a way that they can be validated in cases when they are edited after the fact. The specifics need to provide a collection of credentials that any AI model user can check. The main purpose of the Blockchain Library is for distributed/collaborative learning settings and can also be used for tracking the non-distributed learning process. For undistributed environments, however, it may only be sufficient to record the input data and the final model. On the other hand, to be able to trust the learning process, it is very important to capture multiple perspectives from a different perspective and different participants. The authors discussed a joint AI application, one such use of federated learning that has recently become known. The proposed work describes the tackling of various features of trust in the domain of AI training process by using Blockchain technology; however, for the provision of transparency in governance, the work still requires some governing parameters in respect of using AI and Blockchain technology [37].

2.4 Comparative Analysis

Each qualitative approach has several pros and cons. However, the conceptual framework or conceptual modeling is among the common methods that can be used to introduce a new method in a growing field. Presenting a template or conceptual structure, however, is a demanding task and involves an in-depth knowledge of the subject matter. Research survey-based research, which provides comprehensive interest area knowledge, proposing new research through review-based work is not a successful way, because only comparisons are used with existing research works. On the other hand, an implementation based approach is one of the commonly used approaches where studies use various simulation techniques to validate the research. However, the application-based approach is an efficient research process, but this type of work requires real-world implementation or simulation systems.

3 Selection and Justification of the Preferred Method

For research work with respect to the use of Artificial Intelligence and Blockchain for transparency in Governance, the conceptual framework or conceptual modeling methodology was chosen. The reason for choosing the conceptual structure approach is that it's easy to suggest and demonstrate a new strategy in conceptual work. A growing part of the proposed work can be clearly described with the aid of the conceptual framework. In addition, as regards the use of AI and Blockchain technologies for accountability in governance, there may be different components of the model, and there may also be a data flow, as this is an efficient way to use computational modeling.

On the other hand, the reason for choosing our study's conceptual framework is because it offers the structure of ideas, perceptions, assumptions, beliefs, and related theories that guide and sustain the research work. The most important thing to understand about your theoretical framework is that it is basically a concept or model of what you are studying there that you plan to study, and what is going on with these things. And that's why the temporal theory of phenomena that you are investigating. The purpose of this theory is to evaluate and improve your goals,

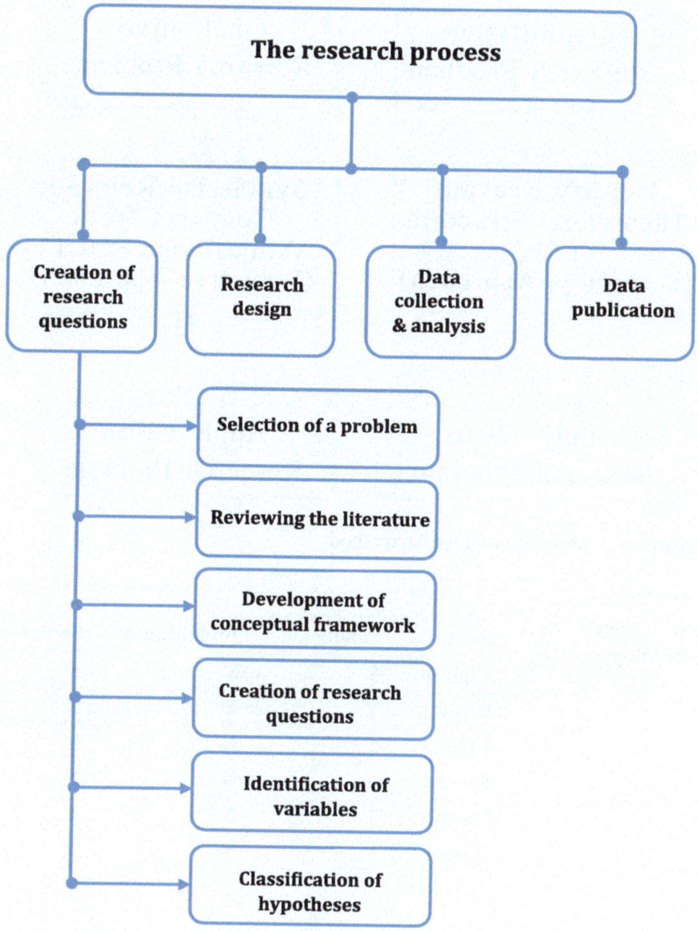

Fig. 1 Conceptual framework research method

promote realistic and relevant research questions, select appropriate approaches, and identify potential authenticity. Figure 1 shows the overall process of the conceptual framework research method [38].

4 Preferred Method Detailed Comparison

The suggested approach, i.e., computational modeling, is an effective way to provide the research paradigm for accountability in governance by using AI and Blockchain technology. Because of this, review-based research is only used to evaluate current approaches, while implementation-based methods require real-world implementing

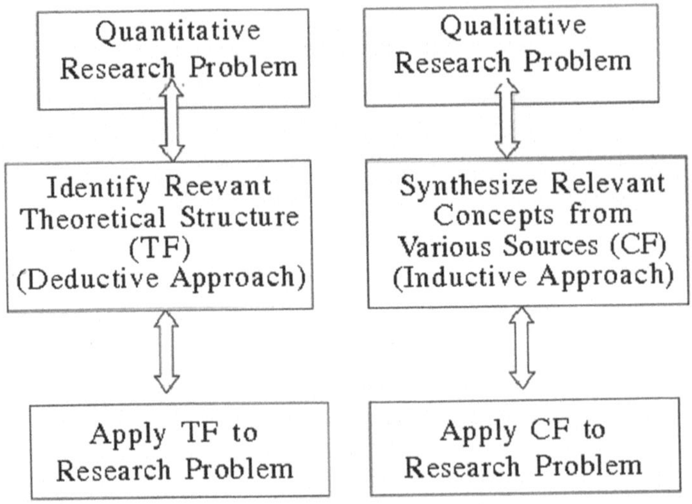

Fig. 2 Comparison of preferred qualitative method

Table 1 Good research practice methodology

Paper	Simulation	Architecture	Review of work
Paper 1	Y	Y	N
Paper 2	Y	Y	N
Paper 3	N	N	N
Paper 4	N	N	N
Paper 5	Y	N	Y
Paper 6	Y	N	Y

tools or simulation platforms to validate and verify the proposed work. Figure 2 shows that the detailed comparison of the preferred qualitative method.

A brief comparison of good research practice methodology for each paper is shown in Table 1.

5 Conclusions

In this work, we have provided a detailed study on the use of AI and Blockchain technology for transparency in governance. We have considered three qualitative approaches for evaluating the research in the proposed area, i.e., conceptual modeling, analysis based work, and implementation based work. For each qualitative approach, we received a detailed summary of two research papers. Based on existing work, we preferred conceptual modeling to the proposed research model with regard to the use of AI and Blockchain technology for governance transparency.

References

1. S.A. Salloum, M. Al-Emran, A.A. Monem, K. Shaalan, A survey of text mining in social media: facebook and twitter perspectives. Adv. Sci. Technol. Eng. Syst. J. **2**(1), 127–133 (2017)
2. S.A. Salloum, M. Al-Emran, K. Shaalan, Mining social media text: extracting knowledge from facebook. Int. J. Comput. Digit. Syst. **6**(2), 73–81 (2017)
3. S.A. Salloum, C. Mhamdi, M. Al-Emran, K. Shaalan, Analysis and classification of arabic newspapers' facebook pages using text mining techniques. Int. J. Inf. Technol. Lang. Stud. **1**(2), 8–17 (2017)
4. S.A. Salloum, M. Al-Emran, S. Abdallah, K. Shaalan, in *Analyzing the Arab Gulf Newspapers Using Text Mining Techniques*, vol. 639 (2018)
5. C. Mhamdi, M. Al-Emran, S.A. Salloum, in *Text Mining and Analytics: A Case Study from News Channels Posts on Facebook*, vol. 740 (2018)
6. S.F.S. Alhashmi, S.A. Salloum, S. Abdallah, in *Critical Success Factors for Implementing Artificial Intelligence (AI) Projects in Dubai Government United Arab Emirates (UAE) Health Sector: Applying the Extended Technology Acceptance Model (TAM)*, vol. 1058 (2020)
7. K.M. Alomari, A.Q. AlHamad, S. Salloum, Prediction of the digital game rating systems based on the ESRB
8. S.A. Salloum, M. Alshurideh, A. Elnagar, K. Shaalan, Mining in educational data: review and future directions, in *Joint European-US Workshop on Applications of Invariance in Computer Vision* (2020), pp. 92–102
9. S.A. Salloum, M. Alshurideh, A. Elnagar, K. Shaalan, Machine learning and deep learning techniques for cybersecurity: a review, in *Joint European-US Workshop on Applications of Invariance in Computer Vision* (2020), pp. 50–57
10. M. Swan, Blockchain thinking: the brain as a decentralized autonomous corporation (commentary). IEEE Technol. Soc. Mag. **34**(4), 41–52 (2015)
11. S.F.S. Alhashmi, S.A. Salloum, C. Mhamdi, Implementing artificial intelligence in the United Arab Emirates healthcare sector: an extended technology acceptance model. Int. J. Inf. Technol. Lang. Stud. **3**(3) (2019)
12. S.F.S. Alhashmi, M. Alshurideh, B. Al Kurdi, S.A. Salloum, A systematic review of the factors affecting the artificial intelligence implementation in the health care sector, in *Joint European-US Workshop on Applications of Invariance in Computer Vision* (2020), pp. 37–49
13. S.A. Salloum, R. Khan, K. Shaalan, A survey of semantic analysis approaches, in *Joint European-US Workshop on Applications of Invariance in Computer Vision* (2020), pp. 61–70
14. R. Shannak, R. Masa'deh, Z. Al-Zu'bi, B. Obeidat, M. Alshurideh, H. Altamony, A theoretical perspective on the relationship between knowledge management systems, customer knowledge management, and firm competitive advantage. Eur. J. Soc. Sci. **32**(4), 520–532 (2012)
15. H. Altamony, M. Alshurideh, B. Obeidat, Information systems for competitive advantage: implementation of an organisational strategic management process, in *Proceedings of the 18th IBIMA conference on innovation and sustainable economic competitive advantage: From regional development to world economic, Istanbul, Turkey*, 9–10 May 2012
16. Z. Alkalha, Z. Al-Zu'bi, H. Al-Dmour, M. Alshurideh, R. Masa'deh, Investigating the effects of human resource policies on organizational performance: An empirical study on commercial banks operating in Jordan. Eur. J. Econ. Financ. Adm. Sci. **51**(1), 44–64 (2012)
17. R. Al-dweeri, Z. Obeidat, M. Al-dwiry, M. Alshurideh, A. Alhorani, The impact of e-service quality and e-loyalty on online shopping: moderating effect of e-satisfaction and e-trust. Int. J. Mark. Stud. **9**(2), 92–103 (2017)
18. H. Al Dmour, M. Alshurideh, F. Shishan, The influence of mobile application quality and attributes on the continuance intention of mobile shopping. Life Sci. J. **11**(10), 172–181 (2014)
19. M. Ashurideh, Customer Service Retention—A behavioural Perspective of the UK Mobile Market (Durham University, 2010)
20. M. Alshurideh, A. Alhadid, B. Al kurdi, The effect of internal marketing on organizational citizenship behavior an applicable study on the University of Jordan employees. Int. J. Mark. Stud. **7**(1), 138 (2015)

21. P. Mamoshina et al., Converging blockchain and next-generation artificial intelligence technologies to decentralize and accelerate biomedical research and healthcare. Oncotarget **9**(5), 5665 (2018)
22. C. Santiso, Can blockchain help in the fight against corruption? in *World Economic Forum on Latin America,* vol. 12 (2018)
23. Z. Zu'bi, M. Al-Lozi, S. Dahiyat, M. Alshurideh, A. Al Majali, Examining the effects of quality management practices on product variety. Eur. J. Econ. Financ. Adm. Sci. **51**(1), 123–139 (2012)
24. A. Ghannajeh et al., A qualitative analysis of product innovation in Jordan's pharmaceutical sector. Eur. Sci. J. **11**(4), 474–503 (2015)
25. Deloitte, Transparency and Responsibility in Artificial Intelligence (2019)
26. J. Sun, J. Yan, K.Z.K. Zhang, Blockchain-based sharing services: What blockchain technology can contribute to smart cities. Financ. Innov. **2**(1), 1–9 (2016)
27. T. Economist, The promise of the blockchain: The trust machine'. Economist **31**, 27 (2015)
28. J.-H. Tseng, Y.-C. Liao, B. Chong, S. Liao, Governance on the drug supply chain via gcoin blockchain. Int. J. Environ. Res. Public Health **15**(6), 1055 (2018)
29. L. Williams, E. McKnight, The real impact of counterfeit medications. US Pharm. **39**(6), 44–46 (2014)
30. R. Qi, C. Feng, Z. Liu, N. Mrad, Blockchain-powered internet of things, e-governance and e-democracy, in *E-Democracy for Smart Cities* (Springer, Berlin, 2017), pp. 509–520
31. S. Nakamoto, A. Bitcoin, A peer-to-peer electronic cash system. *Bitcoin* (2008) URL http://bitcoin.org/bitcoin.pdf
32. J.L. De La Rosa et al., A survey of blockchain technologies for open innovation, in *Proceedings of the 4th Annual World Open Innovation Conference* (2017), pp. 14–15
33. S. Terzi, K. Votis, D. Tzovaras, I. Stamelos, K. Cooper, Blockchain 3.0 smart contracts in E-government 3.0 applications. Preprint at http://arXiv.org/1910.06092 (2019)
34. S. Ølnes, J. Ubacht, M. Janssen, Blockchain in government: Benefits and implications of distributed ledger technology for information sharing (Elsevier, 2017)
35. K. Sarpatwar et al., Towards enabling trusted artificial intelligence via blockchain, in *Policy-Based Autonomic Data Governance* (Springer, Berlin, 2019), pp. 137–153
36. N. Baracaldo, B. Chen, H. Ludwig, J.A. Safavi, Mitigating poisoning attacks on machine learning models: A data provenance based approach, in *Proceedings of the 10th ACM Workshop on Artificial Intelligence and Security* (2017), pp. 103–110
37. S. Schelter, J.-H. Boese, J. Kirschnick, T. Klein, S. Seufert, Automatically tracking metadata and provenance of machine learning experiments, in *Machine Learning Systems Workshop at NIPS* (2017), pp. 27–29
38. G.D. Bouma, R. Ling, L. Wilkinson, *The research process* (Oxford University Press, Oxford, 1993)

Artificial Intelligence Models in Power System Analysis

Hana Yousuf, Asma Y. Zainal, Muhammad Alshurideh, and Said A. Salloum

Abstract The purpose of this chapter is to highlight the main technologies of Artificial Intelligence used in power system where the traditional methods will not be able to catch up all condition of operating and dispatching. Moreover, for each technology mentioned in the chapter there is a brief description where is used exactly power system. Moreover, these methods improve the operation and productivity of the power system by controlling voltage, stability, power-flow, and load frequency. It also permits to control the network such as location, size, and control of equipment and devices. The automation of the power system ensures to support the restoration, fault diagnosis, management, and network security. It is necessary to identify the appropriate AI technique to use it in planning, monitoring, and controlling the power system. Finally the chapter will highlight briefly sustainable side of using AI in power system.

Keywords Power system · Artificial neural network · Fuzzy logic · AI · Genetic algorithms. expert system

---

H. Yousuf
Faculty of Engineering & IT, The British University in Dubai, Dubai, UAE

A. Y. Zainal
Faculty of Business Management, The British University in Dubai, Dubai, UAE

M. Alshurideh
University of Sharjah, Sharjah, UAE

Faculty of Business, University of Jordan, Amman, Jordan

S. A. Salloum (✉)
Machine Learning and NLP Research Group, Department of Computer Science, University of Sharjah, Sharjah, UAE
e-mail: ssalloum@sharjah.ac.ae

© The Editor(s) (if applicable) and The Author(s), under exclusive license to Springer Nature Switzerland AG 2021
A. E. Hassanien et al. (eds.), *Artificial Intelligence for Sustainable Development: Theory, Practice and Future Applications*, Studies in Computational Intelligence 912,
https://doi.org/10.1007/978-3-030-51920-9_12

1 Introduction

In the twenty-first century, Artificial Intelligence has become one of the most advanced technologies employed in various sectors [1–13]. The United Arab Emirates was the first country to launch AI Strategy in the region and world; that shows the adoption of AI in the Federal government's strategic plans is inevitable [14–16]. Several countries, such as China, the USA, UK, and France, adopted AI in their development plan. The key reason behind adopting AI is to integrate various sectors such as healthcare, energy/renewable energy, finance, water, education, and environment.

The current chapter examines the AI in power systems on various dimensions. Because the prevailing traditional methods do not help to give accurate results and reflects the real situation of the system. Artificial Intelligence is by using machines and software development systems that display the intellectual processes and ability of reasoning and thinking as in humans. Power system engineering involves the generation, transmission, distribution, and utilization of electrical power and various electrical devices.

The entrance of renewable energy sources makes the traditional techniques difficult to present different scenarios because of its complexity. Power system analysis must handle a complex, varied, and a large amount of data for its computation, diagnosis, and learning. The sophisticated technology like computers allows handling the difficult issues related to power system planning, operations, design, and diagnosis [17–21]. Henceforth, AI aids in managing the extensive and vast data handling system and gives an accurate and on-time report to make the right decision in resolving power system concerns and improved power systems.

2 AI Techniques: Basic Review

2.1 Expert Systems (ES)

Expert systems are applying Boolean logic, intelligent computer programming. It sanctions to transplant human expertise in specific areas to solve the issues as a human being [3, 5, 11, 13, 22]. That is, it replaces human experts in some particular fields [5, 10, 23–26]. ES is widely used in automation in other sectors; still, its use in power systems and power electronics engineering limited. The fundamental elements of ES embody the knowledge or expertise base, as in Fig. 1. The expert knowledge and database knowledge are the two parts of the knowledge base; that includes the data, facts, and other statements related to expert knowledge [27–36]. So, the location of the program of the human knowledge interface is in the knowledge base. Furthermore, it has data-related to computational methods connected with expert knowledge; it is a rule-based system.

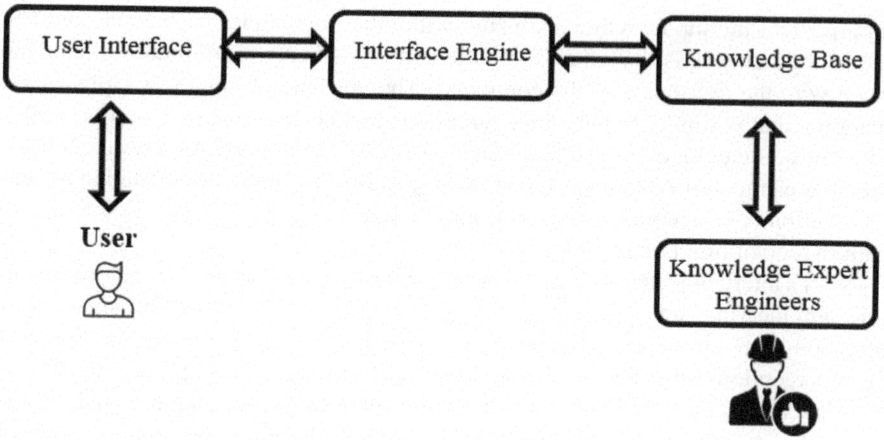

Fig. 1 Elements of an expert system

Application of Expert System in Power system:

- Decision making
- Solving issues based on reasoning, judgment, and heuristics.
- Collection of knowledge.
- In a short time, it handles complex and extensive data and information.

2.2 *Genetic Algorithms (GA)*

As an optimization technique, genetic algorithms (GA) are based on natural selection and genetics. It focuses on the coding of the variables set, not on the actual variables. GA acknowledges the optimal points through a population of possible solution points. It applies the probability transition laws and takes the objective function information [37]. Application of Genetic Algorithms in Power system is:

- Planning—wind turbine positioning, network feeder routing, reactive power optimization, and capacitor placement.
- Operations—maintenance scheduling, hydro-thermal plant co-ordination, reduce loss, load management, controls Facts.
- Analysis—filter design, reduce harmonic distortion, control load frequency, and load flow.

2.3 *Artificial Neural Networks (ANNs to NNWs)*

The artificial neural network is a biologically influenced system because the wiring of neurons converts the inputs to outputs. So, every neuron generates one output as a purpose of the input. While analyzing with other techniques such as FL and ES, the neural network is known as a generic sort of AI because it imitates the human brain with the assistance of the command. The attribute of nonlinear input-output mapping that is similar to patten recognition is in a neurocomputer system (NNW). Therefore, it can mock the human brain's associative memory. As a result, NNS is the vital element of AI that has the efficiency in solving the issues related to pattern recognition or image processing. In traditional methods, it is difficult to solve these pattern recognition issues [38].

NNS's interconnected artificial neurons can resolve various issues related to scientific, engineering, and real-life. ANN is characterized based on the signal flow direction as feedforward or feedback designs. It is quite obvious to use a multi-layer perception (MLP-network or the three-tier feedforward, backpropagation) type. NNS's exhibits the three input signals in the input layer and output signals in the output layer with scaling and descaling. Although the input and output layers of neurons are linear active functions, the hidden middle layer has a nonlinear activation function [38]. Application of Artificial Neural Networks in Power system includes:

- Power system problems related to unspecified nonlinear functions.
- Real-time operations.

2.4 Fuzzy Logic

Fuzzy logic code favors controlling mechanical input. It utilizes software or hardware mode, from simple circuits to mainframes. In power systems, the fuzzy system helps to increase the voltage profile of the power system. It permits to convert the voltage deviation and comparing variables into fuzzy system notions. Fuzzy logic backs to obtain reliable, constant, and clear output because normally, power system investigation employs approximate values and assumptions [37, 38].

The fuzzy interface system (FIS) has five stages in implementation:

- Fuzzification of input variables (Define fuzzy variables).
- Application of a fuzzy operators (AND, OR, NOT) in the IF rule.
- Implies IF and THEN.
- Aggregates the consequences.
- Defuzzification to convert FIS output to value.

Application of Fuzzy Logic in Power system:

- Power system control
- Fault diagnosis
- Stability analysis and improvement
- Assessing security
- Load forecasting

- State estimation
- Reactive power planning and control.

3 AI Applications in Power System

3.1 *AI in Transmission Line*

The fuzzy logic system renders the output of the faulty type based on the fault diagnosis. Whereas, ANN and ES serve to enhance the line performance. The environmental sensors contribute input for the expert system, and it generates an output based on the value of line parameters. Environmental sensors enable ANN to recognize the values of line parameters over the ranges stipulated. The training algorithms of ANN permits to test the neural network and identify the deviation on the performance for each hidden layer [37].

3.2 *Smart Grid and Renewable Energy Systems—Power System Stability*

Probabilistic power system stability assessment involves three components—input variable modeling, computational methods, and determining output indices, as in Fig. 3. The processes and models included in the assessment are represented in Table 1 and Fig. 2 [39].

Figure 3 summarizes that most applied computational methods in different stability analyses based on the reviewed articles. This means that the trends toward using probabilistic computational analysis in the power system increasing rapidly by increasing the complexity and insecurity of the power system in general.

Table 1 Probabilistic assessment of power system stability

Input variable	Computational methods	Output indices
Operational variables	Monte Carlo method	Transient stability
Disturbance variables	Sequential Monte Carlo	Frequency stability
	Quasi-Monte Carlo	Voltage stability
	Markov chain	Small-disturbance stability
	Point estimate	
	Probabilistic collection	

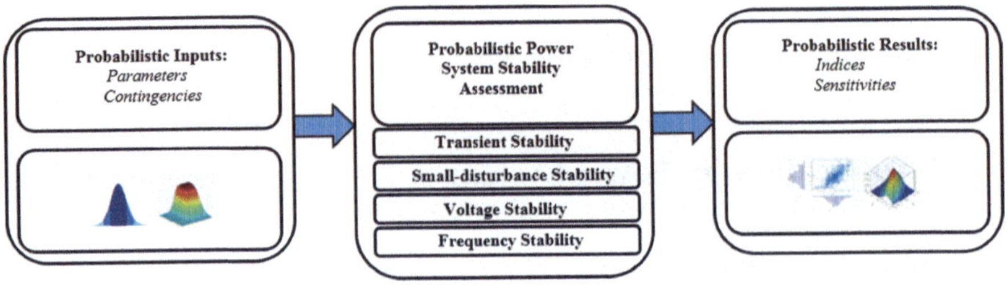

Fig. 2 The framework of the probability assessment of power system stability

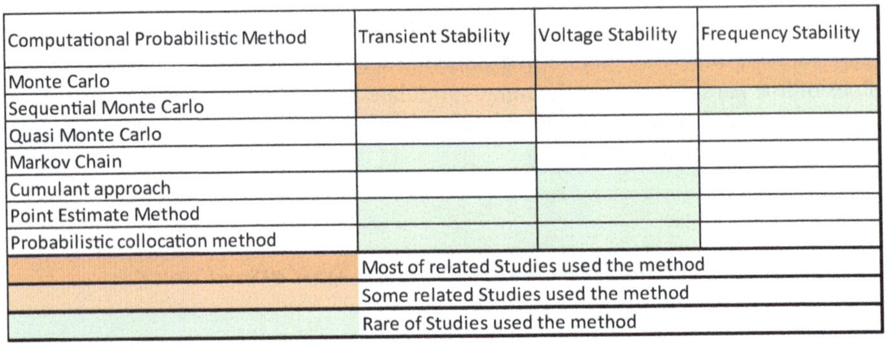

Fig. 3 Computational techniques with stability application

The probabilistic power system analysis comprises of stability, load flow, reliability, and planning [40]. So, it highly supportive during increased uncertainties as in the current situation.

3.3 *Expert System Based Automated Design, Simulation and Controller Tuning of Wind Generation System*

Figure 4 illustrates the ES based software structure for automated design, simulation, and controller turning of the wind generation system. It can be related to the basic ES system because it has the expert system shell that supports the software platform for the ES program. Apart from it, it embeds the knowledge base of the ES where it has rules such as If-Then rules. As per the user specification to the engine interference, the system concludes after it validates with the rules specified in the knowledge base. The design block is responsible for the type of machine, converter system, controller, and other design elements along with the optimum configuration mentioned in the knowledge base. The simulation wing deals with the purpose; it tunes the controller parameters online and also verifies the design power circuit elements of the system. It is necessary to know that simulation is hybrid, so it has plant simulation that is slow

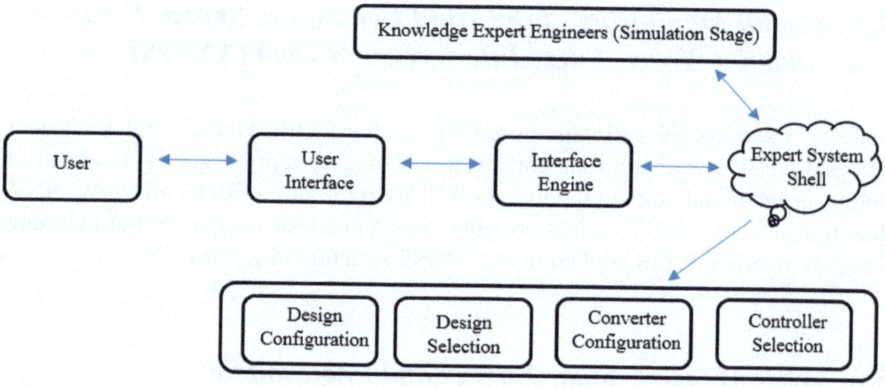

Fig. 4 ES in automated design, simulation and controller of wind generation system

(Simulink/Sim Power System) and the controller simulation (partly in C language and Assembly language) [2].

3.4 Real-Time Smart Grid Simulator-Based Controller

The gird is large and complex, so it is difficult to control, monitor, and protect the smart grid. Yet, centralizing the complete system by integrating the advanced control, information, computer, communication, and other cyber technologies, it is possible to develop ES-based master control. By using a supercomputer-based real-time simulator (RTS) the simplified block diagram as in Figure enables to control SG efficiently [38]. The real-time simulation is extensive and complex, a simulation done in parts with Simulink/SimPowerSystem in correspondence by supercomputers. The outcomes are combined and converted to the C language to improve the speed matching estimate with the real-time operations of the grid. In case of any issues in the centralized control, the regional controller can cancel it. Moreover, in the case of small and autonomous SG, exclusion of regional controllers is possible. Thus, master controller system privileges to know the predictions, demands, actual operating conditions of the grid, equipment usage, depreciation, computes tariff rates, encourages demand-side energy management with smart meters, transient power loading or rejection, stabilizes frequency and voltage, real-time HIL (Hardware-in-the-loop), automated testing, reconfiguration. That is system monitoring, fault protection, and diagnosis of SG.

3.5 Health Monitoring of the Wind Generation System Using Adaptive Neuro-Fuzzy Interference System (ANFIS)

Sensors or sensors less estimation used for monitoring the wind generation system. FIS and NNW are appropriate for the monitoring system because it operates on nonlinear input and output mapping. In ANFIS, it mimics a FIS by an NNW. NNW determines feedforward, so, it is possible to give the desired input-output mapping. The computation of FIS applied in the ANFIS structure (5 layers) [38].

3.6 ANN Models—Solar Energy and Photovoltaic Applications

A photovoltaic generator depends on solar irradiation, humidity, and ambient temperature in generating power. In order to have a maximum generation, the output of the photovoltaic generator must operate in optimal voltage and current, which can be obtained using the Maximum Power Point Tracking (MPPT). One of the most recent algorithms used for MPPT is the ANN-based approach. The main objective of using the ANN-based approach for MPPT is to control the input voltage signal to be as close as the optimal voltage.

3.7 Fuzzy Interference System for PVPS (Photovoltaic Power Supply) System

Figure 5 displays the fuzzy interface system, MLP (Multi-layer perceptron) is applied based on the fuzzy logic rules. So, the input factors include longitude, latitude, and altitude. Whereas, the output includes 12 factors of the mean monthly clearness index [41].

4 Sustainability in Power System Under AI Technology

The world increasing their concerns on environments affect that causes by energy sector where electric energy is produced from the plant (using natural gas and coal) and transmit through the transmission lines to the end user. AI enhances the sustainability in energy sector by combing this technology with renewable energy to produce a clean energy. An example of implementing a sustainable practice in power sector is a distributed panel in the system, which contributes effectively in providing electricity to the system in a clean manner. Many studies conducted recently to measure the improvement of sustainability if AI is fully adopted in the system.

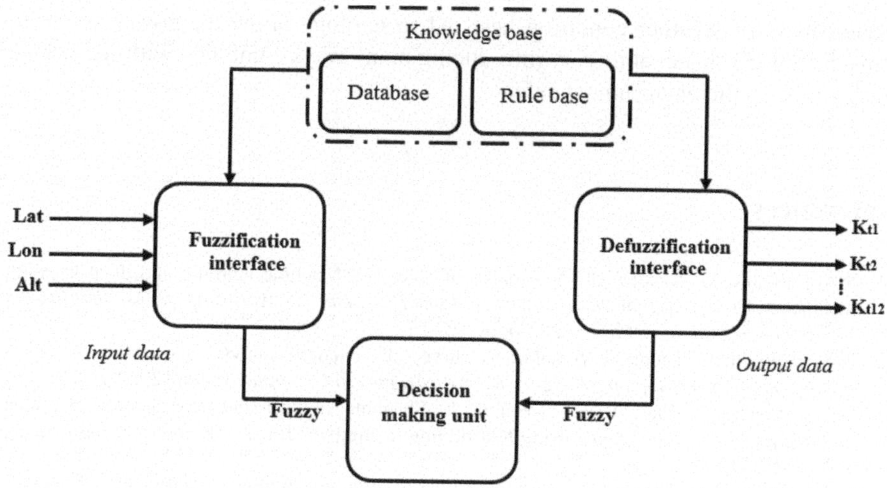

Fig. 5 Fuzzy interface system for PVPS

5 Conclusion and Future Work

In conclusion, traditional methods do not meet the probabilistic condition of power systems. Hence, the implementation of Artificial Intelligence earns more attention. In recent years, various studies, as reviewed here, centered on adopting AI techniques in power systems, comprising smart grids and renewable energy such as solar and wind generators. The four main AI techniques, such as expert systems, fuzzy logic, genetic algorithms, and neural network, are widely adopted. The hybrid of these systems is emerging based on the circumstances. Each technique supports to resolve issues in the power system. The benefits of implementing AI in a power system is a wide range as presented. AI also results in reducing maintenance and operational costs. Besides, AI improves the efficiency of electricity or energy market, includes conventional and renewable energy. AI has a command on planning, control, monitoring, and forecasting activities. It is necessary to identify the appropriate AI technique in planning, monitoring, and controlling the power system.

Moreover, these methods improve the operation and productivity of the power system by controlling voltage, stability, power-flow, and load frequency. It also permits to control the network such as location, size, and control of equipment and devices. The automation of the power system ensures to support the restoration, fault diagnosis, management, and network security. Another essential feature is its accuracy and real-time prediction, estimation and forecast enable the energy sector to manage its resources efficiently and satisfy the demand. Various examples show the use of AI techniques in power systems. Different AI hybrid models will be utilized for forecasting solar energy and it will tested with some enhancement to know which model fits UAE weather condition best. AI technology allow the power sector to move be toward sustainability by introducing many ways combined with renewable energy to keep the environment safe.

References

1. S.A. Salloum, M. Alshurideh, A. Elnagar, K. Shaalan, Machine learning and deep learning techniques for cybersecurity: a review, in *Joint European-US Workshop on Applications of Invariance in Computer Vision* (2020), pp. 50–57
2. S.A. Salloum, R. Khan, K. Shaalan, A survey of semantic analysis approaches, in *Joint European-US Workshop on Applications of Invariance in Computer Vision* (2020), pp. 61–70
3. S.A. Salloum, C. Mhamdi, M. Al-Emran, K. Shaalan, Analysis and classification of arabic newschapters' facebook pages using text mining techniques. Int. J. Inf. Technol. Lang. Stud. **1**(2), 8–17 (2017)
4. S.A. Salloum, M. Al-Emran, K. Shaalan, A survey of lexical functional grammar in the arabic context, Int. J. Com. Net. Tech. **4**(3) (2016)
5. S.A. Salloum, M. Al-Emran, S. Abdallah, K. Shaalan, in *Analyzing the Arab Gulf Newschapters Using Text Mining Techniques*, vol. 639 (2018)
6. S.A. Salloum, M. Alshurideh, A. Elnagar, K. Shaalan, Mining in educational data: review and future directions, in *Joint European-US Workshop on Applications of Invariance in Computer Vision* (2020), pp. 92–102
7. S.F.S. Alhashmi, M. Alshurideh, B. Al Kurdi, S.A. Salloum, A systematic review of the factors affecting the artificial intelligence implementation in the health care sector, in *Joint European-US Workshop on Applications of Invariance in Computer Vision* (2020), pp. 37–49
8. S.F.S. Alhashmi, S.A. Salloum, C. Mhamdi, Implementing artificial intelligence in the United Arab Emirates healthcare sector: an extended technology acceptance model, Int. J. Inf. Technol. Lang. Stud. **3**(3) (2019)
9. K.M. Alomari, A.Q. Alhamad, H.O. Mbaidin, S. Salloum, Prediction of the digital game rating systems based on the ESRB. Opcion **35**(19) (2019)
10. S.A. Salloum, M. Al-Emran, A. Monem, K. Shaalan, A survey of text mining in social media: facebook and twitter perspectives. Adv. Sci. Technol. Eng. Syst. J. **2**(1), 127–133 (2017)
11. S.A. Salloum, M. Al-Emran, A.A. Monem, K. Shaalan, Using text mining techniques for extracting information from research articles, in *Studies in Computational Intelligence*, vol. 740 (Springer, Berlin, 2018)
12. S.A. Salloum, A.Q. AlHamad, M. Al-Emran, K. Shaalan, in *A Survey of Arabic Text Mining*, vol. 740 (2018)
13. C. Mhamdi, M. Al-Emran, S.A. Salloum, in *Text Mining and Analytics: A Case Study from News Channels Posts on Facebook*, vol. 740 (2018)
14. M.T.A. Nedal Fawzi Assad, Financial reporting quality, audit quality, and investment efficiency: evidence from GCC economies. WAFFEN-UND Kostumkd. J. **11**(3), 194–208 (2020)
15. M.T.A. Nedal Fawzi Assad, Investment in context of financial reporting quality: a systematic review. WAFFEN-UND Kostumkd. J. **11**(3), 255–286 (2020)
16. A. Aburayya, M. Alshurideh, A. Albqaeen, D. Alawadhi, I. Ayadeh, An investigation of factors affecting patients waiting time in primary health care centers: An assessment study in Dubai. Manag. Sci. Lett. **10**(6), 1265–1276 (2020)
17. R. Shannak, R. Masa'deh, Z. Al-Zu'bi, B. Obeidat, M. Alshurideh, H. Altamony, A theoretical perspective on the relationship between knowledge management systems, customer knowledge management, and firm competitive advantage. Eur. J. Soc. Sci. **32**(4), 520–532 (2012)
18. H. Altamony, M. Alshurideh, B. Obeidat, Information systems for competitive advantage: Implementation of an organisational strategic management process, in *Proceedings of the 18th*

IBIMA Conference on Innovation and Sustainable Economic Competitive Advantage: From Regional Development to World Economic, Istanbul, Turkey, 9–10 May (2012)
19. M.T. Alshurideh et al., The impact of islamic bank's service quality perception on jordanian customer's loyalty, J. Manage. Res. **9** (2017)
20. B.A. Kurdi, M. Alshurideh, S.A. Salloum, Z.M. Obeidat, R.M. Al-dweeri, An empirical investigation into examination of factors influencing university students' behavior towards elearning acceptance using SEM approach, Int. J. Interact. Mob. Technol. **14**(2) (2020)
21. M. AlShurideh, N.M. Alsharari, B. Al Kurdi, Supply chain integration and customer relationship management in the airline logistics. Theor. Econ. Lett. **9**(02), 392–414 (2019)
22. M.A.R. Abdeen, S. AlBouq, A. Elmahalawy, S. Shehata, A closer look at arabic text classification. Int. J. Adv. Comput. Sci. Appl. **10**(11), 677–688 (2019)
23. A. Ghannajeh et al., A qualitative analysis of product innovation in Jordan's pharmaceutical sector. Eur. Sci. J. **11**(4), 474–503 (2015)
24. A. Alshraideh, M. Al-Lozi, M. Alshurideh, The impact of training strategy on organizational loyalty via the mediating variables of organizational satisfaction and organizational performance: an empirical study on jordanian agricultural credit corporation staff. J. Soc. Sci. **6**, 383–394 (2017)
25. M. Alshurideh, B. Al Kurdi, A. Abumari, S. Salloum, Pharmaceutical promotion tools effect on physician's adoption of medicine prescribing: evidence from Jordan, Mod. Appl. Sci. **12**(11), 210–222 (2018)
26. Z. Zu'bi, M. Al-Lozi, S. Dahiyat, M. Alshurideh, A. Al Majali, Examining the effects of quality management practices on product variety. Eur. J. Econ. Financ. Adm. Sci. **51**(1), 123–139 (2012)
27. H. Al Dmour, M. Alshurideh, F. Shishan, The influence of mobile application quality and attributes on the continuance intention of mobile shopping. Life Sci. J. **11**(10), 172–181 (2014)
28. M. Alshurideh, A. Alhadid, B. Al kurdi, The effect of internal marketing on organizational citizenship behavior an applicable study on the university of Jordan employees. Int. J. Mark. Stud. **7**(1), 138 (2015)
29. M. Alshurideh, A. Shaltoni, D. Hijawi, Marketing communications role in shaping consumer awareness of cause-related marketing campaigns. Int. J. Mark. Stud. **6**(2), 163 (2014)
30. S.A. Salloum, M. Al-Emran, K. Shaalan, The impact of knowledge sharing on information systems: a review, in *13th International Conference, KMO 2018* (2018)
31. R. Al-dweeri, Z. Obeidat, M. Al-dwiry, M. Alshurideh, A. Alhorani, The impact of e-service quality and e-loyalty on online shopping: moderating effect of e-satisfaction and e-trust. Int. J. Mark. Stud. **9**(2), 92–103 (2017)
32. A. ELSamen, M. Alshurideh, The impact of internal marketing on internal service quality: a case study in a Jordanian pharmaceutical company. Int. J. Bus. Manag. **7**(19), 84 (2012)
33. M. Alshurideh, R. Masa'deh, B. Al kurdi, The effect of customer satisfaction upon customer retention in the Jordanian mobile market: an empirical investigation. Eur. J. Econ. Financ. Adm. Sci. **47**(12), 69–78 (2012)
34. M. Ashurideh, Customer service retention—A behavioural perspective of the UK mobile market (Durham University, 2010)
35. G. Ammari, B. Al kurdi, M. Alshurideh, A. Alrowwad, Investigating the impact of communication satisfaction on organizational commitment: a practical approach to increase employees' loyalty. Int. J. Mark. Stud. **9**(2), 113–133 (2017)
36. M. Alshurideh, M. Nicholson, S. Xiao, The effect of previous experience on mobile subscribers' repeat purchase behaviour. Eur. J. Soc. Sci. **30**(3), 366–376 (2012)
37. R.P. Nath, V.N. Balaji, Artificial intelligence in power systems. IOSR J. Comput. Eng. e-ISSN, 661–2278 (2014)
38. B.K. Bose, Artificial intelligence techniques in smart grid and renewable energy systems—some example applications. Proc. IEEE **105**(11), 2262–2273 (2017)
39. K.N. Hasan, R. Preece, J.V. Milanović, Existing approaches and trends in uncertainty modelling and probabilistic stability analysis of power systems with renewable generation. Renew. Sustain. Energy Rev. **101**, 168–180 (2019)

40. K. Meng, Z. Dong, P. Zhang, Emerging techniques in power system analysis (Springer, Berlin, 2010), pp. 117–145
41. R. Belu, Artificial intelligence techniques for solar energy and photovoltaic applications, in *Handbook of Research on Solar Energy Systems and Technologies* (IGI Global, 2013), pp. 376–436

Smart Networking Applications

Internet of Things for Water Quality Monitoring and Assessment: A Comprehensive Review

Joshua O. Ighalo, Adewale George Adeniyi, and Goncalo Marques

Abstract The implementation of urbanisation and industrialisation plans lead to the proliferation of contaminants in water resources which is a severe public challenge. These have led to calls for innovative means of water quality monitoring and mitigation, as highlighted in the sustainable development goals. Environmental engineering researchers are now seeking more intricate techniques conducting real-time monitoring and assessing of the quality of surface and groundwater that is assessable to the human population across various locations. Numerous recent technologies now utilise the Internet of Things (IoT) as a platform in water quality monitoring and assessment. Wireless sensor network and IoT environments are currently being used more frequently in contemporary times. In this paper, the recent technologies harnessing the potentials and possibilities in the IoT for water quality monitoring and assessment is comprehensively discussed. The main contribution of this paper is to present the research progress, highlight recent innovations and identify interesting and challenging areas that can be explored in future studies.

Keywords Actuators · Environment · Internet of things · Sensors · Sustainable development · Water quality

J. O. Ighalo · A. G. Adeniyi
Department of Chemical Engineering, University of Ilorin, P. M. B. 1515, Ilorin, Nigeria
e-mail: oshea.ighalo@yahoo.com

A. G. Adeniyi
e-mail: adeniyi.ag@unilorin.edu.ng

G. Marques (✉)
Instituto de Telecomunicações, Universidade da Beira Interior, 6200-001 Covilhã, Portugal
e-mail: goncalosantosmarques@gmail.com

© The Editor(s) (if applicable) and The Author(s), under exclusive license to Springer Nature Switzerland AG 2021
A. E. Hassanien et al. (eds.), *Artificial Intelligence for Sustainable Development: Theory, Practice and Future Applications*, Studies in Computational Intelligence 912,
https://doi.org/10.1007/978-3-030-51920-9_13

1 Introduction

Water is one of the most abundant natural resources in the biosphere and one that is important for the sustenance of life on earth [1]. The implementation of urbanisation and industrialisation plans lead to the proliferation of contaminants in water resources which is a severe public challenge [2–4]. About 250 million cases of diseases infections are annually reported world-wide due to water pollution-related causes [5]. Therefore, innovative means of monitoring and mitigation water pollution are required [6–8] so that environmental sustainability can be achieved as highlighted in the sustainable development goals (SDGs). Environmental engineering researchers are now developing more intricate techniques for conducting real-time monitoring and assessing of the quality of surface and groundwater that is assessable to the human population across various locations [9, 10]. The internet has powered a lot of technologies and applications which make possible in our time. The Internet of Things (IoT) is an integration of many newly developed digital/information technologies [11].

The IoT now has applications in diverse anthropogenic activities both in the domestic and industrial domain [13]. These include transportation and logistics, healthcare, smart homes and offices [2], water quality assessment [14], tourism, sports, climatology [15], aquaculture [16] and a host of others [17]. More discussion on the IoT can be found elsewhere [18, 19]. Numerous recent technologies now utilise the IoT as a platform in water quality monitoring and assessment [19]. Wireless sensor network and IoT environments are currently being used more frequently in contemporary times. The intricacies of the system require that aspects such as software programming, hardware configuration, data communication and automated data storage be catered for [20].

IoT-enabled AI for Water Quality Monitoring is quite relevant for sustainable development purposes. The presence of clean water to humans is a fundamental part of the sixth (6th) sustainable development goal. It would be difficult to assess which water body and sources is actually clean enough to drink without water quality monitoring. Furthermore, the utilisation of IoT-enabled AI means that any potential water pollution arising from a point or non-point source is quickly identified and mitigated. For 14th sustainable development which emphasises the need to protect life below water, IoT-enabled AI for Water Quality Monitoring would ensure that the quality of water do not go below threshold detrimental to the survival of aquatic flora and fauna.

Within the scope of the authors' exhaustive search, the last detailed review on the subject was published over 15 years ago by Glasgow et al. [21]. In that time frame, a lot has changed in the technology as much advancements and breakthroughs have been made. It would not be out of place to revisit the topic and evaluate recent findings.

In this chapter, the recent technologies harnessing the potentials and possibilities in the IoT for water quality monitoring and assessment is comprehensively discussed. The main contribution of this paper is to present the research progress, highlight recent innovations and identify interesting and challenging areas that can be explored in future studies. After the introduction, the first section discusses the fundamental reasons behind water quality assessment and defines the fundamental indices involved. The next section discusses the importance of IoT in water quality monitoring and assessment. The hardware and software designs for IoT enabled water quality monitoring and assessment for a smart city was discussed in the foregoing section. This is succeeded by an empirical evaluation on the subject matter based on published literature in the past decade and concluded by discussions on knowledge gap and future perspectives.

2 Water Quality Assessment in Environmental Technology

Water quality refers to the physical, chemical and biological characteristics of water [22]. Assessment and monitoring of water quality are essential because it helps in timely identification of potential environmental problems due to the proliferation of pollutants (from anthropogenic activities) [11]. These are usually done both in the short and long term [23]. Monitoring and assessment are also fundamental so that potential regulation offenders can be identified and punished [24]. Technical details as regards the methods for environmental monitoring is discussed by McDonald [25].

There are specific indices used in water quality. A water quality index (WQI) is a dimensionless number used in expressing the overall quality of a water sample based on measurable parameters [26]. Many indices have been developed (as much as 30), but only about seven (7) are quite popular in contemporary times [26]. In all these, the foundational information about the water is gotten from the measurable parameters [27]. The important measurable parameters of water quality are defined below [28].

1. ***Chemical oxygen demand (COD)***: This is the equivalent amount of oxygen consumed (measured in mg/l) in the chemical oxidation of all organic and oxidisable inorganic matter contained in a water sample.
2. ***Biochemical oxygen demand (BOD)***: This is the oxygen requirement of all the organic content in water during the stabilisation of organic matter usually over a 3 or 5 day.
3. ***pH***: This is the measure of the acidity or alkalinity of water. It is neutral (at 7) for clean water and ranges from 1 to 14.
4. ***Dissolved oxygen (DO)***: This is the amount of oxygen dissolved in a water sample (measured in mg/l).
5. ***Turbidity***: This is the scattering of light in water caused by the presence of suspended solids. It can also be referred to as the extent of cloudiness in water measured in nephelometric turbidity units (NTU).
6. ***Electrical conductivity (EC)***: This is the amount of electricity that can flow through water (measured in Siemens), and it is used to determine the extent of soluble salts in the water.

7. ***Temperature***: The is the degree of hotness or coldness of the water and usually measured in degrees Celsius (°C) or Kelvin (K).
8. ***Oxidation-reduction potential (ORP)***: This is the potential required to transfer electrons from the oxidant to the reductant, and it is used as a qualitative measure of the state of oxidation in water.
9. ***Salinity***: This is the salt content of the water (measured in parts per million).
10. ***Total Nitrogen (TN)***: This is the total amount of nitrogen in the water (in mg/l) and is a measure of its potential to sustain and eutrophication or algal bloom.
11. ***Total phosphorus (TP)***: This is the total amount of phosphorus in the water (in mg/l) and is a measure of its potential to sustain and eutrophication or algal bloom.

3 Internet of Things in Water Quality Assessment

Environmental engineering researchers are now seeking more intricate techniques for conducting real-time monitoring and assessing of the quality of surface and groundwater that is assessable to the human population across various locations. Digital communication technologies are now the bedrock of modern society [29] and IoT enabled water quality monitoring and assessment is a vital aspect of that. The traditional method of water quality monitoring requires human personnel taking the readings by instruments and logging the data [30] is considered inefficient, slow and expensive [20]. In this section, the importance of IoT in water quality monitoring and assessment is itemised in light of its advantages over the traditional methods of water sampling and analysis utilised by environmental engineers and scientists when conducting water quality monitoring.

1. The most significant advantage of IoT in water quality monitoring and assessment is the possibility of real-time monitoring. Here, the status of the water quality (based on the different indices) can be obtained at any given time. This is facilitated by the speed of internet communications where data can be transmitted from the sensors in fractions of a second. These incredible speeds are not achievable in traditional water quality monitoring.
2. IoT in water quality monitoring and assessment can be automated. This means that it does not require the presence of human personnel to take readings and log data [31]. Moreover, these IoT systems would require less human resources and eliminate human errors in data logging and computations. Automation is the foundational concept of smart cities and its associated technologies.
3. Alongside the advantage of automation, IoT has led to the use of adaptive and responsive systems in water quality monitoring. These smart-systems can alert authorities or personnel regarding impending danger (such as high water level of an impending flood) or non-optimal conditions (such as in aquaponic systems) [32].

4. IoT in water quality monitoring and assessment is cheaper than hands-on personnel conducting the monitoring and assessment. The cost of human resources is minimised, and an IoT based system would not require.

4 Water Quality Monitoring Systems

IoT aims to provide a continuous presence of distinct cyber-physical systems which incorporate and intelligence capabilities [33, 34]. On the one hand, IoT has changed people daily routine and is today included in most social activities and in particular regarding smart city concept [35]. On the other hand, IoT is a relevant architecture for the design and development of intelligent monitoring systems for water quality assessment.

IoT is currently applied in different kinds of monitoring activities such as thermal comfort [36–40], acoustic comfort [41] and air quality [42, 43]. Moreover, IoT is also applied in agricultural environments such as aquaponics and hydroponics [44–47]. Water quality is crucial in agricultural activities and significantly affect the productivity and efficiency of agricultural ecosystems. IoT systems for enhanced water quality allow to store and compare the water quality data to support the decision making of agricultural plant managers.

The smart city concept associated with multiple strategies which aim to address the most relevant cities challenges using computer science technologies [48]. Currently, cities face crucial challenges regarding their socio-economic goals and the best approaches to meet them and at the same time, improve public health [49]. Water resources are an integral element of cities and are also a crucial challenge regarding their management and quality assessment [50]. Water contamination significantly affects the health and well-being of citizens, and real-time supervisor systems can be used to detect possible contamination scenarios for enhanced public health early.

IoT systems can be located in multiple places and provide a continuous stream of real-time water quality data to various municipal authorities to improve water resources management. The data collected can also be used to plan interventions for enhanced public safety [51].

The technologies used in the design and development of IoT systems in the water management domain are presented in Sect. 4.1.

4.1 Hardware and Software Design

Currently, multiple technologies are available for the design and development of IoT systems. On the one hand, numerous open-source platforms for IoT development such as Arduino, Raspberry Pi, ESP8266 and BeagleBone [52]. These platforms support various short-range communication technologies such as Bluetooth and Wi-Fi but also long-range such as GPRS, UMTS, 3G/4G and LoRA that are efficient methods

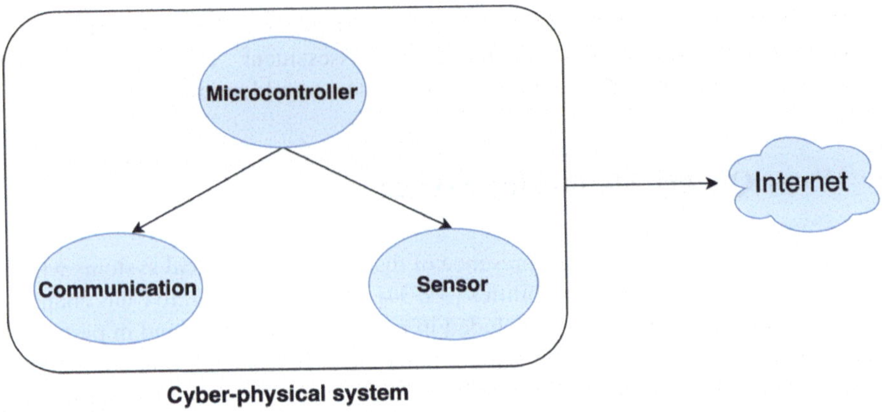

Fig. 1 IoT architecture

for data transmission. Moreover, IoT platforms also support multiple identification technologies, such as NFC and RFID identification technologies [53].

At the hardware level, IoT cyber-physical system can be divided into three elements: microcontroller, sensor and communication (Fig. 1). Commonly, an IoT system is composed by the processing unit, the sensing unit and the communication unit. The processing unit is the microcontroller which is responsible for the interface with the sensor part and can have integrated communication unit or be connected to the communication module for data transmission. The sensor unit is responsible for the physical data collection and is connected to the microcontroller using several interfaces such as analogue input, digital input and I2C. The communication unit is related to the communication technologies used for data transmission. These technologies can be wireless such as Wi-Fi or cabled such as Ethernet.

The data collected using the sensor unit is processed and transmitted to the Internet. These activities are handled using the microcontroller. The analysis, visualization and mineralization of the collected data are conducted using online services and carried by backend services which include more powerful processing units. Multiple low-cost sensors are available with different interface communication and support for numerous microcontrollers which can be applied in the water management domain [54–56].

4.2 Smart Water Quality Monitoring Solutions

Water quality assessment also plays a significant role in multiple agricultural domains such as hydroponics, aquaponics and aquaculture. In these environments water quality must be monitored; however, the main applications involve high priced solutions which cannot be incorporated in the developing countries. Therefore, the cost of water quality monitoring system is a relevant factor for their implementation.

On the one hand, hydroponic applications the nutrients in the water are a crucial factor to be monitored in real-time to provide high-quality products and avoid problems related to contaminations [57]. Therefore, water quality monitoring systems must be incorporated as long with advanced techniques of energy consumption monitoring since hydroponics is associated with high energy consumptions [58, 59]. Moreover, real-time monitoring is essential also in aquaponics since this approach combines the conventional aquaculture methods in the symbiotic environment of plants and depends on nutrient-generators. In aquaponic environments, the excrement produced by animals is used as nitrates that are used nutrient by plants [60]. On the other hand, smart cities require efficient and effective management of water resources [61].

Currently, the availability of low-cost sensors promotes the development of continuous monitoring systems for water monitoring [62]. Furthermore, numerous connectivity methods are available for data transmission of the collected data using wireless technologies [63]. Bluetooth and Zigbee communication technologies can be used to interface multiple IoT units to create short-range networks and be combined with Wi-Fi and mobile networks for Internet connection [64, 65].

Furthermore, smartphones currently have high computational capabilities and support NFC and Bluetooth, which can be used to interface external components such as IoT [66]. In particular, Bluetooth technologies can be used to configure and parametrize IoT water quality monitoring systems and retrieve the collected data in locations where Internet access are not available. On the one hand, mobile devices enable numerous daily activities and provide a high number of solutions associated with data visualization and analytics [67]. On the other hand, people commonly prefer to use their smartphones when compared with personal computers [68, 69].

The current water quality monitoring systems are high cost and do not support data consulting features in real-time. The data collected by these systems are limited since it is not related to the date of data collection and location. The professional solutions available in the literature can be compact and portable. However, that equipment does not provide a continuous data collection and sharing in real-time. Most of these systems only provide a display for data consulting or provide a memory card for data storage. Therefore, the user must extract the information and analyses the results using third-party software.

TDS and conductivity pens are quickly found in the market and are also widely used for water assessment. However, these portable devices do not incorporate data storage or data-sharing features. The user can only check the results using an LCD existent on the equipment. Moreover, this equipment commonly does not have any data storage method.

The development of smart water quality solutions using up to date technologies which provide real-time data access is crucial for the management of water resources (Fig. 2). It is necessary to design architectures which are portable, modular, scalable, and which can be easily installed by the user. The real-time notifications are also a relevant part of this kind of solutions. The real-time notification feature can enable intervention in useful time and consequently address the contamination scenarios in an early phase of development.

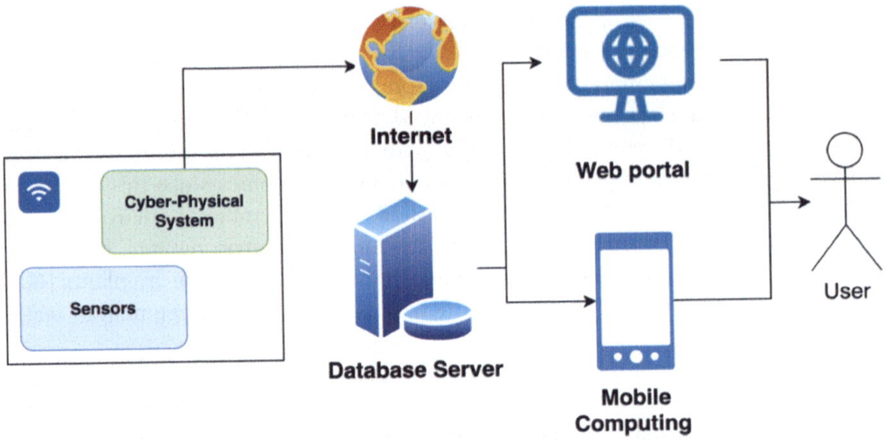

Fig. 2 Smart water monitoring system

5 An Empirical Evaluation of IoT Applications in Water Quality Assessment

In this section, a brief chronological evaluation is made of some of the interesting empirical investigations where IoT enabled technology was utilised in water quality monitoring and assessment. The focus will be not just on studies where an IoT-enabled system was designed for water quality monitoring and assessment but for studies where this technology was applied to specific water bodies within the past decade.

Wang et al. [70] monitored the water quality in the scenic river, Xinglin Bay in Xiamen, China. Their system was divided into three subsystems. There was the data acquisition subsystem, the digital data transmission subsystem and data processing subsystem. The indices monitored were pH, dissolved oxygen (DO), turbidity, conductivity, oxidation-reduction potential (ORP), chlorophyll, temperature, salinity, chemical oxygen demand (COD), NH_4^+, total phosphorus (TP) and total nitrogen (TN). The results of the study were positive as the design was adequate in achieving the set objectives. Furthermore, the water quality was of a good standard as the water had a powerful self-purification ability.

Shafi et al. [71] investigated the pH, turbidity and temperature of surface water across 11 locations in Pakistan, using an IoT enabled system that in-cooperated machine learning algorithms. The four algorithms considered were Support Vector Machine (SVM), k Nearest Neighbour (kNN), single-layer neural network and deep neural network. It was observed from the learning process on the 667 lines of data that deep neural network had the highest accuracy (at about 93%). The model could accurately predict water quality in the future six months.

Saravanan et al. [72] monitored the turbidity, temperature and colour at water pumping in Tamilnadu, India using a Supervisory Control and Data Acquisition

(SCADA) system that is enabled by IoT. The technology was usable in real-time and employed a GSM module for wireless data transfer.

In a quite interesting study, Esakki et al. [73] designed an unmanned amphibious vehicle for pH, DO, EC, temperature, and turbidity of water bodies. The device could function both in air and in water. Part of the mechanical design considerations was in its power requirements, propulsion, hull and skirt material, hovercraft design and overall weight. It was designed for military and civil applications with a mission of time of 25 min, a maximum payload of 7 kg and utilised an IoT based technology.

Liu et al. [74] monitored the drinking water quality at a water pumping station along the Yangtze river in Yangzhou, China. The technology was IoT enabled but incorporated a Long Short-Term Memory (LSTM) deep learning neural network. The parameters assessed were Temperature, pH, DO, Conductivity, Turbidity, COD and NH_3.

Zin et al. [75] utilised wireless sensor network enabled by IoT for the monitoring of water quality in real-time. The system they utilised consisted of Zigbee wireless communication, protocol, Field Programmable Gate Array (FPGA) and a personal computer. They utilised the technology to monitor the pH, turbidity, temperature, water level and carbon dioxide on the surface of the water at Curtin Lake, northern Sarawak in the Borneo island. The system was able to minimise cost and had lesser power requirements. Empirical investigations of IoT applications in water quality monitoring and assessment is summarised in Table 1.

Due to the nature of the sensors, parameters like TDS, turbidity, electrical conductivity, pH and water level are the more popularly studied indices. This was quite apparent from Table 1. It would require a major breakthrough in sensor technology to have portable and cheap sensors that can detect other parameters like heavy metals and other ions. The future of research in this area is likely to be investigations on alternative sensor technologies to determine the wide range of parameters that can adequately describe the quality of water. If this is achievable, then water quality monitoring and assessment would be able to apply correlations of Water Quality Index (WQI) to get quick-WQI values. This would enable rapid determination of the suitability of water sources for drinking.

The current water quality monitoring systems are relatively expensive and do not support data consulting features in real-time. It is predicted that researchers will gradually shift focus from portability in design to affordability. Furthermore, the development of smart water quality solutions using up to date technologies which provide real-time data access is crucial for the management of water resources. It is necessary to design architectures which are portable, modular, scalable, and which can be easily installed by the user. Researchers in the future will likely delve into better real-time monitoring technologies that would incorporate notifications and social media alerts.

Table 1 Summary of IoT applications in water quality monitoring and assessment

Year	Location	Parameters monitored	Ref
2019	Curtin Lake, Borneo island	pH, turbidity, temperature, water level and CO_2	[75]
2019	Pumping station, Yangtze river, Yangzhou, China	Temperature, pH, DO, EC, turbidity, COD and NH_3	[74]
2019	Unspecified location in Bangladesh	pH, turbidity, ORP and temperature	[76]
2018	Pumping station, Tamilnadu, India	Turbidity, temperature and colour	[72]
2018	Unspecified location	pH, DO, EC, temperature, and turbidity	[73]
2018	11 locations in Pakistan	pH, turbidity and temperature	[71]
2018	Unspecified location in India	pH, water level, temperature and CO_2	[13]
2017	Lab setup, India	pH and EC	[1]
2017	Aquaponics system, Manchay, near Lima, Peru	pH, DO and temperature	[77]
2017	Aquaponic system, Chennai, India	pH, water level, temperature and ammonia	[78]
2017	Unspecified location in India	pH, turbidity and EC	[55]
2017	Unspecified location in India	pH, turbidity and water level	[15]
2017	Nibong Tebal, Malaysia	pH and temperature	[79]
2015	Unspecified location in Malaysia	Water level	[12]
2013	Scenic river, Xiamen, China	pH, DO, turbidity, EC, ORP, chlorophyll, temperature, salinity, COD, NH_4^+, TP and TN	[70]
2006	7 locations in South Africa	Unspecified	[80]
2002	Tagus estuary, near Lisbon, Portugal	pH, turbidity and temperature	[81]

6 Conclusions

Urbanisation and industrialisation plans have led to the proliferation of contaminants in water resources which is now a severe environmental challenge. These have led to calls for innovative means of water quality monitoring and mitigation, as highlighted in the SDGs. The recent technologies harnessing the potentials and possibilities in the IoT for water quality monitoring and assessment is comprehensively discussed in this paper. Advantages of IoT in water quality monitoring and assessment are in the possibility of real-time monitoring, automation for smart solutions, adaptive and responsive systems and in a reduction of water quality monitoring costs. A brief chronological evaluation is made of some of the interesting empirical investigations where IoT enabled technology was utilised in water quality monitoring and assessment in the last decade. It was observed that IoT in water quality monitoring and assessment had not been applied to some more sophisticated parameters like heavy metals and other ions. The future of research in this area is likely to be investigations on alternative sensor technologies to determine the wide range of parameters that can adequately describe the quality of water. Cost considerations in the design and real-time data management are also areas of future research interest on the subject matter. The paper was successfully able to present the research progress, highlight recent innovations and identify interesting and challenging areas that can be explored in future studies.

References

1. B. Das, P. Jain, Real-time water quality monitoring system using internet of things, in *2017 International Conference on Computer, Communications and Electronics (Comptelix)*, Jaipur, Rajasthan India, 1–2 July 2017. IEEE
2. J. Shah, An internet of things based model for smart water distribution with quality monitoring. Int. J. Innov. Res. Sci. Eng. Technol. **6**(3), 3446–3451 (2017). http://dx.doi.org/10.15680/IJIRSET.2017.0603074
3. A.G. Adeniyi, J.O. Ighalo, Biosorption of pollutants by plant leaves: an empirical review. J. Environ. Chem. Eng. **7**(3), 103100 (2019). https://doi.org/10.1016/j.jece.2019.103100
4. J.O. Ighalo, A.G. Adeniyi, Mitigation of diclofenac pollution in aqueous media by adsorption. Chem. Bio. Eng. Rev. **7**(2), 50–64 (2020). https://doi.org/10.1002/cben.201900020
5. S.O. Olatinwo, T.H. Joubert, Efficient energy resource utilization in a wireless sensor system for monitoring water quality. EURASIP J. Wireless Commun. Netw. **2019**(1), 6 (2019). https://doi.org/10.1186/s13638-018-1316-x
6. P. Cianchi, S. Marsili-Libelli, A. Burchi, S. Burchielli, Integrated river quality management using internet technologies, in *5th International Symposium on Systems Analysis and Computing in Water Quality Management*, Gent, Belgium, 18–20 Sept 2000
7. J.O. Ighalo, A.G. Adeniyi, Adsorption of pollutants by plant bark derived adsorbents: an empirical review. J Water Process Eng. **35**, 101228 (2020). https://doi.org/10.1016/j.jwpe.2020.101228
8. O.A.A. Eletta, A.G. Adeniyi, J.O. Ighalo, D.V. Onifade, F.O. Ayandele, Valorisation of cocoa (theobroma cacao) Pod husk as precursors for the production of adsorbents for water treatment. Environ. Technol. Rev. **9**(1), 20–36 (2020). https://doi.org/10.1080/21622515.2020.1730983
9. R.G. Lathrop Jr., L. Auermuller, S. Haag, W. Im, The storm water management and planning tool: coastal water quality enhancement through the use of an internet-based geospatial tool. Coastal Manag. **40**(4), 339–354 (2012). https://doi.org/10.1080/08920753.2012.692309
10. J.H. Hoover, P.C. Sutton, S.J. Anderson, A.C. Keller, Designing and evaluating a groundwater quality Internet GIS. Appl. Geogr. **53**, 55–65 (2014). https://doi.org/10.1016/j.apgeog.2014.06.005
11. X. Su, G. Shao, J. Vause, L. Tang, An integrated system for urban environmental monitoring and management based on the environmental internet of things. Int. J. Sustain. Dev. World Ecol. **20**(3), 205–209 (2013). https://doi.org/10.1080/13504509.2013.782580
12. T. Perumal, M.N. Sulaiman, C.Y. Leong, Internet of things (IoT) enabled water monitoring system, in *2015 IEEE 4th Global Conference on Consumer Electronics (GCCE)*, Osaka, Japan, 27–30 Oct 2015. IEEE
13. K. Spandana, V.S. Rao, Internet of things (Iot) based smart water quality monitoring system. Int. J. Eng. Technol. **7**(6), 259–262 (2017)
14. P. Jankowski, M.H. Tsou, R.D. Wright, Applying internet geographic information system for water quality monitoring. Geography Compass. **1**(6), 1315–1337 (2007). https://doi.org/10.1111/j.1749-8198.2007.00065.x

15. P. Salunke, J. Kate, Advanced smart sensor interface in internet of things for water quality monitoring, in *2017 International Conference on Data Management, Analytics and Innovation (ICDMAI)*, Pune, India (24 Feb 2017. IEEE
16. D. Ma, Q. Ding, Z. Li, D. Li, Y. Wei, Prototype of an aquacultural information system based on internet of things E-Nose. Intell. Autom. Soft Comput. **18**(5), 569–579 (2012). https://doi.org/10.1080/10798587.2012.10643266
17. J.J. Caeiro, J.C. Martins, Water Management for Rural Environments and IoT, in *Harnessing the Internet of Everything (IoE) for Accelerated Innovation Opportunities* IGI Global 2019. pp. 83–99. http://dx.doi.org/10.4018/978-1-5225-7332-6.ch004
18. P. Smutný, Different perspectives on classification of the Internet of Things, in *2016 17th International Carpathian Control Conference (ICCC)*, High Tatras, Slovakia, 29 May–1 June 2016. IEEE
19. M.U. Farooq, M. Waseem, S. Mazhar, A. Khairi, T. Kamal, A review on internet of things (IoT). Int. J. Comput. Appl. **113**(1), 1–7 (2015)
20. L. Wiliem, P. Yarlagadda, S. Zhou, Development of Internet based real-time water condition monitoring system, in *Proceedings of the 19th International Congress and Exhibition on Condition Monitoring and Diagnostic Engineering Management*, Lulea, Sweden (12–15 June 2006). Lulea University of Technology Lulea
21. H.B. Glasgow, J.M. Burkholder, R.E. Reed, A.J. Lewitus, J.E. Kleinman, Real-time remote monitoring of water quality: a review of current applications, and advancements in sensor, telemetry, and computing technologies. J. Exp. Mar. Biol. Ecol. **300**(1–2), 409–448 (2004). https://doi.org/10.1016/j.jembe.2004.02.022
22. S.O. Olatinwo, T.-H. Joubert, Energy efficient solutions in wireless sensor system for monitoring the quality of water: a review. IEEE Sens. J. **19**(5), 1596–1625 (2019)
23. K.E. Ellingsen, N.G. Yoccoz, T. Tveraa, J.E. Hewitt, S.F. Thrush, Long-term environmental monitoring for assessment of change: measurement inconsistencies over time and potential solutions. Environ. Monit. Assess. **189**(11), 595 (2017)
24. W.B. Gray, J.P. Shimshack, The effectiveness of environmental monitoring and enforcement: a review of the empirical evidence. Rev. Environ. Econ. Policy. **5**(1), 3–24 (2011). https://doi.org/10.1093/reep/req017
25. T.L. McDonald, Review of environmental monitoring methods: survey designs. Environ. Monit. Assess. **85**(3), 277–292 (2003)
26. A.D. Sutadian, N. Muttil, A.G. Yilmaz, B. Perera, Development of river water quality indices—a review. Environ. Monit. Assess. **188**(1), 58 (2016)
27. X. Yu, Y. Li, X. Gu, J. Bao, H. Yang, L. Sun, Laser-induced breakdown spectroscopy application in environmental monitoring of water quality: a review. Environ. Monit. Assess. **186**(12), 8969–8980 (2014). https://doi.org/10.1007/s10661-014-4058-1
28. A. Bahadori, S.T. Smith, Dictionary of environmental engineering and wastewater treatment. Springer (2016). https://doi.org/10.1007/978-3-319-26261-1_1
29. D. Diamond, *Internet-scale sensing*, ACS Publications (2004)
30. F. Toran, D. Ramırez, A. Navarro, S. Casans, J. Pelegrı, J. Espı, Design of a virtual instrument for water quality monitoring across the Internet. Sensors Actuators B Chem. **76**(1–3), 281–285 (2001). https://doi.org/10.1016/S0925-4005(01)00584-6
31. F. Toran, D. Ramirez, S. Casans, A. Navarro, J. Pelegri, Distributed virtual instrument for water quality monitoring across the internet, in *Proceedings of the 17th IEEE Instrumentation and Measurement Technology Conference* [Cat. No. 00CH37066], Baltimore, MD, USA 1–4 May 2000. IEEE http://dx.doi.org/10.1109/IMTC.2000.848817
32. E.M. Dogo, A.F. Salami, N.I Nwulu, C.O. Aigbavboa, Blockchain and internet of things-based technologies for intelligent water management system, in *Artificial Intelligence in IoT* (Springer 2019), pp. 129–150. http://dx.doi.org/10.1007/978-3-030-04110-6_7
33. D. Giusto, A. Iera, G. Morabito, L. Atzori, *The Internet of things* (Springer, New York, New York, NY, 2010)
34. G. Marques, Ambient assisted living and Internet of things, in *Harnessing the Internet of everything (IoE) for accelerated innovation opportunities*, ed. by P.J.S. Cardoso, et al. (IGI Global, Hershey, PA, USA, 2019), pp. 100–115

35. J. Gubbi, R. Buyya, S. Marusic, M. Palaniswami, Internet of things (IoT): a vision, architectural elements, and future directions. Future Gener. Comput. Syst. **29**(7), 1645–1660 (2013). https://doi.org/10.1016/j.future.2013.01.010
36. G. Marques, R. Pitarma, Non-contact infrared temperature acquisition system based on Internet of things for laboratory activities monitoring. Procedia Comput. Sci. **155**, 487–494 (2019). https://doi.org/10.1016/j.procs.2019.08.068
37. G. Marques, I. Pires, N. Miranda, R. Pitarma, Air quality monitoring using assistive robots for ambient assisted living and enhanced living environments through Internet of things. Electronics **8**(12), 1375 (2019). https://doi.org/10.3390/electronics8121375
38. G. Marques, R. Pitarma, Smartwatch-Based Application for Enhanced Healthy Lifestyle in Indoor Environments, in *Computational Intelligence in Information Systems*, ed. by, S. Omar, W.S. Haji Suhaili, S. Phon-Amnuaisuk, (Springer International Publishing, Cham), pp. 168–177
39. G. Marques, R. Pitarma, Monitoring and control of the indoor environment, in *2017 12th Iberian Conference on Information Systems and Technologies (CISTI)*, Lisbon, Portugal, 14–17 June 2017. IEEE http://dx.doi.org/10.23919/CISTI.2017.7975737
40. G. Marques, R. Pitarma, Environmental quality monitoring system based on internet of things for laboratory conditions supervision, in *New Knowledge in Information Systems and Technologies*, ed. by Á. Rocha, et al. (Springer International Publishing, Cham, 2019), pp. 34–44
41. G. Marques, R. Pitarma, Noise monitoring for enhanced living environments based on Internet of things, in *New Knowledge in Information Systems and Technologies*, ed. by Á. Rocha, et al. (Springer International Publishing, Cham, 2019), pp. 45–54
42. G. Marques, R. Pitarma, Using IoT and social networks for enhanced healthy practices in buildings, in *Information Systems and Technologies to Support Learning*, ed. by Á. Rocha, M. Serrhini (Springer International Publishing, Cham, 2019), pp. 424–432
43. G. Marques, R. Pitarma, An Internet of things-based environmental quality management system to supervise the indoor laboratory conditions. Appl. Sci. **9**(3), 438 (2019). https://doi.org/10.3390/app9030438
44. M. Mehra, S. Saxena, S. Sankaranarayanan, R.J. Tom, M. Veeramanikandan, IoT based hydroponics system using deep neural networks. Comput. Electron. Agric. **155**, 473–486 (2018). https://doi.org/10.1016/j.compag.2018.10.015
45. V. Palande, A. Zaheer, K. George, Fully automated hydroponic system for indoor plant growth. Procedia Comput. Sci. **129**, 482–488 (2018). https://doi.org/10.1016/j.procs.2018.03.028
46. G. Marques, D. Aleixo, R. Pitarma, Enhanced hydroponic agriculture environmental monitoring: an internet of things approach, in *Computational Science—ICCS 2019*, ed. by J.M.F. Rodrigues, et al. (Springer International Publishing, Cham, 2019), pp. 658–669
47. S. Ruengittinun, S. Phongsamsuan, P. Sureeratanakorn P, Applied internet of thing for smart hydroponic farming ecosystem (HFE), in *2017 10th International Conference on Ubi-media Computing and Workshops*, Pattaya, Thailand, 1–4 Aug 2017. IEEE http://dx.doi.org/10.1109/UMEDIA.2017.8074148
48. A. Caragliu, C. Del Bo, P. Nijkamp, Smart cities in Europe. J. Urban Technol. **18**, 65–82 (2011). https://doi.org/10.1080/10630732.2011.601117
49. H. Schaffers, N. Komninos, M. Pallot, B. Trousse, M. Nilsson, A. Oliveira, Smart cities and the future Internet: towards cooperation frameworks for open innovation, in *The Future Internet*, ed. by J. Domingue, et al., (Springer Berlin Heidelberg, 2011). http://dx.doi.org/10.1007/978-3-642-20898-0_31
50. H. Chourabi, T. Nam, S. Walker, J.R. Gil-Garcia, S. Mellouli, K. Nahon, T.A. Pardo, H.J. Scholl, Understanding smart cities: an integrative framework, in *2012 45th Hawaii International Conference on System Sciences (HICSS)*, Maui, Hawaii USA 4–7 July 2012. IEEE http://dx.doi.org/10.1109/HICSS.2012.615
51. S. Talari, M. Shafie-khah, P. Siano, V. Loia, A. Tommasetti, J. Catalão, A review of smart cities based on the internet of things concept. Energies **10**(4), 421 (2017). https://doi.org/10.3390/en10040421
52. K.J. Singh, D.S. Kapoor, Create your own internet of things: a survey of IoT platforms. IEEE Consum. Electron. Mag. **6**(2), 57–68 (2017). https://doi.org/10.1109/MCE.2016.2640718

53. G. Marques, R. Pitarma, M. Garcia, N. Pombo, Internet of things architectures, technologies, applications, challenges, and future directions for enhanced living environments and healthcare systems: a review. Electronics **8**(10), 1081 (2019). https://doi.org/10.3390/electronics8101081
54. A.S. Rao, S. Marshall, J. Gubbi, M. Palaniswami, R. Sinnott, V. Pettigrovet V, Design of low-cost autonomous water quality monitoring system, in *2013 International Conference on Advances in Computing, Communications and Informatics (ICACCI)*, Mysore, India, 22–25 Aug 2013. IEEE http://dx.doi.org/10.1109/ICACCI.2013.6637139
55. S. Geetha, S. Gouthami, Internet of things enabled real time water quality monitoring system. Smart Water. **2**(1), 1 (2016). https://doi.org/10.1186/s40713-017-0005-y
56. N.A. Cloete, R. Malekian, L. Nair, Design of smart sensors for real-time water quality monitoring. IEEE Access. **4**, 3975–3990 (2016). https://doi.org/10.1109/ACCESS.2016.259 2958
57. D.-H. Jung, H.-J. Kim, W.-J. Cho, S.H. Park, S.-H. Yang, Validation testing of an ion-specific sensing and control system for precision hydroponic macronutrient management. Comput. Electron. Agric. **156**, 660–668 (2019). https://doi.org/10.1016/j.compag.2018.12.025
58. T. Gomiero, Food quality assessment in organic vs. conventional agricultural produce: findings and issues. Appl. Soil Ecology. **123**, 714–728 (2018). https://doi.org/10.1016/j.apsoil.2017. 10.014
59. A. Zanella, S. Geisen, J.-F. Ponge, G. Jagers, C. Benbrook, T. Dilli, A. Vacca, J. Kwiatkowska-Malina, M. Aubert, S. Fusaro, M.D. Nobili, G. Lomolino, T. Gomiero, Humusica 2, article 17: techno humus systems and global change—three crucial questions. Appl Soil Ecology. **122**, 237–253 (2018). https://doi.org/10.1016/j.apsoil.2017.10.010
60. C. Maucieri, Z. Schmautz, M. Borin, P. Sambo, R. Junge, C. Nicoletto, Hydroponic systems and water management in aquaponics: a review. Italian J Agron. **13**(1), 1012 (2018). http://dx.doi.org/10.21256/zhaw-3671
61. S. Pellicer, G. Santa, A.L. Bleda, R. Maestre, A.J. Jara, A.G. Skarmeta (2013) A global perspective of smart cities: a survey, in *2013 Seventh International Conference on Innovative Mobile and Internet Services in Ubiquitous Computing (IMIS)*, Massachusetts Ave., NW Washington, DCUnited States: IEEE. http://dx.doi.org/10.1109/IMIS.2013.79
62. L. Parra, S. Sendra, J. Lloret, I. Bosch, Development of a conductivity sensor for monitoring groundwater resources to optimize water management in smart city environments. Sensors **15**(9), 20990–21015 (2015). https://doi.org/10.3390/s150920990
63. I. Ganchev, M.G. Nuno, D. Ciprian, C. Mavromoustakis, R. Goleva, Enhanced living environments: algorithms, architectures, platforms, and systems, in *Lecture Notes in Computer Science*. vol. 11369 (Springer International Publishing, Cham)
64. C. Dobre, Constandinos Mavromoustakis, Nuno Garcia, Rossitza Ivanova Goleva, G. Mastorakis, *Ambient assisted living and enhanced living environments: principles, technologies and control* (Butterworth-Heinemann, Amsterdam; Boston, 2017), p. 499
65. A. Anjum, M.U Ilyas, Activity recognition using smartphone sensors, in *Consumer Communications and Networking Conference (CCNC)*, IEEE: Las Vegas, Nevada, USA 12–13 Jan 2013, pp. 914–919. http://dx.doi.org/10.1109/CCNC.2013.6488584
66. I. Bisio, F. Lavagetto, M. Marchese, A. Sciarrone, Smartphone-centric ambient assisted living platform for patients suffering from co-morbidities monitoring. Commun. Mag. IEEE **53**, 34–41 (2015). https://doi.org/10.1109/MCOM.2015.7010513
67. S. Haug, R.P. Castro, M. Kwon, A. Filler, T. Kowatsch, M.P. Schaub, Smartphone use and smartphone addiction among young people in Switzerland. J. Behav. Addictions **4**(4), 299–307 (2015). https://doi.org/10.1556/2006.4.2015.037
68. D. Kuss, L. Harkin, E. Kanjo, J. Billieux, Problematic smartphone Use: investigating contemporary experiences using a convergent design. Int. J. Environ. Res. Public Health. **15**(1), 142 (2018). https://doi.org/10.3390/ijerph15010142
69. D. Wang, Z. Xiang, D.R. Fesenmaier, Smartphone Use in everyday life and travel. J. Travel Res. **55**(1), 52–63 (2016). https://doi.org/10.1177/0047287514535847
70. S. Wang, Z. Zhang, Z. Ye, X. Wang, X. Lin, S. Chen, Application of environmental internet of things on water quality management of urban scenic river. Int. J. Sustain. Dev. World Ecology. **20**(3), 216–222 (2013). https://doi.org/10.1080/13504509.2013.785040

71. U. Shafi, R. Mumtaz, H. Anwar, A.M.Qamar, H. Khurshid, Surface water pollution detection using internet of things, in *2018 15th International Conference on Smart Cities: Improving Quality of Life Using ICT & IoT (HONET-ICT)*, Islamabad, Pakistan, 8–10h Oct 2018). IEEE
72. K. Saravanan, E. Anusuya, R. Kumar, Real-time water quality monitoring using Internet of Things in SCADA. Environ. Monit. Assess. **190**(9), 556 (2018). https://doi.org/10.1007/s10661-018-6914-x
73. B. Esakki, S. Ganesan, S. Mathiyazhagan, K. Ramasubramanian, B. Gnanasekaran, B. Son, S.W. Park, J.S. Choi, Design of amphibious vehicle for unmanned mission in water quality monitoring using internet of things. Sensors **18**(10), 3318 (2018). https://doi.org/10.3390/s18103318
74. P. Liu, J. Wang, A.K. Sangaiah, Y. Xie, X. Yin, Analysis and prediction of water quality using LSTM deep neural networks in IoT environment. Sustainability **11**(7), 2058 (2019). https://doi.org/10.3390/su11072058
75. M.C. Zin, G. Lenin, L.H.Chong, M. Prassana, Real-time water quality system in internet of things, in *IOP Conference Series: Materials Science and Engineering*, vol 495, no 1, p. 012021 (2019). http://dx.doi.org/10.1088/1757-899X/495/1/012021
76. M.S.U. Chowdury, T.B. Emran, S. Ghosh, A. Pathak, M.M. Alam, N. Absar, K. Andersson, M.S. Hossain, IoT based real-time river water quality monitoring system. Procedia Comput. Sci. **155**, 161–168 (2019). https://doi.org/10.1016/j.procs.2019.08.025
77. S Abraham, A. Shahbazian, K. Dao, H. Tran, P Thompson, An Internet of things (IoT)-based aquaponics facility, in *2017 IEEE Global Humanitarian Technology Conference (GHTC)*, San Jose, California, USA, 2017. IEEE
78. M. Manju, V. Karthik, S. Hariharan, B. Sreekar, Real time monitoring of the environmental parameters of an aquaponic system based on Internet of Things, in *2017 Third International Conference on Science Technology Engineering and Management (ICONSTEM)*, Chennai, India, 23–24 Mar 2017. IEEE
79. K.H. Kamaludin, W. Ismail, Water quality monitoring with internet of things (IoT), in *2017 IEEE Conference on Systems, Process and Control (ICSPC)*, Malacca, Malaysia, 15–17h Dec 2017. IEEE
80. P. De Souza, M. Ramba, Wensley A, E. Delport, Implementation of an internet accessible water quality management system for ensuring the quality of water services in South Africa, in *WISA Conference*, Durban, South Africa. Citeseer (2006)
81. O. Postolache, P. Girao, M. Pereira, H. Ramos, An internet and microcontroller-based remote operation multi-sensor system for water quality monitoring. Sensors **2**, 1532–1536 (2002)

Contribution to the Realization of a Smart and Sustainable Home

Djamel Saba, Youcef Sahli, Rachid Maouedj, Abdelkader Hadidi, and Miloud Ben Medjahed

Abstract Home automation is the set of connected objects that make the house itself connected. We sometimes even speak of an automated or intelligent home. However, connected objects allow the house to react automatically according to one or more events. This document presents a contribution to the realization of a wireless smart and also sustainable home. The house concerned by the construction is powered by a renewable, clean, and free energy source which is photovoltaic energy. Then, the house management system is based on an Arduino and embedded microprocessor-based systems. This work is a hybridization between several disciplines such as computer science, electronics, electricity, and mechanics. Next, a smart and sustainable home is characterized by many benefits like resident comfort, security, and energy-saving. The first part of this project focuses on building a model with the modules used (sensors, actuators, Wi-Fi interface, etc.). The second part is reserved for the implementation of the system and to make it controllable via a smartphone or a computer.

Keywords Home automation · Arduino · Photovoltaic energy · Artificial intelligence · Internet of thing · Sensors and actuators

---

D. Saba (✉) · Y. Sahli · R. Maouedj · A. Hadidi · M. B. Medjahed
Unité de Recherche en Energies Renouvelables en Milieu Saharien, URER-MS, Centre de Développement des Energies Renouvelables, CDER, 01000 Adrar, Algeria
e-mail: saba_djamel@yahoo.fr

Y. Sahli
e-mail: sahli.sofc@gmail.com

R. Maouedj
e-mail: ra_maouedj@yahoo.fr

A. Hadidi
e-mail: hadidiabdelkader@gmail.com

M. B. Medjahed
e-mail: benmedjahed_78@yahoo.fr

© The Editor(s) (if applicable) and The Author(s), under exclusive license to Springer Nature Switzerland AG 2021

A. E. Hassanien et al. (eds.), *Artificial Intelligence for Sustainable Development: Theory, Practice and Future Applications*, Studies in Computational Intelligence 912,
https://doi.org/10.1007/978-3-030-51920-9_14

Abbreviations

AI	Artificial Intelligence
IoT	Internet of Things
MEMS	Micro-Electro Mechanical Systems
M2M	Machine to Machine
RFID	Radio Frequency Identification
WSN	Wireless Sensor Network

1 Introduction

Home automation is the set of connected objects and applications that transform a house into a smart home, ready to simplify life in all areas of everyday life [1]. In addition, the various elements of a smart home (heating, lighting, multiple sockets, alarms, video surveillance devices, etc.) can be controlled from mobile applications, available on smartphones or tablets [2]. Making a house smart is first and foremost about providing comfort and security to the occupants. Using equipment that can be controlled remotely, it is possible to change the temperature, control the lighting or verify that a person does not enter the house during the absence of residents of the dwelling [3]. Comfort is ensured by a more intuitive use of the devices. A connected house becomes 100% multimedia accommodation: radio or music follows you in all rooms, and you can launch an application using voice (thanks to smart speakers) [4]. Housing is also becoming more energy efficient. When the heating is modulated using an intelligent thermostat and the lights go out when the rooms are empty, this saves significant electricity savings [5, 6]. The Smarter connected thermostat makes it possible to adapt the management of the heating to the pace preferred by the occupant of the house, to program scenarios, manage the unexpected, and control the heating on a touch screen or, remotely. It is easy to create and modify scenarios adapted to the needs of the occupant, for example for a day or the week, depending on the rhythms and habits of the whole family: work and school periods/vacation periods/weekends, etc., depending on specific needs: teleworking day, cocooning… Once the programming is done, the management system takes care of everything.

The first level for demonizing a house is that of connected objects, which work individually through an application and are connected via Wi-Fi to the box of the house. It can be a thermostat which manages the radiators of the house and gives indications on energy expenditure [7]; a robotic vacuum cleaner that is triggered remotely to find a clean home on the way home from work [8]; a fridge that indicates the food to be bought and indicates the expiry dates [9]; a video surveillance system with facial recognition [10]; an entry door with biometric lock [11]; a robotic lawn-mower [12]; a pool water analysis system [13]. The home automation system, for its part, is fully connected: shutters, alarm, air conditioning, video system, heating, IT… Everything is centralized and works with the house's electrical network, by radio. But a more reliable solution is to create a 27 V "bus" network specifically dedicated to home automation [14]. Then, the smart home is home automation, which also anticipates needs. Brightness sensors tell the shutters when to close; presence sensors automatically lower the heating and turn off the lights; an outdoor anemometer and rain gauge close the blinds, windows, and shutters when it is too windy or raining, etc. The smart home makes everyday life easier and manages all connected objects [15]. These new possibilities not only improve comfort but also help reduce costs. Home automation allows you to configure the devices so that they are only activated when necessary: the light turns on when someone enters the apartment, the heater turns on in the morning and turns off automatically, resulting in significant savings. Some devices even go so far as to recognize who is in front of the mirror, and to adjust the lighting according to each person's preferences. However, automated home offers many amenities. Some may seem superfluous, and not everything is necessary for everyone. The elderly or those experiencing health problems, however, benefit from the advantages that these new techniques bring them. For example, to better support deaf or hard of hearing people, the doorbell can be paired with a light signal. Likewise, an acoustic signal makes it possible to ask for help during a fall, and the electric blinds can close automatically without the need to make a physical effort. Also, companies have embraced the smart revolution fairly quickly, deploying devices such as alarms, sensors, monitors, and cameras. These devices have become much easier to install and more affordable, which greatly facilitated their entry into the home. They are also becoming more advanced, so the possibilities are now limitless, and all homes can benefit from smart technology.

The "smart home" is a familiar concept, even if consumers only know it for its ability to control their heating remotely from their phone or to turn on or off a lamp plugged into a smart outlet. But this is only the beginning of the enormous potential of home automation techniques. Smart devices as we know them are becoming much more important. Cameras will be at the center of our future smart homes as they evolve. Smart cameras with super sensors that can process contextual and behavioral analysis will enter our lives soon [16].

Individually, a smart outlet or smart thermostat can save you time or energy, and a CCTV camera can add a layer of security to your home. But industry experts estimate that the typical home could hold more than 500 smart devices by 2022 [17]. People will benefit more from smart devices by connecting them, wirelessly, to trigger each other, and share and use all the information generated. Consumers' homes will speak to them. This year, people will start to think more about how the devices can be used together, the only limit being their imagination. However, intelligent techniques will be used as differentiating factors in traditional residential construction, as well as in-home renovations and commercial redevelopments. There is certainly evidence to show that smart technology, including cameras, smart alarms, thermostats, doors, and windows with sensors, could make properties more attractive to buyers.

Thanks to intelligent technology, more and more advanced functionalities, and many more devices are now within our reach. It's extremely exciting, but one could easily reach a point where smart technology exceeds the limits of the home network infrastructure that consumers have in place. Each device needs access to the Internet and generates data. To adapt, people need to get back to basics and find the right platform to lay the foundation for the smart home. In other words, they need to make sure that they have a high-performance router and robust Internet connectivity at home to be able to handle data traffic. Without it, a fast and smooth intelligent experience will not be possible. Finally, the main functions that we will be able to program for a smart home focus on three sectors:

- Security: to ensure better protection of your home. It will be possible to automate certain tasks (example: triggering the alarm at a fixed time, closing the shutters remotely, switching on presence detectors, controlling video surveillance, etc.);
- Communication: everything related to leisure can also be automated. You can start the television from a distance, play music or receive certain data at fixed times necessary for medical monitoring via a computer;
- Energy management in the home: home automation still makes it possible to adjust the thermostat in the various rooms, to close the shutters at certain hours which save a few degrees in winter when the cold drops at the same time as at night, etc.

The smart home has many advantages, including:

- Better time management: by scheduling repetitive tasks such as opening or closing the shutters, setting off the alarm at fixed times, opening your portal from your smartphone.
- Greater security: a home automation system is often better protected against burglary.
- One way to limit energy expenditure: home automation offers the possibility of adjusting the thermostat according to the hours of the day and according to the rooms and to benefit from a constant temperature. This avoids overheating in winter or using the air conditioning at full speed in summer.

The remainder of this paper is organized as follows. Section 2 presents artificial intelligence. Section 3 explains the Internet of Things. Section 4 explains the Smart Home. Section 5 details Home automation technologies. Section 6 presents Home Automation Software. Section 7 clarifies Home automation and photovoltaic energy. Section 8 provides an introduction to the implementation of the project. Finally, Sect. 9 concludes the paper.

2 AI

AI refers to systems or machines that mimic human intelligence to perform tasks and that can improve based on the information collected through iteration [18, 19]. AI is characterized by advantages that are linked to the process and the ability to think and analyze in-depth data to the maximum to a particular format or function. Although artificial intelligence conjures up images of high-performance robots resembling

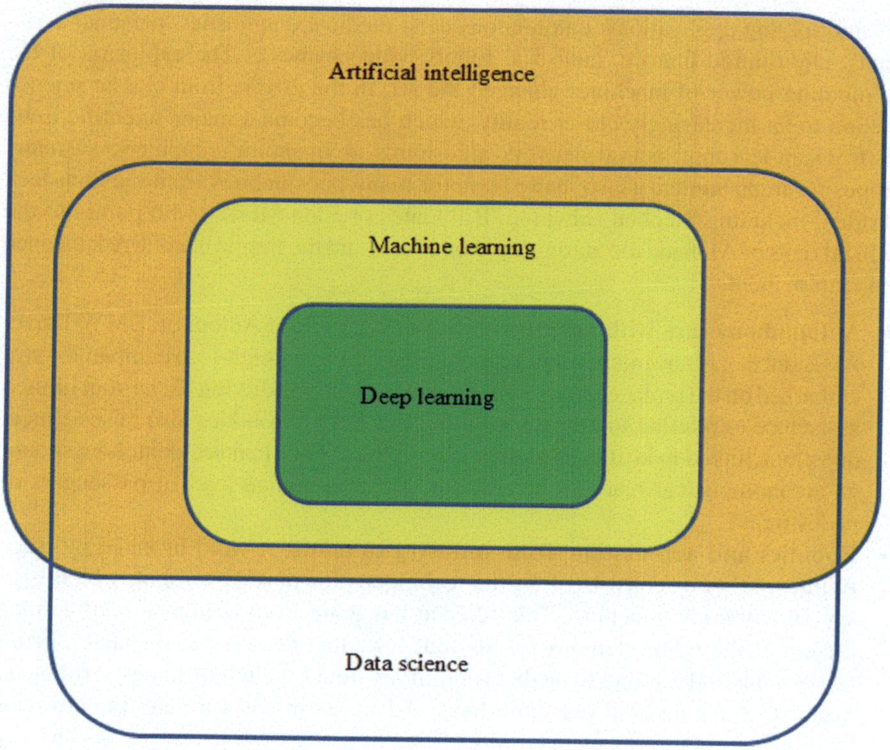

Fig. 1 Domains that make up AI

humans and invading the world, artificial intelligence is not intended to replace us. It aims to significantly improve human capacities and contributions. This makes it a very valuable business asset.

AI has become a catch-all term for applications that perform complex tasks that previously required human intervention, such as communicating with customer's online or playing chess. The term is often used interchangeably with the fields that make up AI such as machine learning and deep learning (Fig. 1). There are differences, however. For example, machine learning focuses on creating systems that learn or improve their performance based on the data they process. It's important to note that, even though all of the machine learning is based on artificial intelligence, it's not just about machine learning.

2.1 Some Applications of AI

The emerging technology of AI, crosses several techniques simulating human cognitive processes. Existing since the 1960s, research has recently developed to the point of multiplying applications: autonomous cars, medical diagnostics, personal assistants, algorithmic finance, industrial robots, video games… The explosion of the computing power of machines changed the AI, in the 2010s, from classic science fiction to an increasingly closer reality, which has become a major scientific issue [20]. Deep-learning, neural network algorithms or quantum computers: so many hopes for trans humanists, so many fears for many personalities from the high-tech world—including Stephen Hawking, Bill Gates or Elon Musk—who point out the ethical risks of AI made too autonomous or aware, and the fragile benefit-risk balance on employment.

- **Autonomous cars without drivers**: Google Car, Tesla Autopilot, BMW Driver Assistance… Many international car manufacturers and high-tech companies have embarked on the replacement—by software—of human driving [21]. From limited assistance to piloting to total automation, this field of robotics also raises ethical questions linked to artificial intelligence: in the event of an accident, how should an autonomous car react to the risk posed on the human lives of passengers or pedestrians?
- **Robotics and automation, from industry to home**: A word invented by Isaac Asimov, robotics consists of all the techniques for making a robot, which is to say an automatic machine. This science has gone from fiction to reality since the era of the writer, famous for his four laws that dictate their behavior. From heavy industrial robotics to medical or military fields, including domestic robotics, machines have invaded our daily lives. Advances in software and the artificial intelligence controlling them are already making it possible to replace us—at least in part—on many increasingly complex tasks, such as driving with autonomous cars, or too dangerous, like driving exploration of another planet or radioactive site [22]. The possibility of the development of autonomous offensive weapons, or "killer robots", worries the UN, however, like many scientists.
- **Science fiction**: a literary and cinematographic genre, science fiction often envisions an imaginary future, filled with extraterrestrials from another planet. Some scientists are trying to assimilate it to real science like the cosmologist Stephen Hawking or this team of Slovak researchers who exhibited a prototype flying car in January 2014 [23]. Others try to dissociate them. Chinese researchers have, for example, demonstrated that a trip back in time is almost impossible, thus demonstrating the potential existence of the famous DeLorean car from the film Back to the Future.
- **Quantum physics**: after Einstein's theory of relativity, quantum physics is a new way of looking at science dating back to the twentieth century. It is based on the principle of the black body discovered by Max Planck and Werner Heisenberg. The experience of Schrödinger's cat makes the field of quantum mechanics more complex. Researchers use it to design computers or quantum calculators, like IBM or D-Wave, which interest large organizations like the NSA [24]. They even plan to make quantum teleportation possible.

2.2 *AI Methodological Approaches*

There are two different methodological approaches: symbol processing and the neural approach, symbolic artificial intelligence, neural artificial intelligence.

2.2.1 Symbolic AI

Symbolic AI is the classic approach we have to artificial intelligence. It is based on the idea that human intelligence can be reconstructed at a conceptual, logical, and orderly level, regardless of concrete empirical values: this is called a top-down approach [25]. Knowledge, including spoken and written languages, is represented in the form of abstract symbols. Thanks to the manipulation of symbols and based on algorithms, machines learn to recognize, understand, and use these symbols. The intelligent system obtains information from expert systems, within which data and symbols are classified in a specific way, most of the time in a logical and interconnected way. The intelligent system can rely on these databases to compare their content with their own. Then, typical applications of symbolic AI include word processing and speech recognition, as well as other logical disciplines, such as chess. Symbolic AI works according to strict rules and makes it possible to solve extremely complex problems thanks to the development of computer capacities. This is how Deep Blue, the IBM computer with symbolic artificial intelligence, defeated world chess champion, Garri Kasparov, in 1996 [26].

Symbolic AI performance depends on the quality of expert systems but is also inherently limited. The developers had high hopes in these systems: thanks to advances in technology, intelligent systems could also become more and more powerful, and the dream of artificial intelligence seemed at hand. However, the limits of symbolic AI are becoming more and more obvious. The degree of complexity of the expert system, therefore, does not matter, because symbolic artificial intelligence remains relatively inflexible in comparison. Indeed, the system based on strict rules is difficult to manage when it is faced with exceptions, variations, or uncertainties. On the other hand, symbolic AI has great difficulty in acquiring autonomous knowledge.

2.2.2 Neural AI

It was Geoffrey Hinton and two of his colleagues who, in 1986, developed the concept of neural artificial intelligence, and at the same time revitalized the field of AI [27]. They further developed the back propagation of the gradient. This laid the groundwork for deep learning, used today by almost all artificial intelligence technologies. With this learning algorithm, deep neural networks can learn continuously and develop independently of each other. This represents a great challenge, which the symbolic AI was unable to meet.

Neural artificial intelligence (also called subsymbolic AI) is therefore distinguished from the principles of symbolic representation of knowledge. As with human intelligence, knowledge is segmented into small functioning units, artificial neurons, which are linked to ever-growing groups. This is called a bottom-up approach. The result is a rich and varied network of artificial neurons.

Neuronal artificial intelligence aims to imitate as precisely as possible the functioning of the brain and to artificially simulate a neural network. Unlike symbolic AI, the neural network is stimulated and trained to progress; in robotics, for example, this stimulation is done using sensory and motor data. It is thanks to these experiences that AI itself generates knowledge that is constantly growing. Herein lies the major innovation: although the training itself requires a lot of time, it allows the machine to learn on its own in the longer term. We sometimes talk about learning machines. This is what makes neural AI-based machines very dynamic systems with adaptive abilities, which sometimes are no longer fully understood by humans.

3 IoT

The IoT is "a network that connects and combines objects with the Internet, following the protocols that ensure their communication and exchange of information through a variety of devices" [28]. Then, the IoT can also be defined as "a network which allows, via standardized and unified electronic identification systems, and wireless mobile devices, to directly and unambiguously identify digital entities and objects and thus to be able to recover, store, transfer and process data without discontinuity between the physical and virtual worlds" [29]. There are several definitions on the concept of IoT, but the most relevant definition is that proposed by Weill and Souissi who defined IoT as "an extension of the current Internet towards any object which can communicate directly or indirectly" [30]. With electronic equipment they connected to the Internet. This new dimension of the Internet is accompanied by strong technological, economic and social games, in particular with the major savings that could be achieved by adding technologies that promote the standardization of this new field, especially in terms of communication, while ensuring the protection of individual rights and freedoms.

3.1 The IoT History

The IoT has not existed for a very long time. However, there have been visions of machines communicating with each other since the early 1800s. Machines have provided direct communications since the telegraph (the first landline) was developed in the 1830s and 1840s. Described as "wireless telegraphy", the first radio voice transmission took place on June 3, 1900, providing another element necessary for the development of the IoT. The development of computers began in the 1950s.

The Internet, itself an important component of the IoT, started in 1962 as part of the DARPA (Defense Advanced Research Projects Agency) and evolved into ARPANET in 1969 [31]. In the 1980s, commercial service providers began to support public use of ARPANET evolved into our modern Internet. Global positioning satellites (GPS) became a reality in early 1993, with the Department of Defense providing a stable and highly functional system of 24 satellites. This was quickly followed by the launch of private commercial satellites into orbit. Satellites and landlines provide basic communications for much of IoT. An additional and important element in the development of a functional IoT was the remarkably intelligent decision of IPV6 to increase the address space. Steve Leibson of the Computer History Museum says: "The expansion of address space means that we could assign an IPV6 address to each atom on the surface of the Earth, and still have enough addresses to make another one hundred lands. This way, we will not run out of Internet addresses anytime soon" [32]. In addition, the IoT, as a concept, was not officially named until 1999. One of the first examples of the Internet of Things dates back to the early 1980s and was a Coca-Cola machine, located at Carnegie Melon University. Local programmers would connect to the refrigerator on the Internet and vary if there was an available drink, and if it was cold, before making the trip. Then, in 2013, the Internet of Things became a system using multiple technologies, ranging from the Internet to wireless communication and from MEMS to embedded systems. Traditional areas of automation (including building and home automation), wireless sensor networks, GPS, control systems, and more, all support IoT.

3.2 Operation

The IoT allows the interconnection of different smart objects via the Internet. Thus, for its operation, several technological systems are necessary. "The IoT designates various technical solutions (RFID, TCP/IP, mobile technologies, etc.) which make it possible to identify objects, to capture, store, process, and transfer data in physical environments, but also between contexts physical and virtual universes" [33]. Indeed, although there are several technologies used in the operation of Ido, we only focus on a few that are, according to Han and Zhanghang, the key techniques of Ido. These techniques are: RFID, WSN, and M2M.

- **RFID**: the term RFID includes all technologies that use radio waves to automatically identify objects or people [34]. It is a technology that makes it possible to memorize and retrieve information remotely thanks to a label that emits radio waves. It is a method used to transfer data from labels to objects or to identify objects remotely. The label contains electronically stored information that can be read remotely.
- **WSN**: it is a set of nodes that communicate wirelessly and which are organized in a cooperative network [35]. Each node has a processing capacity and can contain different types of memories, an RF transceiver, and a power source, as it can also

take into account the various sensors and actuators [6]. As its name suggests, the WSN then constitutes a network of wireless sensors which can be a technology necessary for the functioning of the IoT.

- **M2M**: it is "the association of information and communication technologies with intelligent objects to give them the means to interact without human intervention with the information system of an organization or company" [36].

3.3 Areas of Application

The IoT concept is exploding as we have an increasing need in everyday life for intelligent objects capable of making it easier to achieve objectives. Thus, the fields of application of the IoT can be varied. Several areas of application are affected by the IoT. Gubbiet et al. Have classified the applications into four areas [37]: personal, transportation, environment, infrastructure, and public services… etc. (Fig. 2).

The fields of application of IoT are multiple. Industry, health, education, and research are cited as examples. However, it will be possible in the future to find the

Fig. 2 IoT areas of application

IoT concept anywhere, anytime, and available to everyone. "IoT consists of a world of (huge) data, which, if used correctly, will help to address today's problems, particularly in the following fields: aerospace, aviation, automotive, telecommunications, construction, medical, the autonomy of people with disabilities, pharmaceuticals, logistics, supply chain management, manufacturing and life cycle management of products, safety, security, environmental monitoring, food traceability, agriculture and livestock" [38].

3.4 Relationship Between IoT and IA

Among the technological advances that fascinate, Artificial Intelligence (AI) and the Internet of Objects (Ido) are taking center stage. This enthusiasm testifies to an unprecedented transformation in our history, bringing man and machine together for the long term.

The combination of these two technologies offers many promises of economic and social progress, affecting key sectors such as health, education, and energy.

Artificial intelligence boosts productivity and redistributes the cards in terms of required skills. The analysis of the very numerous sensors present on connected objects increases the efficiency, reliability, and responsiveness of companies.

They thus transform the link they maintain with their consumers and, by extension, their culture. As such, the concept of digital twin offers new opportunities to better control the life cycle of products, revolutionize the concept of predictive maintenance or the design of innovative solutions. So many innovations at the service of humans, if it is placed at the heart of this interaction.

Many challenges accompany the development of these two technologies. In addition to the issue of cyber security and the management of an ever-increasing volume of data, there is the complexity of the evolution of imagined solutions [39].

4 Smart Home

The first home automation applications appeared in the early 1980s. They were born from the miniaturization of electronic and computer systems. The development of electronic components in household products has improved performance while reducing the energy consumption costs of equipment [40]:

- An approach aimed at bringing more comfort, security, and conviviality in the management of housing thus guided the beginnings of home automation.
- Home automation has been bringing innovations to the market for more than 20 years now. But it is only since the 2000s that home automation seems to be more interesting. Some research and industry institutions are working on a smart home concept that could potentially spawn new technologies and attract more consumers.

- Currently, the future of home automation is assured. Home automation is attracting more and more individuals who want to better manage the many features of their home.
- One of the hopes on which home automation professionals rely is to make this concept the best possible support for carrying out daily tasks. Since 2008, scientists and specialists have been thinking, for example, about robots guiding people daily.

4.1 Home Automation Principles

Home automation, everyone talks about it without really knowing what it is about. You only need to consult the manufacturers' catalogs to be convinced of this. Dictionaries are full of more or less similar definitions. The Hachette encyclopedic dictionary, edition of 1995, tells us that home automation is "computer science applied to all the systems of security and regulation of household tasks intended to facilitate daily life" Vast program! Where does it stop? electricity, where do the automations stop, where does home automation start?

If there is still about fifteen years, an electrician was enough to carry out all the electrical installation of a building, it is quite different today. The skills required are multiple—electronics, automation, security, thermal, energy—because all household equipment is closely coupled. All this equipment is connected by specialized wired or wireless links.

The central unit can manage all of this equipment and may be connected to a computer network either directly over IP over Ethernet, or via a telephone modem over a switched residential network.

We can summarize by saying that home automation is the technological field that deals with home automation, hence the etymology of the name which corresponds to the contraction of the terms "home" (in Latin "Domus") and "automatic" [41]. It consists of setting up networks connecting different types of equipment (household appliances, HiFi, home automation equipment, etc.) in the house. Thus, it brings together a whole set of services allowing the integration of modern technologies in the home.

4.2 Definition of the Smart Home

The definitions of the smart home sometimes cause ambiguities, mainly the confusion between the terms "Home Automation", and "Smart Home". Home automation (home automation) today this term is rather replaced by that of a smart home which means a paradigm that positions itself as a successor of the home automation, profiting from the advances in ubiquitous computing which one also calls ambient computing, integrating including the Internet of Things. In addition to the dominant dimension of IT, the smart home as represented in the 2010s also wants to be more user-centered, moving away from the technophile approach characteristic of home automation of the 1990s [42]. The principle of operation of a house smart is to centralize the control. Unlike a conventional electrical installation, the control and power circuits are separate. It thus becomes possible to establish links between the control members and the command receivers, which usually belong to independent subsystems. In 2009, the first home automation boxes appeared on the market. Unlike previous wire solutions, which are often very expensive, home automation boxes use the power of the Internet and wireless. With or without a subscription, they allow open use and can be controlled from a computer, smartphone, or touchpad. The installation is very simple and is done in a few minutes by an inexperienced user.

4.3 Ambient Intelligence

Due to technological developments and miniaturization, the computer no longer has activity only focused on its end-user, IT also takes into account the environment thanks to information from sensors that can communicate with each other. The environment can be very diverse: the house, the car, the office, the meeting room. Before the development of the ambient intelligence concept, the term ubiquitous computing was proposed by Weiser in 1988 [43] in the Xerox PARC project to designate a model of human-machine interaction in which the processing of information relating to the activities of the everyday life was integrated into the objects.

The vision of ubiquitous computing was made possible by the use of a large number of miniaturized, economical, robust, and network-connected devices that we could, therefore, install in places where people carry out their daily activities. Then, Uwe Hansmann of the company IBM proposed the term "pervasive" computer to describe this technology. In the literature, there is great confusion regarding the use of these two terms and that of ambient intelligence. They are often used indistinctly. However, we consider more relevant the distinction proposed by Augusto and Mccullagh [44] who affirms that the concept of ambient intelligence covers a wider field than the simple ubiquitous availability of resources and that it requires the application of artificial intelligence to achieve their goal, which is to be wise and show initiative. So unlike the other two concepts, ambient intelligence calls on contributions from other fields such as machine learning, multi-agent systems, or robotics.

4.4 Communicating Objects

M2M (Machine-to-Machine) brings together solutions allowing machines to communicate with each other or with a central server without human intervention. It is a recent booming market, driven by favorable technological and economic dynamics and its development will vary depending on the sector of activity [36]. However, we can already identify that of safety and health which are among the most promising… At the home automation level, we naturally observe the same development of these communicating objects. The objects of daily life are equipped with communication solutions, such as boilers (Thermibox by ELM Leblanc which draws on M2M expertise from Webdyn), household appliances, etc. If the primary functionalities of our household equipment of yesterday remain the same, they see their capacities increased tenfold because of this "interconnection". But we also observe many specialized objects that appeared a short time ago.

5 Home Automation Technologies

The essence of a home automation installation is communication between its different elements. For this, many protocols were born, because each manufacturer has achieved its communication protocol, which has led to a very complex situation. The protocols presented below are not proprietary: most of them are standardized and/or open.

5.1 Wireless Protocols

Wireless protocols are very popular today, the great freedom of placement of the sensors and switches that they bring allows them to be placed in places that are sometimes improbable, very often in what is called "last meters", these places where information is needed, but where it is relatively expensive to wire a dedicated Fieldbus. They also allow you not to have to wire certain parts, so that you can renovate/rearrange them more easily in the future. These protocols sometimes require the use of batteries, the main defect is, therefore, the lifespan of the latter, in some cases; it drops to a few months, which is very restrictive. The short-range (free space: 300 m, one dwelling: around 30 m) of these facilities means they are used for well-defined purposes, but in the case of a single-family home, the limitations are mostly acceptable [45].

The protocols presented below use the frequencies 868 MHz for Europe and 315 MHz for North America [46]:

5.1.1 EnOcean

Is a radio technology (868 MHz) standardized IEC (ISO/IEC 14543-3-10) promoted by the EnOcean Alliance and by the company EnOcean.

The purpose of this protocol is to make various devices communicate using the surrounding energy harvest. EnOcean equipment is therefore cordless and battery-free! Energy harvested from the environment can come from various physical principles:

- Reverse piezoelectric effect;
- Photoelectric effect;
- Seebeck effect.

Research is underway to recover energy from vibrations or energy from the surrounding electromagnetic field. It is, of course, obvious that the energy optimization which had to be carried out is very advanced, in order to be able to support radio transmissions with so little energy. A super—capacity is often added within this equipment so that it can emit even in the event of a shortage of their primary energy; and some display several months of autonomy under these conditions.

Communication between devices takes place via prior manual pairing; then each device can address up to 20 other devices. This standard is free in terms of implementation; however, many players join the EnOcean Alliance in order to be able to benefit from licenses to the energy harvesting patents held by the Alliance.

5.2 *802.15.4*

802.15.4 is an IEEE standard for wireless networks of the LR-WPAN (Low Rate Wireless Personal Area Network) family. On the OSI model, this protocol corresponds to the physical layers and data link and allows the creation of mesh or star type wireless networks. It is relatively easy to find 802.15.4 transceivers from specialized resellers including microprocessors and 128 KB of onboard RAM to implement all kinds of applications above 802.15.4.

5.2.1 6LoWPAN

Is an abbreviation of IPv6 Low power Wireless Personal Area Networks. This IETF project aims to define the encapsulation and header compression mechanisms of IPv4 protocols and especially IPv6 for the 802.15.4 standard.

This project, although already having products on the market, is not yet as mature as the other solutions presented above. It should reach maturity in the medium term, and receives for the moment a very good reception by the actors of the medium, which should give it a bright future.

The integration of the 6LowPAN stack has been done in the Linux kernel since version 3.3 and work continues on this subject.

5.2.2 Z-Wave

Z-Wave was developed by the Danish company Zen-Sys which was acquired by the American company Sigma Designs in 2008 and communicates using low power radio technology in the 868.42 MHz frequency band. The Z-Wave radio protocol is optimized for low-bandwidth exchanges (between 9 and 40 kbps) and battery-powered or electrically powered devices, as opposed to Wi-Fi for example, which is intended for high-speed exchanges and on electrically powered devices only.

Z-Wave operates in the sub-gigahertz frequency range, which depends on the regions (868 MHz in Europe, 908 MHz in the US, and other frequencies according to the ISM bands of the regions). The range is around 50 m (more outdoors, less indoors). The technology uses mesh technology (Mesh Network) to increase range and reliability.

5.2.3 ZigBee

Is a free protocol governed by the ZigBee Alliance. The ZigBee protocol generally works above 802.15.4, it implements the network and application layers of the OSI model. This implementation makes it possible to take advantage of the advantages of the 802.15.4 standard in terms of communication. The main additions are the addition of network and application layers which, among other things, allow each to carry out message routing; the addition of ZDO (ZigBee Device Object) governed by the specification; and the addition of custom objects by the manufacturers.

This protocol still suffers from certain problems, the most important being an interoperability problem. As seen above, the protocol gives manufacturers the possibility of defining their application objects. The manufacturers are of course not deprived of it, which causes total incompatibilities, some manufacturers having re-implemented their undocumented protocols above ZigBee. The integration of the ZigBee/802.15.4 stack has been performed in the Linux kernel since version 2.6.31. ZigBee begins its transformation to an IP network via the Smart Energy Profile version 2.0 specification.

5.3 Carrier Currents

Protocols using carrier currents are very popular today because they reduce wiring and supposedly do not use radio frequencies. They nevertheless have disadvantages, they are very quickly disturbed by the electrical environment (radiators, dimmers, etc.), they do not pass, or very poorly, the electrical transformers and the electromagnetic radiation of the cables through which they pass make them very good radio transmitters.

X10 is an open communication protocol for home automation, mainly used on the continent present on the west side of the Atlantic Ocean [47]. This protocol was born in 1975 and uses the principle of the carrier current. This protocol is not recommendable at present for a new installation; it offers very low bit rates which cause high latencies (of the order of a second for sending an order). Many other limitations are present and detailed on the Web.

5.4 *Wired Protocols*

Wireless protocols are often supported by a field bus that extends the overall capabilities of the installation. Among the wired protocols, there are two main families, the centralized ones, those which use an automaton or a server to govern the whole installation; and the other category, decentralized protocols where sensors and actuators interact directly with each other, without a central point [48]. Each approach has its advantages and disadvantages.

5.4.1 Modbus

Is an old protocol placed in the public domain operating on the master-slave application layer mode. It works on different media: RS-232, RS-485, or Ethernet. This protocol requires centralization because of its use of a master. It supports up to 240 slaves [48]. Its use for home automation is now anecdotal or reserved for economic construction projects.

5.4.2 DALI

DALI (Digital Addressable Lighting Interface) is a standardized IEC 60929 and IEC 62386 protocol that allows you to manage a lighting installation via a two-wire communication bus. It is the successor of 0–10 V for the variation of the light intensity. Its capacities are limited (64 luminaires divided into a maximum of 16 groups per bus), but managers can be used that can interconnect several buses (PLCs, for example) [48]. DALI is often used in the tertiary (office) or, to a lesser extent, in residential housing.

5.4.3 DMX512

Lighting control methods include several well-defined and long-used standards. This is the case with DMX512, more commonly known as DMX (Digital Multiplexing). It is mainly used in the world of the stage (concerts, TV sets, sound & light shows), for controlling dynamic lighting. The DMX 512 is, to date, the most widespread and most universal protocol, it is used everywhere and by all manufacturers of scenic lighting equipment, which makes it possible to find dimmer power blocks capable of varying several pieces of equipment, at very affordable prices [49]. These blocks can also support powers higher than what one could do in DALI. The DMX 512 uses an RS-485 serial link to control 512 channels by assigning them a value between 0 and 255.

5.5 *1-Wire*

It is a communication bus, very close to the operation of the I^2C bus [50]. This bus is currently not used much for home automation, although some installations remain.

5.5.1 KNX

It is an open standard (ISO/IEC 14543-3) born from the merger of three protocol specifications: EIB, EHS, and Bâtibus. It is mainly used in Europe. KNX is described by a specification written by the members of the KNX Association, which also takes care of the promotion and the reference configuration software (proprietary ETS software) [51]. Different physical layers can be implemented. room for KNX, twisted pair, and Ethernet are the most widespread, but others can also be encountered, although very marginal: infrared, carrier current, and radio transmission. These physical layers are very slow (except Ethernet) and penalize the protocol for large networks.

In use, this protocol proves to be decentralized, the sensors communicate directly to the actuators that they must control, without going through a central point. The configuration of a network is done with dedicated proprietary ETS software (designed by the KNX association), other software exists but has very low visibility compared to the ETS juggernaut. When there is a change in the behavior of the network, the operation of the protocol requires a total reload of the firmware (firmware) of the equipment (s) concerned (sensor or actuator).

The implementation is relatively complex and the protocol reveals possibilities that are quite low and which depend very much on the equipment's firmware. Again, an installation with only one brand of equipment is preferable, to take full advantage of the possibilities of these.

The scalability of the installation of this type is very low unless you have kept all the configuration in place (including firmware, which can quickly be cumbersome), and the operating logic is quite complex to understand. for a non-regular.

5.5.2 LonWorks

It is a field-level building network historically created by the Californian company Echelon, which now provides the basic equipment (chips with the onboard LonTalk protocol). This network uses the ANSI/CEA-709.1-B standard LonTalk protocol, and free use. It is widely used as a field bus to control HVAC equipment (heating, ventilation, air conditioning), as well as for lighting control [52]. Geographically, it is mainly used in the United States, England, and several countries in Asia; probably making it the network of its kind with the most facilities around the world.

In the same way as KNX, LonWorks is a decentralized type network. This allows it to communicate over very long distances with a speed of 78 kb/s. The speed depends on the physical layer used, among which are: twisted pair, carrier current, optical fiber, and radio waves.

LonWorks has several advantages, but one of the most important is interoperability. The use of SNVT (Standard Network Variable Type), standardized network variables, for communication between nodes, requires integrators to carry out their configurations. In addition, manufacturers are strongly encouraged to create their new products while respecting the use of SNVTs, which ensures maximum compatibility between brands.

LonMark International is the association dedicated to the promotion and development of equipment compatible with LonWorks. It manages and maintains the standards linked to development between the manufacturers who are part of this association; it also manages to advertise of the standard and products, certifications, cancellation/creation of SNVT, etc.

There is some software to implement a LonWorks network: NL220, LonMaker, among others. The software flexibility offered by LonWorks to manufacturers is such that anyone can develop software capable of starting up a network of this type, by complying with the documentation. In addition to these advantages, there is the possibility of creating "LNS Plugins", software that allows the configuration of a product or a network through a graphical interface independently of the software used to create the network. The network configuration is saved in the "LNS Database", a very small database that defines the entire network and which is common to all configuration software. Projects on LonWorks and Linux are implemented, such as lonworks4linux, but they are not yet well defined.

5.5.3 Protocol xPL

The xPL project aims to provide a unified protocol for the control and supervision of all household equipment. This protocol aims to be simple in its structure while providing a large range of functionalities. It has, for example, auto-discovery and auto-configuration functions that allow it to be "Plug and Play", unlike many other home automation protocols [53]. Due to its completely free origins, it is implemented in many free home automation software, but it is very hard to find compatible hardware to equip a home. It is simple to implement and is part of devices incorporating the "plug and use" principle. His motto: "Light on the cable by design". In a local network, it uses the UDP protocol.

5.5.4 BACnet

BACnet (Building Automation and Control networks) is a communication protocol specified by ASHRAE and is an ISO and ANSI standard. It is a network layer protocol that can be used over several links and physical layer technologies, such as LonTalk, UDP/IP….

BACnet also integrates the application layer thanks to a set of dedicated objects. These objects represent the information managed and exchanged by the devices [54]. Its object approach as close as possible to the application layers makes it a good candidate as a high-level protocol in a BMS or home automation installation. BACnet is often seen as the protocol that will unify everyone else, thanks to its advanced features. He is therefore very much appreciated for supervision in Technical Building Management.

6 Home Automation Software

6.1 OpenHAB

OpenHAB (Open Home Automation Bus) is a home automation software written in Java with OSGi and published under the GPL v3 license [55]. OpenHAB has a web server for its interface, it also has applications for iOS and Android. The software can be ordered in a specific way: by sending orders via XMPP (Protocol of Communication). Development is active, many modules for communication with other protocols should arrive in later versions.

6.2 FHEM

It is a home automation server written in Perl under GPL v2 license. This German software allows you to manage the FS20, 1-Wire, X10, KNX, and EnOcean protocols. Its documentation and forums, mostly in German, are a negative point for many users [56].

6.3 HEYU

Is a home automation program usable from the command line. This program is written in C and is licensed under the GPL v3 (older versions have a special license). HEYU is specifically intended for the X10 protocol, to communicate with this network, the preferred interface is the CM11A of XA0 Inc. This project has recently been very active, its late opening and its exclusive use of X10 have undoubtedly caused its surrender.

DomotiGa: is a home automation software for GNU/Linux written in Gambas and under GPL v3 license, its origins are Dutch. This software is compatible with 1-Wire, KNX, X10, xPL, Z-Wave, and many more.

MisterHouse: is a multi-platform software written in Perl under the GPL license. This software is aging and no longer seems to be maintained, it is nevertheless regularly returned to the fore during research on free home automation. Due to its American roots, this software allows you to manage X10, Z-Wave, EIB, 1-Wire networks.

6.4 Domogik

It is software written in Python under the GPL v3 + license. It was born on the forum ubuntu-fr.org between several people who wanted home automation software. It is in active development and currently allows basic habitat management. Its architecture is based on the internal xPL protocol [57]. It is gradually extending its functionality to the protocols most used in home automation. For the moment, the following protocols are supported: x10, 1-Wire, LPX800, Teleinfo, RFID, Wake on LAN/Ping. The software has a web interface and an Android application.

6.5 Calaos

It is a commercial home automation solution based on a Modbus PLC and a GNU/Linux home automation server. The majority of the applications code is free under the GPL v3 license. The solution is intended primarily for new houses with strong integration at the time of the construction study. A Calaos installation uses automata that control all the electrical elements in the house, as well as acquire switches, temperature probes, presence sensors, etc. The controller is capable of interacting with field buses such as DALI or KNX. Then comes the home automation server, which will control the automaton and thus manage all the rules of the house (such as pressing a switch which generates the launch of a scenario). It also gives access to the home in different forms of interfaces: Web, touch (based on EFL), mobile applications, etc. A Calaos system is also capable of managing IP cameras, as well as audio multi-room.

6.6 OpenRemote

The goal of OpenRemote is to create a free software stack that manufacturers can integrate at very low cost into their products, to create control surfaces for the building. OpenRemote supports a large number of protocols including X10, KNX, Z-Wave, ZigBee, 6LoWpan, etc. The idea is to reuse the screens already present in places of life such as smartphones, tablets, and desktop computers. So currently supported: Android, iOS, as well as GWT for web applications. All of the code is licensed under the AGPL license.

6.7 LinuxMCE

It is a GNU/Linux distribution that aims to provide complete and integrated multimedia management, home automation, telephony, and security solution for a home. It is based on a lot of free software, such as VLC, Asterisk, Xine, etc. All of this software is implemented jointly to create a coherent whole. Lots of additional code allows realizing the various configuration and control interfaces. This distribution in development makes it possible to manage the following home automation protocols: KNX, 1-Wire, DMX, X10, EnOcean… It is based on old versions (10.04, for the development version) of Kubuntu. It is probably the most successful free solution currently, its developers compare it to proprietary solutions at more than USD 100,000. Unfortunately, it requires the use of a dedicated distribution, the more resourceful can peel it to extract the components that interest them and recreate a home automation server on one of their home server.

7 Home Automation and Photovoltaic Energy

Home automation brings together all the techniques used to control, program, and manage certain actions remotely in a house. The domotized home, connected house, or even smart home aims to improve the comfort of these inhabitants, as well as their security. But that's not all. Home automation also saves on bills, helping you control your consumption.

All the devices in your house can be connected by Wi-Fi or using a network cable, allowing you to remotely control your heating, your shutters, or even your alarm system.

Today, when we talk about home automation, we are essentially talking about saving energy. It is for this reason that the association of home automation with self-consumption is obvious to us, and to many people who are interested in it, simply because it makes it possible to optimize the energy savings made possible by progress home automation.

In this work, we speak of energy self-consumption for individuals. However, the term self-consumption here refers to the production of energy through the installation of photovoltaic panels. This allows you to consume your own produced energy and therefore considerably reduce your electricity bills. By opting for the use of home automation coupled with photovoltaic installations, it allows to optimize energy consumption and reduce energy costs.

Thanks to advances in home automation and intelligent energy systems, it is now possible to better coordinate the production and consumption of renewable electricity. With the confirmed dynamics of the self-consumption market, the synergies between these two fields seem more and more evident.

Intelligent energy management allows consumers to make the most of their solar production to save on energy bills. Linking the photovoltaic installation to the household electrical appliances, the device measures energy production and consumption in real-time. Then thanks to intelligent energy management algorithms [58]. They learn habits of energy consumption. Coupled with weather forecasts, they determine when is the best time to trigger household appliances to make the most of the solar energy produced during the day: triggering of the hot water tank and programming of the washing machine or washing machine dishes during the day to allow the exploitation of photovoltaic electricity production.

8 Implementation

This section is devoted to an introduction to the implementation of the smart home equipped with solar energy. Many of the important points that we will address, which are important in the realization of this project, such as the means and programs used, the prices of the materials and the means used to know the total cost of realization.

8.1 Cost of Home Automation

It should be remembered that the price of a home automation installation can vary depending on the desired application. There are many types of installation and the prices of home automation systems are variable according to your request. Here is a price order for different elements that can make up a home automation installation (Table 1) [59].

There are many types of installation and the prices of home automation systems vary according to demand. Here is a price order for different elements that can make up a home automation installation (Table 2) [59].

For this work, the budget will depend on the number of peripherals that we have used (Table 3).

Table 1 Applications for a home automation installation

Desired application	Description
Optimizing energy management	Programming of heating and lighting, switching off devices on standby
Reinforce security	Alarms, remote monitoring, centralized opening and closing, assistance to people
Comfort and home automation	Home automation heating and lighting adapted to your needs, control of HiFi devices

Table 2 Home automation systems prices

Elements of the home automation system	Price
Home automation alarm	Between €200 and €1000
Electric gate	Between €500 and €1500
Electric shutters	Between €150 and €800
Home automation sensors	From €20
Home automation switches	Between €50 and €150
Home automation remote control	About €350
Home automation control screen	Between €100 and €500
Home automation system	Between €500 and €1500
Management software	Often free
Home automation wiring	Around €2000

Table 3 Peripherals used

Modules	Numbers
PIR motion sensor	X1
MQ6 gas sensor	X1
Flame sensor	X1
Humidity sensor	X1
Arduino Mega	X1
ESP8266 12E	X1
9 g servo motor	X1
LCD 1602 I2C	X1
LED 5 mm	X5
Buzzer	X1

8.2 *Hardware and Software Used*

8.2.1 Arduino Mega Board

Arduino designates a free ecosystem comprising cards (Arduino Uno, Arduino Leonardo, Arduino Méga, Arduino Nano…), software (notably the Arduino IDE), or even libraries. These programmable electronic systems make it easy to build projects and to approach both the electronic approach and the software approach.

The Arduino Méga 2560 board is a microcontroller board based on an ATmega2560 (Fig. 3). It contains everything necessary for the operation of the microcontroller; To be able to use it and get started, simply connect it to a computer using a USB cable (or power it with an AC adapter or battery, but this is not essential, 1 power supplied by the USB port) (Table 4).

Fig. 3 Arduino mega 2560 board

Table 4 Characteristic summaries

Microcontroller	ATmega2560
Operating voltage	5 V
Supply voltage (recommended)	7–12 V
Supply voltage (limits)	6–20 V
Clock speed	16 MHz
EEPROM memory (non-volatile memory)	4 KB
SRAM memory (volatile memory)	8 KB

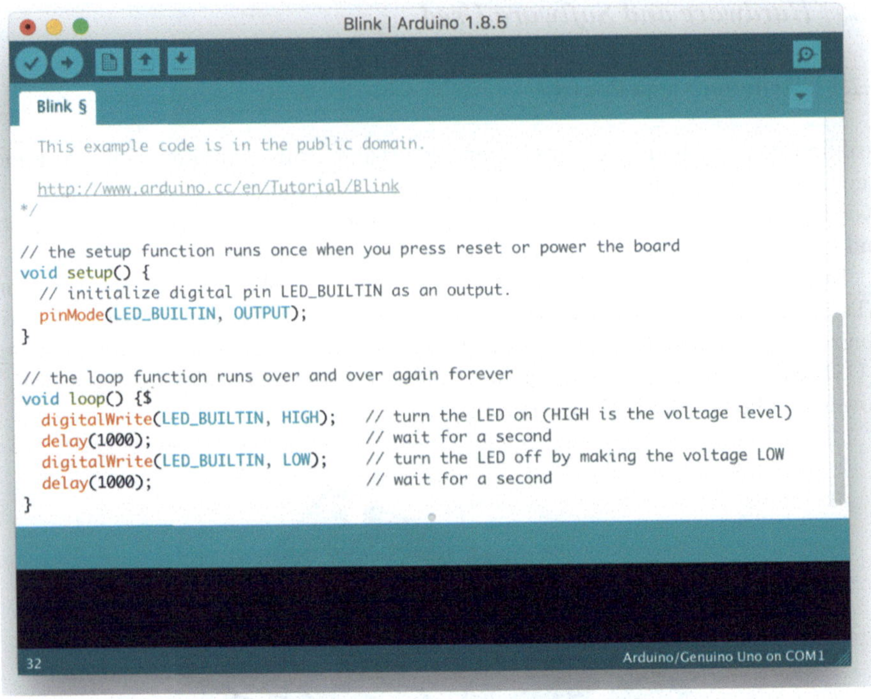

Fig. 4 Arduino software (IDE)

8.2.2 Arduino Software (IDE)

The open-source Arduino Software (IDE) makes it easy to write code and transfer it to the board. It works on Windows, Mac OS X, and Linux. The environment is written in Java and based on processing and other open-source software (Fig. 4). This software can be used with any Arduino board.

9 Conclusion

The United Nations Environment Program (UNEP) is the highest environmental authority in the United Nations system. Its mission is to lead the way and encourage partnerships in the protection of the environment while being a source of inspiration and information for people and a facilitating instrument enabling them to improve the quality of their life without compromising that of future generations.

A habitat is a place of great importance for everything, of its nature it is the place where one stays and returns. All people, and especially the elderly, spend a lot of their time at home, hence the considerable influence of the home on the quality and nature of life. Improving the feeling of comfort and security in the home, therefore, appears to be quite important from a social point of view.

Not long ago, computer science was applied to the creation of smart homes to improve people's living conditions while at home and provide them with reliable remote control. Such a house is a residence equipped with ambient computer technologies intended to assist the inhabitant in the various situations of domestic life. The so-called smart houses increase the comfort of the inhabitant through natural interfaces to control lighting, temperature, or various electronic devices. In addition, another essential goal of applying information technology to habitats is the protection of individuals. This has become possible through systems capable of anticipating and predicting potentially dangerous situations or of reacting to events endangering the inhabitant. The beneficiaries of such innovations can be autonomous individuals but also more or less fragile people with limited movement capacity. Intelligent systems can remind residents, among other things, of their medication, facilitate their communication with the outside world, or even alert relatives or emergency services.

IoT promises to be an unprecedented development. Objects are now able to communicate with each other, to exchange, to react, and to adapt to their environment on a much broader level. Often described as the third wave of the new information technology revolution, following the advent of the Internet in the 1990s, then that of Web 2.0 in the 2000s, the Internet of Things marks a new stage in the evolution of cyberspace. This revolution facilitates the creation of intelligent objects allowing advances in multiple fields; one of the fields most affected by the emergence of IoT is home automation. Indeed, the proliferation of new means of communication and new information processing solutions are upsetting living spaces. Housing, which has become an intelligent living space, must not only be adapted to the people who live there, to their situations and needs but also be ready to accommodate new systems designed to relieve daily life, increase the possibilities and reach a higher level services and comfort (Internet access, teleworking, monitoring of consumption, research of information, etc.). But despite the involvement of many companies in the field, it is clear that few applications are now operational and widely distributed. Commercial solutions in the home automation market are dominated by smart control gadgets like automatic light or smart thermostat, but the complete home automation solution called smart home remains inaccessible to the common consumer because of the costs and incompatibility of most of these solutions with houses already built.

In recent years, the rate of energy consumption has increased considerably, which is why the adoption of an energy management system (EMS) is of paramount importance. The energy supply crisis caused by the instability of oil prices and the compulsory reduction of greenhouse gases is forcing governments to implement energy-saving policies. As residences consume up to 40% of the total energy of a developed country, an energy management system of a residence, using information and communications technologies, becomes more and more important, and necessary to set up. However, several projects have been proposed to design and implement an efficient energy management system in the building sector using IoT technology.

Finally, the research carried out constitutes an important database that attracts researchers in this field. It is rich in information on smart homes with clean solar energy. The solar smart home project, therefore, offers many advantages such as the comfort of the population, the protection of property as well as the rational and economical management of electric energy.

This work still needs to be developed on several points:

- Updating of information, whether related to physical means or programs.
- It is also possible to collaborate with experts in construction and materials to develop the house, such as the materials used for the installation of the walls, the materials used for the interior ventilation of the house. All of these things can make your home more energy-efficient.
- The use of other optimization algorithms.
- The development of a more robust user interface which allows the introduction of all comfort parameters.

References

1. D. Saba, Y. Sahli, B. Berbaoui, R. Maouedj, Towards smart cities: challenges, components, and architectures, in eds. by A.E. Hassanien, R. Bhatnagar, N.E.M. Khalifa, M.H.N. Taha, *Studies in Computational Intelligence: Toward Social Internet of Things (SIoT): Enabling Technologies, Architectures and Applications*. (Springer, Cham, 2020), pp 249–286
2. D. Saba, Y. Sahli, F.H. Abanda et al., Development of new ontological solution for an energy intelligent management in Adrar city. Sustain. Comput. Inform. Syst **21**, 189–203 (2019). https://doi.org/10.1016/J.SUSCOM.2019.01.009
3. D. Saba, F.Z. Laallam, H.E. Degha et al., Design and development of an intelligent ontology-based solution for energy management in the home, in *Studies in Computational Intelligence*, 801st edn., ed. by A.E. Hassanien (Springer, Cham, Switzerland, 2019), pp. 135–167
4. D. Saba, R. Maouedj, B. Berbaoui, Contribution to the development of an energy management solution in a green smart home (EMSGSH), in *Proceedings of the 7th International Conference on Software Engineering and New Technologies—ICSENT 2018* (ACM Press, New York, USA, 2018), pp. 1–7
5. D. Saba, H.E. Degha, B. Berbaoui, R. Maouedj, *Development of an ontology based solution for energy saving through a smart home in the City of Adrar in Algeria* (Springer, Cham, 2018), pp. 531–541
6. H.E. Degha, F.Z. Laallam, B. Said, D. Saba, Onto-SB: human profile ontology for energy efficiency in smart building, in *2018 3rd International Conference on Pattern Analysis and Intelligent Systems (PAIS)* (IEEE, Larbi Tebessi University Algeria, Tebessa, Algeria, 2018)
7. D. Saba, H.E. Degha, B. Berbaoui et al., Contribution to the modeling and simulation of multiagent systems for energy saving in the habitat, *International Conference on Mathematics and Information Technology (ICMIT 2017)* (IEEE, Adrar, Algeria, 2018), pp. 204–208
8. T.B. Asafa, T.M. Afonja, E.A. Olaniyan, H.O. Alade, Development of a vacuum cleaner robot. Alexandria Eng. J. (2018). https://doi.org/10.1016/j.aej.2018.07.005
9. L. Xie, B. Sheng, Y. Yin et al., Fridge: an intelligent fridge for food management based on RFID technology, in: *UbiComp 2013 Adjunct-Adjunct Publication of the 2013 ACM Conference on Ubiquitous Computing* (2013)
10. A. Beghdadi, M. Asim, N. Almaadeed, M.A. Qureshi, Towards the design of smart video-surveillance system, in *2018 NASA/ESA Conference on Adaptive Hardware and Systems, AHS* (2018)

11. J. Baidya, T. Saha, R. Moyashir, R. Palit, Design and implementation of a fingerprint based lock system for shared access, in *2017 IEEE 7th Annual Computing and Communication Workshop and Conference, CCWC* (2017)
12. A.V. Proskokov, M.V. Momot, D.N. Nesteruk et al., Software and hardware control robotic lawnmowers. J. Phys.: Conf. Ser. (2018)
13. F. Bu, X. Wang, A smart agriculture IoT system based on deep reinforcement learning. Futur. Gener Comput. Syst **99**, 500–507 (2019). https://doi.org/10.1016/J.FUTURE.2019.04.041
14. G.M. Toschi, L.B. Campos, C.E. Cugnasca, Home automation networks: a survey. Comput. Stand. Interfaces (2017). https://doi.org/10.1016/j.csi.2016.08.008
15. P.S. Nagendra Reddy, K.T. Kumar Reddy, P.A. Kumar Reddy et al., An IoT based home automation using android application, in *International Conference on Signal Processing, Communication, Power and Embedded System, SCOPES 2016 Proceedings* (2017)
16. T.H.C. Nguyen, J.C. Nebel, F. Florez-Revuelta, Recognition of activities of daily living with egocentric vision: a review. Sensors (2016)
17. I. Khajenasiri, A. Estebsari, M. Verhelst, G. Gielen, A review on internet of things solutions for intelligent energy control in buildings for smart city applications. Energy Procedia 770–779 (2017)
18. D. Saba, F.Z. Laallam, A.E. Hadidi, B. Berbaoui, Contribution to the management of energy in the systems multi renewable sources with energy by the application of the multi agents systems "MAS". Energy Procedia **74**, 616–623 (2015). https://doi.org/10.1016/J.EGYPRO.2015.07.792
19. D. Saba, F.Z. Laallam, A.E. Hadidi, B. Berbaoui, Optimization of a multi-source system with renewable energy based on ontology. Energy Procedia **74**, 608–615 (2015). https://doi.org/10.1016/J.EGYPRO.2015.07.787
20. M. Flasiński, History of artificial intelligence, in *Introduction to Artificial Intelligence* (2016)
21. S. Hunter, Google self-driving car project. Google X (2014). https://doi.org/10.1017/CBO9781107415324.004
22. V.R. Prasath Kumar, M. Balasubramanian, S. Jagadish Raj, Robotics in construction industry. Indian J. Sci. Technol. (2016). https://doi.org/10.17485/ijst/2016/v9i23/95974
23. W. Shatner, C. Walter, Star trek. I'm working on that : a trek from science fiction to science fact. Pocket Books (2004)
24. J. Mehra, Einstein, physics and reality (2010)
25. R. Sun, Artificial intelligence: connectionist and symbolic approaches, in *International Encyclopedia of the Social & Behavioral Sciences* 2nd edn
26. T. Munakata, Thoughts on deep blue vs. kasparov. Commun. ACM (1996) https://doi.org/10.1145/233977.234001
27. G.E. Hinton, How neural networks learn from experience. Sci. Am. (1992). https://doi.org/10.1038/scientificamerican0992-144
28. S. Li, XuL Da, S. Zhao, The internet of things: a survey. Inf. Syst. Front. **17**, 243–259 (2015). https://doi.org/10.1007/s10796-014-9492-7
29. E. Borgia, The internet of things vision: key features, applications and open issues. Comput. Commun. **54**, 1–31 (2014). https://doi.org/10.1016/J.COMCOM.2014.09.008
30. R. Saad, in *Modèle collaboratif pour l'Internet of Things (IoT)* (2016)
31. D.G. Perry, S.H. Blumenthal, R.M. Hinden, The ARPANET and the DARPA internet. Libr. Hi Tech (1988)
32. P.V. Paul, R. Saraswathi, The internet of things—a comprehensive survey, in *6th International Conference on Computation of Power, Energy, Information and Communication, ICCPEIC 2017* (2018)
33. R. Khan, S.U. Khan, R. Zaheer, S. Khan, Future internet: the internet of things architecture, possible applications and key challenges, in *Proceedings of the 10th International Conference on Frontiers of Information Technology, FIT 2012* (2012)
34. L. Identificaci, R. Frecuencia, R.F. Identification, RFID: TECNOLOGÍA, APLICACIONES Y PERSPECTIVAS. Rfid Tecnol. Apl. Y Perspect. (2010)

35. S. Srivastava, M. Singh, S. Gupta, Wireless sensor network: a survey, in *2018 International Conference on Automation and Computational Engineering, ICACE 2018* (2018)
36. P.K. Verma, R. Verma, A. Prakash et al., Machine-to-machine (M2M) communications: a survey. J. Netw. Comput. Appl. (2016)
37. I. Lee, K. Lee, The internet of things (IoT): applications, investments, and challenges for enterprises. Bus. Horiz. **58**, 431–440 (2015). https://doi.org/10.1016/J.BUSHOR.2015.03.008
38. W.L. Wilkie, E.S. Moore, Expanding our understanding of marketing in society. J. Acad. Mark. Sci. (2012). https://doi.org/10.1007/s11747-011-0277-y
39. J. Roy, Cybersecurity. Public Adm. Inf. Technol. (2013)
40. Y. Liu, B. Qiu, X. Fan et al., Review of smart home energy management systems. Energy Procedia (2016)
41. Climamaison Domotique: Définition (2019). https://www.climamaison.com/domotique/definition.htm. Accessed 2 Jan 2019
42. M. Alaa, A.A. Zaidan, B.B. Zaidan et al., A review of smart home applications based on Internet of Things. J. Netw. Comput. Appl. **97**, 48–65 (2017). https://doi.org/10.1016/J.JNCA.2017.08.017
43. P. Remagnino, G.L. Foresti, Ambient intelligence: a new multidisciplinary paradigm. IEEE Trans. Syst. Man Cybern. Part A: Syst. Hum **35**, 1–6 (2005)
44. J. Augusto, P. Mccullagh, Ambient intelligence: concepts and applications. Comput. Sci. Inf. Syst. (2007). https://doi.org/10.2298/csis0701001a
45. A. Boukerche, Protocols for wireless sensor (2009)
46. J. Haase, *Wireless Network Standards for Building Automation* (Springer, New York, NY, 2013), pp. 53–65
47. A. Kailas, V. Cecchi, A. Mukherjee, A survey of communications and networking technologies for energy management in buildings and home automation. J. Comput. Netw. Commun. (2012)
48. S. Al-Sarawi, M. Anbar, K. Alieyan, M. Alzubaidi, Internet of things (IoT) communication protocols: review, in *ICIT 2017 8th International Conference on Information Technology, Proceedings* (2017)
49. L.E. Frenzel, DMX512. in *Handbook of Serial Communications Interfaces* (2016)
50. L.A. Magre Colorado, J.C. Martínez-Santos, *Leveraging 1-wire communication bus system for secure home automation* (Springer, Cham, 2017), pp. 759–771
51. D.-F. Pang, S.-L. Lu, Q.-Y. Zhu, Design of intelligent home control system based on KNX/EIB bus network, in *2014 International Conference on Wireless Communication and Sensor Network* (IEEE, 2014), pp. 330–333
52. U. Ryssel, H. Dibowski, H. Frank, K. Kabitzsch, Lonworks. in *Industrial Communication Systems* (2016)
53. S. Huang, B. Li, B. Guo et al., Distributed protocol for removal of loop backs with asymmetric digraph using GMPLS in P-cycle based optical networks. IEEE Trans. Commun. **59**, 541–551 (2011). https://doi.org/10.1109/TCOMM.2010.112310.090459
54. S. Tang, D.R. Shelden, C.M. Eastman et al., BIM assisted building automation system information exchange using BACnet and IFC. Autom. Constr. (2020). https://doi.org/10.1016/j.autcon.2019.103049
55. openHAB Foundation eV. openHAB (2017). https://www.openhab.org/. Accessed 1 Apr 2017
56. M. Vukasovic, B. Vukasovic, Modeling optimal deployment of smart home devices and battery system using MILP, in *2017 IEEE PES Innovative Smart Grid Technologies Conference Europe, ISGT-Europe 2017 Proceedings* (2017)
57. D.M. Siffert, Pilotage d'un dispositif domotique depuis une application Android (2014)
58. N.C. Batista, R. Melício, J.C.O. Matias, J.P.S. Catalão, Photovoltaic and wind energy systems monitoring and building/home energy management using ZigBee devices within a smart grid. Energy (2013). https://doi.org/10.1016/j.energy.2012.11.002
59. Exemple devis domotique pour une maison connectée. https://www.voseconomiesdenergie.fr/travaux/domotique/prix. Accessed 10 May 2020

Appliance Scheduling Towards Energy Management in IoT Networks Using Bacteria Foraging Optimization (BFO) Algorithm

Arome Junior Gabriel

Abstract Modern life is almost impossible without electricity, and there is an explosive growth in the daily demand for electric energy. Furthermore, the explosive increase in the number of Internet of Things (IoT) devices today has led to corresponding growth in the demand for electricity by these devices. Serious energy crisis arises as a consequence of these high energy demand. One good solution to this problem could be Demand Side Management (DSM) which involves scheduling of consumers' appliances in a fashion that will ensure peak load reduction. This ultimately will ensure stability of the Smart Grid (SG) networks, minimization of electricity cost, as well as maximization of user comfort. In this work, we adopted Bacteria Foraging (BFA) Optimization technique for the scheduling of IoT appliances. Here, the load is shifted from peak hours toward the off peak hours. The results show that BFA optimisation based scheduling technique caused a reduction in the total electricity cost and peak average ratio.

Keywords Demand side management · Internet of things network · Smart grid · Electricity optimization · Meta-heuristic optimization algorithms

1 Introduction

An Internet of Things (IoT) based Smart Grid (SG) is a more efficient form of the traditional power grid, and is often called the next generation power grid system. SG improves the reliability, efficiency and effectiveness of the traditional power grid by riding on a collection of several technologies and applications working together as the fundamental structure [called Advanced Metering Infrastructure (AMI)], to provide a 2-way communication mechanism for exchange of information (about current electricity status, pricing data and control commands in real-time) between utility (electric energy suppliers) and consumers/users.

A. J. Gabriel (✉)
School of Computing, Federal University of Technology, Akure, Nigeria
e-mail: ajgabriel@futa.edu.ng

© The Editor(s) (if applicable) and The Author(s), under exclusive license to Springer Nature Switzerland AG 2021

A. E. Hassanien et al. (eds.), *Artificial Intelligence for Sustainable Development: Theory, Practice and Future Applications*, Studies in Computational Intelligence 912,
https://doi.org/10.1007/978-3-030-51920-9_15

The information exchanged between consumers and utility through the AMI is used for energy optimization. Indeed, energy optimization has become a huge necessity in today's world especially due to the explosive increase in demand for electric power for running modern homes, businesses and industries. DSM is one of the most important aspects of SG energy optimization. It provides balance between demand and supply. Through DSM users are encouraged to shift their load from on peak hours to off peak hours. Demand response (DR) and load management are two main functions of DSM [1]. In load management, the focus is on the efficient management of energy. It reduces the possible chances of distress and blackouts. It also plays an important role in reducing peak to average ratio (PAR), power consumption and electricity cost. Load management involves scheduling of appliances. The shifting of appliance load is done either via task scheduling or energy scheduling. Task scheduling involves switching appliances on/off depending on the prevailing electricity price at a given time interval. Energy scheduling on the other hand entails reducing appliances' length of operational time (LoT) and power consumption.

According to Rasheed et al. [2], DR refers to steps taken by consumers in reaction to dynamic price rates announced by utility. Truly, changes in grid condition can result in corresponding change in electricity demand level. This rapid change creates an imbalance of demand and supply, and this imbalance, within a short time can pose great threat to the stability of the power grid. DR helps provide flexibility at relatively low rates. It is beneficial for both utility and consumers. It aims at educating the consumers to consume maximum of their energy requirements during off peak hours. It also results in reduction of PAR, which is beneficial to the utility.

The relationship between demand and supply are better captured using dynamic pricing rates than flat rate pricing schemes. Some of the dynamic pricing tariffs are day ahead pricing (DAP), time of use (TOU), RTP, inclined block rate (IBR) and critical peak pricing (CPP). These encourage consumers to shift high load appliances from on-peak hours to off-peak hours, resulting in minimization of electricity cost and reduction in PAR.

Several DSM strategies have been proposed in recent years to achieve above said objectives. In [1–4,], formal non-stochastic techniques like integer linear programming (ILP), non-integer linear programming (NILP), and mixed integer non-linear programming (MILP), were adopted for energy consumption and electricity cost minimization. However these methods cannot handle efficiently the stochastic nature of price signals and consumer energy demand. Other researchers proposed stochastic schemes for overcoming the limitations of the non-stochastic methods. The daily rapid increase in demand for energy from residential homes has resulted in so much research interest been directed at home appliances scheduling. In this work, we have adopted bacteria foraging optimization algorithm for scheduling of appliances in order to minimize consumption and consumer electricity bills.

The rest of this work is organised as follow: Section 2 presents review of related literature. In Sect. 3, the proposed system model was discussed. Section 4 contains discussion on the BFO algorithm adopted in this study. Report on simulation and results is contained in Sect. 5. Section 6 shows the conclusion.

2 Review of Related Literature

In recent times, several researches have been carried out on ways to develop optimal strategies for home energy management with regards to the smart grid. The most fundamental objectives of these proposals are minimization of cost, and load balancing. In this subsection, we present some of the existing related literature, highlighting the objective(s) and limitation(s) of their approaches.

A Mixed Integer Linear Programming (MILP) based HEM system modelling was proposed in [5]. The author carried out evaluation of a smart household using MILP framework, putting into consideration the operation of a smart household that owns a PV, an ESS that consists of a battery bank and also an EV with V2H. Two-way energy exchange is allowed through net metering. The energy drawn from the grid has a real-time cost, while the energy sold back to the grid is considered to be paid a flat rate. A cost reduction of over 35% was reportedly achieved. Increase in the size of the population however, leads to very high computational time requirement.

Mixed integer linear programming based algorithm was proposed by [6] which schedules the home appliances.

The main focus of this work was minimization of cost with allowance through multi-level preference, where job switching for the appliances can be possible on lower cost according to demand.

Mixed Integer Non Linear Programming (MINLP) approach was adopted in both [7] and [8]. The authors of Moon and Lee [7] worked on the problem of multi residential electricity load scheduling with the objective of ensuring user satisfaction given budget limits. This problem was formulated and solved as a Mixed Integer Non Linear.

Programming (MINLP) problem. PL-Generalized benders algorithm was also for solving this problem in a distributed manner. The authors reported optimized scheduling of electricity load, plus a reduction in trade-off between user-comfort and cost. In [8], a HEM was proposed towards ensuring optimal scheduling of residential home appliances based on dynamic pricing scheme. Although the authors achieved their goal of cost reduction as shown in their simulation results where 22.6 and 11.7% reduction in cost for peak price and normal price scheme respectively, they incurred a different type of cost in terms of high computational complexity and time. Indeed, the formal approach used in both [7] and [8] incurs high computational overhead with respect to time, especially as the number of home appliances in consideration increases.

Ogunjuyigbe et al. proposed load satisfaction algorithm in [9], with the goal of ensuring cost minimization and maximization of user-comfort. The authors reported achievement of minimum and maximum cost and user comfort respectively in their simulation results. It was also discovered through their sensitivity analysis carried out on different user budgets, that users' satisfaction is directly proportional to user's budget. However, Peak Average Ratio (PAR) an important metric was completely ignored.

Recursive models have also been proposed for evaluating peak demand under different power demand scheduling scenarios. Vardakas et al. in [10], adopted recursive method for calculating peak demand in compressed, delay and postponement request settings and compared same with the default non-scheduled default scenario using real time pricing scheme for an infinite number of appliances in the residential home management system. The authors also considered the involvement of consumers in energy management program together with RES integration. Their simulations result reveals satisfactory accuracy while analytical models calculate peak demand results in very small computational time. However, their assumption of infinite number of appliances results in overestimation of power consumption. In order to handle this limitation in [10], the authors in [11] proposed four scenarios for a finite number of appliances. They also considered the participation of consumers in HEM so as to ensure social welfare. The analytical results produce low timely results which are essential for near real-time decisions.

In [3], authors propose an optimization power scheduling scheme to implement DR in a residential area. In this paper electricity price is announced in advance. Authors formulate the power scheduling problem as an optimization problem to obtain the optimal schedules. Three different operation modes are considered in this study. In first mode, consumer does not care about discomfort and considers only electricity cost. Whereas, in the second operation mode consumer only cares about discomfort. In third operation mode consumer cares about both discomfort and electricity cost. Results show that proposed scheduling scheme achieves significant trade-off between discomfort and electricity cost.

The Authors of Rasheed et al. [2] proposed an optimal stopping rule based opportunistic scheduling algorithm. They categorised consumers into active consumers, semi active consumers, passive consumers based on energy consumption pattern. Authors proposed two scheduling algorithms. Modified first come first serve (MFCFS) algorithm for reduction of electricity cost, and priority enable early deadline first (PEEDF) algorithm for maximizing consumer comfort. The authors in their simulation results demonstrated the effectiveness of the proposed algorithms in the target objectives. 22.77% reduction in PAR and 22.63% reduction in cost was achieved. However, installation and maintenance of RES which can be quite expensive was completely ignored.

Muralitharan et al. used a multi-objective evolutionary algorithm to reduce consumer cost and waiting time in SG [12]. The authors have applied threshold policy in order to avoid peak and balance load. The penalty in form of additional charges has been incorporated in their proposed model if consumers exceed price threshold limits. The simulation results minimize both electricity cost and waiting time.

In [13], renewable energy generation and storage models were proposed. Day ahead load forecasting (DLF) model based on artificial neural network (ANN) was also presented. The authors made use of energy consumption patterns of two previous days to forecast the nature of demand load for the next day. 38% improvement in speed of execution and 97% improvement in confining non confining non-linearity in load demand curve of previous days. Were reportedly achieved. However their forecast were not error free.

Genetic Algorithm (GA) based solutions for DSM were proposed in [1] towards achieving residential load balancing. The specific objectives of this work was to ensure increase in electricity cost savings and user comfort. Appliances were grouped as regular, user-aware, thermostatically-controlled, elastic, and inelastic appliances.

Scheduling of appliances was done using intelligent programmable communication thermostat (IPCT) and conventional programmable communication thermostat. The simulation results show that proposed algorithms achieved 22.63 and 22.77% reduction in cost and PAR respectively. However, their technique incurred high increase in system complexity.

In [14], the authors proposed an energy management system based on multiple users and load priority algorithm. The major objectives of this proposal was to reduce electricity consumption and cost. The strategic scheduling was based on multiple used influence and load priority for TOU energy pricing.

The authors of Wen et al. [15] proposed a reinforcement learning based on Markov decision process model where Q-learning algorithm was adapted for the design of the scheduler. This algorithm does not require a predefined function for the consumer dissatisfaction in case of job rescheduling.

The article in [16] suggested a double cooperative based game theory technique for the development of a HEM in a bid to ensure cost minimization for consumers. Deliberated utilities were considered using cooperative game theory.

In [17] Hybrid differential evolution with harmony search (DE-HS) algorithm is proposed. This paper proposed the generation scheduling of micro grid consisting of traditional generators, photovoltaic (PV) systems, generation of wind energy, battery storage and electric vehicles (EV). EV act in two ways, as a load demand and also used as storage device. Proposed hybrid DE-HS algorithm is used to solve scheduling problem. The paper also modelled the uncertainty of wind PV systems towards ensuring that the stability of micro grid is maintained. Their results as presented reveals that the proposed hybrid technique requires minimum investment cost. The paper considered two scenarios; scheduling of micro grid (MG) with storage system and EV and without storage system and EV for comparison purpose. The proposed method performed better in the first scenario (with storage and EV) as it took 7.83% less cost as compared to the other case.

The authors in [18] considered power flow constraints when they proposed a hybrid harmony search algorithm (HAS) with differential evolution (DE) day-ahead model based scheduling in a micro-grid. Their main goals were: to minimize the total generation and operation cost of PV, wind turbine (WT), diesel generator (DG) as well as batteries. A comparative analysis of the proposed model and technique with other techniques like DE, hybrid DE and GA (GADE), modified DE (MDE) and hybrid particle swarm optimization with DE (PSODE) was carried out in order to evaluate their proposed HSDE. Simulation results indicated that in terms of better convergence (low cost with minimum CPU time), the proposed technique performed well compared to the other techniques. In order to further demonstrate the robustness of the proposed technique, both normal and fault operation modes are considered in test micro grid.

Reduction in energy consumption, monetary cost and PAR were reported achieved in [19]. In order to achieve these goals. Appliances with low priority were switched off. Priorities were assigned to appliances as consumer wants.

Beyond Smart Grid, nature-inspired optimization algorithms have been used in other domains, with huge successes. For instance, success in the use of genetic tabu search algorithm for optimized poultry feed formulation was reported in literature [20].

Indeed, existing literature reveals the superiority of meta-heuristic techniques over other ones with respect to handling of large and complex scenarios, while enjoying less time of execution. BFA is proposed in our work for evaluating our objectives due to its ability to perform efficiently even in the face of increasing population size. Besides, BFA also has self-healing, self-protection and self-organization capabilities.

Table 1 presents a summary of the related works in literature.

3 System Model

In this work, we consider a single household scenario. A home is made up of several appliances.

3.1 Category of Loads

The appliances in a given home can be categorised into manageable and non-manageable loads. Due to the high nature of its energy consumption and predictability in its operation, most research efforts as obtainable in existing literature, are directed at manageable loads. Manageable loads include appliances like refrigerator, water heater, dish washer and even washing machine. Non-manageable loads on the other hand include appliances like TV, laptops, phones and lights. These are home appliances having insignificant loads compared with the manageable load examples. Besides, these appliances are interactive and have little scheduling flexibilities [4].

In this work, we focus on the manageable loads. We have considered two major sub-categories; Shift-able and non-shift-able appliances. The system model in Fig. 1 captures a summary of the workings of the proposed system.

Table 1 Summary of related research works

References	Technique(s)	Objectives	Limitations
Impacts of small-scale generating and storage units presented in [5]	MLP	Electricity cost reduction	PAR is ignored increased complexity
Residential demand response scheduling presented in [6]	MLP	Cost minimization with allowance for job switching for the appliances	High computation time
Multi-residential demand response scheduling presented in [7]	MINLP and PL generalized benders algorithm	Reduce trade-off between user comfort and electricity consumption cost	Increased system complexity
Optimal residential appliance scheduling under dynamic pricing scheme via HEMDAS presented in [8]	MINLP	User satisfaction within given budget limit constraint	High computational overhead with regards to time
User satisfaction-induced demand side load management in residential buildings with user budget constraint presented in [9]	Load satisfaction algorithm	Cost minimization and user-comfort maximization	PAR ignored
Performance evaluation of power demand scheduling scenarios in a smart grid environment presented in [10]	Recursive models	Calculation of peak demand	Assumption of infinite number of appliances resulted in overestimation of power consumption
Power demand control scenarios for smart grid applications with finite number of appliances presented in [11]	Recursive methods	Peak demand calculation	Only finite set of appliances were considered
Residential power scheduling for demand response in smart grid presented in [3]	ILP	Minimization of cost and consumer discomfort	PAR is neglected, RES integration not considered
Priority and delay constrained demand side management in real-time price environment with renewable energy source presented in [2]	OSR, MFCFS and PEEDF	Cost and energy-consumption minimization via RESs	Individual appliance scheduling was ignored. High cost of installation and maintenance of RES was also not considered

(continued)

Table 1 (continued)

References	Technique(s)	Objectives	Limitations
Multi-objective optimization technique for demand side management with load balancing approach in smart grid presented in [12]	Multi-objective evolutionary algorithm	Minimization of cost and user delay time	Dominant energy scheduling of an appliance was not considered
A modified feature selection and artificial neural network-based day-head load forecasting model for a smart grid presented in [13]	DLF-based ANN	Load forecasting	Presence of errors in their forecasts
Real time information based energy management using customer preferences and dynamic pricing in smart homes presented in [1]	GA, IPCT, CPCT	Maximise cost savings and reduction in PAR	High system complexity incurred
Optimal operation of micro-grids through simultaneous scheduling of electrical vehicles and responsive loads considering wind and PV units uncertainties. Presented in [17]	DE-HS	Minimization of cost	Hazardous emission of pollutants was not considered
Optimal day ahead scheduling model is presented in [18]	HS-DE	Minimization of total generation and operation cost	Increased system complexity
Priority based scheduling is used in [19]	Priority-based Scheduling	Reduction in energy consumption and cost	Appliances with low priority may face starvation
Appliance scheduling optimization in smart home networks presented in [4]	Mixed integer programming	Minimization of electricity cost	Not scalable without occurring increased computation time

3.2 *Specific Objectives of This Work*

The specific objective of this work is to develop a BFA based scheduling system towards achieving load balancing, cost and PAR reduction and also measure consumer comfort.

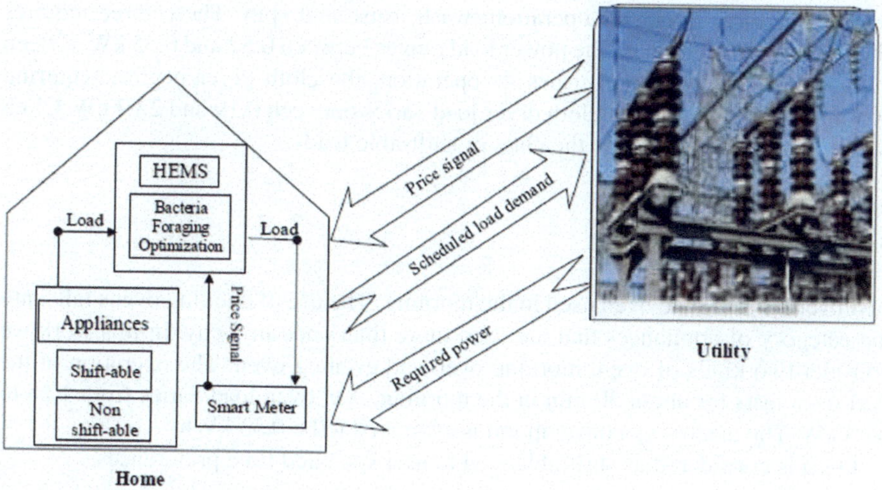

Fig. 1 Proposed system model

3.3 *Description of Major Home Appliances Considered*

In this work, we consider a single household with the following key electricity consuming appliances: dishwasher, clothes washer and dryer, morning oven, evening oven, electric vehicle (EV), refrigerator and air conditioner (AC).

It is common knowledge that various appliances have fixed timing for the completion of their cycles. This implies that they have fixed power rating which can be determined from the appliance specifications or by conducting experiments to that effect. The time of use (TOU) price tariffs was used in this work. The following section presents a description of each of the appliances considered in this work.

3.3.1 Dish Washer

The dishwasher has three major operating cycles: wash, rinse and dry. Completing all the cycles requires about 105 min to complete. The load varies between a minimum of 0.6 kW and a maximum of 1.2 kW as the dish washer runs. The dish washer is belongs to the class of shift-able loads. The energy consumption is about 1.44 kWh for one complete cycle of dish washer.

3.3.2 Cloth Washer and Cloth Dryer

These two appliances work in sequence. This implies that the cloth washer runs its course to completion, only then the cloth dryer comes on and take over. The cloth washer has three cycles of operation; wash, rinse and spin. These three requires about 45 min to complete. The power load ranges between 0.52 and 0.65 kW. Fifteen minutes after the washer finishes its operation, the cloth dryer begins, requiring 60 min to finish operation. Cloth dryer load varies between 0.19 and 2.97 kW. Cloth washer and dryer belongs to the class of shift-able loads.

3.3.3 AM Oven

AM oven refers to the oven used in the morning. The use of cooking ovens falls into the category of appliances that are used more than once in a day. In this work we consider two kinds of oven, morning oven and evening oven. The operation of the AM oven lasts for about 30 min in the morning. AM oven load varies from 1.28 to 0.83 kW. The electricity consumption is estimated to be 0.53 kWh.

Oven is considered as shift-able load to user specified time preferences.

3.3.4 PM Oven

The PM oven is the oven used in the evening. The PM oven lasts longer in its operation and in this case, two burners are used. The evening oven runs for 1.5 h, with load varying between 0.75 and 2.35 kW. The electricity consumption is 1.72 kWh.

3.3.5 Electric Vehicle (EV)

The manufacture of EVs is on the rise in today's world. Hybrid vehicles that works both on gas and electric batteries are becoming common now. These batteries are charged via home electricity. The EV takes 2.5 h to charge fully at a constant 3 kW load and immediately tapers off to zero. The consumption of EV is estimated to be 7.50 kWh. EV falls into the class of loads that is shift-able to user preferred time when TOU tariff is the lowest like between 7 p.m. and 7 a.m.

3.3.6 Refrigerator

Refrigerator falls in the category of appliances which works 24 h a day. The only time compressor rests is when the inside temperature is lower or equal to the set temperature of the refrigerator. The compressor also rests when defrost heating starts. The maximum and minimum load during the operation of the refrigerator is 0.37 kW and 0 kW respectively. The electricity energy consumption is 3.43 kWh/day. Refrigerator belongs to the class of continuous non-shift-able load.

3.3.7 Air Conditioner (AC)

The load profile of the air-conditioner (AC) considered here varies between 0.25 kW when its compressor is switched off and peaks at 2.75 kW when compressor of is working. The compressor goes off when the room temperature inside the room is equal or below the set room temperature. However, air fan continues to work for circulation of air. The energy consumption of 2.5 ton AC is around 31.15 kWh per day. AC belongs to the class of continuous non-shift-able load, and its usage could be based on prevailing weather condition.

3.3.8 Non-Manageable Appliances

These are other appliances available in a typical household setting. Examples are televisions, transistor radios, lights, clock, phones, and even personal computers. As highlighted earlier, their loads compared to the major load discussed above are insignificant and not power consuming. Besides, these appliances have little scheduling flexibilities, and as such are not considered in this work.

3.4 Length of Operation Time

In this work, we consider a day of 24 h as divided into 96 time slots. All the time slots are represented by their starting times. The starting slot on a given day is 6:00 a.m., while the ending time slot is 5:45 a.m. the next day. This implies that each of the 96 time slot denotes an interval of 15 min. The end time of an individual slot is obtained by adding 15 min to the starting time. For example, for time slot 2, the starting time is 06:15 a.m. and ending time is 06:30 a.m.

3.5 Appliances Scheduling Optimization Problem Formulation

Several researches in literature have focused on optimizing users' electricity consumption pattern, towards achieving stability of the grid and reduction in electricity demand at on-peak hours, electricity cost, and PAR. This indeed, is difficult due to the random nature of users' consumption pattern and electricity prices. The authors of Qayyum et al. [4] formulated and solved the appliance scheduling problem using a mixed integer programming (MIP) approach for which decision variables and support binary decision variables were specified and applied. The specific target of their work was reduction of electricity. The performance evaluation of this proposed deterministic technique was carried out using the time of use (TOU) pricing tariff.

Table 2 Properties of appliances considered in this experiment

Category	Appliances	Power rating (kwh)	Daily usage (h)
Shiftable Loads	Dish washer	0.7	1.75
	Cloth washer	0.62	0.75
	Cloth dryer	1.8	1.0
	Morning (a.m.) oven	1.2	0.5
	Evening (p.m.) oven	2.1	1.5
	Electric vehicle	2.0	2.5
Non-shiftable loads	Refrigerator	0.3	24
	Air conditioner	2.0	24

This work however, has the problem of scalability. As attempts at scaling up the number of appliances in consideration results in overall system complexities.

Our work aims specifically at ensuring optimization of electric energy consumption patterns via appliances scheduling, in order to achieve load balancing, cost minimization and reduction in PAR. In order to achieve these objectives, we propose bacteria foraging optimization algorithm (BFA) for scheduling home appliances in smart grid. Our technique is stochastic and meta-heuristic in nature and is able to overcome the limitation of the work in [4] and other works that used formal deterministic techniques. TOU pricing signals is adopted for the computation of electricity cost. Scheduling of appliances is carried out over 96 time slots of 15 min each for a given day, based on the TOU pricing signal.

A given household is equipped with an advanced metering infrastructure (AMI) which enables a bidirectional communication between the home energy management system (HEMS) and utility. Appliances classification, life time and power ratings are shown in Table 2.

3.5.1 Problem Definition

Our optimization problem is formally is defined as minimizing daily electricity cost as captured in Eq. (1), subject to Eqs. 2–7:

$$\min \sum_{i=1}^{N} \sum_{j=1}^{n} P_{i,j}^{k} X_{i,j}^{k} \tag{1}$$

$$L(k) = E_s(k) + E_{ns} \tag{2}$$

$$L(k) < \mu\text{th} \tag{3}$$

$$\sum L_A = \sum L_B \tag{4}$$

$$\sum \text{Cost}_A < \sum \text{Cost}_B \tag{5}$$

$$k_\propto < k < k_\beta \tag{6}$$

$$X_{i,j}^k \in [0, 1] \tag{7}$$

The objective function is as stated in Eq. 1. This is subject to the constraints defined in Eqs. 2–7. Equation 2 represents the total energy consumption of all the appliances at time interval *t*. The implication of Eq. 3 is that energy consumption in a particular time slot should be less than or equal to specified threshold. This aids the reduction of PAR. Equation 4 shows the power consumption constraint. Its implication is that the total unscheduled load (power consumption before scheduling) must be equal to the total scheduled load. It also ensures that length of operation of each appliance must not be affected by scheduling. Equation 5 shows that total cost of scheduled load should be less of unscheduled load. Equation 6 represents start time ($k_\propto$) and ending time (k_β) of appliances. The current ON and OFF status of appliances is given in Eq. 7 as 1 or 0 respectively.

4 BFA Meta-Heuristic Optimization Technique

This section presents the BFA meta-heuristic optimization approach which we adopted in our HEMS for home appliances scheduling towards achieving effective DSM. The scheduling done covers 8 of such appliances.

The Bacterial Foraging Optimization Algorithm was motivated by an activity called "chemotaxis" shown by bacteria when they forage (search for food). These bacteria drive themselves around by rotation of the flagella. In order to allow for forward movement, the flagella is rotated in an anti-clockwise manner, allowing the bacterium to "swim" or "run". To change direction or "tumble", the flagellum is rotated in a clockwise manner. Once the bacteria "tumble" randomly into a desired direction, it then "swims" again. Repeating these two ("swimming" and "tumbling") actions alternatively aids the organism's search for nutrients in indiscriminate directions. Once the organism is approaching food, swimming becomes more frequent. In contrast, as the organism moves away from one food in search of another nutrient source, tumbling (change in direction) becomes more frequent than swimming. Chemotaxis is fundamentally a complex combination of swimming and tumbling that helps bacteria stay in locations where nutrients are available in higher concentrations.

Fundamentally, the BFO system is made up of three (3) major mechanisms, viz., chemotaxis, reproduction, and elimination-dispersal. The following sub-section presents briefs on these processes;

4.1 Chemotaxis

Here, a "tumble" indicates a unit walk with random direction, while if in contrast, the unit walk (in the last step) is in the same direction, we talk about a "run". Assuming a bacterium at jth chemotactic, kth reproductive, and lth elimination-dispersal step is represented as $\theta^i(j, k, l)$, then given that, the run-length unit parameter, $C(i)$, stands for the chemotactic step size during each run or tumble. Then, in each computational chemotactic step, the movement of the ith bacterium can be represented as

$$\theta^i(j+1, k, l) = \theta^i(j, k, l) + C(i)\frac{\Delta(i)}{\sqrt{\Delta^T(i)\Delta(i)}} \tag{8}$$

where $\Delta(i)$ stands for the direction vector of the jth chemotactic step.

if the bacterial movement is *run*, then, $\Delta(i)$ is the same with the last chemotactic step; otherwise, $\Delta(i)$ is a random vector whose elements lie in $(-1, 1)$.

With the activity of run or tumble taken at each step of the chemotaxis process, a step fitness, denoted as $J(j, k, l)$, can be evaluated.

4.2 Reproduction

Reproduction procedure is quite significant. At this stage, the health status of each of the organism is computed as the sum of the step fitness during its life. After that, the entire population of organisms (bacteria) are arranged in reverse order of their health status. Now, only the first half of population are selected (survives), each of the selected bacterium splits into two duplicate bacteria, thus, preserving the total number of bacteria under consideration.

4.3 Elimination and Dispersal

Although the chemotaxis provides a basis for local search, and the reproduction process speeds up the convergence, these two (2) alone are not sufficient requirements for global optima searching. This for instance could be because, bacteria may get trapped at the local optima (initial positions). The elimination and dispersal procedure helps with eliminating this probability of bacteria being trapped at the local optima. So, after certain amount of reproduction procedures, some bacteria are chosen according to some criteria, to be killed and moved to another position within the environment.

In summary, the BFA approach is nature inspired optimization technique that works based on the natural bacteria foraging steps. In this optimization algorithm, the cell is allowed to stochastically and collectively swarm for best solution. These BFA steps include: chemotaxis step, reproduction step and the elimination-dispersal step. During the chemotaxis step, the length of life of a bacteria is measured based on the number of chemotactic steps. Here the cost (fitness) j_i of bacteria is calculated by the proximity to other bacteria's new position θ_i after a tumble along the manipulated cost surface one at a time by adding the proximity to other bacteria's new position θ_i after a tumble along the manipulated cost surface one at a time by adding step size F_i in the tumble direction lie between $[-1, 1]$. Random direction vector (i) is generated to represent the tumble. In reproduction cell, it is basically selection phase of this algorithm. In this step, bacteria cells performed well over their life duration are selected for next generation. Elimination dispersal step is based on fitness function. Here the previous expired cells are discarded and new population is inserted.

5 Experiment Results and Discussion

In this section, the performance evaluation of the proposed approach is evaluated and results as discussed demonstrates how effective the proposed BFA based appliances scheduling approach is our specific objectives

in this simulation was balancing of load for Time of Use (TOU) price signals, minimization of cost as well as reduction of PAR. The units of measurement of cost, load and waiting time for both shift-able and non-shift-able load categories is cents, kWh and hour respectively.

In this paper, we consider 15 min as the Operation Time Interval (OTI). That is each hour of a day is divided into 4 slots of 15 min each. This implies for a whole day, we have 96 slots all together.

The performance appraisal of the proposed approach to DSM is carried out in terms of *user comfort*, *electricity cost*, *total load* and *PAR*.

5.1 User Comfort

In this work, user comfort is seen in terms of waiting time. Waiting time is the time a user will wait before an appliance gets turned on. User Comfort is inversely proportional to electricity cost or price, this implies that, in order to minimize cost of electricity consumed, users will have to trade off their comfort as they will wait for off-peak hours when their appliances gets turned on by the scheduler. Nevertheless,

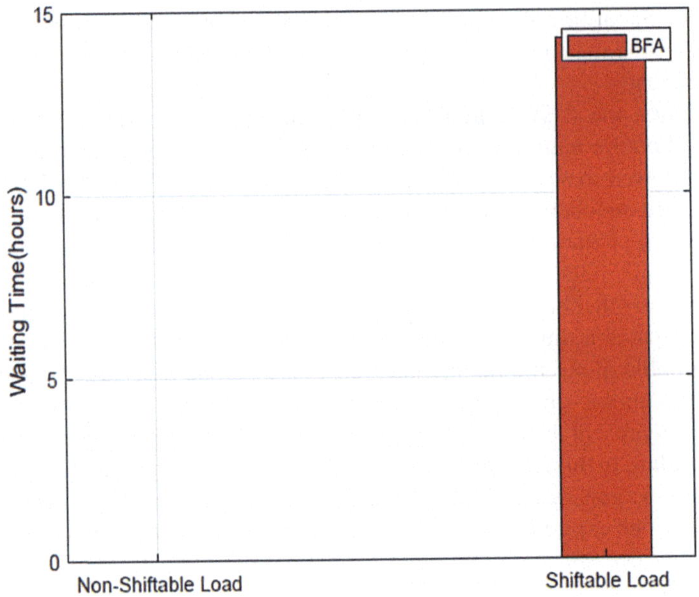

Fig. 2 Total users' waiting time

if comfort is preferred by the user (i.e. they don't want to wait for or delay their operations), then they must compromise on cost (i.e. they must pay high cost). Figure 2 shows the total waiting time of shift-able appliances.

5.2 Electricity Cost

Figure 3a shows the graph of electricity cost per slot as per before and after scheduling with BFA. Results reveals that cost paid during on peak hours is greatly reduced than what is obtainable in the unscheduled scenario. This is because bulk of the load during on peak hour has been shifted to off-peak hours.

The bar plot in Fig. 3b clearly further show the effectiveness of the BFA based scheduling approach in terms of total cost reduction. BFA scheduling caused a 24% reduction of the total cost from 148 to 112%.

5.3 Load or Electricity Consumption

Figure 4 shows the Load or electricity consumption at different time slots of a given day. It is clear that the BFA-scheduling succeeded in shifting or spreading the bulk of the load at on-peak hours to off-peak hours. As a result of this shifting of loads

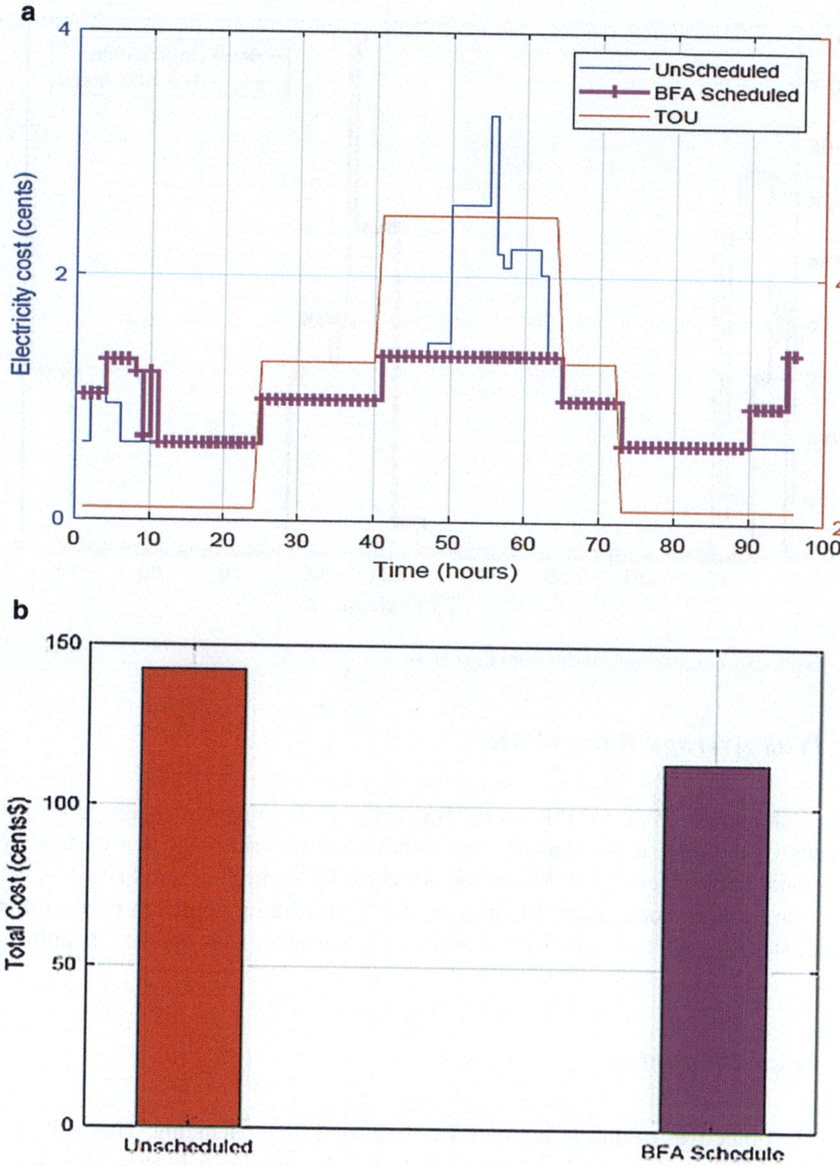

Fig. 3 **a** Graph of cost of electricity per time slot. **b** Reduced total cost given in cents

to time slots where electricity price is quite low, minimization of cost or electricity bills is achieved.

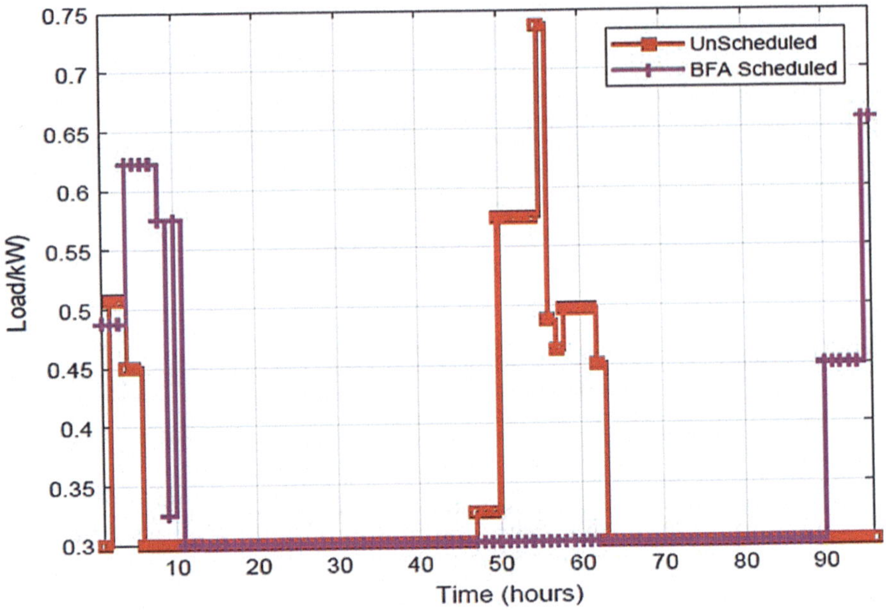

Fig. 4 Electricity consumption at different time slots

5.4 Peak Average Ratio (PAR)

Figure 5 shows the PAR for this work. Basically, PAR improves its capacity and efficiency of the grid. It also ensures the stability of the grid. PAR is also directly proportional to electricity cost. Therefore, a reduced PAR implies a reduction in the bills of the users or consumers. Clearly, the BFA scheduling approach used in this work has significantly reduced the PAR from what was obtainable before scheduling.

5.5 Load Balancing

One of the objectives of this assignment was to ensure load balancing. This will also ensure the stability of the grid. As shown in Fig. 6 the Load is balanced, since total load before and after scheduling is equal.

6 Conclusion

This work presented an IoT devices scheduling model that is based on BFA. BFA is a nature inspired meta-heuristic algorithm. Performance evaluation of the proposed

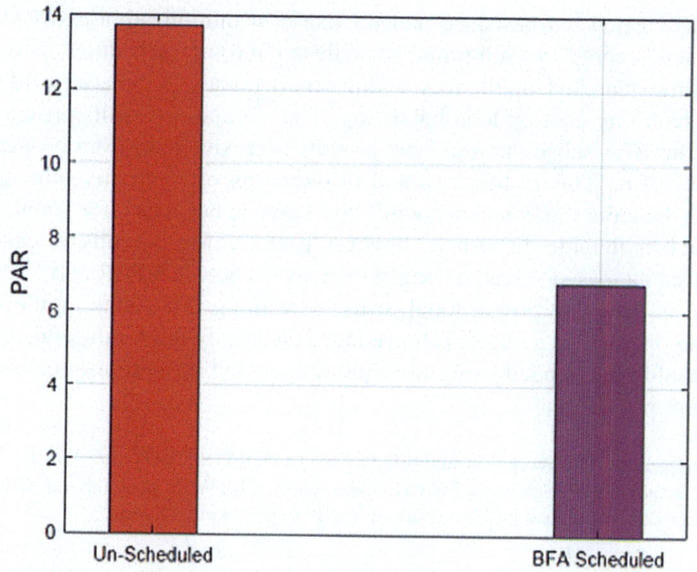

Fig. 5 Plot of the peak average ratio

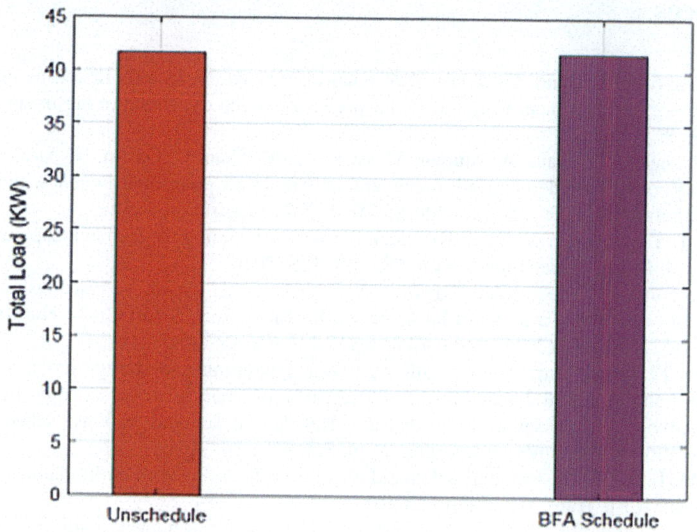

Fig. 6 Balanced total load before and after scheduling

method was carried out based on metrics like cost minimization and PAR reduction. Efficient energy consumption is achieved through scheduler, which helps in scheduling smart/IoT appliances within smart homes. A Smarter grid helps in reducing electricity cost by load balancing. The simulation results presented indicates that the BFA scheduler was able to shift excessive load from on-peak hours to off-peak hours. This in turn resulted in reduction of electricity bills and PAR. The former benefits the consumer while the latter is necessary for stability of the grid, hence beneficial to the utility. However, user comfort in terms of waiting time of appliances increased. Thus, a compromise exists between PAR and UC, as electricity cost is inversely proportional to users' waiting time. This implies that the waiting time increase is a trade-off for reduced electricity bill for the consumer. This research could be extended using multiple homes, hybrid optimisation algorithms and different price signals.

Acknowledgements This work was supported by the COMSATS University Islamabad, Pakistan, The World Academy of Science (TWAS) under the CIIT-TWAS postdoctoral fellowship of 2016/2017, as well as the Federal University of Technology, Akure, Nigeria.

References

1. M.B. Rasheed, N. Javaid, M. Awais, Z.A. Khan, U. Qasim, N. Javaid., Real time information based energy management using customer preferences and dynamic pricing in smart homes. Energies **9**(7), 542 (2016)
2. M.B. Rasheed, N. Javaid, A. Ahmad, M. Awais, Z.A. Khan, U. Qasim, N. Alrajeh, Priority and delay constrained demand side management in real time price environment with renewable energy source. Int. J. Energy Res. **40**(14), 2002–2021 (2016)
3. K. Ma, T. Yao, J. Yang, X. Guan, Residential power scheduling for demand response in smart grid. Int. J. Electr. Power Energy Syst. **78**, 320–325 (2016)
4. F.A. Qayyum, M. Naeem, A.S. Khwaja, A. Anpalagan, L. Guan, V. Venkatesh, Appliance scheduling optimization in smart home networks. Spec. Sect. Smart Grids: Hub Interdiscip. Res. **3**(1), 2179–2190 (2015). https://doi.org/10.1109/access.2015.2496117
5. O. Edic, Economic impacts of small-scale own generating and storage units, and electric vehicles under different demand response strategies for smart households. Appl. Energy
6. R. Jovanovic, A. Bousselham, I.S. Bayram, Residential demand response scheduling with consideration of consumer preferences. Appl. Sci. **6**(1), 16 (2016)
7. S. Moon, J. Lee, Multi-residential demand response scheduling with multi-class appliances in smart grid. IEEE Trans. Smart Grid (2016)
8. E. Shirazi, J. Shahram, Optimal residential appliance scheduling under dynamic pricing scheme via HEMDAS. Energy Build. **93**, 40–49 (2015)
9. A.S.O. Ogunjuyigbe, T.R. Ayodele, O.A. Akinola, User satisfaction-induced demand side load management in residential buildings with user budget constraint. Appl. Energy **187**, 352–366 (2017)
10. J.S. Vardakas, N. Zorba, V.V. Christos, Performance evaluation of power demand scheduling scenarios in a smart grid environment. Appl. Energy **142**, 164–178 (2015)
11. John S. Vardakas, Nizar Zorba, Christos V. Verikoukis, Power demand control scenarios for smart grid applications with finite number of appliances. Appl. Energy **162**, 83–98 (2016)

12. K. Muralitharan, S. Rathinasamy, S. Yan, Multiobjective optimization technique for demand side management with load balancing approach in smart grid. Neurocomputing **177**, 110–119 (2016)
13. A. Ahmad, N. Javaid, N. Alrajeh, Z.A. Khan, U. Qasim, A. Khan, A modified feature selection and artificial neural network-based day-ahead load forecasting model for a smart grid. Appl. Sci. **5**(4), 1756–1772 (2015)
14. J. Abushnaf, A. Rassau, and W. Grnisiewicz, W. (2015). Impact on electricity use of introducing time of use pricing to a multiuser home energy
15. Z. Wen, D. O'Neill, H. Maei, Optimal demand response using device-based reinforcement learning. Smart Grid IEEE Trans. On **6**(5), 2312–2324 (2015)
16. B. Gao, X. Liu, W. Zhang, Y. Tang, Autonomous household energy management based on a double cooperative game approach in the smart grid. Energies **8**(7), 7326–7343 (2015)
17. S.S. Reddy, J.Y. Park, J.Y. Jung, Optimal operation micro-grid using hybrid differential evolution and harmony search algorithm. Front. Energy **10**(3), 355–362 (2016)
18. J. Zhang, Y. Wu, Y. Guo, B. Wang, H. Wang, H. Liu, A hybrid harmony search algorithm with differential evolution for day-ahead scheduling problem of a microgrid with consideration of power flow constraints. Appl. Energy **183**, 791–804 (2016)
19. M. Rastegar, M. Fotuhi-Firuzabad, H. Zareipour, Home energy management incorporating operational priority of appliances. Int. J. Electr. Power Energy Syst. **74**, 286–292 (2016)
20. S.A. Oluwadare, A.J. Gabriel, O.G. Ogunride, Tabu-genetic algorithm-based model for poultry feed formulation. Int. J. Sustain. Agric. Res. **6**(2), 94–109 (2019). https://doi.org/10.18488/journal.70.2019.62.94.109

The manufacturer's authorised representative in the EU is Springer Nature Customer Service Centre GmbH, Europaplatz 3, 69115 Heidelberg, Germany. If you have any concerns regarding our products, please contact ProductSafety@springernature.com

Printed and bound by CPI Group (UK) Ltd, Croydon, CR0 4YY
22/09/2026
02221060-0012

Springer Earth System Sciences

Series editors

Philippe Blondel, Bath, UK
Eric Guilyardi, Paris, France
Jorge Rabassa, Ushuaia, Argentina
Clive Horwood, Chichester, UK

More information about this series at http://www.springer.com/series/10178

Matías C. Ghiglione
Editor

Geodynamic Evolution of the Southernmost Andes

Connections with the Scotia Arc

Springer

Editor
Matías C. Ghiglione
National Council of Argentina (CONICET)
Universidad de Buenos Aires
Buenos Aires
Argentina

ISSN 2197-9596 ISSN 2197-960X (electronic)
Springer Earth System Sciences
ISBN 978-3-319-39725-2 ISBN 978-3-319-39727-6 (eBook)
DOI 10.1007/978-3-319-39727-6

Library of Congress Control Number: 2016941068

© Springer International Publishing Switzerland 2016
This work is subject to copyright. All rights are reserved by the Publisher, whether the whole or part of the material is concerned, specifically the rights of translation, reprinting, reuse of illustrations, recitation, broadcasting, reproduction on microfilms or in any other physical way, and transmission or information storage and retrieval, electronic adaptation, computer software, or by similar or dissimilar methodology now known or hereafter developed.
The use of general descriptive names, registered names, trademarks, service marks, etc. in this publication does not imply, even in the absence of a specific statement, that such names are exempt from the relevant protective laws and regulations and therefore free for general use.
The publisher, the authors and the editors are safe to assume that the advice and information in this book are believed to be true and accurate at the date of publication. Neither the publisher nor the authors or the editors give a warranty, express or implied, with respect to the material contained herein or for any errors or omissions that may have been made.

Printed on acid-free paper

This Springer imprint is published by Springer Nature
The registered company is Springer International Publishing AG Switzerland

Foreword

This new volume of the "Springer Earth System Science" deals with the attractive Southernmost Andes–Antarctica connection issue, which has called the attention of several generations of field geologists, geophysicists, and oceanologists. A new generation of researchers, but already experienced young scientists, is presenting the results of several years of investigation in these remote regions of very difficult access. It is interesting to compare the early work on land of Katz (1964)[1] and the studies of Dalziel et al. (1974)[2] in the Rocas Verdes marginal basin and the Scotia Arc, to see how much we have advance in the knowledge and the timing of the tectonic processes. This book will be a milestone for understanding the new connections of the Patagonian–Fuegian Andes with the Scotia Arc and related areas.

Special attention should be paid to the work of Matías C. Ghiglione and coworkers, who is based on an extensive and detailed fieldwork. His research started in the Tierra del Fuego Island during his Ph.D. thesis (1999–2003), to continue in the Patagonian Andes of Argentina and Chile over the last years. As a result of this extensive fieldwork, together with available industrial seismic lines, he has proposed new details and processes on the geodynamic evolution of the southernmost Andes and its relation with subsidence and sedimentation in the Argentine continental margin.

The progress in the knowledge of the metamorphic conditions (pressure, temperature, and timing), the geochemical characteristics, and age of the metamorphic complexes is another piece of new data that allow setting thoughtful paleogeographic schemes for the Paleozoic and Mesozoic evolutions. Calderon and coworkers present new solid hypotheses, which will deserve further research and consideration.

[1]Katz HR (1964) Some new concepts on geosynclinal development and mountain building at the southern end of South America. XXII Intern Geol Congr, India, Proceedings 4:242–255.

[2]Dalziel IWD, De Wit MF, Palmer KF (1974) Fossil marginal basin in the southern Andes. Nature 250:291–294.

Most of the Andes have evidence in recent years of important variations in the present and old subduction geometries of the Benioff zones. The Southernmost Andes is not an exception, and the data presented by González Guillot show firm evidence of these variations based on petrological and spatial changes in the magmatic front through time.

The review of the Scotia arc development, at different time windows, and their connections with the Patagonian–Fuegian Andes through time, profits of the comprehensive experience of Graeme Eagles on this subject. The different schemes of Eagles show didactically and rigorously fine details on the processes involved in this complex evolution.

Seismicity in the southernmost South America and its complex relation with the geodynamic context of the Scotia Arc is an important but little understood topic. Sue and Ghiglione undertake this subject from the view of brittle deformation, fault kinematics, and related seismicity, to elucidate deformation pattern in the southernmost Andes and surrounding southern oceans.

The analysis of the Patagonian Orocline, a classic problem of the last 50 years in the Southernmost Andes since the early proposal of Warren Carey, clearly shows that the continental margin bend is not a proper orocline. Several paleomagnetic studies have shown the complexity of this evolution and how rotations vary on time and space through the orogen. The proposal of Marco Maffione, based on the new and old paleomagnetic and structural data, for changing the Patagonian Orocline to a Patagonian Arc, deserves attention. This is not just a change in the denomination, if not an identification of previous paleogeography and further modifications, which could be a clue to the final understanding of this important feature.

As a final remark, I want to express that the reading of this book brought to my memory nice scenes of navigation in the small zodiacs, landing in abrupt fjords, and walking in wet pit plains, which arouse my envy to these young researchers that can continue to enjoy these beautiful landscapes, even under the rain and intense snowfall, but full of excellent geology. Congratulations to the authors and especially to the editor for the initiative and for the robust final product obtained.

Victor A. Ramos
Emeritus Professor
Universidad de Buenos Aires

Contents

An Introduction to the Geodynamic Evolution of the Southernmost Andes: Connections with the Scotia Arc 1
Matías C. Ghiglione

Tectonic Evolution of Paleozoic and Mesozoic Andean Metamorphic Complexes and the Rocas Verdes Ophiolites in Southern Patagonia 7
Mauricio Calderón, Francisco Hervé, Francisco Fuentes, Julie C. Fosdick, Fernando Sepúlveda and Gonzalo Galaz

Magmatic Evolution of the Southernmost Andes and Its Relation with Subduction Processes 37
Mauricio González Guillot

Tectonic Reconstructions of the Southernmost Andes and the Scotia Sea During the Opening of the Drake Passage 75
Graeme Eagles

The Relation Between Neogene Denudation of the Southernmost Andes and Sedimentation in the Offshore Argentine and Malvinas Basins During the Opening of the Drake Passage 109
Matías C. Ghiglione, Christian Sue, Miguel E. Ramos, Jonathan E. Tobal and Rocío E. Gallardo

Wrenching Tectonism in the Southernmost Andes and the Scotia Sea Constrained from Fault Kinematic and Seismotectonic Overviews 137
Christian Sue and Matías C. Ghiglione

Kinematic Evolution of the Southern Andean Orogenic Arc 173
Marco Maffione

Index 201

Contributors

Mauricio Calderón Carrera de Geología, Universidad Andres Bello, Santiago, Chile

Graeme Eagles Helmholtz Centre for Marine and Polar Research, Alfred Wegener Institute, Bremerhaven, Germany

Julie C. Fosdick Department of Geological Sciences, Indiana University, Bloomington, USA

Francisco Fuentes Carrera de Geología, Universidad Andres Bello, Santiago, Chile

Gonzalo Galaz Departamento de Geología Regional, Instituo de Geología, Universidad nacional Autónoma de México, México D.F., México

Rocío E. Gallardo Departamento de Ciencias Geológicas, Instituto de Estudios Andinos "Don Pablo Groeber" CONICET, FCEN—Universidad de Buenos Aires, Buenos Aires, Argentina

Matías C. Ghiglione Departamento de Ciencias Geológicas, Instituto de Estudios Andinos "Don Pablo Groeber" CONICET, FCEN—Universidad de Buenos Aires, Buenos Aires, Argentina

Mauricio González Guillot Centro Austral de Investigaciones Científicas CADIC-CONICET, and Instituto de Ciencias Polares, Ambiente y Recursos Naturales, Universidad Nacional de Tierra del Fuego, Tierra del Fuego, Ushuaia, Argentina

Francisco Hervé Carrera de Geología, Universidad Andres Bello, Santiago, Chile; Departamento de Geología, Universidad de Chile, Santiago, Chile

Marco Maffione Department of Earth Sciences, Utrecht University, Utrecht, The Netherlands

Miguel E. Ramos Departamento de Ciencias Geológicas, Instituto de Estudios Andinos "Don Pablo Groeber" CONICET, FCEN—Universidad de Buenos Aires, Buenos Aires, Argentina

Fernando Sepúlveda Servicio Nacional de Geología y Minería, Santiago, Chile

Christian Sue Bourgogne Franche-Comté University, Laboratoire Chrono-environnement, Besançon, France

Jonathan E. Tobal Departamento de Ciencias Geológicas, Instituto de Estudios Andinos "Don Pablo Groeber" CONICET, FCEN—Universidad de Buenos Aires, Buenos Aires, Argentina

An Introduction to the Geodynamic Evolution of the Southernmost Andes: Connections with the Scotia Arc

Matías C. Ghiglione

1 Introduction

The backbone of this book is the analysis of the relation between the tectonic evolution of the Southernmost Andes and its southern neighbors: Antarctica and the Scotia Sea. The Southernmost Andes comprise the southern bent of the Andean mountains, where they progressively change their strike from the N–S-oriented Southern Patagonian Andes to the E–W-trending Fuegian Andes (Fig. 1). These Andean segments and their corresponding basement, sedimentary basins, magmatic rocks, and batholithic belts are linked, even if sometimes elusively, to the evolution of the Antarctic Peninsula and of the islands spread in the Southern ocean.

The Scotia Sea separating South America and the Antarctic Peninsula (Fig. 1) is only ~30 Myr old, very young, considering a joint geological history of the Southernmost Andes, Patagonia, and Antarctica than can be traced back to late Paleozoic times (i.e., 300 Myr or more). Analysis of marine geophysical data in the Scotia Sea (cf. Eagles et al. 2006) and geological data from Tierra del Fuego (Ghiglione et al. 2008) demonstrate that the disconnection of South America and Antarctica started at about 50 Ma. However, oceanic spreading producing the final opening of the Drake Passage, and the consequent strike-slip movements along the North and South Scotia ridges (Fig. 1), occurred just since the Oligocene. Therefore, there is a joint history of Antarctica and South America that was shattered by the opening of the Scotia Sea; the remaining pieces of the puzzle, how can

M.C. Ghiglione (✉)
Departamento de Ciencias Geológicas, Instituto de Estudios Andinos "Don Pablo Groeber"
CONICET, FCEN—Universidad de Buenos Aires, Buenos Aires, Argentina
e-mail: matias@gl.fcen.uba.ar

© Springer International Publishing Switzerland 2016
M.C. Ghiglione (ed.), *Geodynamic Evolution of the Southernmost Andes*,
Springer Earth System Sciences, DOI 10.1007/978-3-319-39727-6_1

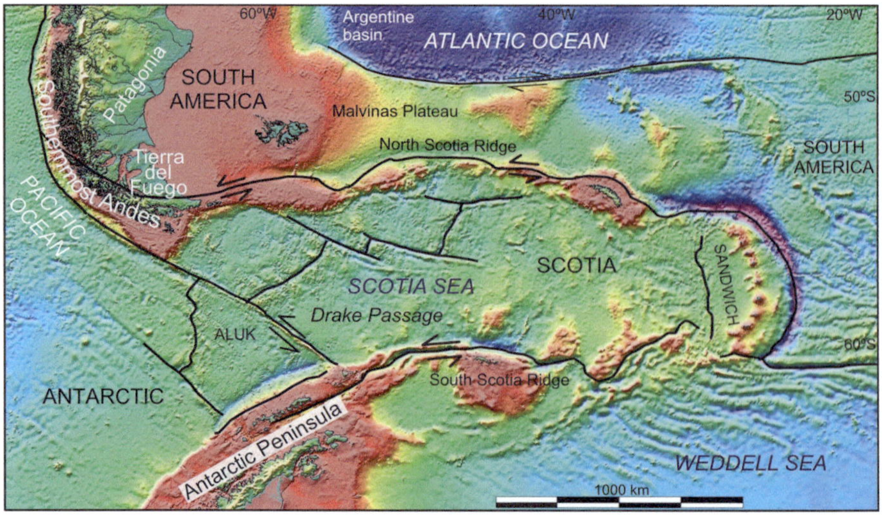

Fig. 1 Major tectonic plates (*black uppercase letters*) and some other geological and geographical features studied in this book, enclosing the Southernmost Andes, Scotia Sea, and the Antarctic Peninsula region

they fit together, and how were they separated from each other are the subjects of this book.

The overview presents state-of-the-art reviews intended for all scientist and professionals, as well as PhD and Master students working in the region. Since each chapter presents the necessary background and a historical appraisal, it is anticipated to be a useful tool for interested readers trying to understand the main geodynamic framework and debates of the Southernmost Andes–Scotia Arc region.

The following chapters summarize the works and thoughts of young Earth Science researchers that have worked on the geological and geophysical exploration of the region for over the last two decades. Each of the main authors have all finish their PhD's around 10 years ago on a number of multidisciplinary subjects in the region and since then have keep conducting field works, sampling, and analyzing the area. Multiple point of views and methods are therefore presented to address update summaries of new and long-lasting debates, including the complex evolution of the Rocas Verdes marginal basin (Dalziel et al. 1974) and the Patagonian Batholith (Katz 1964), plate kinematic reconstructions, and fault kinematics of the southern South America–Scotia Sea region, paleomagnetism and kinematics of the *Patagonian Orocline* (Carey 1958), orogenic growth in the Southernmost Andes and basin dynamics. The impact of the studied data in a number of global topics, such as the Paleozoic geodynamic context of Patagonia and Antarctica within the *Terra Australis* orogen (*sensu* Cawood 2005), the Eocene–Oligocene global climatic change, offshore sedimentation in the Atlantic platform, active seismicity, kinematics of deformation along the North and South Scotia ridges, and the opening of the Drake Passage, is also addressed.

The authors started from scratch exhaustive data compilations and present them in tables and summary figures. It should be highlighted the consistency and coherence of the data, although most information in the region is scarce and scattered due to the remoteness of islands and fiords, in a polar context of weather conditions that are bearable only in the summer months. Indeed, the compiled information is the product of multiple campaigns over 30 years, data collected with several methods, different logistics, and multiple countries involved. After they were filtered for modern requisites and compiled in regional databases, they are, however, very consistent and stress the thoughtfulness of the referenced researchers.

Therefore, one of the main outputs of this book is the presentation of extensive synthetic databases and their thoughtful analysis. The tectonic history of the Southernmost Andes, and their rich and complex relation with their southern neighbors is, however, too big to be left without personal views and perceptions. The following chapters present not only a review of theories and databases, but they address new perspectives, and potential paradigms that will lead the future research on the region.

2 Book Structure

The backbone used to structure the book is a chronological order of studied topics that are also arranged in a comprehensible way to understand the context. However, each chapter addresses the necessary background to understand the analyzed subjects and methodologies. The following chapters start with the evolution of Paleozoic and Mesozoic metamorphic complexes (second chapter), and the Mesozoic–Cenozoic magmatic evolution of the Patagonian and Fuegian batholiths (third chapter) that constitute the basement of the region. The book continues with Cenozoic plate kinematic reconstructions from geological, marine, and satellite geophysical data (fourth chapter), the late Cenozoic foreland basin dynamics and sediment dispersal during the opening of the Drake Passage (fifth chapter) and a seismotectonic overview (sixth chapter), to finally present a disquisition about the kinematic evolution of the Patagonian Orocline (seventh chapter).

Calderón et al. (2016) summarize the tectonic and magmatic evolution of Paleozoic and Mesozoic metamorphic complexes in southern Patagonia and the Rocas Verdes ophiolites in the second chapter, on the base of combined petrological, geochemical, and geochronological datasets. This chapter explains the evolution of magmatic belts, marginal basins, and continental fragments that once formed part of the peripheral realm of southwestern Gondwanan and gives details on their connections with the Antarctic Peninsula. These authors revisit the evolution of the Rocas Verdes basin, explaining its relation with the Patagonian and Fuegian batholiths and the Magallanes–Austral basin, and present schematic reconstructions of its Jurassic to Cretaceous tectonic evolution. Additionally, the possibility that the Antarctic Peninsula was located west of the present margin of southern Patagonia (cf. Jokat et al. 2003) is also addressed.

The Jurassic through Miocene episodic magmatic growth of the main arc and rear arc magmatic suites in the Southernmost Andes is discussed in the third chapter by Guillot (2016). This author presents a detailed and updated report of field features, lithology, geochemistry and age of the Fuegian Batholith and satellite plutons, and a thoughtful comparison among the different plutonic suites. An account of variations in the composition of magmas and shifts in the position of the volcanic front in the Patagonian–Fuegian batholiths are shown in sketches and maps, and are associated to fluctuating subduction parameters. The peculiar tectonic setting that affected the magmatic arc construction includes the opening and closure of the Rocas Verdes marginal basin and oroclinal bending in the latest Cretaceous–Paleogene. A mid-Cretaceous slab-flattening event, which caused a cratonward migration of the volcanic front, the widening of the arc, and inception of potassic magmatism in the rear arc is also proposed.

In the fourth chapter, Eagles (2016) reviews geological data exposed around the Scotia Sea margins, and the patterns of fracture zones and magnetic reversal isochrons from marine and satellite geophysical data. This author reviews what can be interpreted of the plate tectonic history of the region by using these resources and presents a detailed and quantitative approach to its plate kinematics. A set of tectonic reconstructions of the Southernmost Andes and the Scotia Sea during the opening of the Drake Passage from 6 to 50 Ma are shown in this fourth chapter. Additionally, main changes in the plate tectonic configuration and in the geodynamic setting are related to deformation and kinematics in the Southernmost Andes. This author presents an overview an debate of pre-Scotia times in which the established proximity of South Georgia to Tierra del Fuego in Early Cretaceous times is disputed.

Ghiglione et al. (2016) present in the fifth chapter an overview of Neogene orogenic growth and denudation in the Southernmost Andes, and related sediment dispersal in the Magallanes–Austral-Malvinas foreland basin during the opening of the Drake Passage. These authors propose that great volumes of sediments produced by Miocene–Pliocene denudation of the cordillera bypassed Patagonia and reached the Argentinean continental margin, where they were accommodated in thick sequences with high sedimentation rates. The evidence summarized in this fifth chapter shows a sequence of geodynamic events and paleogeographic changes leading to that output, including the approximation and collision of the Chile Ridge against the Pacific margin, the dynamic uplift of extra-Andean Patagonia, and the propagation of Antarctic waters masses into the Argentine Basin after the deepening of the Drake Passage.

In the sixth chapter, Sue and Ghiglione (2016) present a synthesis of the fault system associated with major fault zones in the Southernmost Andes and the Scotia Sea and review the available kinematic databases from fault slip data and GPS-measured deformation. These authors establish a current stress pattern stable since Oligocene times, reflecting that the global left-lateral motion between Antarctica and South America plate is the main driving force for the entire area.

Discrepancies between long-term and short-term slip velocities along the Magallanes–Fagnano fault system are also evaluated, suggesting a complex mechanics that makes calculating a recurrence time history for the characteristic earthquakes difficult.

The seventh and final chapter by Maffione (2016) gives an structural and paleomagnetic account of the regional-scale curvature developed in the Patagonian–Fuegian Andes, namely Patagonian Orocline (Carey 1958). In this chapter, a historical, and comprehensive review of the up-to-date paleomagnetic, structural, and magnetic fabric data is presented, together with a discussion on current models and future challenges. This author propose that the orogenic curvature of the Southernmost Andes does not specifically fit in any of the classic definitions of curved mountain belts, and it rather represents a "hybrid" curved belt. For this reason, the more generic term of "Patagonian Arc" is introduced, to signify a slip-partitioning mechanism along preexisting faults that controlled the kinematic evolution of the bent.

References

Calderón M, Hervé F, Fuentes F, Fosdick JC, Sepúlveda F, Galaz G (2016) Tectonic evolution of Paleozoic and Mesozoic andean metamorphic complexes and the Rocas Verdes ophiolites in southern Patagonia. In: Ghiglione MC (ed) Geodynamic Evolution of the Southernmost Andes. Springer Earth System Sciences, pp 7–36

Carey SW (1958) The tectonic approach to continental drift. Continental drift: a symposium: Hobart. Geology Department, University of Tasmania, Tasmania, pp 177–355

Cawood PA (2005) Terra Australis orogen: Rodinia breakup and development of the pacific and Iapetus margins of Gondwana during the Neoproterozoic and Paleozoic. Earth-Sci Rev 69:249–279

Dalziel IWD, De Wit MF, Palmer KF (1974) Fossil marginal basin in the southern Andes. Nature 250:291–294

Eagles G (2016) Tectonic reconstructions of the Southernmost Andes and the Scotia Sea during the opening of the Drake Passage. In: Ghiglione MC (ed) Geodynamic Evolution of the Southernmost Andes. Springer Earth System Sciences, pp 75–108

Eagles G, Livermore RA, Morris P (2006) Small basins in the Scotia Sea: The Eo-cene drake passage gateway. Earth Planet Sci Lett 242:343

Ghiglione MC, Yagupsky D, Ghidella M, Ramos VA (2008) Continental stretching preceding the opening of the Drake Passage: evidence from Tierra del Fuego. Geology 36:643–646

Ghiglione MC, Sue C, Ramos ME, Tobal JE, Gallardo RE (2016) The relation between neogene denudation of the Southernmost Andes and sedimentation in the offshore Argentine and Malvinas basins during the opening of the Drake Passage. In: Ghiglione MC (ed) Geodynamic Evolution of the Southernmost Andes. Springer Earth System Sciences, pp 109–135

Guillot MG (2016) Magmatic evolution of the southernmost Andes and its relation with subduction processes. In: Ghiglione MC (ed) Geodynamic Evolution of the Southernmost Andes. Springer Earth System Sciences, pp 37–74

Jokat W, Boebel T, König M, Meyer U (2003) Timing and geometry of early Gondwana breakup. J Geophys Res 108:2428

Katz HR (1964) Some new concepts on geosynclinal development and mountain building at the southern end of South America. XXII Intern Geol Congr India Proc 4:242–255

Maffione M (2016) Kinematic evolution of the Southern Andean orogenic arc. In: Ghiglione MC (ed) Geodynamic Evolution of the Southernmost Andes. Springer Earth System Sciences, pp 173–200

Sue C, Ghiglione MC (2016) Wrenching tectonism in the Southernmost Andes and the Scotia Sea constrained from fault kinematic and seismotectonic overviews. In: Ghiglione MC (ed) Geodynamic Evolution of the Southernmost Andes. Springer Earth System Sciences, pp 137–171

Tectonic Evolution of Paleozoic and Mesozoic Andean Metamorphic Complexes and the Rocas Verdes Ophiolites in Southern Patagonia

Mauricio Calderón, Francisco Hervé, Francisco Fuentes, Julie C. Fosdick, Fernando Sepúlveda and Gonzalo Galaz

Abstract The petrology and geochronology of Paleozoic and Mesozoic metamorphic and ophiolitic complexes in southern Patagonia constrain the evolution of magmatic belts, marginal basins, and continental fragments that once formed part of the peripheral realm of the southwestern Gondwanan margin. The upper Paleozoic metasedimentary rocks exposed along the eastern slopes of the Andean ranges contain prominent zircon components of Devonian and earliest Carboniferous age. These strata were deposited in marine basins fed by sediment derived from almost coeval felsic magmatic belts and recycled units with Ordovician and Cambrian components; some of which have been identified to the east, in Argentina. Permian to Triassic metasedimentary complexes, mainly exposed along the western slope of the Southern Patagonian Andes and in the Antarctic Peninsula, show conspicuous peaks of Permian detrital zircons with sparse older age components, supporting their deposition near active magmatic belts in a convergent margin setting. It is proposed that rapid northward migration of Gondwana (between 330 and 270 Ma) promoted crustal extension and the establishment of an arc and back-arc marginal basin configuration during the Permian. Subsequent closure of the marginal basins occurred during the Late Triassic–Early Jurassic Chonide orogeny. Middle Jurassic fragmentation of Gondwana involved continental extension, synrift silicic and bimodal magmatism, and seafloor-spreading, leading to the oceanic-type Late

M. Calderón (✉) · F. Hervé · F. Fuentes
Carrera de Geología, Universidad Andres Bello, Sazie 2119, Santiago, Chile
e-mail: mccaldera@gmail.com

F. Hervé
Departamento de Geología, Universidad de Chile, Plaza Ercilla 803, Santiago, Chile

J.C. Fosdick
Department of Geological Sciences, Indiana University, 1001 East 10th Street, Bloomington, USA

F. Sepúlveda
Servicio Nacional de Geología y Minería, Av. Santa María 0104, Santiago, Chile

G. Galaz
Departamento de Geología Regional, Instituo de Geología, Universidad nacional Autónoma de México, México D.F. 04510, México

© Springer International Publishing Switzerland 2016

M.C. Ghiglione (ed.), *Geodynamic Evolution of the Southernmost Andes*, Springer Earth System Sciences, DOI 10.1007/978-3-319-39727-6_2

Jurassic to Early Cretaceous Rocas Verdes basin. Once the Antarctic Peninsula microplate migrated southward, subduction was re-established in southern Patagonia in Early Cretaceous time, involving processes of subduction erosion and high-P/T metamorphism in the forearc region. The Rocas Verdes basin started to close in the latest Early Cretaceous involving underthrusting and subduction of its oceanic seafloor. These convergent processes culminated with the Late Cretaceous Patagonian and Fuegian orogeny involving the tectonic emplacement of ophiolitic complexes, collision of drifted crustal slivers against the continent, and development of the Magallanes–Austral foreland basin.

Keywords Metamorphic complexes · Rocas Verdes ophiolites · Patagonian Andes · Fuegian Andes · Antarctic Peninsula · Magallanes–Austral basin

1 Introduction

The southernmost edge of South America, hereafter referred to as southern Patagonia, encompasses igneous and metamorphic complexes that represent one interconnected assemblage of diachronous magmatic belts and accretionary complexes that once formed part of the southwestern margin of Gondwana (Hervé et al. 2003; Pankhurst et al. 2003; Eagles 2016), also known as the Terra Australis Orogen (Cawood 2005). Some of these units composed the continental basement, in which the Upper Jurassic to Lower Cretaceous Rocas Verdes marginal basin was developed (Katz 1964; Dalziel et al. 1974; Suárez and Pettigrew 1976; Dalziel 1981; Stern and de Wit 2003; Calderón et al. 2007a), following the onset of Gondwana fragmentation in Middle Jurassic times (Bruhn et al. 1978; Pankhurst et al. 2000). Today, the rift and oceanic-type seafloor remnants of the Rocas Verdes basin are preserved as discontinuous belts of supracrustal bimodal igneous suites and ophiolitic complexes tectonically juxtaposed onto the cratonic margin of southern Patagonia during the late Mesozoic Andean orogeny (Dott et al. 1977; Dalziel 1981; Cunningham 1994; Stern and de Wit 2003; Calderón et al. 2013) (Fig. 1).

The intricate internal architecture of metamorphic and ophiolitic complexes, coupled with the lack of fossil records in complexly deformed metasedimentary rocks, has fostered the development and application of several analytical techniques (geochemistry, geothermobarometry, igneous and detrital U–Pb zircon geochronology, low-temperature thermochronology) to resolve temporal and genetic relationships among them and provide an improved framework for paleogeographic analysis. The understanding of these complexes has been further established with the combination of U–Pb SHRIMP (Sensitive High-Resolution Ion Microprobe) geochronology and Lu-Hf isotopic data for individual zircon grains, providing timing of formation and sediment deposition and the nature of host magmas and provenance areas (Hervé et al. 2003, 2010a, b; Pankhurst et al. 2003; Augustsson et al. 2006; Augustsson and Bahlburg 2008; Fanning et al. 2011; Castillo et al. 2015). Furthermore, the geological evolution of some of these

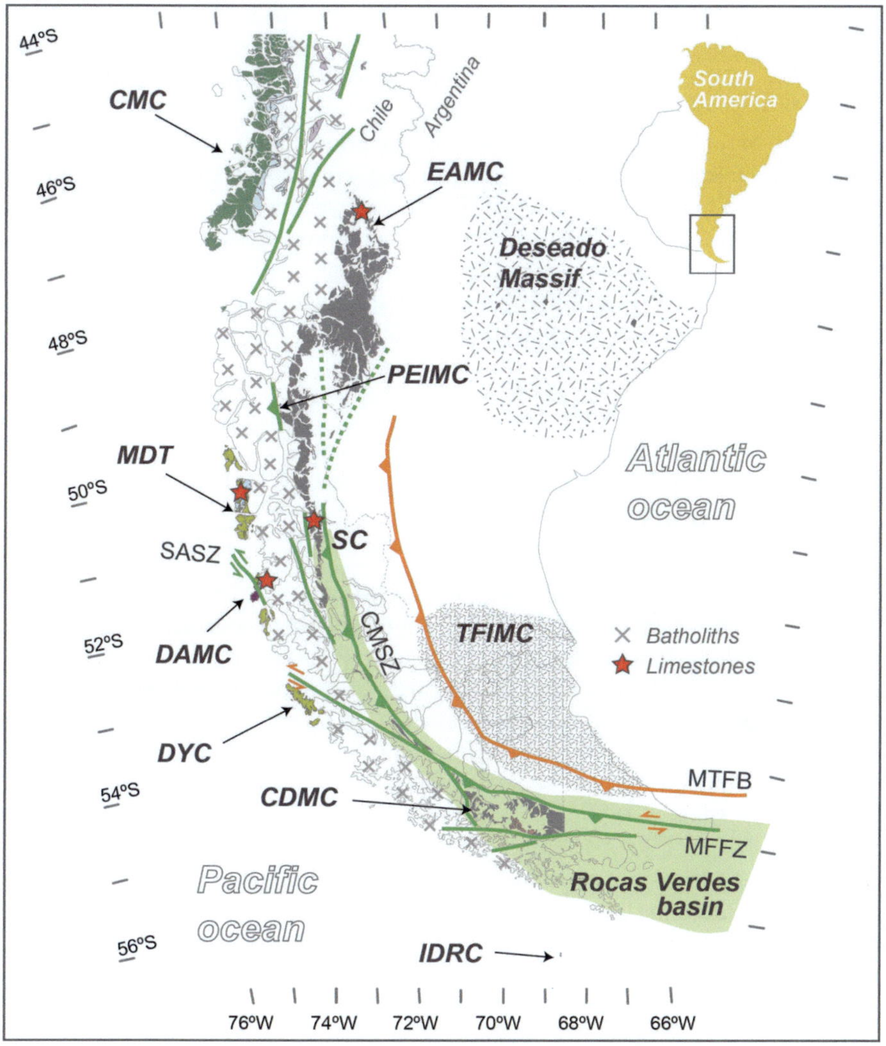

Fig. 1 Geological sketch map of the Paleozoic and Mesozoic igneous and metamorphic complexes in southern South America (after Sernageomin 2003). The contour of the Rocas Verdes basin ophiolites and related volcanic and sedimentary rocks is illustrated. Major fault zones are illustrated. Abbreviations of geological units are: *CDMC* Cordillera Darwin Metamorphic Complex; *CMC* Chonos Metamorphic Complex; *DAMC* Diego de Almagro Metamorphic Complex; *DC* Denaro Complex; *DRMC* Diego Ramirez Metamorphic Complex; *DYC* Duque de York Complex; *EAMC* Eastern Andean Metamorphic Complex; *MDT* Madre de Dios Terrane; *MRMC* Main Range Metamorphic Complex; *PEIMC* Puerto Edén Igneous and Metamorphic Complex; *SC* Staines Complex; *TL* Tarlton Limestones; *TFIMC* Tierra del Fuego Igneous and Metamorphic Complex. Abbreviations of major fault zones are: *CMSZ* Canal de las Montañas Shear Zone, *MFFZ* Magallanes-Fagnano fault zone, *MTFB* Magallanes Thrust and Fold Belt; *SAZS* Seno Arcabuz Shear Zone

complexes has been proposed using determinations of metamorphic pressure (P)–temperature (T) conditions and low-temperature thermochronology (Willner et al. 2000, 2004, 2009; Ramírez-Sánchez et al. 2005; Calderón et al. 2007b, 2012; Sepúlveda et al. 2008; Maloney et al. 2011).

In light of already-published data from heterochronous petrotectonic assemblages, the paleogeographic evolution of the continental margin of southern Patagonia is discussed in terms of shared geodynamic scenarios, such as the emplacement of magmatic belts (arc and/or rifts) and related sedimentary basins on older basement complexes, and development of oceanic-type marginal basins above subduction zones. The possibility that the Antarctic Peninsula was located west of the present margin of southern Patagonia (cf. Jokat et al. 2003; Hervé et al. 2006) is also addressed.

2 Igneous and Metamorphic Complexes of Patagonia and Tierra del Fuego

The Southern Patagonian and Fuegian Andes are a relatively low-elevation mountain belt made up in part by the upper Paleozoic to Mesozoic metamorphic complexes and by Jurassic to Cretaceous supra-subduction zone ophiolitic complexes of the Rocas Verdes basin. Most of the metamorphic and ophiolitic complexes crop out west and east of the continuous Mesozoic to Cenozoic Patagonian Batholith (Hervé et al. 1984, 2007a; Pankhurst et al. 1999; Guillot 2016) with portions of continental basement rocks buried beneath the Jurassic volcaniclastic deposits of the Chon Aike Silicic Large Igneous Province and younger sedimentary infill of the Magallanes–Austral basin (Pankhurst et al. 2003; Hervé et al. 2008) (Fig. 1). The older continental rocks in southern Patagonia are in the Deseado Massif area (Pankhurst et al. 2003) and below the Jurassic to Cenozoic strata in southern Patagonia and Tierra del Fuego (Söllner et al. 2000; Hervé et al. 2010a). In this section, we present a summarized description of metamorphic and ophiolitic complexes, detailing lithologies, U–Pb zircon geochronology, metamorphic characteristics, and geodynamic significance.

Two main divisions are recognized and employed to arrange the descriptions: Upper Paleozoic to Mesozoic **Andean metamorphic complexes** (Sect. 2.1) and **Ophiolitic complexes of the Rocas Verdes basin** from Jurassic to Cretaceous supra-subduction zone (Sect. 2.2). The **Mesozoic Mylonitic Belts** of the Seno Arcabuz (SASZ) and Canal de las Montañas (CMSZ) shear zones acting as main tectonic boundaries (Figs. 1 and 2) are described separately (Sect. 2.3).

2.1 Andean Metamorphic Complexes

The Andean metamorphic complexes are mainly composed by polydeformed metasedimentary rocks, local bodies of limestone, and metamorphosed mafic rocks. Deformation and metamorphism precludes conventional stratigraphic logging and preservation of biostratigraphic records. However, U–Pb SHRIMP geochronology of detrital zircon grains in different metasedimentary rocks provides constraints on the timing of sediment deposition of their protoliths and their possible

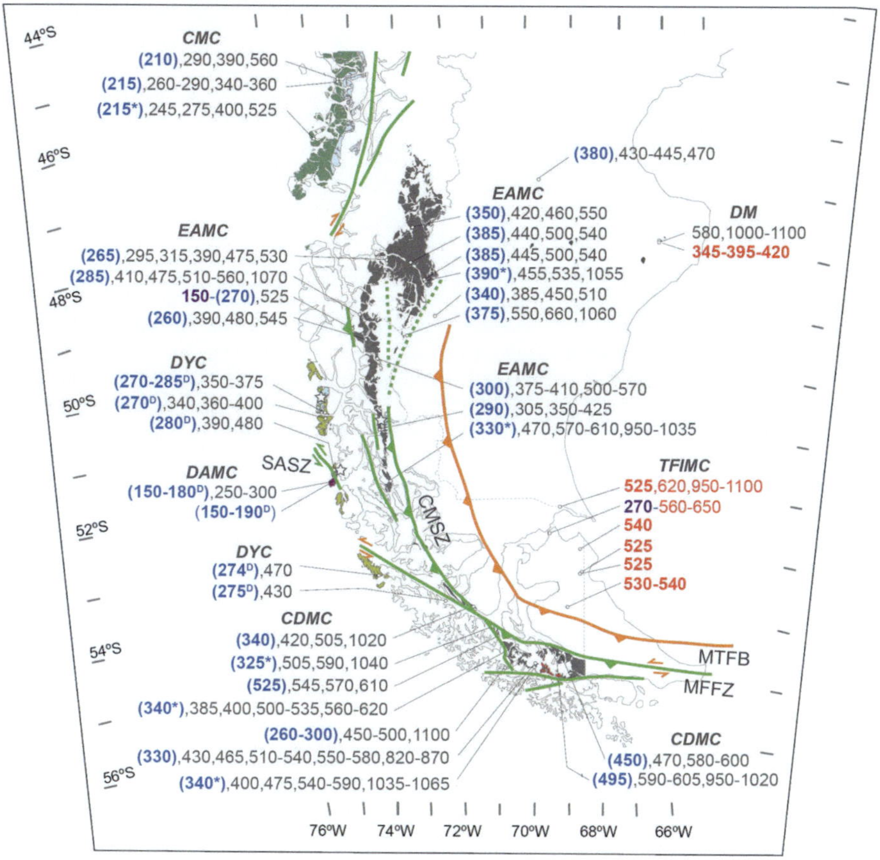

Fig. 2 Geological sketch map with published SHRIMP U–Pb zircon ages (Ma). In metamorphic rocks, the detrital zircon ages in parenthesis (*blue*) are the youngest obtained detrital age calculated with 3 zircon grains that can be interpreted as maximum sedimentation ages (*asterisks* indicate the age of 1 or 2 grains); older ages indicate principal provenance grouping of zircons. In situ crystallization ages in plutonic and gneissic rocks are indicated in *red*; older ages indicate inherited crustal components involved during their formation. Ages in *purple* correspond to the metamorphic age obtained in zircon overgrowths from migmatitic gneisses. For a detailed analysis of the data, reference should be made to the original papers (Hervé et al. 2003, 2010a, b; Pankhurst et al. 2003; Augustsson et al. 2006; Augustsson and Bahlburg 2008; Permuy Vidal et al. 2014; Malkowski et al. 2015; Castillo et al. 2015). Abbreviations are the same than those in Fig. 1

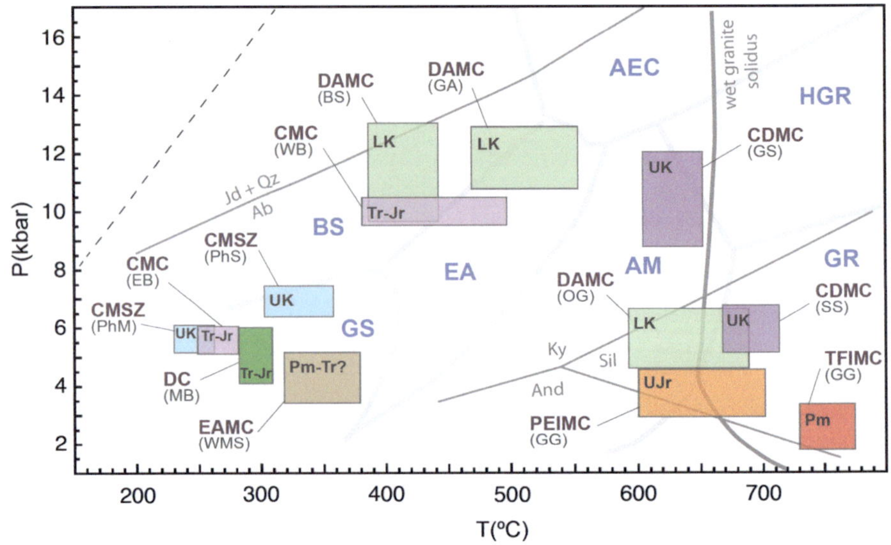

Fig. 3 Pressure–temperature diagram showing metamorphic conditions calculated in metamorphic rocks and the age of metamorphism. The overview of general metamorphic facies, stability field of andalusite-sillimanite-kyanite and wet granite solidus are also shown (Spear et al. 1999; Ernst 2009). For a detailed analysis of the data, reference should be made to the original papers (Ramírez-Sánchez et al. 2005; Calderón et al. 2007b, 2012; Willner et al. 2000, 2004, 2009; Sepúlveda et al. 2008; Hervé et al. 2010a; Maloney et al. 2011). Metamorphic facies abbreviations are: *GS* greenschist; *BS* blueschist; *EA* epidote amphibolite; *AM* amphibolite; *GR* sillimanite-granulite; *HGR* kyanite-granulite; *AEC* amphibolite-eclogite. Abbreviations are the same than those in Fig. 1

paleogeographic settings. The compilation of U–Pb detrital and igneous zircon ages of igneous and metamorphic complexes is shown in Fig. 2. A P–T diagram indicating the P–T conditions and age of metamorphism rocks belonging to different metamorphic complexes is illustrated in Fig. 3. Here, we describe the igneous and metamorphic complexes from the oldest to the youngest.

2.1.1 Tierra del Fuego Igneous and Metamorphic Complex (TFIMC)

The TFIMC corresponds to the Fuegian basement of the Magallanes–Austral basin (Söllner et al. 2000; Pankhurst et al. 2003; Hervé et al. 2010a), buried beneath km-thick successions of Jurassic volcaniclastic and younger sedimentary strata. This complex is dominated by foliated peraluminous biotite granitoids, and some containing cordierite and garnet, variably transformed to orthogneisses. Gneisses are predominately composed by coarse-grained aggregates of quartz, plagioclase, and K-feldspar, separated by finer bands of deformed and partly chloritized green biotite. The attitude of the foliation is difficult to determine, but is usually at high angles to the core length, suggesting that a low-dipping foliation is predominant. Some of the gneisses contain the mineral assemblage cordierite-sillimanite-garnet-biotite-plagioclase-orthoclase-quartz.

The plutonic zircon age spectra in samples of this complex (Hervé et al. 2010a) display: (1) a Cambrian igneous component (525–540 Ma); (2) a Cambrian migmatitic gneiss with peaks of inherited zircons at ~950–1100 and 560–650 Ma; and (3) a Permian high-grade anatectic event (of ~270 Ma) recorded in metamorphic zircon overgrowths on precursor rocks with noticeable Cambrian zircon components. The P–T metamorphic conditions calculated in the Permian cordierite-sillimanite-garnet migmatitic gneisses varies between 2–3 kbar and 730–770 °C, respectively (Fig. 3).

The oldest rocks in the TFIMC can be related to the sedimentary source of phyllitic metasedimentary rocks at Dos Hermanos locality (Deseado Massif, Fig. 1) characterized by a younger peak of ~565 Ma in detrital zircons (Pankhurst et al. 2003). Meanwhile, the Permian age for the upper crustal anatexis in the TFIMC is comparable with the age of high-grade metamorphism (~260 Ma) in gneisses located at Adie inlet (eastern margin of Antarctic Peninsula), formed from precursor metasedimentary rocks with Cambrian ages (Millar et al. 2002). These indicate similar and contemporaneous crustal anatectic processes in southern Patagonia and western Antarctic Peninsula (Hervé et al. 2010a).

2.1.2 Eastern Andean Metamorphic Complex (EAMC)

The EAMC includes the previously defined Cochrane and Lago General Carrera units (Lagally 1975) as well as the Staines Complex at the latitude of Puerto Natales (Forsythe and Allen 1980). Metamorphic complexes located in the Andean range consist mainly of polydeformed turbiditic successions, with minor limestone bodies and metabasites. The EAMC has been correlated with other units in Argentina such as the Devonian to lower Carboniferous Bahía de la Lancha and Río Lácteo Formations, and the Cerro-Negro mica schists (Permuy Vidal et al. 2014).

On the basis of detrital zircon U–Pb analyses (Hervé et al. 2003; Augustsson et al. 2006; Augustsson and Bahlburg 2008), the EAMC includes sedimentary components deposited between the late Devonian and the early Carboniferous (eastern domain), as well as younger turbidites deposited during the Permian and Triassic (western domain) (Fig. 2). Metamorphosed turbidites from the eastern domain shows Ordovician, Devonian, and early Carboniferous peaks of detrital zircons, and the western domain shows peaks of Permian zircons with variable proportions of Carboniferous, Devonian, Ordovician, and Cambrian grains.

Provenance studies based on petrographic and geochemical data (Hervé et al. 1998; Faúndez et al. 2002; Ramírez-Sánchez 2002; Augustsson and Bahlburg 2003; Lacassie 2003) suggest that turbidites of the eastern domain record deposition in a passive continental margin and were derived from a cratonic source, which possibly had undergone a complex and extended sedimentary recycling history. However, Augustsson and Bahlburg (2008) suggested that the eastern domain was fed from metasedimentary country rocks of an evolving magmatic arc. In contrast, detrital zircon U–Pb ages in the sedimentary protoliths of the Cerro Negro schists yield a statistically well-defined Devonian maximum depositional age (of ~380 Ma), supporting a forearc setting for its deposition (Permuy Vidal et al. 2014). The detrital zircon spectra, occurrence of metaplutonic lithic fragments, detrital biotite, and the major element composition of the metasedimentary strata of the western domain indicate that the basin probably was fed from the arc proper and their country rocks (Augustsson and Bahlburg 2008). Potential sedimentary source areas for the Ordovician, Devonian, and early Carboniferous detrital zircons in the eastern domain are located to the east, in the Deseado Massif, and in northern Patagonia (Figs. 1 and 2) (Pankhurst et al. 2003, 2006; Moreira et al. 2013).

Potential sources of Permian detritus for the western domain are the plutonic complexes in the northern Patagonia interpreted as the result of voluminous and widespread silicic plutonism and volcanism linked to break-off of the subducted plate between the Deseado Massif and northern Patagonia (Fig. 1) (Pankhurst et al. 2006). Magmatic belts of Permian and Triassic age are not known at present in southern Patagonia, with the exception of punctual Permian migmatites and granites described in the TFIMC (Hervé et al. 2010a). Similarly, Permian and Triassic granitoids crop out sporadically along the Antarctic Peninsula, and many of them are peraluminous (Millar et al. 2002).

The metamorphic grade of the EAMC is in the greenschist facies or lower, and higher-grade rocks occur only in the contact aureoles of Mesozoic to Cenozoic intrusions (Ramírez-Sánchez 2002; Calderón 2000; Valdés 2005). P–T calculations in low-grade metapelites of the EAMC (Cochrane unit), consisting of quartz, albite, and white mica, yielded P–T conditions of 4 ± 1.2 kbar and 300–390 °C (Fig. 3). The rocks of this complex were affected by a Permian metamorphic event (Thomson and Hervé 2002; Augustsson et al. 2006) under lower P/T metamorphic conditions than those that are typical of subduction accretionary complexes (Hervé et al. 1998; Ramírez-Sánchez et al. 2005). At the Puerto Edén area, the EAMC underwent Late Jurassic low P/T amphibolite facies metamorphism and anatexis (Watters 1964; Calderón 2000; Hervé et al. 2003).

2.1.3 Cordillera Darwin Metamorphic Complex (CDMC)

The CDMC consists of metasedimentary, metaplutonic and metavolcanic units, of supposed late Paleozoic to early Mesozoic age, which have a unique late Mesozoic metamorphic imprint among the metamorphic basement complexes of the Southern Patagonian and Fuegian Andes, which is characterized by the generation of biotite, staurolite, kyanite, and sillimanite zones (Nelson et al. 1980; Hervé et al. 1981a; Kohn et al. 1995; Klepeis et al. 2010). The main lithologies of this complex are: (1) pelitic schists with quartz, plagioclase, muscovite, biotite, chlorite, garnet, staurolite, ilmenite, kyanite, and sillimanite (fibrolite); (2) metabasites with garnet, plagioclase, hornblende, biotite, chlorite, quartz, epidote, titanite, and ilmenite; (3) garnet-bearing orthogneisses; and (4) quartz-mica schists generated by dynamic metamorphism of precursor Jurassic silicic volcaniclastic rocks (cf. Hervé et al. 2010b; Klepeis et al. 2010; Maloney et al. 2011).

The detrital zircon U–Pb age spectra in most metasedimentary rocks of the CDMC are characterized by Mississippian peaks ($\sim$330–340 Ma) with variable proportions of Devonian, Ordovician, Cambrian, and Neo- to Meso-proterozoic components (Fig. 2) (Hervé et al. 2010b). These age peak signatures are comparable with detrital zircon patterns from the eastern domain of the EAMC. The paucity of Cambrian detrital zircons indicates that the TFIMC was not the main source of detritus for the protolith of the CDMC, suggesting the existence of a pre-Jurassic suture zone between both metamorphic complexes (Hervé et al. 2010b).

Metamorphic P–T conditions determined by Maloney et al. (2011) in garnet-bearing schists constrain a P–T path dominated by decompression from 12 to 9 kbar at $\sim$620 °C (Fig. 3), culminating with staurolite and kyanite growth at 75 Ma (U–Th–Pb in situ dating of monazite). Additionally, these authors indicate that late contact aureoles developed adjacent to granite intrusions include sillimanite-bearing migmatites formed at P of 6 kbar and at T higher than 670 °C. The metamorphic petrology and field relationships are consistent with processes of continent-ward underthrusting, imbrications, and crustal thickening during the closure of the Rocas Verdes basin (Klepeis et al. 2010) with later exhumation controlled by thrusting (Maloney et al. 2011) that is an alternative to the model that considers the CDMC as an extensionally exhumed metamorphic core complex (Dalziel and Brown 1989; Kohn et al. 1993).

2.1.4 Madre de Dios Terrane (MDT)

The MDT is composed of three tectonically interleaved lithostratigraphic units: the Tarlton Limestone, the Denaro, and the Duque de York complexes (Forsythe and Mpodozis 1979, 1983). The Tarlton Limestone (TL) is a massive pelagic limestone body that was deposited in an oceanic carbonate platform during late Carboniferous–early Permian times (Douglass and Nestell 1976) overlying a broadly contemporaneous (Ling et al. 1985) oceanic substrate (the Denaro Complex). The Denaro Complex (DC) represents the oceanic floor composed of pillow basalts with enriched (E) and normal (N) mid-oceanic ridge basalts (MORB) geochemical signatures, radiolarian and Mn- and Fe-bearing cherts probably representing an oceanic ridge environment distal from the continental influence of Gondwanaland. This exotic terrane was later accreted to the continental margin during the early Mesozoic. The metabasalts of the DC preserve relic igneous phases, augite, and chromite and comprise metamorphic mineral assemblages of the pumpellyite-actinolite facies consisting of albite, chlorite, epidote, pumpellyite, stilpnomelane, titanite, garnet (grandite), white mica (phengite), actinolite, titano-magnetite, and quartz (Sepúlveda 2004; Sepúlveda et al. 2008). Peak metamorphic conditions of $\sim$290–310 °C and 4–6 kbar (Fig. 3) were estimated by Willner et al. (2009). These authors suggest that the Madre de Dios Accretionary Complex was frontally accreted in a subduction zone environment.

The Duque de York Complex (DYC) is a turbiditic succession, which was deposited unconformably over the TL and the DC when they reached the vicinity of the continental margin (Forsythe and Mpodozis 1979, 1983). The DYC includes palynomorphs of Permian age (Sepúlveda et al. 2010), as well as radiolarian cherts which indicate an early Permian deposition age (Quezada 2010). Metapelites exhibit porphyroblasts of cordierite and biotite in contact with the batholith (Parra 2015).

Detrital zircon U–Pb age data from rocks of the DYC show a dominant peak of Permian zircons (~270–290 Ma) (Hervé et al. 2003; Sepúlveda et al. 2010; Castillo et al. 2015), consistent with the biostratigraphic age. These rocks also comprise subordinate Carboniferous and Ordovician zircon populations, the latter becoming an important contribution in the southernmost outcrops of this unit (e.g., Desolation Island, Castillo et al. 2015). Combination of petrographic and geochemical analyses in rocks of the DYC (Lacassie 2003) indicates that this unit was derived from a geochemically intermediate igneous source, interpreted to have originated on a deeply dissected continental magmatic arc. The DYC basin was probably adjacent to the continental crust of Gondwanaland, in an active margin tectonic setting (Faúndez et al. 2002; Lacassie et al. 2006). Zircon fission track data indicate that the DYC (and probably the TL and DC units) was metamorphosed before or during the earliest Jurassic under low-grade metamorphic conditions (Thomson and Hervé 2002).

2.1.5 Chonos Metamorphic Complex (CMC)

The CMC consists predominantly of metaturbidites (Pimpirev et al. 1999), with more restricted occurrences of metabasites and metacherts. The CMC has a Late Triassic depositional age, as indicated by fossil fauna (Fang et al. 1998) and by detrital zircon SHRIMP U–Pb age determinations (Hervé and Fanning 2000). The complex was differentiated into two north–south-trending belts by Hervé et al. (1981b). The eastern belt has well-preserved primary sedimentary and volcanic structures, which are progressively obliterated when passing into the more pervasively deformed and recrystallized western belt rocks. Detrital zircons show characteristic peaks of the Upper Triassic detrital zircons with variable proportions of Permian, Carboniferous, Devonian, and Cambrian components (Hervé et al. 2003)

The lithologies of this complex correspond to psammopelitic rocks in the eastern belt with detrital quartz, plagioclase, K-feldspar, and muscovite in a very fine matrix consisting of quartz, albite, illite, chlorite (clinochlore), and white mica (muscovite and phengite). Accessory phases are titanite, rutile, zircon, tourmaline, apatite, and epidote. Metabasites with relict pillow structures and subophitic textures display mineral assemblages of quartz, albite, amphibole (actinolite), pumpellyite, chlorite, stilpnomelane, titanite, epidote, and veinlets of phengitic white mica. Meta-ironstones, usually interleaved with tectonic lenses of metachert, are dark finely banded carbonate rocks consisting of siderite, chlorite, and graphite. The metamorphic mineral assemblage of metapsammopelitic schists in the western belt consists of quartz, albite, white mica (generally phengite), chlorite, titanite, graphite with epidote, calcite, tourmaline, and Fe–Ti oxides as common accessory phases. Foliated metabasites or greenschists are made up by amphibole (magnesium-hornblende), epidote, quartz, albite, titanite, ilmenite, and rare apatite. Exceptional garnet-bearing metabasites also occur. Meta-ironstone in the western belt consists of rocks formed by quartz-rich bands with abundant stilpnomelane and subordinate chlorite, amphibole, calcite, and white mica. Willner et al. (2000) resolved that maximum P–T conditions are 4.5–6 kbar and 250–280 °C for the eastern belt and 8–10 kbar and 380–550 °C for the western belt (EB and WB, respectively, in Fig. 3).

2.1.6 Puerto Edén Igneous and Metamorphic Complex (PEIMC)

The PEIMC consists of medium- to high-grade metamorphic rocks, migmatites, and plutonic rocks. Schists are composed of quartz, muscovite, plagioclase, biotite, pinnitized cordierite, andalusite (some with inclusions of staurolite), sillimanite, and K-feldspar. Accessory minerals are tourmaline, zircon, apatite, corundum, graphite, and Fe–Ti oxides. Also occur medium- to coarse-grained leucocratic gneisses, sillimanite gneisses, and stromatitic migmatites. Metamorphic assemblages, textural relations, and geothermobarometry suggest a low P/T and nearly isobaric metamorphic and partial melting event (~600–700 °C and 3–4.5 kbar) superimposed on the greenschist facies metamorphic rocks of the EAMC (Calderón 2006; Calderón et al. 2007b).

Metamorphic overgrowths on zircons in sillimanite paragneisses record a Late Jurassic (~150 Ma) age taken as evidence of local gneiss formation under in situ anatectic conditions during the emplacement of the Jurassic components of the South Patagonian Batholith (Hervé et al. 2003) formed during continental rifting in a supra-subduction setting (Calderón et al. 2007b).

2.1.7 Diego de Almagro Metamorphic Complex (DAMC)

The DAMC is composed of two subunits of differing metamorphic imprints, one composed of garnet amphibolites and blueschists, and the other of quartz-mica schists and an orthogneiss. The contact between them has not been observed in the field. SHRIMP U–Pb ages in zircons from the orthogneiss and a quartz-rich spessartine-bearing schist interleaved in the blueschists (Hervé and Fanning 2003) have yielded Middle Jurassic ages that are interpreted as the crystallization age of their igneous forerunners, a muscovite-garnet-bearing granite and a rhyolitic rock. This complex is in tectonic contact (Forsythe 1981, 1982) with the DYC along the mid-crustal sinistral strike-slip Seno Arcabuz shear zone (Olivares et al. 2003; Willner et al. 2004).

Quartz-rich mica schist and a mylonitized granite contain Late Jurassic zircons of ~166 and 170 Ma, respectively (Hervé and Fanning 2003). According to these authors, these rocks were metamorphosed during the Cretaceous in a subduction zone environment, which developed blueschist assemblages in metabasalts interleaved with the mica schist.

The mineralogy of the DAMC metamorphic rocks is well described by Willner et al. (2004) and Hyppolito (2010). The amphibolites consist of Ca-amphibole (hornblende and actinolite), epidote, chlorite, albite, quartz, phengite, titanite, and subordinate garnet (partially altered to chlorite), calcite, ilmenite, apatite, and pyrite. Orthogneisses and schists within the Seno Arcabuz shear zone consist of quartz, plagioclase, muscovite, biotite, apatite, and tourmaline. Hypidioblastic garnet with helicitic inclusion trails of graphite, biotite, and white mica occur in the schists. These rocks were overprinted by discrete shear band cleavage with newly grown quartz, phengite, chlorite, stilpnomelane, epidote, and albite. Amphibolites in the shear zone consist of hornblende, quartz, plagioclase, epidote, and titanite. Willner et al. (2004) established a converging P–T path followed by blueschists (initially crystallized at 9.5–13.5 kbar and 380–450 °C) and amphibolites (first crystallized at 11.2–13.2 kbar and 460–565 °C) into the P–T space defined between 6.3–9.6 kbar and 320–385 °C, respectively (Fig. 3). Orthogneisses crystallized first at 4.9–6.5 kbar and 580–690 °C and their retrograde phengite, chlorite, and stilpnomelane assemblage crystallized at a minimum pressure of $\sim$5.7 kbar at 300 °C. The high P/T metamorphism in the rocks of the DAMC developed in a subduction zone during the Cretaceous (Hervé et al. 1999; Willner et al. 2004).

2.1.8 Diego Ramírez Metamorphic Complex (DRMC)

The DRMC is located in the Diego Ramirez Island, in the southern extreme of South America, and consists of pillow basalts and metasedimentary rocks that form a crush mélange (Wilson et al. 1989) with glaucophane-bearing metamorphic assemblages. A Middle Jurassic Rb–Sr whole-rock errorchron in the metasedimentary rocks (Davidson et al. 1989) suggests that these rocks can be correlated with those of the DAMC. Both complexes are envisaged as the remnants of mafic oceanic floor developed in a Jurassic marginal basin located between southern Patagonia and Antarctic Peninsula (Hervé and Fanning 2003).

2.2 Rocas Verdes Ophiolitic Complexes

The Rocas Verdes ophiolitic complexes preserve an incomplete "ophiolite pseudostratigraphy" (lacking the ultramafic rocks) and are particularly well suited for examining the tectonic and petrological transition from rift to back-arc basin evolution and subsequent closure and orogeny in the Southern Patagonian Andes. The rocks are foliated and metamorphosed in a discontinuous way; most parts of the large complexes preserve original textures and structures with characteristic secondary mineralogy of non-deformative seafloor hydrothermal metamorphism (Stern et al. 1976; Elthon and Stern 1978; Allen 1982; Calderón 2006; Avendaño 2008; Prades 2008; Calderón et al. 2013). In this section, the lithological, geochronological, geochemical, and isotopic information of the main ophiolitic complexes is presented. The location and geochronological data of ophiolitic complexes and

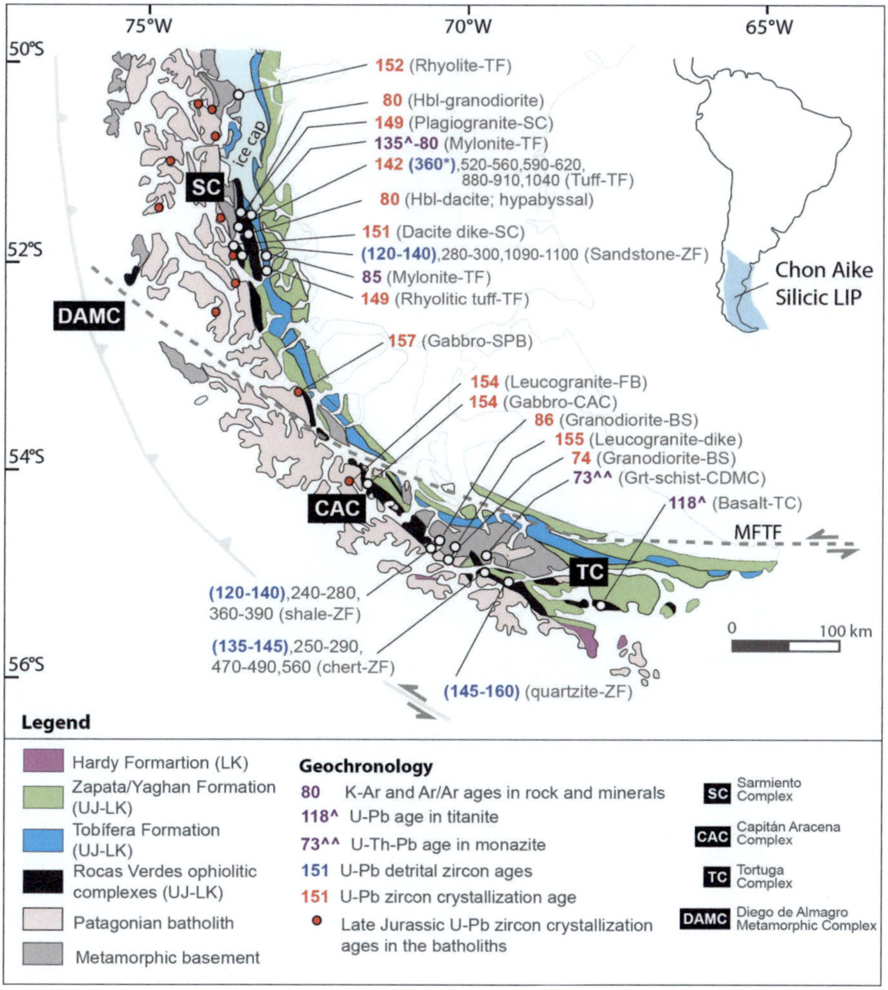

Fig. 4 Geological sketch map of Paleozoic and Mesozoic lithostratigraphic units in the Southernmost Andes of South America, including the Rocas Verdes ophiolites, folded and sheared volcanic and sedimentary successions of the Tobífera, Zapata, Yahgán Formations, the Patagonian and Fuegian batholiths, and metamorphic complexes. In sedimentary rocks, detrital zircon ages in parenthesis (*blue*) are the youngest obtained detrital age grains (*asterisks* indicate the age of 1 or 2 grains); older ages indicate principal provenance grouping of zircons. In situ crystallization ages in volcanic and plutonic rocks are indicated in *red*. Ages in *purple* correspond to the metamorphic age obtained by several methods as indicated in the figure. For a detailed analysis of the data, reference should be made to the original papers (Hervé et al. 2007a; Klepeis et al. 2010; Maloney et al. 2011; Calderón et al. 2007a, 2012, 2013). Abbreviations are: *CAC* Capitán Aracena Complex; *DAMC* Diego de Almagro Metamorphic Complex; *SC* Sarmiento Complex; *TC* Tortuga Complex

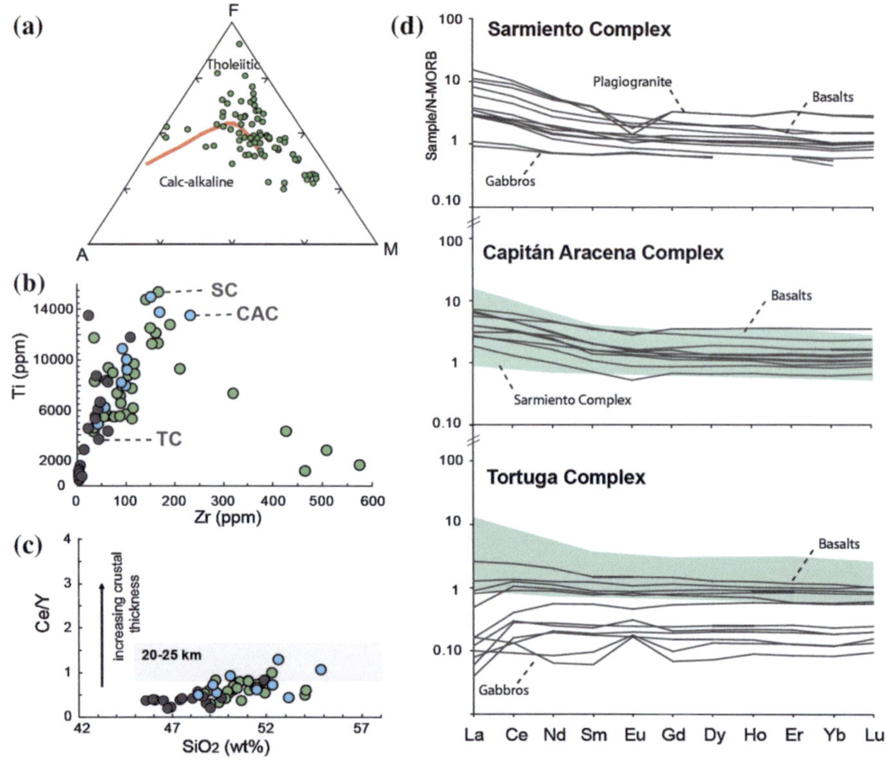

Fig. 5 Geochemical and petrological features of pillow basalts, basaltic dikes, diabases, and gabbroic rocks of the different Rocas Verdes ophiolitic complexes shown in diagrams. **a** AFM, **b** Ti versus Zr, **c** Ce/Y versus SiO_2 [values of crustal thickness are from Mantle and Collins (2008)]. **d** Normal mid-ocean ridge basalt (N-MORB) normalized Rare Earth Element diagram of the ophiolitic complexes of the Rocas Verdes basin, using Sun and McDonough (1989) normalization values. For a detailed analysis of the geochemical data, reference should be made to the original papers (Stern 1979, 1980; Saunders et al. 1979; Fildani and Hessler 2005; Calderón et al. 2013)

related granitic rocks are presented in Fig. 4. The available major and trace element geochemical composition and Nd-isotopic data for the ophiolitic rocks and other geological units in the region are illustrated in Figs. 5 and 6.

2.2.1 Sarmiento Complex (SC)

The Sarmiento Complex occurs in a north–south-trending 10- to 20-km-wide belt bound by the Canal de las Montañas Shear Zone in the east. Three main lithological units are distinguished (Calderón et al. 2007c): a thick mafic extrusive layer of pillow basalts and breccias with intercalations of radiolarian chert; a mafic–felsic

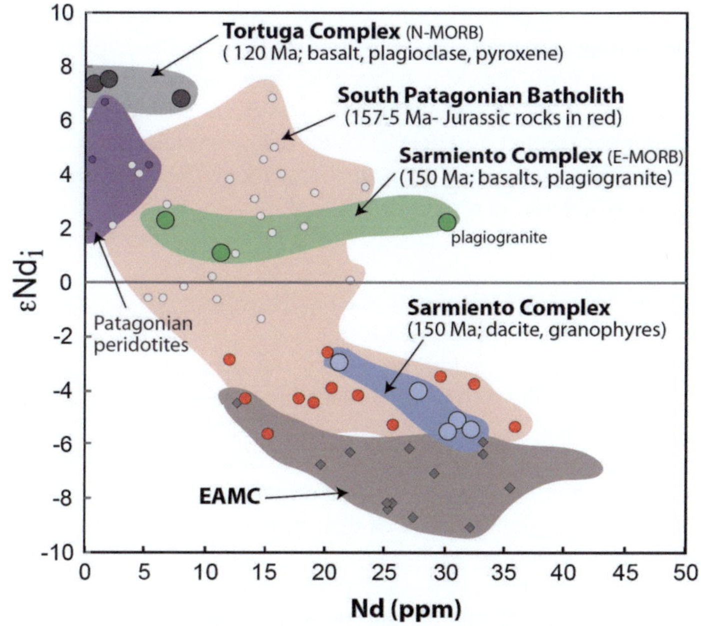

Fig. 6 Comparison of εNd_i versus Nd values of: igneous rocks from the Sarmiento and Tortuga ophiolitic complexes, and the South Patagonian Batholith; metasedimentary rocks, schists, and gneisses of the Eastern Andean Metamorphic Complex (EAMC); and peridotite xenoliths in extra Andean Patagonia (Stern 1991; Augustsson and Bahlburg 2003; Schilling et al. 2005; Hervé et al. 2007a; Calderón et al. 2007c). Noticeable are the similarities between felsic rocks of bimodal association in the Sarmiento Complex and Jurassic granites of the batholith, indicating the involvement of crustal components during their formation. It is also obvious the cogenetic origin between the basalts and plagiogranites, the last as product of extreme fractional crystallization

extrusive layer consisting predominantly of layers of pillow basalts, including intercalations of rhyolitic tuffs, hyaloclastites, and rhyolitic and dacitic dikes, which is in turn cut by mafic intrusive bodies; and a mafic–felsic intrusive layer consisting mainly of medium-grained granophyre cut by fine-grained gabbro and late sub-horizontal plagiogranite dikes. The base of this unit is composed of metagabbro and amphibolite. These rocks are exposed in two north–south-trending steeply dipping thrust sheets located in the central and western portions of the Sarmiento Cordillera (Allen 1982; Calderón et al. 2007a; Rapalini et al. 2008).

A dacite dyke crosscutting pillow-basalt successions and a plagiogranite dyke in sheeted dikes of the basal Sarmiento Complex contain ~150 Ma zircon crystals, indicating that mafic submarine volcanism had started prior to or during the Late Jurassic (Calderón et al. 2007a). The neighbor silicic pyroclastic rocks of the Tobífera Formation, intercalated with synrift deposits along fault-bounded grabens within Paleozoic metasedimentary rocks, were dated at 148 and 142 Ma, and thus coeval with ophiolitic magmatism. A fine-grained sandstone layer between the hemipelagic shales of the Zapata Formation located at c. 1000 m above its depositional contact over pillow basalts reveals 12 ages between 132 and 143 Ma (a prominent peak that constrains an apparent maximum depositional age) and a Permian population (260–300 Ma) defined by four grains. Pre-tectonic grains of titanite from phyllonites of the basal and low-grade metamorphic sole of the Sarmiento Complex yielded a SHRIMP U–Pb mean crystallization age of roughly 135 Ma that probably constrain the age of seafloor-like metamorphism in the back-arc setting (Calderón et al. 2013).

2.2.2 Capitán Aracena Complex (CAC)

To the south of and in the middle portion of the main trace of the Magallanes–Fagnano Fault Zone, the CAC crops out in a northwest–southeast-trending 20-km-wide belt located at the northeast of Cordillera Darwin (Prades 2008). It forms a body elongated in a southeast–northwest direction that is thrust northward over metasedimentary rocks of the Yahgán Formation (Otzen 1987). The rocks are associated with diverse granitoid bodies that intrude them and with rhyolitic rocks of the Tobífera Formation.

Basaltic pillow lavas are occasionally cut by mafic dikes and preserve primary variolitic, amygdaloidal, porphyritic, and glomerophyric textures. The fine-grained seriate, intersertal, microgranular, and ophitic textures are characteristic in mafic dikes. There are also short tracts of green aphanitic foliated rocks, as their occasional intense foliation appears in the field to be mafic dike complexes. Coarse-grained gabbros were observed at several localities and, at some of these, were close to pillow basalts with no apparent discontinuity between them.

A medium-grained gabbro collected 100 m below pillow basalts having no discontinuity with the underlying rocks yielded a weighted mean age of ~154 Ma (Calderón et al. 2013). A similar age of ~155 Ma was obtained in a leucogranite intruded by ophiolitic mafic dikes. Both indicate that the early rifting stage of the Rocas Verdes basin development at the latitudes of the CAC was accompanied by bimodal magmatism.

2.2.3 Tortuga Complex

The Tortuga ophiolite is exposed in an east–west-trending 20-km-wide belt extending from the southwestern part of the Isla Navarino to the region south of the Canal Beagle. The southern border of the ophiolite is in fault contact with a tightly folded and weakly foliated package of turbiditic rocks. To the north, the possible depositional contact between the Tortuga ophiolite and the sedimentary rocks is obscured by subsequent thrust-fold deformation and earliest Late Cretaceous magmatism (Hervé et al. 1984). The Tortuga Complex is remarkable for its exceptional exposures of mafic oceanic-like crustal pseudostratigraphy, which grades upward from cumulate gabbros to pillow and sheet-flow basalts (Godoy 1978; Stern 1979).

The extrusive unit of the Tortuga ophiolite consists of a more than 2000-m-thick succession of submarine volcanic rocks, including massive and pillowed basalts, volcanic breccias, hyaloclastites, and intercalated horizons of chert and siltstones. The basalts bear plagioclase and clinopyroxene, and some contain subordinate pseudomorphs after discernible primary orthopyroxene and olivine. Pillow basalts preserve a fine-grained variolitic texture, with skeletal plagioclase laths and fans of dendritic clinopyroxene. Massive lava flows, or diabase sills, with characteristic columnar structures have medium-grained ophitic and variolitic textures of plagioclase intergrown with pyroxene granules. Massive diabase screens within a 1-km-thick level of sheeted dikes consist of medium-grained plagioclase and clinopyroxene and orthopyroxene and have ophitic, subophitic, and intergranular textures. The exposed base of the Tortuga ophiolite consists of an igneous complex made up of slightly altered massive and layered gabbros, most of which are two pyroxene and olivine gabbros, leucogabbros, and clinopyroxene troctolites intruded by dikes of basalt and fine diabase with chilled margins.

A cherty layer directly overlying pillow basalts yielded two dominant groups of zircon ages, about 136 and 143 Ma (that constrains an apparent maximum depositional age), with subordinate zircon grains of Permian, Cambrian, and latest Neoproterozoic ages (Calderón et al. 2013).

U–Pb LA-MC-ICP-MS in situ analyses in titanite, which is in association with amphibole, epidote, chlorite, plagioclase, and quartz in veins within the margins of pillow basalts yielded and Aptian mean crystallization age of ~118 Ma (Calderón et al. 2013). This can be interpreted as the age of greenschist facies seafloor metamorphism in the back-arc basin.

2.2.4 Geochemistry and Nd-Isotope Compositions

The mafic rocks of the Rocas Verdes basin ophiolites crystallized from tholeiitic magmas that are characterized by increasing Ti contents during the early stages of differentiation followed by decreasing contents in the later stages because of fractionation of magnetite (Fig. 5) (Stern 1979, 1980; Miller et al. 1994; Calderón et al. 2013). Geochemistry is also compatible with the concept of a thinned crust resulting from extension in a rift or back-arc environment.

Rare Earth Elements (REE) diagrams normalized to the composition of normal mid-oceanic ridge basalts show the similarity between the Sarmiento and the Capitán Aracena complexes; these are more enriched in large-ion lithophile elements than is the Tortuga Complex, which suggests that the former have a greater affinity with E-MORB magma. In the Tortuga Complex, mafic rocks crystallized from N-MORB magma probably generated by decompression melting of the sub-back-arc mantle and record the most advanced evolutionary stage of the Rocas Verdes basin opening to a mid-ocean ridge-type setting (cf., Saunders et al. 1979; Stern 1979; Calderón et al. 2013). The contrasting affinities of E- and N-type MORB compositions are also suggested by the Nd-isotopic compositions of mafic rocks of the Sarmiento and Tortuga complexes (Fig. 6) (Stern 1991; Calderón et al.

2007c). These features reflect that mantle-derived tholeiitic magmas were generated by melting of progressively more depleted sources during the interplay between back-arc mantle convection and the release of stress across the continental plate boundary (Stern and de Wit 2003).

2.3 Mesozoic Mylonitic Belts

2.3.1 Seno Arcabuz Shear Zone (SASZ)

The SASZ is a NNW-trending low-grade mylonitic belt that juxtaposes different structural levels at Isla Diego de Almagro (DAMC and DYC) and record two deformation events (Olivares et al. 2003): an early mylonitic foliation, and a late low-strain foliation originated by folding of the older fabric. Ductile deformation and sinistral-reverse kinematic indicators are typical of transpressional shear zones. According to the syntectonic mineralogy, deformation occurred at greenschist metamorphic grade. Zircon fission track ages from rocks of the SASZ are of ~65 Ma and record the end of shearing that juxtaposed all rocks in the middle crust (Willner et al. 2004).

2.3.2 Canal de las Montañas Shear Zone (CMSZ)

The CMSZ is a low-grade mylonite belt generated from felsic ignimbritic, pelitic and basaltic protoliths of the Upper Jurassic–Lower Cretaceous Rocas Verdes basin, interpreted as the metamorphic sole thrust of the Sarmiento Complex (Calderón et al. 2012). Highly strained rocks of the CMSZ display a reverse, continent-ward tectonic transport, with a minor dextral component of shearing. Transitional pumpellyite–actinolite and the upper greenschist facies metamorphic conditions at ~5–6 kbar and 230–260 °C indicate that the primary shearing event occurred in a subduction zone setting. In situ 40Ar/39Ar laserprobe chronology yielded ages of ~85 Ma on syntectonic phengite.

3 Discussion

In this section, we summarize the Paleozoic and Mesozoic tectonic and magmatic evolution of Southern Patagonian Andes based on the combined petrological, geochemical, and geochronological datasets described herein. Our tectonic synthesis is considered within the paleogeographic context provided by the PLATES program (The University of Texas Institute for Geophysics) in which the southwestern continental margin of Gondwana was nearly WNW-trending until Triassic time.

In these models, three discrete episodes of rapid drifting of the Gondwanan margin occurred between 420–390 Ma (to the south), 390–360 Ma (to the west), and 330–300 Ma (to the north). These changes, attributed to underlying mantle dynamics and changes in the configuration of tectonic plates, were likely a major control on the tectonic evolution of the continental margin and sediment routing of clastic detritus shed into marginal basins, as reflected by detrital zircon provenance data discussed below (Fig. 2). Additionally, we revisit the tectonic evolution of the Rocas Verdes basin in light of new geochronology and petrology of different supracrustal units (ophiolites, metamorphic complexes, mylonite belts) metamorphosed under different P–T conditions, at different times.

3.1 Devonian-Carboniferous

The occurrence of lower Devonian and Mississippian plutonic rocks in the Deseado Massif and the southward migration of Gondwana (420–390 Ma) suggest that subduction processes operated along the continental margin before the accretion of the Antarctic Peninsula to southern Patagonia (Fig. 7). However, given the persisting lack of information about the tectonic setting in which these magmatic rocks have formed, we prefer to conservatively refer to them as magmatic belts associated with either to subduction-related magmatic arcs or to volcanic rift zones.

Detrital zircon U–Pb geochronology of turbidite protoliths to the upper Paleozoic metamorphic rocks of the eastern domain of the EAMC and the CDMC points to the existence of Cambrian, Ordovician, Devonian, and Carboniferous magmatic belts in southern Patagonia and/or neighboring continental areas that have since been removed or buried (cf. Hervé et al. 2003, 2010a; Augustsson et al. 2006; Augustsson and Bahlburg 2008). Most of the Cambrian, Devonian, and Carboniferous igneous and metamorphic rocks are found in the Deseado Massif and in the TFIMC. However, detrital zircons of Ordovician age are recovered from some of these rocks and no evidence of plutonic/volcanic complexes of this age is known in southern Patagonia (Fig. 2). Plutonic complexes of Ordovician age crop out in the North Patagonian Massif (Pankhurst et al. 2006) and the evidence of a similar belt possibly concealed in southern Patagonia is indirectly provided by Middle Ordovician granitic cobbles in basal clastic successions of the Jurassic Tobífera Formation (to the north of the CDMC; Hervé et al. 2010b) and within conglomerates of the Permian Golondrina Basin, east of the Deseado Massif (Fig. 1) (Pankhurst et al. 2003).

According to paleogeographic models, the Antarctic Peninsula microplate transited from east to west, parallel to the westward drift of Gondwana (390–360 Ma) until the microplate became attached to the south part of southern Patagonia in the Mid-Carboniferous (Fig. 7) (cf. Ramos 2010). Bell and Suarez (2000) suggested that the main deformation and metamorphic events in the eastern belt of the EAMC occurred in pre-late Carboniferous times probably resulting from microplate interactions, rather than forearc accretion. A reasonable scenario that we consider is that, during the middle–late Devonian, a transform boundary was established along the continental margin with episodic magmatism and transpressional deformation in precursor basins.

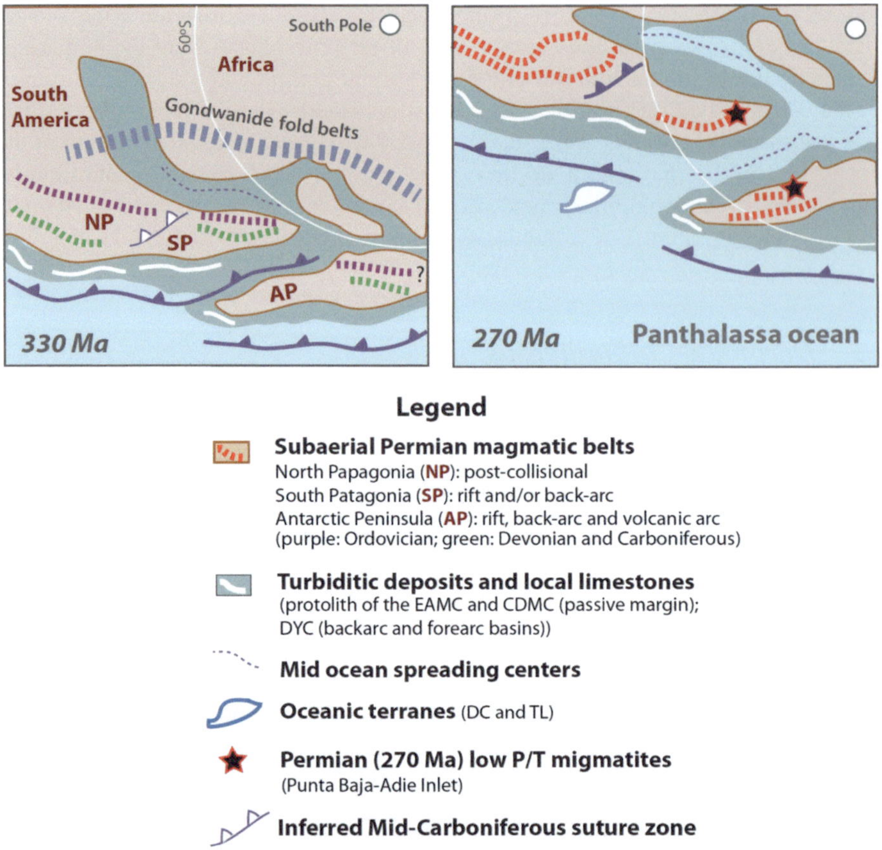

Fig. 7 Schematic reconstructions of the late Mississippian to early Permian paleogeographic and tectonic evolution of Gondwana margin at southern South America based on the geodynamic significance of Paleozoic metamorphic and igneous complexes in southern Patagonian and Antarctic Peninsula. The location of Antarctic Peninsula is based on paleogeographic maps of the Institute for Geophysics, The University of Texas at Austin

3.2 *Permian-Triassic*

Detrital zircon U–Pb geochronology of metasedimentary rocks of the western belt of the EAMC and the DYC indicates that their sedimentary protoliths contain predominantly Permian zircons and subordinate older grains. A wide variety of detrital components are recognized in the western belt of the EAMC, with zircons of Cambrian, Ordovician, Devonian, and Pennsylvanian ages. In contrast, the Duque de York Complex (DYC) yields a dominant peak of Permian zircons (~270–280 Ma; Hervé et al. 2003; Fanning et al. 2011; Castillo et al. 2015) with minor but discernable peaks of zircons with Devonian age and few Ordovician grains (Fig. 2). Cambrian components, such as those igneous rocks in the TFIMC and detrital components in the EAMC and CDMC, are absent. This detrital signature is consistent with a depositional setting of the DYC far from the influence of Cambrian basement rocks and tectonic evolution and recycling of Patagonian Andes metasedimentary complexes.

The Trinity Peninsula Group (TPG) in the northern Antarctic Peninsula yields a similar detrital zircon U–Pb age distribution compared to the DYC, but with the addition of pronounced Ordovician zircon populations (Hervé et al. 2006; Barbeau et al. 2010; Bradshaw et al. 2012; Castillo et al. 2015). Cobbles of Ordovician and Devonian igneous rocks are also found in Permo-Carboniferous conglomerates of the TPG (Bradshaw et al. 2012) suggesting that hidden magmatic belts, such as those identified in Patagonia, were proximal to the TPG (Fig. 7). Amphibolite-grade metamorphic rocks in Antarctic Peninsula were formed during episodic events of Ordovician, Permian, and Triassic metamorphism linked to tectono-magmatic activity in this block (cf. Pankhurst et al. 2003; Wendt et al. 2008; Riley et al. 2012).

Provenance analyses in turbidites of the DYC and TPG suggest deposition in forearc basins of the Gondwana convergent margin (Faúndez et al. 2002; Lacassie et al. 2006; Castillo et al. 2015), despite lack of known bedrock exposures of Permian igneous sources. In contrast, previous workers postulate that Permo-Triassic turbidites in Antarctic Peninsula were also deposited in a passive margin environment and/or within back-arc basins (cf. Burton-Johnson and Riley 2015). Taking this into consideration, we propose that the apparent southward jump of the magmatic activity in the region was related to the fast northward drift of the supercontinent at ~330 Ma that promoted the relative retreat of the continental margin and crustal extension in the supra-subduction setting with the consequent establishment of an arc and back-arc basin configuration during the Permian and Early Triassic (Fig. 7). Such as in retreating convergent margins (e.g., the modern western Pacific), contemporaneous development of passive margins, back-arc, and forearc basins are present, and this complex setting may help elucidate the variety of sedimentary environments interpreted for the western belt of the EAMC, the DYC, and the TPG. Afterward, the marginal basins were closed during the Late Triassic–Early Jurassic Chonide orogeny. Dalziel et al. (2000) proposed that this orogenic event was related to subduction of buoyant oceanic lithosphere that induced shallow flat subduction of the downgoing lithosphere.

3.3 *Jurassic-Cretaceous*

During the initial stage of Gondwana breakup, continental extension was accompanied by synrift silicic and bimodal magmatism and oceanic-type spreading in the overriding and attenuated continental crust (cf. Bruhn et al. 1978; Pankhurst et al. 2000; Stern and de Wit 2003). Paleogeographic models show that at the end of the Triassic, the southwestern Gondwana margin was NW-trending at that at 180 Ma Antarctic Peninsula was detached from southern Patagonia resulting in the development of marginal basins separating both continental blocks, and the emplacement of the Upper Jurassic—Lower Cretaceous Rocas Verdes basin in southern Patagonia (Dalziel 1981).

During the incipient stages of the Rocas Verdes basin evolution (Fig. 8), repetitive injection of mantle-derived mafic magmas triggered high-grade metamorphism, anatexis, and granite magma generation at mid- to upper crustal levels, resulting in the generation of bimodal magmatic associations and granites (Calderón et al. 2007a, b, c). Granophyres and dacitic dikes in the Sarmiento Complex are isotopically similar to the Upper Jurassic granitoids in the South Patagonian Batholith (Fig. 6) and thus probably represent the earliest components of the composite batholith that formed during the early rifting phase of the Rocas Verdes basin evolution, in a supra-subduction setting.

The back-arc stage of the Rocas Verdes basin started once the Antarctic Peninsula microplate was translated southward in Early Cretaceous time, and subduction was re-established along the margin of southern Patagonia. Subduction activity is recorded in blueschist and epidote-amphibolite facies metamorphic rocks of the DAMC (Fig. 5), involving processes of subduction erosion and high-P/T metamorphism of the forearc segment (Willner et al. 2004). These processes were coeval with the greater extension and ocean ridge-type magmatism in the back-arc basin suggesting that both domains, the forearc and back-arc, were tectonically decoupled.

The Rocas Verdes basin started to close at the end of the Early Cretaceous (Aptian–Albian), involving underthrusting and subduction of its oceanic seafloor until Late Cretaceous time (Calderón et al. 2013), coeval with the early development of the Magallanes–Austral foreland basin signaled by distal turbiditic deposits in the sedimentary record (Fildani et al. 2003; Fosdick et al. 2011). A subduction environment for the closure of the Rocas Verdes basin has been determined in felsic schists and mylonites of the CMSZ (Hervé et al. 2007b; Calderón et al. 2012). These processes culminated with collision or accretion of the drifted magmatic arc and/or peripheral crustal blocks against the continent (Cunningham 1995; Kohn et al. 1995; Klepeis et al. 2010; Fosdick et al. 2011; Maloney et al. 2011; Calderón et al. 2012, 2013; Betka et al. 2015; Guillot 2016), in part recorded in high-P and high-T metamorphic rocks of the CDMC.

Lithospheric shortening, thickening, and resulting thrust-loading of the retroarc Patagonian fold and thrust belt caused deepening of the Magallanes–Austral foreland basin during the Late Cretaceous. The sedimentary infill of the Magallanes–Austral basin records additional detrital geochronological evidence of lower Paleozoic magmatic arc activity (Fildani and Hessler 2005; Barbeau et al. 2009; Romans et al. 2010; Fosdick et al. 2015), although variable deformation and exhumation across the magmatic arc and hinterland have obscured much of the Paleozoic outcrop record.

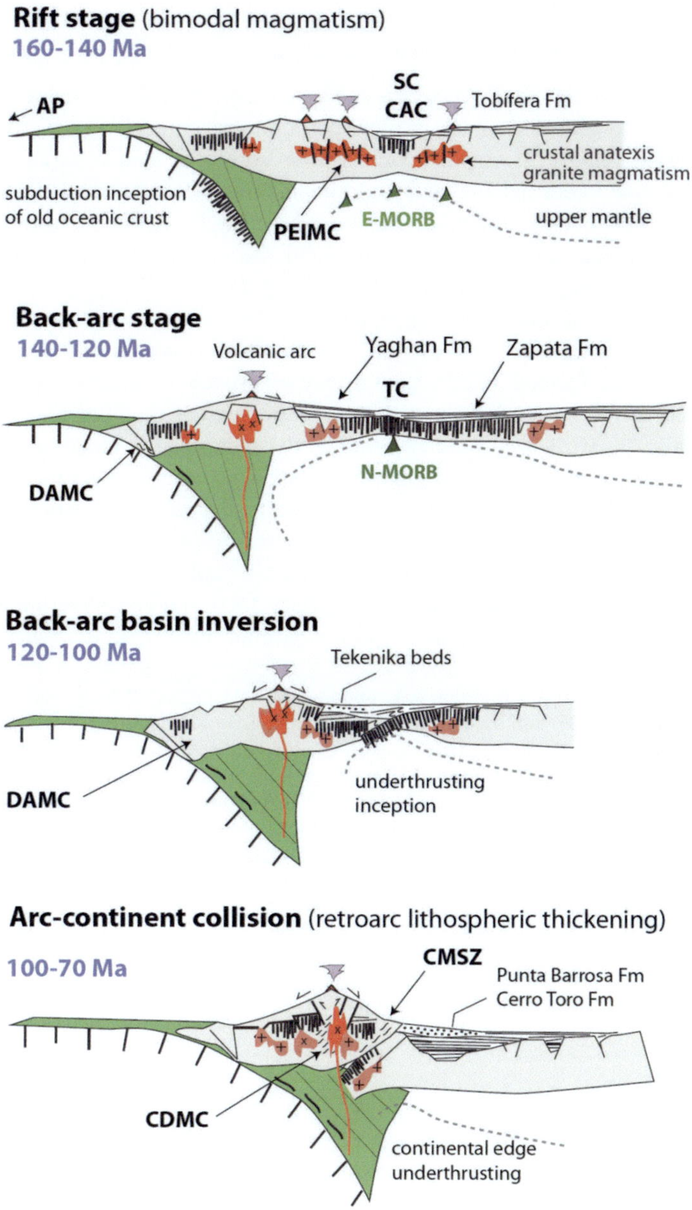

Fig. 8 Schematic reconstructions of the Jurassic to Cretaceous tectonic evolution of the Rocas Verdes basin after Calderón et al. (2013). The time span coincides with the relative separation of Antarctic Peninsula from South America. Abbreviations of metamorphic and ophiolitic complexes are presented in the caption of Figs. 1 and 4

4 Conclusions

The southern edge of South America records a complex Phanerozoic geological history related to global changes in plate configuration, including several episodes of magmatism related to crustal extension or contraction in a supra-subduction setting. Because older basins and sedimentary successions are overprinted by younger tectonic events, much of the history must be deduced from combined petrological and geochronological studies and the application of actualistic plate tectonic models and paleotectonic reconstruction models. We propose that extensional fault zones and spreading ridges, developed during breakup and/or dispersion of peripheral microplates, can evolve to transform zones, and vice versa. The closure of basins, formed within such zones of crustal weakness, causes internal deformation of the basin in-fill and formation of low- or medium-grade metamorphic complexes, with well-defined, frequently tectonically reactivated margins. Whether these zones host voluminous injections of mantle-derived magmas during continental extension, processes will contribute to the lateral growth of the continental crust and reworking via anatexis and generation of granitic magmas. An integrated four-dimensional analysis of igneous and metamorphic complexes, including ophiolites and batholiths, is necessary for more complete understanding of the entire geological evolution of the Patagonian and Antarctic regions.

Acknowledgements This contribution was supported by the CONICYT Anillo Antártico Projects (ARTG-04 and ACT-105) and by the FONDECYT projects to FH (1980741, 1010412, 7010412, 1050431) and to MC (11075000, 1161818), which allowed the study of the Patagonian Andes in recent years. Captains Conrado Alvarez, Gilles Rigaud, and Hugo Cárdenas in their boats Penguin, Foam, Explorador, Morgane, Chonos, and Marypaz II took us safely to remote areas. C. Mark Fanning has guided the SHRIMP U–Pb zircon age determinations and participated in most field campaigns. Valuable discussions and fieldwork with Elisa Ramirez, Cecilia Herrera, Augusto Rapalini, Umberto Cordani, Hans Massonne, Cristóbal Ramirez de Arellano, Bob Pankhurst, Andrea Fildani, and Brian Romans are greatly acknowledged.

References

Allen RB (1982) Geología de la Cordillera Sarmiento, Andes Patagónicos, entre los 51°00′ y 52° 15′ Lat. S, Magallanes, Chile. Servicio Nacional de Geología y Minería, Boletín 38:1–46

Augustsson C, Bahlburg H (2003) Active or passive continental margin? Geochemical and Nd isotope constraints of metasediments in the backstop of a pre-Andean accretionary wedge in southernmost Chile (46°30′–48°30′S). Geol Soc London Spec Publ 208:253–268

Augustsson C, Münker C, Bahlburg H, Fanning CM (2006) Provenance of late Palaeozoic metasediments of the SW South American Gondwana margin: a combined U–Pb and Hf-isotope study of single detrital zircons. Geol Soc London 163:983–995

Augustsson C, Bahlburg H (2008) Provenance of late Palaeozoic metasediments of the Patagonian proto-Pacific margin (southernmost Chile and Argentina). Int J Earth Sc 97:71–88

Avendaño V (2008) Petrología del Complejo Ofiolítico Tortuga, Magallanes Chile: evidencias de un metamorfismo Cretácico Inferior. Graduation thesis, Departamento de Geología, Universidad de Chile, 128 p

Barbeau DL, Olivero EB, Swanson-Hysell NL, Zahid KM, Murray KE, Gehrels GE (2009) Detrital-zircon geochronology of the eastern Magallanes foreland basin: implications for Eocene kinematics of the northern Scotia Arc and Drake Passage. Earth Planet Sc Lett 284:489–503

Barbeau DL, Davis JT, Murray KE, Valencia V, Gehrels GE, Zahid KM, Gombosi J (2010) Detrital-zircon geochronology of the metasedimentary rocks of northwestern Graham Land. Antarctic Sci 22:65–78

Betka P, Klepeis K, Mosher S (2015) Along-strike variation in crustal shortening and kinematic evolution of the base of a retroarc fold and thrust belt: Magallanes, Chile 53°S–54°S. Geol Soc Am Bull B31130-1

Bradshaw JD, Vaughan APM, Millar IL, Flowerdew MJ, Trouw RAJ, Fanning CM, Whitehouse MJ (2012) Permo-Carboniferous conglomerates in the Trinity Peninsula Group at View Point, Antarctic Peninsula: sedimentology, geochronology and isotope evidence for provenance and tectonic setting in Gondwana. Geol Mag 149(4):626–644

Bruhn RL, Stern CR, de Wit MJ (1978) Field and geochemical data bearing on the development of a Mesozoic volcano-tectonic rift zone and back-arc basin in southernmost South America. Earth Planet Sci Lett 41:32–46

Burton-Johnson A, Riley TR (2015) Autochthonous v. accreted terrane development of continental margins: a revised in situ tectonic history of the Antarctic Peninsula. Geol Soc London 172:822–835

Calderón M (2000) Metamorfismo de contacto en el margen oriental del batolito Sur-Patagónico (48°–50°30′S), Magallanes, Chile. Graduation thesis, Departamento de Geología, Universidad de Chile, 70 pp

Calderón M (2006) Petrogenesis and tectonic evolution of Late Jurassic bimodal magmatic suites (Sarmiento Complex) and migmatites (Puerto Edén Igneous Metamorphic Complex) in the Southern Patagonian Andes, Chile. Doctoral thesis, Universidad de Chile, Santiago

Calderón M, Fildani A, Hervé F, Fanning CM, Weislogel A, Cordani U (2007a) Late Jurassic bimodal magmatism in the northern seafloor remnant of the Rocas Verdes basin, Southern Patagonian Andes. Geol Soc London 164:1011–1022

Calderón M, Hervé F, Massonne H-J, Tassinari CG, Pankhurst RJ, Godoy E, Theye T (2007b) Petrogenesis of the Puerto Eden Igneous and Metamorphic Complex, Magallanes, Chile: late Jurassic anatexis of meta-greywackes and granitoid magma genesis. Lithos 93:17–38

Calderón M, Hervé F, Cordani U, Massonne H-J (2007c) Crust-Mantle interactions and the generation of silicic melts: insights from the Sarmiento Complex, southern Patagonian Andes. Rev Geol Chile 34:249–275

Calderón M, Fosdick JC, Warren C, Massonne H-J, Fanning CM, Fadel Cury L, Schwanethal J, Fonseca PE, Galaz G, Gaytán D, Hervé F (2012) The low-grade Canal de las Montañas Shear Zone and its role on the tectonic emplacement of the Sarmiento Ophiolitic Complex and Late Cretaceous Patagonian Andes orogeny, Chile. Tectonophysics 524–525:165–185

Calderón M, Prades CF, Hervé F, Avendaño V, Fanning CM, Massonne H-J, Theye T, Simonetti A (2013) Petrological vestiges of Late Jurassic-Early Cretaceous transition from rift to back-arc basin in southernmost Chile: new age and geochemical data from the Capitán Aracena, Carlos III and Tortuga Ophiolitic complexes. Geoch J 47:201–217

Castillo P, Fanning CM, Hervé F, Lacassie JP (2015) Characterisation and tracing of Permian magmatism in the south-western segment of the Gondwana margin; U–Pb age, Lu–Hf and O isotopic compositions of detrital zircons from metasedimentary complexes of northern Antarctic Peninsula and western Patagonia. Gondwana Res (in press)

Cawood PA (2005) Terra Australis Orogen: Rodinia breakup and development of the Pacific and Iapetus margins of Gondwana during the Neoproterozoic and Paleozoic. Earth Sci Rev 69:249–279

Cunningham WD (1994) Uplifted ophiolitic rocks on Isla Gordon, southernmost Chile: implications for the closure history of the Rocas Verdes marginal basin and the tectonic evolution of the Beagle Channel region. J South Am Earth Sci 7:135–147

Cunningham WD (1995) Orogenesis at the southern tip of the Americas: the structural evolution of the Cordillera Darwin metamorphic complex, southernmost Chile. Tectonophysics 244:197–229

Dalziel IWD, de Wit MJ, Palmer KF (1974) Fossil marginal basin in the southern Andes. Nature 250:291–294

Dalziel IWD (1981) Back-arc extension in the southern Andes: a review and critical reappraisal. Philos T Roy Soc A 300:319–335

Dalziel IWD, Brown RL (1989) Tectonic denudation of the Darwin metamorphic core complex in the Andes of Tierra del Fuego, southernmost Chile: implications for Cordilleran orogenesis. Geology 17:699–703

Dalziel IWD, Lawver LA, Murphy JB (2000) Plumes, orogenesis, and supercontinental fragmentation. Earth Planet Sci Lett 178:1–11

Davidson J, Mpodozis C, Godoy E, Hervé F, Muñoz N (1989) Jurassic accretion of a high buoyancy guyot in southernmost South America: the Diego Ramírez Islands. Rev Geol Chile 16:247–251

Dott RH, Winn RD, de Wit MJ, Bruhn RL (1977) Tectonic and sedimentary significance of Cretaceous Tekenika Beds of Tierra del Fuego. Nature 266:620–622

Douglass RC, Nestell MK (1976) Late Paleozoic foraminifera from Southern Chile. US Geological Survey Special Paper 858, 49 pp

Eagles G (2016) Tectonic reconstructions of the southernmost Andes and the Scotia Sea during the opening of the Drake Passage. In: Ghiglione MC (ed) Geodynamic Evolution of the Southernmost Andes. Springer Earth System Sciences, pp 75–108

Elthon D, Stern CR (1978) Metamorphic petrology of the Sarmiento ophiolite complex, Chile. Geology 6:464–468

Ernst WG (2009) Subduction-zone metamorphism, calc-alkaline magmatism, and convergent-margin crustal evolution. Gondwana Res 18:8–16

Fang Z, Boucot A, Covacevich V, Hervé F (1998) Discovery of late Triassic fossils in the Chonos Metamorphic Complex, southern Chile. Rev Geol Chile 25:165–173

Fanning CM, Hervé F, Pankhurst RJ, Rapela CW, Kleiman LE, Yaxley GM, Cas-tillo P (2011) Lu–Hf isotope evidence for the provenance of Permian detritus in accretionary complexes of western Patagonia and the northern Antarctic Peninsula region. J South Am Earth Sci 32:485–496

Faúndez V, Hervé F, Lacassie JP (2002) Provenance studies of prelate Jurassic metaturbidite successions of the Patagonian Andes, southern Chile. N Z J Geol Geophys 45:411–425

Fildani A, Hessler AM (2005) Stratigraphic record across a retroarc basin inversion: Rocas Verdes–Magallanes basin, Patagonian Andes, Chile. Geol Soc Am Bull 117:1596–1614

Fildani A, Cope TD, Graham SA, Wooden JL (2003) Initiation of the Magallanes foreland basin: timing of the southernmost Patagonian Andes orogeny revised by detrital zircon provenance analysis. Geology 31:1081–1084

Forsythe R (1981) Geological investigations of pre-Late Jurassic terranes in the Southernmost Andes. Doctoral thesis, Columbia University, 298 pp

Forsythe R (1982) The late Paleozoic and early Mesozoic evolution of southern South America: a plate tectonic interpretation. Geol Soc London 139:671–682

Forsythe R, Allen RB (1980) The basement rocks of Península Staines, Región XII, Province of Ultima Esperanza, Chile. Rev Geol Chile 10:3–15

Forsythe RD, Mpodozis C (1979) El Archipiélago de Madre de Dios, Patagonia Occidental, Magallanes: rasgos generales de la estratigrafía y estructura del basamento pre-Jurásico Superior. Rev Geol Chile 7:13–29

Forsythe RD, Mpodozis C (1983) Geología del Basamento pre-Jurásico Superior en el Archipiélago Madre de Dios, Magallanes, Chile. Servicio Nacional de Geología y Minería, Boletín 39, 63 p

Fosdick JC, Romans BW, Fildani A, Bernhardt A, Calderón M, Graham SA (2011) Kinematic evolution of the Patagonian retroarc fold-thrust belt and Magallanes foreland basin, Chile and Argentina, 51°30′S. Geol Soc Am Bull 123:1679–1698

Fosdick JC, Grove M, Graham SA, Hourigan JK, Lovera O, Romans BW (2015) Detrital thermochronologic record of foreland burial heating, sedimentary provenance, and orogenesis in Patagonia. Basin Res 27:546–572

Godoy E (1978) Observaciones en el Complejo Ofiolítico de Isla Milne Edwards-Cerro Tortuga (Isla Navarino) Magallanes-Chile. VII Congreso Geológico Argentino, Neuquén, Actas, II:625–636

Guillot MG (2016) Magmatic evolution of the southernmost Andes and its relation with subduction processes. In: Ghiglione MC (ed) Geodynamic Evolution of the Southernmost Andes. Springer Earth System Sciences, pp 37–74

Hervé F, Fanning CM (2000) Late Triassic zircons in meta-turbidites of the Chonos Metamorphic Complex, southern Chile. Rev Geol Chile 28:91–104

Hervé F, Fanning CM (2003) Early Cretaceous subduction of continental crust at the Diego de Almagro archipelago, Southern Chile. Episodes 26:285–289

Hervé F, Nelson E, Kawashita K, Suárez M (1981a) New isotopic ages and the timing of orogenic events in the Cordillera Darwin, southernmost Chilean Andes. Earth Planet Sci Lett 55:257–265

Hervé F, Mpodozis C, Davidson J, Godoy E (1981b) Observaciones estructurales y petrográficas en el basamento metamórfico del Archipiélago de los Chonos entre el Canal King y el Canal Ninualac, Aisén. Rev Geol Chile 13(14):3–16

Hervé M, Suárez M, Puig A (1984) The Patagonian batholith S of Tierra del Fuego, Chile: timing and tectonic implications. Geol Soc London 141:909–917

Hervé F, Aguirre L, Godoy E, Massonne H-J, Morata D, Pankhurst RJ, Ramírez E, Sepúlveda V, Willner A (1998) Nuevos antecedentes acerca de la edad y las condiciones P-T de los Complejos Metamórficos en Aysén, Chile. X Congreso Latinoamericano de Geología, Buenos Aires, II:134–137

Hervé F, Prior D, López G, Ramos VA, Rapalini A, Thomson S, Lacassie JP, Fanning CM (1999) Mesozoic blueschists from Diego de Almagro, Southern Chile. II South American Symposium on Isotope Geology, Córdoba, Actas, (Extended Abstract):318–321

Hervé F, Fanning CM, Pankhurst RJ (2003) Detrital zircon age patterns and provenance in the metamorphic complexes of Southern Chile. J South Am Earth Sci 16(107):123

Hervé F, Miller H, Pimpirev C (2006) Patagonia-Antarctica connections before Gondwana break-up. In: Fütterer DK, Dmaske D, Kleinschmidt G, Miller H, Tessensohn F (eds) Antarctica: contributions to global earth sciences. Springer, Berlin, Heidelberg, New York, pp 217–228

Hervé F, Pankhurst RJ, Fanning CM, Calderón M, Yaxley GM (2007a) The South Patagonian batholith: 150 my of granite magmatism on a plate margin. Lithos 97:373–394

Hervé F, Massonne H-J, Calderón M, Theye T (2007b) Metamorphic PT conditions of Late Jurassic rhyolites in the Magallanes fold and thrust belt, Patagonian Andes, Chile. J Iberian Geol 33:5–16

Hervé F, Calderón M, Faundez V (2008) The metamorphic complexes of the Patagonian and Fueguian Andes. Geol Acta 6:43–53

Hervé F, Calderón M, Fanning CM, Kraus S, Pankhurst RJ (2010a) SHRIMP chronology of the Magallanes basin basement, Tierra del Fuego: Cambrian plutonism and Permian high-grade metamorphism. Andean Geology 37(2):253–275

Hervé F, Fanning CM, Pankhurst RJ, Mpodozis C, Klepeis K, Calderón M, Thomson SN (2010b) Detrital zircon SHRIMP U–Pb age study of the Cordillera Darwin Metamorphic Complex of Tierra del Fuego: sedimentary sources and implications for the evolution of the Pacific margin of Gondwana. Geol Soc London 167:555–568

Hyppolito T (2010) Termobarometria, evolução tectono-metamórfica, e geoquímica de xistos azuis, rochas eclogíticas e litotipos associados da ilha Diego de Almagro, Patagônia Chilena. Master Thesis, Instituto de Geociências, Universidade de São Paulo, São Paulo.

Jokat W, Boebel T, König M, Meyer U (2003) Timing and geometry of early Gondwana breakup. J Geophys Res 108:2428

Katz HR (1964) Some new concepts on geosynclinal development and mountain building at the southern end of South America. In: Proceedings of 22nd international geological congress, India New Delhi. vol 4, pp 242–255
Klepeis K, Betka P, Clarke G, Fanning CM, Hervé F, Rojas L, Mpodozis C, Thomson SN (2010) Continental underthrusting and obduction during the Cretaceous closure of the Rocas Verdes rift basin, Cordillera Darwin, Patagonian Andes. Tectonics 29:TC3014
Kohn MJ, Spear F, Dalziel IWD (1993) Metamorphic P–T paths from Cordillera Darwin, a core complex in Tierra del Fuego, Chile. J Petrol 34:519–542
Kohn MJ, Spear FS, Harrison TM, Dalziel IWD (1995) 40Ar/39Ar geochronology and P-T-t paths from the Cordillera Darwin Metamorphic Complex, Tierra del Fuego, Chile. J Metam Geol 13:251–270
Lacassie JP (2003) Estudio de la Proveniencia Sedimentaria de los Complejos Metamórficos de los Andes Patagónicos (46°–51°S), mediante la aplicación de redes neuronales e isótopos estables. Doctoral thesis, Universidad de Chile, 119 pp
Lacassie JP, Hervé F, Roser B (2006) Sedimentary provenance study of the post-Early Permian to pre-early cretaceous metasedimentary duqye de York Complex, Chile. Rev Geol Chile 33:199–219
Lagally U (1975) Geologische Untersuchungen mi Gebiet Lake General Carrera - Lake Cochrane, Prov. Aysen/Chile unter besorender Berücksichtigung des Grundgebirges und seiner tektonik. Doctoral thesis, Universität München, 143 pp
Ling HY, Forsythe RD, Douglass CR (1985) Late Paleozoic microfaunas from southernmost Chile and their relation to Gondwanaland forearc development. Geology 13:357–360
Malkowski MA, Grove M, Graham SA (2015) Unzipping the Patagonian Andes—Long-lived influence of rifting history on foreland basin evolution. Lithosphere. doi:10.1130/L489.1
Maloney KT, Clarke GL, Klepeis KA, Fanning CM, Wang W (2011) Crustal growth during back-arc closure: cretaceous exhumation history of Cordillera Darwin, southern Patagonia. J Metam Geol 29:649–672
Mantle GW, Collins WJ (2008) Quantifying crustal thickness variations in evolving orogens: correlation between arc basalt composition and Moho depth. Geology 36:87–90
Millar IL, Pankhurst RJ, Fanning CM (2002) Basement chronology of the Antarctic Peninsula: recurrent magmatism and anatexis in the Palaeozoic Gondwana margin. Geol Soc London 159:145–157
Miller CA, Barton M, Hanson R, Flemming T (1994) An Early Cretaceous volcanic arc-marginal basin transition zone, Peninsula Hardy, southernmost Chile. J Volc Geoth Res 63:33–58
Moreira P, Fernández R, Hervé F, Fanning CM, Schalamuk IA (2013) Detrital zircons U–Pb SHRIMP ages and provenance of La Modesta Formation, Patagonia Argentina. J South Am Earth Sci 47:32–46
Nelson E, Dalziel IWD, Milnes AG (1980) Structural geology of the Cordillera Darwin: collision style orogenesis in the southernmost Chilean Andes. Eclogae Geologicae Helveticae 73:727–751
Olivares B, Cembrano J, Hervé F, López G, Prior D (2003) Geometría y cinemática de la Zona de Cizalle Seno Arcabuz, Andes patagónicos, Chile. Rev Geol Chile 30:39–52
Otzen G (1987) Geología Regional de la isla Capitán Aracena y Clarence, Magallanes, Chile. Informe Técnico ENAP (inedit report)
Pankhurst RJ, Weaver SD, Hervé F, Larrondo P (1999) Mesozoic-Cenozoic evolution of the North Patagonian Batholith in Aysén, Southern Chile. Geol Soc London 156:673–694
Pankhurst RJ, Riley TR, Fanning CM, Kelley SP (2000) Episodic silicic volcanism in Patagonia and the Antarctic Peninsula: Chronology of magmatism associated with the break-up of Gondwana. J Petrol 41:603–625
Pankhurst RJ, Rapela CW, Loske WP, Márquez M, Fanning CM (2003) Chronological study of the pre-Permian basement rocks of southern Patagonia. J South Am Earth Sci 16:27–44
Pankhurst RJ, Rapela CW, Fanning CM, Márquez M (2006) Gondwanide continental collision and the origin of Patagonia. Earth Sci Rev 76:235–257

Parra F (2015) Análisis de proveniencia y ambiente de sedimentación de las un-idades sedimentarias de la isla Diego de Almagro (51°30′S). Graduation thesis, Departamento de Geología, Universidad de Chile, 85 pp

Permuy Vidal C, Moreira P, Guido DM, Fanning CM (2014) Linkages between the southern Patagonia pre-Permian basements: new insights from detrtital zircons U-Pb SHRIMP ages from the Cerro Negro District. Geol Acta 12:137–150

Pimpirev C, Miller H, Hervé F (1999) Preliminary results on the lithofacies and palaeoenvironmental interpretation of the Paleozoic turbidite sequence in Chonos Archipielago, Southern Chile. Comunicaciones 48–49:3–12

Prades CF (2008) Petrología y metamorfismo de las rocas basálticas en la isla Capitán Aracena, isla Carlos III y Estero La Pera, Region de Magallanes, Chile. Graduation thesis, Departamento de Geología, Universidad de Chile, 135 p

Quezada A (2010) Proveniencia sedimentaria y ambiente deposicional de unidades atribuidas a los complejos Duque de York y Denaro, Patagonia XII Región, Chile. Graduation thesis, Departamento de Geología, Universidad de Chile, 126 pp

Ramírez-Sánchez E (2002) Geotermobarometría en metapelitas de complejos metamórficos de Aysen, Chile. Doctoral thesis, Universidad de Chile, 143 pp

Ramírez-Sánchez E, Hervé F, Kelm U, Sassi R (2005) P–T conditions of metapelites from metamorphic complexes in Aysen, Chile. J South Am Earth Sci 19:373–386

Ramos V (2010) The Grenville-age basement of the Andes. J South Am Earth Sci 29:77–91

Rapalini AE, Calderón M, Singer S, Hervé F, Cordani U (2008) Tectonic implications of a paleomagnetic study of the Sarmiento Ophiolitic Complex, southern Chile. Tectonophysics 452:29–41

Riley TR, Flowerdew MJ, Whitehouse MJ (2012) U–Pb ion-microprobe zircon geochronology from the basement inliers of eastern Graham Land, Antarctic Peninsula. Geol Soc London 169:381–393

Romans BW, Fildani A, Graham SA, Hubbard SM, Covault JA (2010) Importance of predecessor basin history on sedimentary fill of a retroarc foreland basin: provenance analysis of the Cretaceous Magallanes basin, Chile (50°52′S). Basin Res 22:640–658

Saunders AD, Tarney J, Stern CR, Dalziel IWD (1979) Geochemistry of Mesozoic marginal basin floor igneous rocks from southern Chile. Geol Soc Am Bull 90:237–258

Schilling M, Conceição RV, Mallmann G, Koester E, Kawashita K, Hervé F, Mo-rata D, Motoki A (2005) Spinel-facies mantle xenoliths from Cerro Redondo, Argentine Patagonia: petrographic, geochemical, and isotopic evidence of interaction between xenoliths and host basalt. Lithos 82:485–502

Sepúlveda FA (2004) Metamorpfismo de bajo grado en rocas del Complejo Denaro, archipiélagoMadre de Dios, XII Región. Graduation thesis, Departamento de Geología, Universidad de Chile, 49 pp

Sepúlveda FA, Hervé F, Calderón M, Lacassie JP (2008) Petrology of igneous and metamorphic units from the allochthonous Madre de Dios Terrane, Magallanes, Chile. Gondwana Res 13:238–249

Sepúlveda FA, Palma-Heldt S, Hervé F, Fanning CM (2010) Permian depositional age of metaturbidites of the Duque de York Complex, southern Chile: U-Pb SHRIMP data and palynology. Andean Geol 37:375–397

Sernageomin (2003) Mapa Geológico de Chile: versión digital. publicación geológica digital, No. 4, 2003. CDROM, versión 1.0, 2003. Base Geológica escala 1:1.000.000. Gobierno de Chile, Servicio Nacional de Geología y Mi-nería, Subdirección Nacional de Geología

Söllner F, Miller H, Hervé M (2000) An Early Cambrian granodiorite age from the pre-Andean basement of Tierra del Fuego (Chile): the missing link between South America and Antarctica? J South Am Earth Sci 13:163–177

Spear FS, Kohn MJ, Cheney JT (1999) P-T paths from anatectic pelites. Contrib Miner Petrol 134:17–32

Stern CR (1979) Open and closed system igneous fractionation within two Chilean ophiolites and the tectonic implications. Contrib Miner Petrol 68:243–258

Stern CR (1980) Geochemistry of Chilean ophiolites: evidence for the compositional evolution of the mantle source of back-arc basin basalts. J Geophys Res 85:955–966
Stern CR (1991) Isotopic composition of Late Jurassic and Early Cretaceous mafic igneous rocks from the Southernmost Andes: implications for sub-Andean mantle. Rev Geol Chile 18:15–23
Stern CR, de Wit MJ (2003) Rocas Verdes ophiolites, southernmost South America: remnants of progressive stages of development on oceanic-type crust in a continental margin back-arc basin. In: Dilek Y, Robinson PT (eds) Ophiolites in earth history, vol 218, pp 1–19 (Geol Soc London Sp Pub)
Stern CR, de Wit MJ, Lawrence JR (1976) Igneous and metamorphic processes associated with formation of Chilean ophiolites and their implications for ocean floor metamorphism, seismic layering and magnetism. J Geophys Res 81:4370–4380
Suárez M, Pettigrew TH (1976) An upper Mesozoic Island-arc-back-arc system in the southern Andes and South Georgia. Geol Mag 113:305–328
Sun S, McDonough W (1989) Chemical and isotopic systematics of oceanic basalts: implications for mantle composition and processes. Geol Soc London Spec Publ 42:313–345
Thomson SN, Hervé F (2002) New time constraints for the age of metamorphism at the ancestral Pacific Gondwana margin of Southern Chile. Rev Geol Chile 29:255–271
Valdés A (2005) Petrología de la aureola metamórfica de contacto del plutón monzogranítico Río Murta, XI región de Aysén, Chile. Graduation thesis, Universidad de Concepción, 143 pp
Watters W (1964) Geological work at Puerto Edén, Wellington Island, southern Chile. Transac Roy Soc NZ (Geology) 2:155–168
Wendt AS, Vaughan APM, Tate A (2008) Metamorphic rocks in the Antarctic Peninsula region. Geol Mag 145:655–676
Willner A, Hervé F, Massonne H-J (2000) Mineral chemistry and pressure–temperature evolution of two contrasting high-pressure–low-temperature belts in the Chonos Archipelago, Southern Chile. J Petrol 4:309–330
Willner AP, Hervé F, Thomson SN, Massonne H-J (2004) Converging PT-paths of Mesozoic HP-LT metamorphic units (Diego de Almagro Island, Southern Chile, 51°30′S): evidence for juxtaposition during late shortening of an active continental margin. Miner Petrol 81:43–84
Willner AP, Sepúlveda FA, Hervé F, Massonne H-J, Sudo M (2009) Conditions and timing of pumpellyite–actinolite-facies metamorphism in the Early Mesozoic frontal accretionary prism of the Madre de Dios Archipelago (Latitude 50°20′S; Southern Chile). J Petrol 50:2127–2155
Wilson TJ, Hanson RE, Grunow AM (1989) Multistage melange formation within an accretionary complex, Diego Ramírez Islands, Southern Chile. Geology 17:11–14

Magmatic Evolution of the Southernmost Andes and Its Relation with Subduction Processes

Mauricio González Guillot

Abstract The magmatic arc in the southernmost Andes is represented by the Fuegian Batholith and satellite plutons, plus minor volcanic and volcaniclastic sequences. Magmatism was active from the Late Jurassic to the Miocene in response to subduction of the Pacific plates beneath South America. The batholith was constructed episodically and includes a calc-alkaline trench-side arc and shoshonitic rear-arc suites. The oldest components are minor S-like granitoids; recent volcanism is very scarce. The volcanic front shifted, and the composition of magmas changed several times due to a variable subduction configuration. Major tectonic events included the opening and closure of the Rocas Verdes marginal basin in the Jurassic–Cretaceous and oroclinal bending during the latest Cretaceous–Paleogene. A well-established slab shallowing event initiated in the mid-Cretaceous, causing a cratonward migration of the volcanic front, the widening of the arc, and inception of potassic magmatism in the rear-arc. This event was associated with marginal basin inversion, regional peak metamorphism, and orogenic deformation. Different timing and position of rear-arc suites with respect to the volcanic front in the Fuegian and Southern Patagonian Andes may have been influenced by differential development of the marginal basin, which was wider to the south. The vanishing of magmatism in the Miocene may have responded to highly oblique convergence plus buoyant subduction.

Keywords Fuegian Batholith · Patagonian Batholith · Trench-side arc · Rear-arc · Subduction parameters · Fuegian Andes · Rocas Verdes basin

---

M.G. Guillot (✉)
Centro Austral de Investigaciones Científicas CADIC-CONICET,
and Instituto de Ciencias Polares, Ambiente y Recursos Naturales,
Universidad Nacional de Tierra del Fuego, Tierra del Fuego, Ushuaia, Argentina
e-mail: g_guillot@cadic-conicet.gob.ar

© Springer International Publishing Switzerland 2016
M.C. Ghiglione (ed.), *Geodynamic Evolution of the Southernmost Andes*,
Springer Earth System Sciences, DOI 10.1007/978-3-319-39727-6_3

1 Introduction

The Fuegian Andes constitutes the segment of the Andean Cordillera that runs east–west at the southernmost tip of South America (Fig. 1). The Mesozoic–Cenozoic magmatic arc that forms part of this belt encompasses several suites related to subduction of the Pacific plates beneath South America.

The main arc is composed of calc-alkaline plutons that make up the bulk of the Fuegian Batholith, along with volcanic rocks of the Hardy and Packsaddle Formations (Fig. 1) (Hervé et al. 1984; Puig et al. 1984; Suárez et al. 1985; Miller et al. 1994). To the rear-arc in Tierra del Fuego, small isolated, shoshonitic plutons crop out, grouped in the Fuegian Potassic Magmatism (FPM) (González Guillot et al. 2009), and other minor high-K, calc-alkaline hypabyssal intrusions, the Ushuaia Peninsula Andesites (UPA) (González Guillot et al. 2011). The main arc and rear-arc units represent magmas generated in a trench-side and back-arc side, respectively, similar to those formed in dual volcanic chains (c.f. Tatsumi and Eggins 1995).

The construction of the batholithic backbone was episodic in both the Fuegian and Patagonian Andes (e.g., Hervé et al. 2007a, 1984). However, while magmatism on the latter remained approximately on the same general axis (Hervé et al. 2007a), the Fuegian Batholith is wider and registers a more evident shift of the volcanic front during time (Hervé et al. 1984) as a consequence of changing subduction parameters. For instance, in the mid-Cretaceous, an episode of slab flattening took place (Stern et al. 1991; Stern and De Wit 2003; González Guillot et al. 2009, 2011) which produced a landward migration of the calc-alkaline volcanic front and the inception of potassic, rear-arc magmatism (Fig. 1). This process was associated with crustal deformation during the closure of the Rocas Verdes marginal basin (e.g., Dalziel et al. 1974; Stern et al. 1991). Accordingly, Folguera and Ramos (2011) proposed that events of slab flattening and steepening of the subducted plate have been repetitive along discrete sectors in the southernmost Andes since the last 100 Myr.

Hence, the Fuegian Andes provides an excellent scenario to study across-arc chemical and temporal variations of magmatic rocks in continental arc settings, from calc-alkaline units at the volcanic front, to potassic plutons at the farthest rear-arc. It has the additional attribute of a marginal basin developed during the early stages of arc construction. However, no thorough research on the evolution of arc magmatism has been done so far in this segment of the Andes, mainly because of the remoteness and inaccessibility of the region.

Fig. 1 Regional geological map of Tierra del Fuego and South Patagonia. Abbreviations: *MFFS* Magallanes–Fagnano fault system, *FTB* fold and thrust belt front, *B* Burney volcano, *C* Cook volcano, *PA* Punta Arenas, *P* Porvenir, *RG* Río Grande, *U* Ushuaia. The numbers refer to rear-arc intrusions: *1* and *2* UPA at Ushuaia Peninsula and Trapecio Hill, respectively, 3–6 FPM plutons: Ushuaia (*3*), Moat (*4*), Jeu-Jepén (*5*), and Kranck (*6*). *7* corresponds to a monzonitic pluton correlated with the FPM. *Sources* Hervé et al. (1984), Suárez et al. (1985), Sernageomin (2003), Olivero and Malumián (2007), González Guillot et al. (2009), Anguita (2010), Klepeis et al. (2010), and McAtamney et al. (2011). The *black dots* and *white squares* and *diamonds* are from Hervé et al. (2007a) ▶

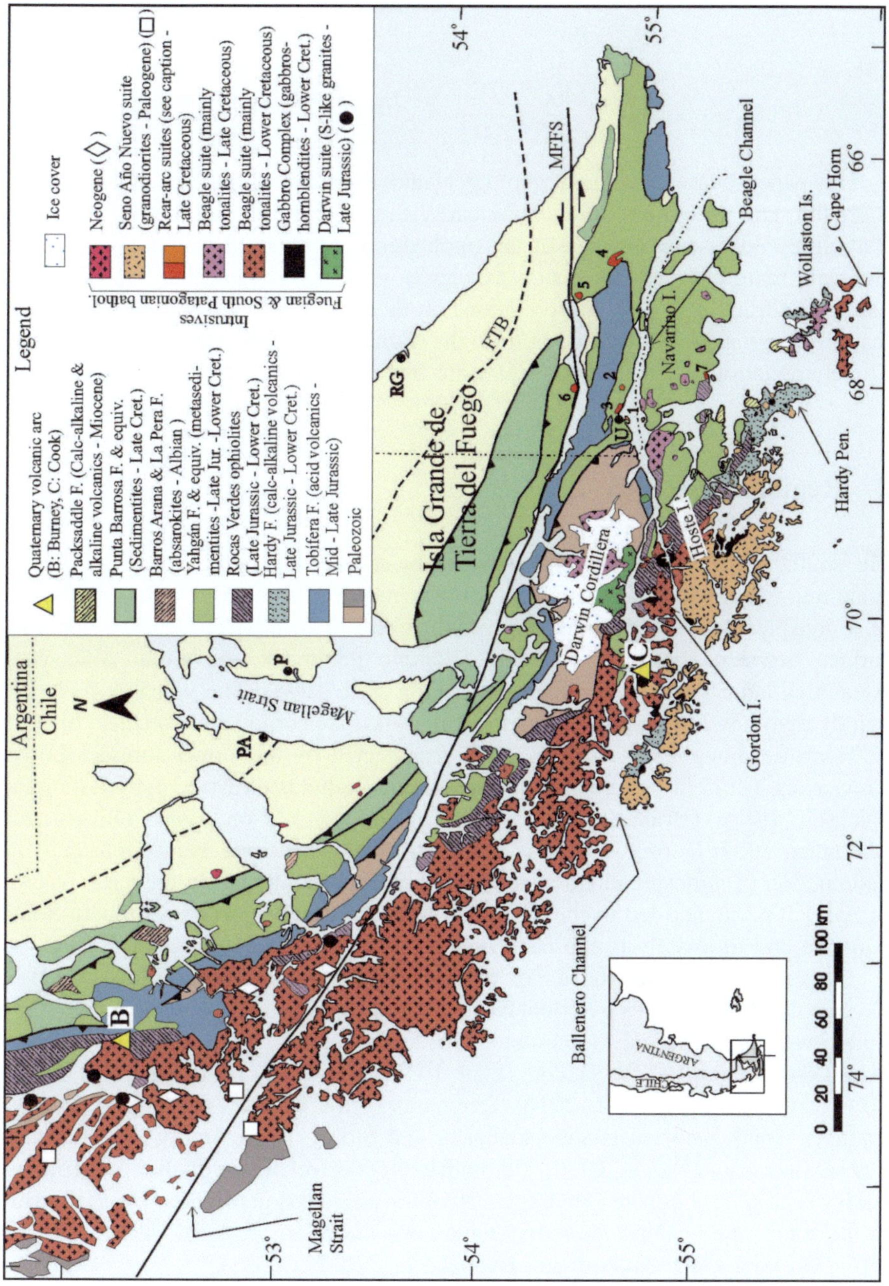
Legend
Quaternary volcanic arc (B: Burney, C: Cook)
Packsaddle F. (Calc-alkaline & alkaline volcanics - Miocene)
Punta Barrosa F. & equiv. (Sedimentites - Late Cret.)
Barros Arana & La Pera F. (absarokites - Albian)
Yahgán F. & equiv. (metasedimentites -Late Jur. -Lower Cret.)
Rocas Verdes ophiolites (Late Jurassic - Lower Cret.)
Hardy F. (calc-alkaline volcanics - Late Jurassic - Lower Cret.)
Tobífera F. (acid volcanics - Mid - Late Jurassic)
Paleozoic
Ice cover
Fuegian & South Patagonian bathol. Intrusives
Neogene (◇)
Seno Año Nuevo suite (granodiorites - Paleogene) (□)
Rear-arc suites (see caption - Late Cretaceous)
Beagle suite (mainly tonalites - Late Cretaceous)
Beagle suite (mainly tonalites - Lower Cretaceous)
Gabbro Complex (gabbros-hornblendites - Lower Cret.)
Darwin suite (S-like granites - Late Jurassic) (●)
Argentina
Chile
N
Magellan Strait
PA
P
RG
Isla Grande de Tierra del Fuego
FTB
MFFS
U
Darwin Cordillera
Hoste I.
Navarino I.
Beagle Channel
Gordon I.
Hardy Pen.
Wollaston Is.
Cape Horn
Ballenero Channel
Magellan Strait
CHILE
ARGENTINA
0 20 40 60 80 100 km
53°
54°
55°
74°
72°
70°
68°
66°
1
2
3
4
5
6
7

Table 1 List of abbreviations used in the text

FB	Fuegian Batholith
MPF	Fuegian Potassic Magmatism
PB	Patagonian Batholith
SPB	Southern Patagonian Batholith
UPA	Ushuaia Peninsula Andesites

This paper reviews the major geological and geochemical features of the Fuegian Batholith and satellite plutons, compares them with the Southern Patagonian Batholith, and discusses some of the implications of changing subduction parameters and marginal basin inversion to magma genesis, emplacement, and differentiation. Additionally, new interpretations on the timing and tectonic setting during rear-arc magmatism are provided, in light of the new U/Pb zircon ages available. The abbreviations used in this chapter are listed in Table 1.

2 Regional Geology

The southernmost Andes are characterized by several magmatic episodes occurred since mid-Mesozoic times (see Fig. 1 for location of lithological units). Remnants of a conspicuous bimodal volcanic province are present in the surface and subsurface between Patagonia and the Antarctic Peninsula, the Chon Aike–West Antarctica large igneous province (Pankhurst et al. 1998). The volcanic rocks are mainly rhyolites and basalts, but flows of intermediate composition are also present. In Tierra del Fuego, this magmatism is represented by the Upper Jurassic–Lower Cretaceous Tobífera or Lemaire Formation and includes effusive, pyroclastic, and epiclastic facies (Bruhn et al. 1978; Hanson and Wilson 1991; Olivero and Malumián 2007; González Guillot et al. 2016). This volcanic event represents the culmination of widespread extensional deformation that began in the Late Triassic or Early Jurassic and led to the formation of a number of basins in southern South America and ultimately to the fragmentation of Gondwana (Kay et al. 1989).

The extension led to the development of the ensialic, Upper Jurassic–Lower Cretaceous Rocas Verdes marginal basin, floored by oceanic-type mafic rocks, along a narrow strip bordering the continental margin between 51°S and 56°S (Dalziel et al. 1974; Suárez 1977b; Saunders et al. 1979; Stern 1979, 1980; Stern and De Wit 2003; Calderón et al. 2007, 2016). Remnants of the Rocas Verdes basin are also found in South Georgia Island (Alabaster and Storey 1990; Storey and Alabaster 1991; Mukasa and Dalziel 1996), although a direct correlation with the southernmost Andes is motive of debate (see Eagles 2016 for a discussion on the subject). Further to the south, the oceanic floor may have been linked to the Weddell Sea (de Wit 1977; Grunow 1993; Mukasa and Dalziel 1996; Barker 2001).

The magmatic arc is represented by the Patagonian and Fuegian batholiths and the Hardy Formation volcanics (Suárez et al. 1985; Miller et al. 1994; Hervé et al. 2007a). Other scarce volcanic units in Tierra del Fuego, the Miocene Packsaddle

Formation and the Recent Cook deposits, are also related to subduction (Puig et al. 1984; Suárez et al. 1985). The Packsaddle volcanics include calc-alkaline (21 Ma) and alkaline (18 Ma) terms.

The Hardy Formation consists of Upper Jurassic–Lower Cretaceous pyroclastic and volcaniclastic rocks and lavas of calc-alkaline affinity (Suárez et al. 1985; Miller et al. 1994). Toward the back-arc, this unit interfingers with the contemporaneous Yahgán Formation (Fig. 1), which represents the volcaniclastic infill of the marginal basin (Suárez and Pettigrew 1976). Northward, the Yahgán Formation is laterally replaced by black mudstones of the Beauvoir-La Paciencia Formations (Olivero and Martinioni 2001; Olivero and Malumián 2007).

The PB extends from 39°S to 56°S and is usually subdivided into three segments: the North Patagonian Batholith, north of 46°30'S, the Southern Patagonian Batholith (SPB, 46°30′–53°00′S), and the Fuegian Batholith (FB), south of 53°S. The FB and SPB were emplaced episodically between 157 and 15 Ma (Halpern 1973; Hervé et al. 1984, 2007a; Suárez et al. 1985, 1986; Bruce et al. 1991; Martin et al. 2001), while the locus of magmatism changed many times during its construction.

The closure of the marginal basin encompassed crustal deformation, peak regional metamorphism, and uplift (Dalziel et al. 1974; Dott et al. 1977; Bruhn 1979; Hervé et al. 1981; Nelson 1982; Dalziel 1986; Cunningham 1995). Several authors (e.g., Gealey 1980; Nelson et al. 1980; Dalziel 1981; Cunningham 1995; Kraemer 2003; Fildani and Hessler 2005; Galaz et al. 2005; Hervé et al. 2007b) suggested, and recently Klepeis et al. (2010) and Maloney et al. (2011) emphasized, a model of basin inversion by underthrusting of the leading edge of the continent under the magmatic arc, while subduction of the proto-Pacific margin under South America was still active. This occurred prior to 86 Ma in Cordillera Darwin (Klepeis et al. 2010) or 80 Ma at 52°S (Calderón et al. 2012).

Deformation and landward propagation of the Andean fold and thrust belt were mainly controlled by thrusting in a compressional regime (Klepeis 1994a; Klepeis et al. 2010; Maloney et al. 2011; Likerman et al. 2013); although alternative interpretations hold a transpressional closure of the Rocas Verdes basin during a period of oblique convergence, with partitioned strike-slip component localized along the Canal Beagle (Cunningham 1993, 1995; Ghiglione and Ramos 2005).

The $\sim$90° bending of the southernmost Andes was attributed either to true oroclinal rotation (*Patagonian Orocline* sensu Carey 1955) initiated in the Late Cretaceous (Cunningham et al. 1991; Kraemer 2003; Rapalini et al. 2015), or to an original curved orogen (Diraison et al. 2000), or to a progressive arc initiated by rotation during the closure of the Rocas Verdes basin and culminated mainly as a primary arc during the final stages of orogenic deformation (Ghiglione and Cristallini 2007; Poblete et al. 2014; Torres Carbonell et al. 2014; Maffione et al. 2010, 2015).

In the late Paleogene, convergence in the southernmost Andes declined and sinistral strike-slip and oblique-slip faulting dominated (Cunningham 1993, 1995; Klepeis 1994b; Klepeis and Austin 1997; Diraison et al. 2000; Ghiglione and Ramos 2005; Gombosi et al. 2009; Klepeis et al. 2010). Wrenching initiated along the Canal Beagle and in the Neogene shifted northward to the Magallanes–Fagnano fault system (Fig. 1) (e.g., Klepeis 1994b; Menichetti et al. 2008). This transition coincides with the formation of the South America—Scotia transform boundary (Lodolo et al. 2003; Rossello 2005; Eagles 2016). Associated with this change in plate boundary configuration, magmatism in the arc vanished during the Neogene in Tierra del Fuego, reappearing only as minor manifestations in the Quaternary. The present volcanic arc is composed of six Quaternary volcanoes, all of them but the southernmost one, located to the east of the batholith (Fig. 1) (Puig et al. 1984; Stern and Kilian 1996). According to Rubio et al. (2000), subduction between 52°S and 57°S is still active or has recently ceased and is highly oblique, with a relative convergence rate of 1.3 mm/a (Pelayo and Wiens 1989).

3 The Fuegian Batholith and Satellite Plutons

The FB is a multiple intrusion dominated by granodiorites and tonalities (Kranck 1932; Suárez 1977a; Hervé et al. 1984; Suárez et al. 1985; Nelson et al. 1988). It can be divided into (i) the S-like granitoids of Cordillera Darwin (Sect. 3.1), (ii) the I-type calc-alkaline plutons of the main arc (Sect. 3.2), and (iii) the shoshonitic and other potassic intrusions of the rear-arc (Sect. 3.3).

3.1 The Jurassic S-like Granitoids

These granitoids are grouped in the Darwin suite, which is dominated by orthogneisses intruded in Paleozoic basement rocks and Mesozoic cover sequences on the continental flank of the batholith in the Fuegian and Patagonian Andes (Nelson et al. 1980; Hervé et al. 1981a, 2007a) (Fig. 1). The orthogneisses consist of biotite granites, granodiorites, and quartz monzonites, with minor garnet, muscovite, and amphibole. Myrmekite and accessory allanite are also common. The rocks range from weakly foliated to mylonitic and are intruded by mafic dykes of the ophiolitic suite and the undeformed Beagle suite granitoids (Nelson et al. 1980). Associated amphibole gabbros provide the suite a bimodal nature (Hervé et al. 2007a).

The granites are Upper Jurassic, with U/Pb zircon ages of 164–153 Ma (Mukasa and Dalziel 1996; Klepeis et al. 2010) and an initial $^{87}Sr/^{86}Sr$ ratio of 0.7087 (Hervé et al. 1981). In the SPB, these plutons were dated in 157–144 Ma, with initial $^{87}Sr/^{86}Sr$ ratios of 0.7068–0.7092 and negative εNdt values of −4.2 to −7.0 (Hervé et al. 2007a). The age and isotopic composition suggest that they are the plutonic equivalents of the voluminous Tobífera rhyolites (Bruhn et al. 1978; Hervé et al. 1981, 2007a).

The silicic (76 % SiO_2), low CaO (0.6 %) and Sr (11–58 ppm), and peraluminous nature of the orthogneisses, together with their high initial $^{87}Sr/^{86}Sr$ ratios, are compatible with an origin either by (i) anatexis of the lower crust triggered by the heat provided from under-plated mantle-derived basalts or by (ii) crustal contamination of I-type magmas (Bruhn et al. 1978; Nelson et al. 1980; Suárez et al. 1990; Hervé et al. 2007a). According to Suárez et al. (1990), they are somewhat intermediate in their nature between true S-type and I-type granites generated in an extensional setting during fragmentation of Gondwana, adjacent to a subduction zone. Furthermore, the rhyolites of the Tobífera Formation are high K_2O, calc-alkaline, and peraluminous, a distinctive feature of rhyolites of extensional environments involving anatexis of continental crust (Gust et al. 1985).

3.2 *The Main Arc*

Included here are the I-type, calc-alkaline plutons that form the bulk of the batholith. These plutons, along with the Hardy Formation and Packsaddle volcanics, represent the trench-side suites of the magmatic arc.

3.2.1 Lithology

The plutons south of Canal Beagle were divided into three different suites considering age, petrography, fabric, and contact relationships with the hosting rock (Fig. 1) (Hervé et al. 1984; Suárez et al. 1985): (i) a Gabbroic assemblage (Lower Cretaceous), (ii) the Beagle suite (Lower to Upper Cretaceous), and (iii) the Seno Año Nuevo suite (Paleogene).

The Gabbroic assemblage was separated from other suites rather arbitrarily to group gabbros, gabbronorites, diorites, and hornblendites exposed as relicts within later granitoids. They form bodies of up to 30 km^2 scattered along the FB. Contact relationships with other granitoids indicate that they are older in all cases. The suite consists mainly of hornblende gabbro, exhibits cumulate textures, and shows no penetrative foliation, although mylonitic structures are occasionally found. The mineralogy includes combinations of labradorite, hornblende, clinopyroxene, orthopyroxene, and rare olivine and biotite (Figs. 2 and 3).

These rocks have yielded hornblende K/Ar ages from 141 to 103 Ma, i.e., embracing the pretectonic to syntectonic stages of the Late Cretaceous Andean orogeny (Hervé et al. 1984). Basic synplutonic dykes and inclusions in the Beagle granitoids, interpreted as part of the Gabbroic assemblage, would extend this basic magmatism to the Late Cretaceous.

The Beagle suite dominates the FB (Fig. 1) and consists of tonalites and granodiorites, with minor quartz monzodiorites and quartz diorites (Fig. 4a). Except at Wollaston Islands, the granitoids show synmagmatic penetrative foliation formed during emplacement. Mineral composition is andesine, alkali feldspar, quartz, hornblende (8 %), and biotite (7 %) (Fig. 2). Clinopyroxene is extremely scarce and appears mantled by hornblende. Accessory phases are apatite, titanite, opaque, zircon, and allanite.

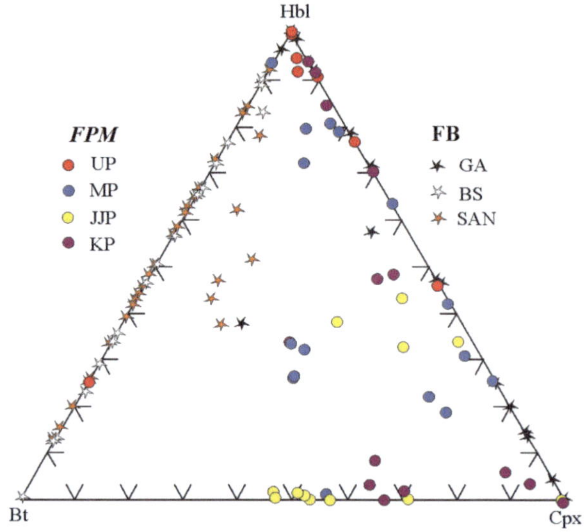

Fig. 2 Mafic mineral composition of trench-side arc and rear-arc suites in Tierra del Fuego (Suárez et al. 1985; González Guillot 2009; González Guillot et al. 2012). *UP* Ushuaia pluton, *MP* Moat pluton, *JJP* Jeu-Jepén pluton, *KP* Kranck pluton, *GA* Gabbroic assemblage, *BS* Beagle suite, *SAN* Seno Año Nuevo suite

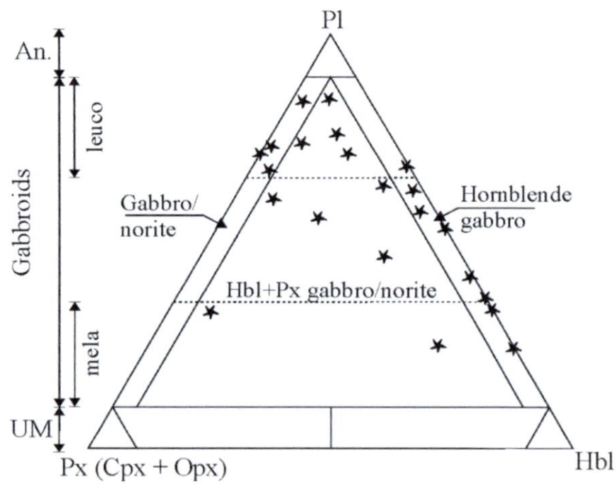

Fig. 3 Modal composition of the Gabbroic assemblage (Suárez et al. 1985)

K/Ar and Rb/Sr mineral ages, and scarce $^{40}Ar/^{39}Ar$ and U/Pb ages range between 117 and 74 Ma (Halpern and Rex 1972; Halpern 1973; Hervé et al. 1984; Suárez et al. 1985, 1986; Mukasa et al. 1988; Kohn et al. 1995; Klepeis et al. 2010; McAtamney et al. 2011). Lower and Upper Cretaceous plutons are concentrated to the south and north of the FB, respectively (Fig. 1).

The Seno Año Nuevo suite is exposed along the Pacific side of the batholith (Fig. 1). The plutons are non-foliated and range from quartz diorites to granites, with tonalite, granodiorite, and quartz monzodiorites as the most common lithologies (Fig. 4a). They are petrographically comparable with the Beagle suite, but somewhat more K-feldspar-rich (Fig. 2) and younger. K/Ar and Rb/Sr mineral ages are in the range 60–34 Ma (Hervé et al. 1984; Suárez et al. 1985).

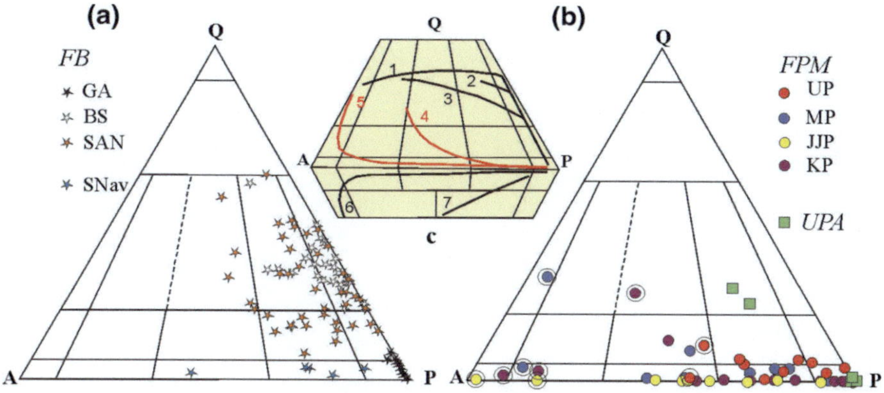

Fig. 4 Modal composition of trench-side arc and rear-arc. **a** Gabbroic assemblage (*GA*), Beagle (*BS*), and Seno Año Nuevo (*SAN*) suites (Suárez et al. 1985) and a monzonitic pluton of south Navarino Island (*SNav*) (Katz and Watters 1966). **b** FPM plutons (*UP* Ushuaia, *MP* Moat, *JJP* Jeu-Jepén, *KP* Kranck) (González Guillot 2009; González Guillot et al. 2012), and UPA (González Guillot et al. 2011). The *encircled* samples in the FPM correspond to late-stage veins. **c** Magmatic trends of Lameyre (1987), Lameyre and Bowden (1982), and Lameyre and Bonin (1991): *1* tholeiitic, *2* calc-alkaline tonalitic–trondhjemitic, *3* calc-alkaline granodioritic, *4* monzonitic, *5* alkaline (silica-saturated), *6* alkaline (silica-undersaturated), and *7* alkaline (silica strongly undersaturated)

3.2.2 Geochemistry

The geochemical data of the FB are scarce and restricted to major elements (Suárez 1977a; Suárez et al. 1985) and to an incomplete suite of trace elements obtained ~40 years ago by means of XRF (Suárez 1977a).

The three suites described are subalkaline (Fig. 5a) and metaluminous, typical of Cordilleran batholiths. The Beagle and Seno Año Nuevo suites define medium- to high-K, calc-alkaline trends, whereas the Gabbroic assemblage has a more ambiguous composition, straddling the low-K tholeiitic and medium-K dividing line (Fig. 6). The K_2O/Na_2O ratio varies on average from 0.24 (Gabbroic assemblage), through 0.35 (Beagle) to 0.59 (Seno Año Nuevo) (Table 2). Suárez et al. (1985) indicated the Gabbroic assemblage represents an immature stage of the arc and that the other two suites are progressively more evolved.

Trace elements normalized to MORB show serrated patterns characteristic of arc magmas, with high large ion lithophile element (LILE) content relative to N-MORB, K and Sr peaks, and Nb troughs (Fig. 7). The LILE/HFSE (high-field-strength elements) ratios are relatively high, with Rb/Zr averaging 0.34 (Table 2). LREE are enriched with respect to chondrite and Y, with La_N/Y_N of 1.81–14.35 (Table 2).

The Hardy Formation This unit represents the volcanic counterpart of the Mesozoic arc. The lavas range from basalt to dacite, with medium-K, calc-alkaline geochemistry (Figs. 5a and 6) (Suárez et al. 1985; Miller et al. 1994). Normalized

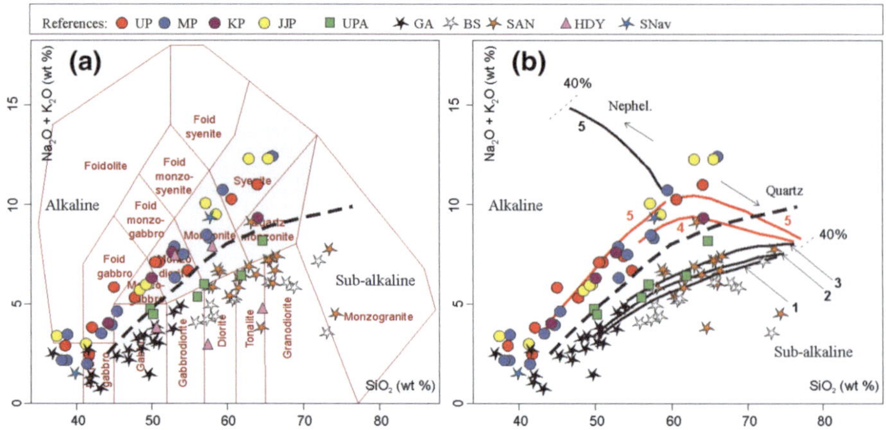

Fig. 5 Total alkali silica (*TAS*) diagrams for the *FB* and satellite plutons. **a** Trench-side arc suites (FB and Hardy Fm (HDY); Suárez et al. 1985 and Miller et al. 1994, respectively), and rear-arc suites (FPM; Acevedo et al. 2002; González Guillot 2009; González Guillot et al. 2012; and UPA plus a monzonitic pluton of south Navarino Island (*SNav*); González Guillot et al. 2011 and Kranck 1932, respectively). Fields of Middlemost (1991), the shading marks the mildly alkaline (transalkali) field. Alkaline and subalkaline divide of Miyashiro (1978). **b** The same diagram with trends of Lameyre (1987) and Lameyre and Bonin (1991): *1* calc-alkaline tonalitic–trondhjemitic, *2* tholeiitic, *3* calc-alkaline granodioritic, *4* monzonitic, and *5* alkaline. Normative *Q* and *Ne* contents are indicated. Same references as in Fig. 4

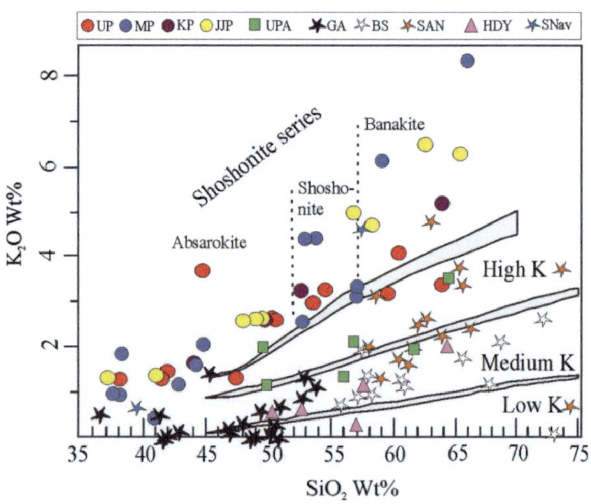

Fig. 6 K_2O versus SiO_2 diagram for the FB and satellite plutons. Division lines of Rickwood (1989). Same references as in Fig. 4

Table 2 Summarized chemical composition and ratios of trench-side arc and rear-arc suites

Suite	GA		BS*		SAN		Hardy		UPA		FPM	
	Range	Aver. (*n*)	Range	Aver. (*n*)	Range	Aver. (*n*)	Range	Aver. (*n*)	Range	Aver. (*n*)	Range	Aver. (*n*)
SiO_2	36.7–56.7	47.4(19)	48.4–73.1	62.1(40)	54.4–74.2	62.8(18)	50.5–64.5	56.6(5)	49.8–64.6	56.6(6)	37.2–65.86	50.44(37)
K_2O/Na_2O	0.00–1.32	0.24(19)	0.02–1.09	0.44(40)	0.17–1.07	0.59(18)	0.10–0.76	0.26(5)	0.33–0.74	0.50(7)	0.25–1.78	0.82(39)
K_{51}**		0.58		0.76		1.00		0.38		1.46		3.04
Rb/Zr	–	–	0.05–0.70	0.34(28)	–	–	0.02–0.38	0.21(5)	0.29–0.81	0.49(6)	0.20–1.73	0.92(30)
La_N/Y_N***	–	–	1.85–14.31	6.50(28)	–	–	1.79–3.79	2.75(5)	5.51–23.07	12.25(6)	1.49–14.46	7.19(30)
$La_N/Y_{N(51)}$***	–	–		4.92	–	–		2.37		6.93		7.46
Th	–	–	2.00–28.00	8.61(28)	–	–	0.60–6.60	3.06(5)	7.40–23.00	12.96(7)	0.30–21.60	7.39(32)
Th/La	–	–	0.11–0.80	0.41(28)	–	–	0.10–0.44	0.26(5)	0.32–0.49	0.42(6)	0.04–0.62	0.27(30)
Ce/Pb	–	–	0.77–14.50	3.66(28)	–	–	3.96–7.82	5.38(5)	5.61–9.71	8.20(5)	4.72–47.00	22.02(16)
Pb/Th	–	–	0.50–3.00	1.49(28)	–	–	0.94–4.33	2.03(5)	0.43–0.68	0.60(5)	0.19–5.33	1.15(16)

GA Gabbroic assemblage, *BS* Beagle suite, *SAN* Seno Año Nuevo suite (Suárez et al. 1985). Data of Hardy Fm from Miller et al. (1994)
FPM Fuegian Potassic Magmatism (Acevedo et al. 2002; González Guillot 2009; González Guillot et al. 2012)
UPA Ushuaia Peninsula Andesites (Acevedo et al. 2002; González Guillot et al. 2011)
*Includes data from Suárez (1977a)
**K_2O at 51 % SiO_2 from linear regression lines
***Norm. factors: CI chondrite (Toylor and McLennen 1985)

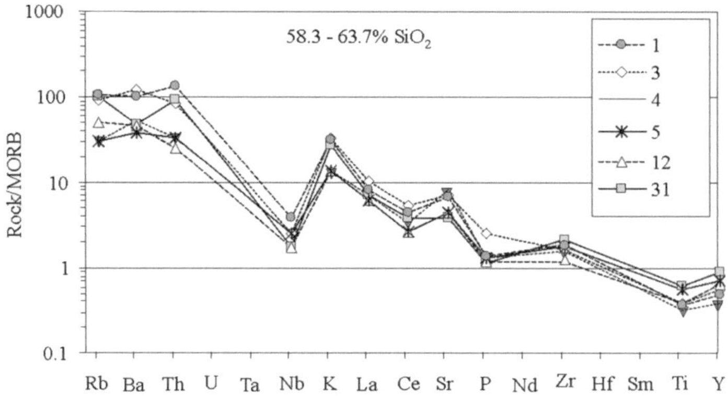

Fig. 7 Normalized trace multielement diagram of the FB (Beagle suite, Suárez 1977a). Normalization factors from Sun and McDonough (1989)

trace element plots display patterns typical of magmatic arcs, such as enrichment in LILE relative to N-MORB and negative HFSE anomalies relative to LILE, P, and LREE. Chondrite normalized data indicate moderate LREE enrichment (Table 2). Fossils in the volcaniclastic sequences indicate a Tithonian–Valanginian age (Suárez et al. 1985), whereas intercalated rhyolite intrusions yielded ages of 120 and 107 Ma (U/Pb zircon and $^{40}Ar/^{39}Ar$ in hornblende, respectively) (Miller et al. 1994).

3.3 The Rear-Arc

The rear-arc in Tierra del Fuego is composed of isolated plutons with mildly alkaline, potassic composition, the Fuegian Potassic Magmatism (FPM) (González Guillot et al. 2009), and other minor intrusions of high-K, calc-alkaline composition, the Ushuaia Peninsula Andesites (González Guillot et al. 2011). Both suites appear as scattered intrusions north of Canal Beagle, east of Cordillera Darwin, separated from the rest of the batholith (Fig. 1). Other potassic units within the axis of the FB and in the Southern Patagonian Andes may be related to the FPM and are briefly described below.

3.3.1 The Fuegian Potassic Magmatism (FPM)

Field Features and Lithology

This suite is composed of four small (25-3 km^2), composite plutons (Ushuaia, Moat, Jeu-Jepén, and Kranck) (Fig. 1), each one encompassing a broad lithological spectrum from ultramafic to felsic facies (Figs. 4b and 8). Lithologies include clinopyroxenites, hornblendites, gabbros, diorites, monzogabbros, monzodiorites, monzonites, syenites, and alkali feldspar syenites. Quartz varieties of the mafic to felsic facies may be present in some plutons, as well as late lamprophyre dykes.

Pyroxenites and hornblendites are abundant in the Ushuaia and Moat plutons, where they form massive bodies or autoliths in more felsic rocks. Contrarily, in the Jeu-Jepén and Kranck plutons, they are scarce, restricted only to small (decameter scale) lenses and clusters of autoliths. Associated pegmatitic veins or pockets of anorthosite–gabbro and thin aplitic dykes of meladiorite/malagabbro are common.

Gabbros, diorites, monzogabbros, and monzodiorites form also massive bodies and show variable textures (granular, cumular, porphyritic, traquitic). Hornblende or biotite are almost absent in these lithologies in Jeu-Jepén and Ushuaia plutons, respectively. The other plutons may contain pyroxene–hornblende, pyroxene–biotite, and/or pyroxene–hornblende–biotite varieties (Fig. 2).

Monzonites form massive bodies and minor dykes in all but the Ushuaia pluton. Late-stage magmatic pulses are rectilinear to anastomosing veins and veinlets, and breccia infill of syenite and K-feldspar syenite (Fig. 4b), or monzonite in the Ushuaia pluton.

Scarce veinlets of alkali feldspar granite and monzogranite occur in the Moat and Kranck plutons, respectively (Fig. 4b). The Ushuaia and Jeu-Jepén plutons also contain border facies of quartz monzodiorite and tonalite with biotite + garnet ± fibrolite, which appear to be hybrids between dioritic or monzodioritic magmas and anatectic melts from the country metapelites (Fig. 8) (González Guillot and Acevedo 2009).

Crosscut relationships indicate a general temporal trend of emplacement and crystallization from the ultramafic, through mafic to felsic facies and finally lamprophyres, accompanied by progressive changes in mineral composition, e.g., an increase in Ab content in plagioclase and Fe content in mafic minerals, consistent with magmatic differentiation processes.

Igneous layering between ultramafic and mafic facies is common, and accompanied by igneous foliation parallel to it. Evidences of mingling between mafic and ultramafic facies are also present. Additionally, the Ushuaia pluton shows concentric compositional zonation with a pyroxenite core and more evolved lithologies toward the margins (Fig. 8). This and other features, including positive anomalies in platinum group elements, led González Guillot and Schalamuk (2009) to compare it to Alaskan-type complexes.

The suite lacks metamorphism and penetrative deformation.

Age

There is a large variability in the ages reported for the FPM, even from a single pluton, depending on the lithologies sampled and the methods employed. The most reliable ages are U/Pb zircon ages from the Ushuaia and Jeu-Jepén plutons, a late-stage felsic dyke of 75 ± 2 Ma (Barbeau et al. 2009) and a monzonite of 72 ± 0.8 Ma (Cerredo et al. 2011a), respectively. These are consistent with

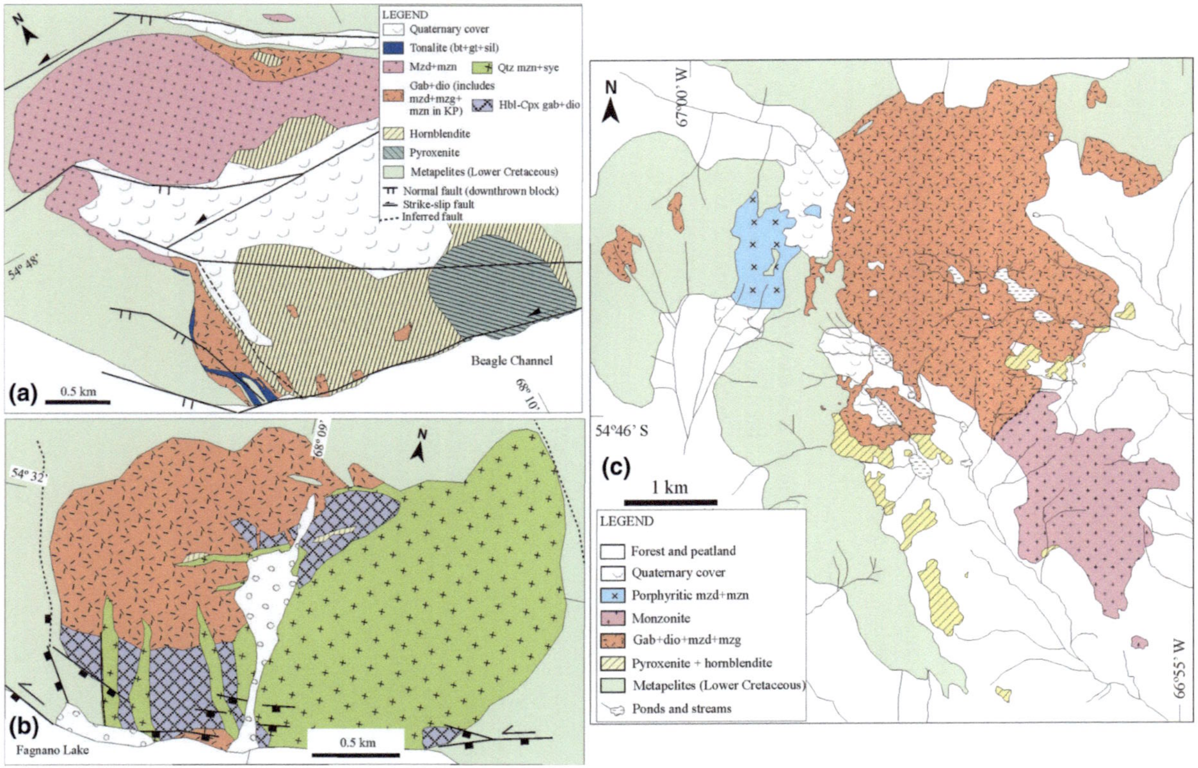

Fig. 8 Geological maps of FPM plutons. **a** Ushuaia pluton (modified from González Guillot 2009). **b** Kranck pluton (modified from González Guillot et al. 2012). **c** Moat pluton (modified from González Guillot et al. 2009)

unpublished U/Pb zircon ages of mafic to felsic facies from the Ushuaia and Kranck plutons in the range of 78–68 Ma (M. Martins Pimentel, pers. com.).

Older lower-resolution results extend this magmatism to the Early Cretaceous. These are K/Ar ages (hornblende or whole rock) from the Ushuaia, Jeu-Jepén, and Kranck plutons in the range of 113–93 Ma (Acevedo et al. 2000, 2002; Peroni et al. 2009; Cerredo et al. 2011b), or a Rb/Sr isochron age of 115 ± 3 Ma in the Moat pluton (González Guillot et al. 2009). Since the plutons lack penetrative subsolidus fabric and metamorphism, they should postdate the main Andean phase occurred prior to 86 Ma (see Section "Regional Geology"). Consequently, all the K/Ar and Rb/Sr ages must be erroneous.

Geochemistry

All plutons show a mildly alkaline (transalkali) trend in a TAS diagram (Fig. 5a) and have high-K_2O contents, the samples occupying the shoshonitic field on a K_2O–SiO_2 plot (Fig. 6). The K_2O/Na_2O ratio is high, ≥ 0.6 in most samples and ≥ 1 in rocks with silica content >55 % (Table 2). The content in TiO_2 is low (<1.4 %), except in ultramafics with Ti-magnetite and ilmenite cumulates. These are typical chemical characteristics that distinguish the shoshonitic magma series (Morrison 1980).

Trace elements normalized to MORB show the characteristic serrated pattern of arc magmas, with high LILE content, K and Sr spikes, and troughs in Nb and Ta (Fig. 9). LILE/HFSE ratios are high, with Rb/Zr from 0.4 to 1.7 in most samples (Table 2). Rare earth elements (REE) normalized to chondrite show enrichment of LREE (light) relative to HREE (heavy) and a rather horizontal trajectory of the latter (Fig. 10). La/Yb varies from 8.4 to 27.8, except in the ultramafic rocks, which show a strong depletion in LREE, a trend that diminishes gradually toward the most evolved rocks.

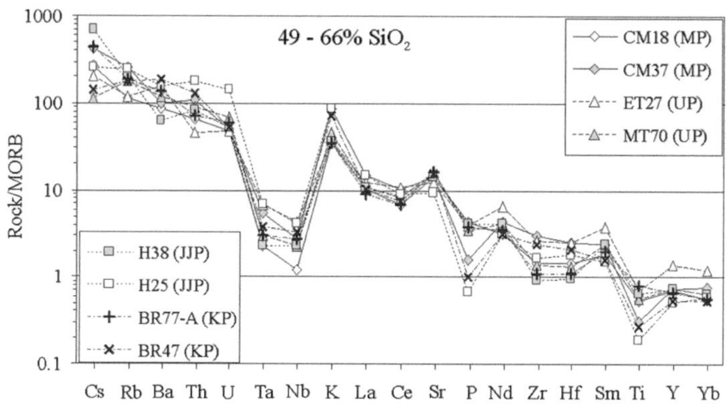

Fig. 9 Normalized trace multielement diagram of the FPM (González Guillot 2009; González Guillot et al. 2012). Normalization factors from Sun and McDonough (1989)

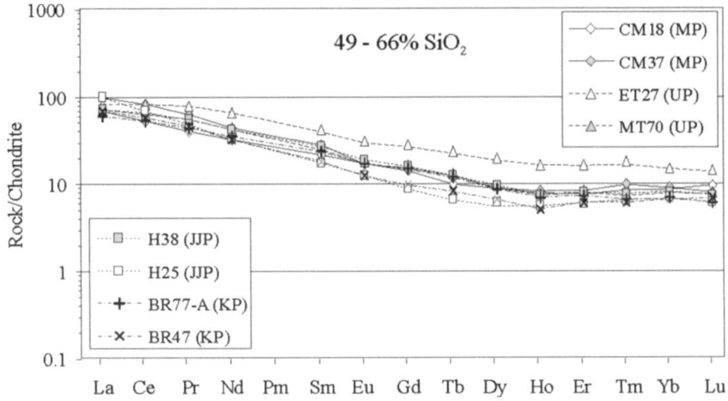

Fig. 10 Normalized REE patterns of the FPM (González Guillot 2009; González Guillot et al. 2012). Normalization factors from Taylor and McLennan (1985)

The suite is typically I-type, metaluminous, except the biotite–garnet–fibrolite border facies, which are peraluminous.

The monzonites of Navarino Island Kranck (1932) and Katz and Watters (1966) described scarce plutons composed of gabbro to monzonite in southern Navarino Island (Fig. 1). Chemically, they show mildly alkaline and shoshonitic trends in the TAS and K_2O–SiO_2 diagrams, respectively (Figs. 5a and 6). Their petrography and chemistry are similar to that of the FPM. Additionally, some hornblendites with mildly alkaline composition from Gordon Island (Fig. 1), of 110–106 Ma (K/Ar in hornblende, Suárez et al. 1985), may also be included in the FPM (Figs. 5a and 6).

The absarokites of Barros Arana and La Pera Formations These are rear-arc units that crop out to the east of the SPB (Fig. 1). They consist of a sequence of mafic dykes, lavas, and volcaniclastic breccias occurring within the sedimentary infill of the Rocas Verdes basin (Soffia 1988; Stern et al. 1991) and formed before the Andean deformation, in the late or post-Albian (Stern et al. 1991; Mpodozis et al. 2007). The geochemistry indicates that they were originated by subduction (Stern et al. 1991; Prades 2008; Anguita 2010). The age of the Barros Arana Formation is 104 Ma (hornblende K/Ar, Stern et al. 1991), whereas the maximum age of the La Pera Formation is estimated as 102 Ma (detrital zircon, Mpodozis 2006). The fossil content in the Erezcano and Canal Bertrand Formations, where the basalts intercalate, confirms an Albian age (Fuenzalida and Covacevich 1988; Castelli et al. 1992). The remarkable characteristic of these units is the high total alkalis, K_2O, LILE, and LREE contents and high LILE/HFSE and LREE/HREE ratios, typical of the mafic members of the shoshonite association (absarokites). In these respects, they are equivalent to the FPM, although the absarokites are older and less radiogenic in terms of $^{143}Nd/^{144}Nd$ and $^{87}Sr/^{86}Sr$ ratios. The differences in isotopic composition may be due to a greater time of residence within the crust of the FPM magmas, in comparison with the basalts, related to the mode of emplacement (intrusive vs. extrusive, respectively) and the tectonic regime (see Sect. 5) (González Guillot et al. 2009).

3.3.2 The Ushuaia Peninsula Andesites (UPA)

The UPA consist of hypabyssal intrusions intruded in Lower Cretaceous units that crop out mainly in the Ushuaia Peninsula, the type locality, and on the top of the Trapecio Hill (Fig. 1). At Rodríguez Hill, 4 km to the north of Kranck pluton, strongly altered lamprophyric and andesitic dykes are present, dated in 104 Ma (K/Ar whole rock), which trend N–S, i.e., orthogonal to the Andean structures (Martinioni et al. 1999). Dykes similar in composition, orientation, and position toward the rear-arc were reported at 52°S and dated in 79 Ma ($^{40}Ar/^{39}Ar$ on hornblende, Calderón et al. 2012).

Field Features and Lithology

Irregular sheet-like intrusions and plugs crop out at both the Ushuaia Peninsula and Trapecio Hill, ranging 0.1–10 m thick (Fig. 11). The rocks have no metamorphism, are undeformed, and crosscut the slaty cleavage and folding of the country rock. Depending on the composition and texture, they vary among quartz meladiorites, granodiorites, andesites, dacites, and lamprophyres (Fig. 4b) (González Guillot et al. 2011).

The quartz meladiorites form dykes or small apophysis of dark-colored rocks (M > 50 %) with (hornblende) megacrystic to seriate texture. The granodiorites form thin dykes and breccia infill crosscutting quartz meladiorites. The andesites and dacites form dykes and sills of variable thickness, from few millimeters up to 10 m wide, and cut all the other lithologies. They are light-colored (M 5–20 %) and have subvolcanic textures (porphyritic, seriate, and microgranular), with parts showing autobreccia texture. They bear enclaves of hornblendite. All the sequence is cut by 10–30-cm-thick lamprophyre dykes (Fig. 11).

Age

K/Ar data from the Ushuaia Peninsula yielded an age range of 104–77 Ma (Ramos et al. 1986; Acevedo et al. 2002; Elsztein 2004; Peroni et al. 2009). The youngest ages correspond to a "granitic porphyry" (Ramos et al. 1986) and two lamprophyre dykes (Elsztein 2004) that yielded ~77 and ~88 Ma, respectively. These younger ages are roughly coincident with a U/Pb zircon age of 83 ± 0.93 Ma from an andesite sill on that locality (unpublished, M. Martins Pimentel, pers. com.) and indicate the rocks are post-tectonic, in agreement with the lack of metamorphism and penetrative deformation. The more reliable U/Pb age also evidences that the UPA is older than the plutons of the FPM, contrarily to what was previously suggested based on the available K/Ar ages (González Guillot et al. 2011).

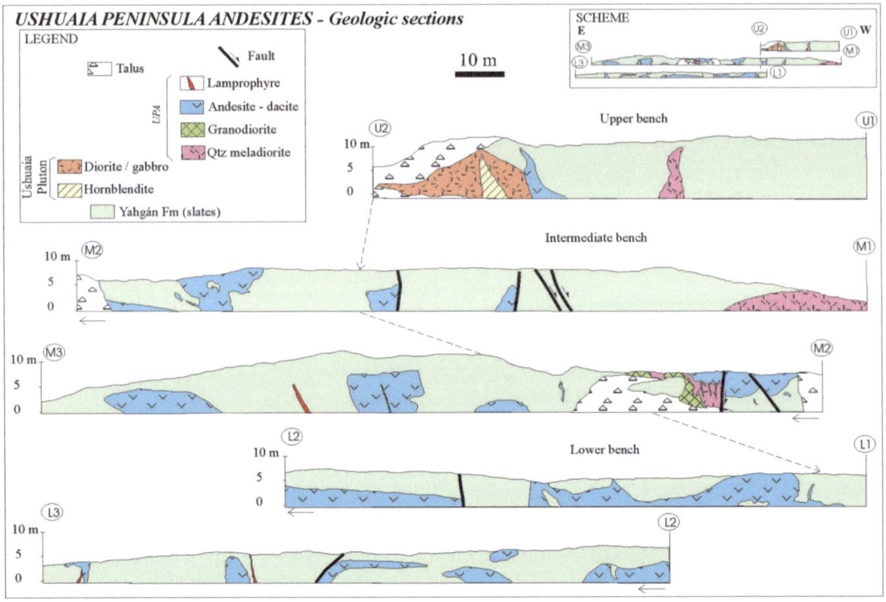

Fig. 11 Geological profiles of the UPA from a quarry at Ushuaia Peninsula. Modified from González Guillot et al. (2011)

Geochemistry

The UPA is subalkaline, metaluminous (A/CNK from 0.66 to 1.02), characterized by a calc-alkaline trend with medium to high-K_2O content (Figs. 5a and 6). The K_2O/Na_2O ratio varies from 0.34 to 0.74 (Table 2).

Normalized trace element plots (Fig. 12) show a typical serrate pattern, with Nb, Ta, and Ti troughs, K and Sr spikes, and high LILE/HFSE ratios, that characterizes subduction-related magmas. Total abundances of LILE and HFSE are high, especially Ba (505–1434 ppm), Sr (752–1351 ppm), Rb (38-89 ppm), Nb (7–13 ppm), Ta (0.5–1.1 ppm), and Th (7–23 ppm). All (except Rb) show good positive correlation with SiO_2.

The REE normalized to chondrite show a steep arrangement of LREE and a roughly flat to concave pattern of HREE (Fig. 13) as is typical of subduction-related rocks with significant hornblende fractionation. High LREE and slightly depleted HREE cause a high LREE/HREE ratio, with La/Yb ranging from 11 to 38. The Sm/Yb ratio varies between 2.2 and 3.1.

The low HREE (1.26–1.73 ppm Yb) and Y (9.0–18.9 ppm) contents of the UPA and the high LREE/HREE are similar to magnesian andesites from western Aleutians (Yogodzinski et al. 1994, 1995). The high LILE and HFSE contents, and the low HREE contents and U/Th (0.18–0.30) and Ba/Th (36–116) ratios,

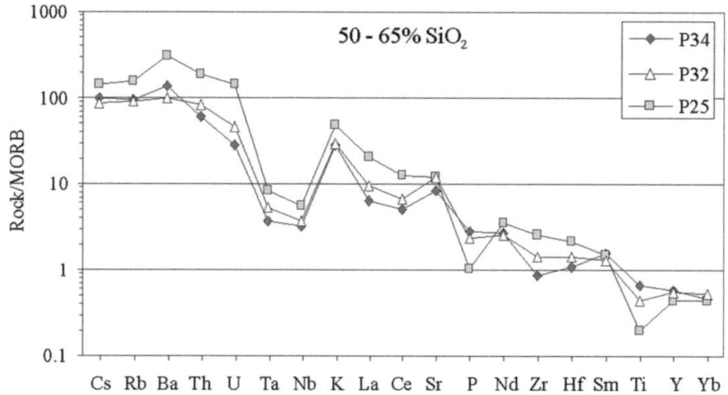

Fig. 12 Normalized trace multielement diagram of the UPA (González Guillot et al. 2011). Normalization factors from Sun and McDonough (1989)

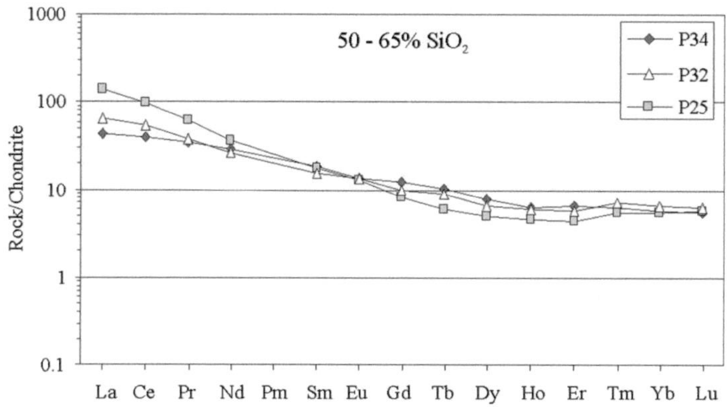

Fig. 13 Chondrite normalized REE patterns of the UPA (González Guillot et al. 2011). Normalization factors from Taylor and McLennan (1985)

especially, resemble also the northern lavas of the Holocene Austral Volcanic Zone (AVZ) in the Andes (49°S–54°S), which is considered a typical example of adakitic magmatism (Stern and Kilian 1996). This chemical signature of the UPA is interpreted as caused by contribution to the mantle source either of (i) subducted sediment partial melts, generated in the garnet stability field, or (ii) of partial melts of the top of the subducted basaltic crust (González Guillot et al. 2011).

3.4 Comparison Among the Different Plutonic Suites

Following the modal criteria of Lameyre and Bowden (1982), Lameyre (1987), and Lameyre and Bonin (1991) (Fig. 4), the suites of the FB define a calc-alkaline tonalitic–trondhjemitic trend (Beagle suite) and a calc-alkaline tonalitic–trondhjemitic to granodioritic trend (Seno Año Nuevo suite), whereas the Gabbroic assemblage could be defined either as tholeiitic or calc-alkaline. On the other hand, the rear-arc suites define a silica-saturated alkaline trend and to a lesser extent a monzonitic trend (FPM) and a calc-alkaline granodioritic trend (UPA). The Jurassic S-like granitoids correspond to the granitoids of crustal origin.

These compositional trends were also distinguished previously using geochemical parameters (Fig. 5a). On the other hand, the Hardy Formation does not follow a clear trend in Fig. 5a, with samples straddling the subalkaline–alkaline divide, although its medium-K, calc-alkaline composition is depicted in the K_2O–SiO_2 diagram (Fig. 6). The distinction of these trends may be improved by comparison with the compilations of Lameyre (1987) and Lameyre and Bonin (1991) (Fig. 5b). The mildly alkaline trend of the FPM follows a silica-saturated alkaline trend, although few samples divert to a monzonitic trend (equivalent to the volcanic shoshonitic associations, c.f. Pagel and Leterrier 1980) (curves 5 and 4 in Fig. 5b, respectively). The plutons of southern Navarino Island and some hornblendites of the Gabbroic assemblage also show a silica-saturated alkaline trend. These and the FPM have K_2O contents typical of the shoshonite series, whereas the main of the FB and the Hardy Formation show lower K_2O content, which increases with time from the low to the high-K_2O fields in the K_2O–SiO_2 diagram (Fig. 6). The composition of the UPA straddles the divide between medium- and high-K_2O calc-alkaline fields.

Monzonitic series, as defined by Lameyre and Bowden (1982) and Lameyre and Bonin (1991), include gabbros, diorites, monzogabbros, monzodiorites, monzonites, quartz monzonites, monzogranites, and scarce syenogranites. On the other hand, the silica-saturated alkaline series are bimodal, since they lack the intermediate lithologies, and are dominated by the granitic end members. This bimodal character is not present in the FPM, nor is there an abundance of felsic end members (actually intermediate members dominate). Thus, it follows that the silica-saturated alkaline trend of the FPM is merely a transitional variation of the monzonitic trend.

Trace elements provide more differences between the trench-side arc and rear-arc units. The FPM has higher LILE and P, compared to the calc-alkaline batholith (Fig. 14). Additionally, it also shows higher LILE/HFSE (e.g., Rb/Zr 0.57–3.28) and La_N/Y_N (3.62–14.65) ratios compared to the FB (Rb/Zr 0.23–0.60, La_N/Y_N 4.54–10.39) (Table 2).

4 Spatial and Temporal Distribution of Magmatic Suites

Figure 15 summarizes the information given in Sects. 3.1–3.4. It evidences that magmatism in the Fuegian Andes was episodic and that the batholith grew incrementally, including the rear-arc intrusions. In Tierra del Fuego, the magmatic suites are arranged with an ESE–WNW orientation, and trends progressively become more N–S toward the north, parallel to the present trench (Fig. 1). The oldest FB

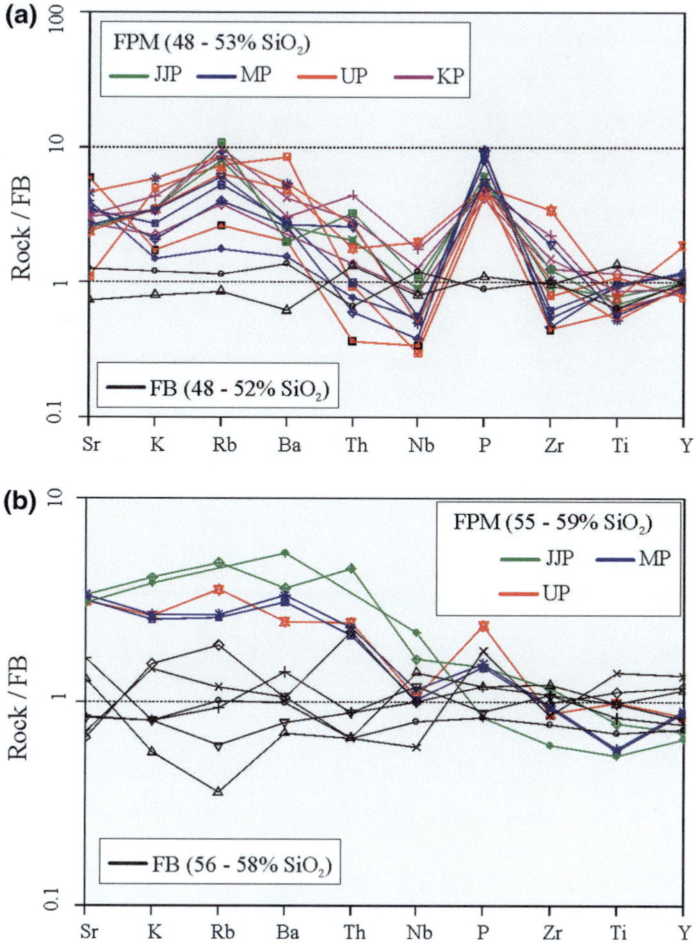

Fig. 14 Trace multielement diagram of the FPM (González Guillot 2009; González Guillot et al. 2012) and FB (Suárez 1977a), normalized to the average of the samples of the FB in each case. Abbreviations as in Fig. 2. **a** Basic samples. **b** Intermediate samples

plutons were emplaced along the cratonic margin of the batholith during the Late Jurassic (Fig. 15). Afterward, a first oceanward shift in the volcanic front is documented, as Lower Cretaceous plutons and volcanics outcrop toward the Pacific side (Fig. 15) (Suárez et al. 1985, 1986; Miller et al. 1994; Hervé et al. 2007a). During the Late Cretaceous, magmatism of the Beagle suite shifted cratonward (Fig. 15), since Upper Cretaceous plutons crop out to the NNE of the Lower Cretaceous ones. Additionally, there seems to be a decrease in ages from south to north of the Canal Beagle, as most plutons to the south have ages in the range 98–80 Ma (Halpern 1973; Hervé et al. 1984; Suárez et al. 1986; Kohn et al. 1995; Mukasa and Dalziel 1996), whereas ages in Cordillera Darwin range from 86 to

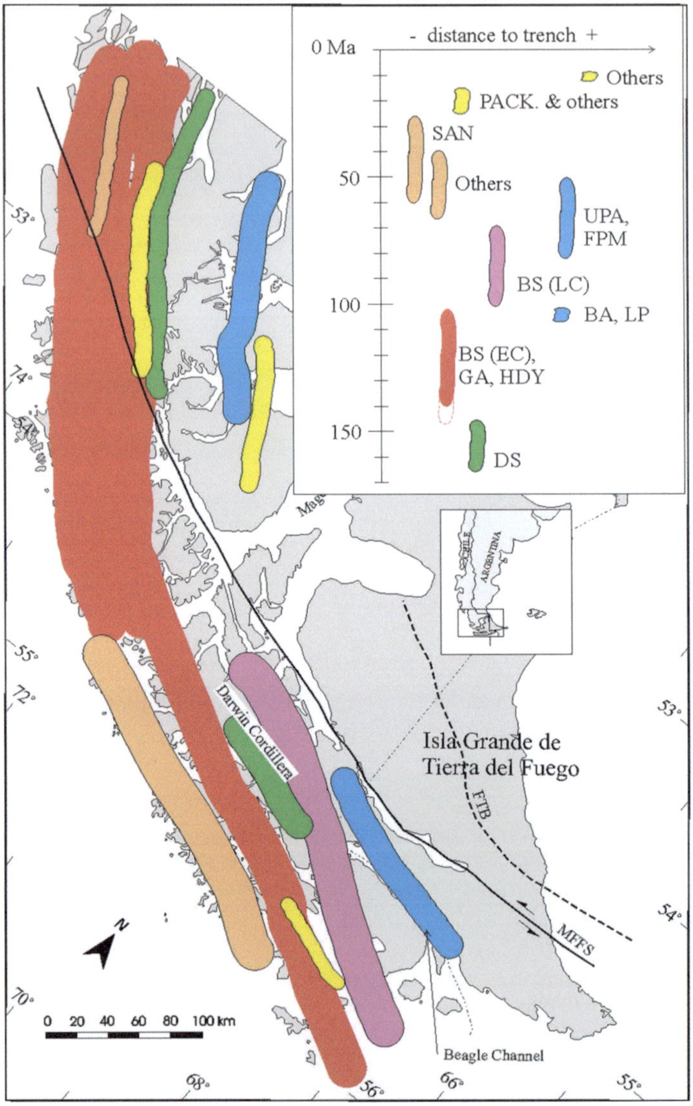

Fig. 15 Simplified map showing the geographic distribution of trench-side arc and rear-arc suites in Tierra del Fuego and Southern Patagonia. The inset emphasizes the time distribution and relative distance to the paleo-trench. Data from Hervé et al. (1984, 2007a), Suárez et al. (1985, 1986), Stern et al. (1991), Miller et al. (1994), Klepeis et al. (2010), Barbeau et al. (2009), and Cerredo et al. (2011a), this review. *DS* Darwin suite, *EC* Early Cretaceous, *LC* Late Cretaceous, *BA* Barros Arana Fm, *LP* La Pera Fm, PACK Packsaddle Fm. Others refer to unnamed plutons in the SPB. Other abbreviations as in Table 1 and Figs. 1, 2 and 5

73 Ma (Mukasa and Dalziel 1996; Klepeis et al. 2010; McAtamney et al. 2011). Rear-arc intrusions in Tierra del Fuego are coeval with the Beagle suite in Cordillera Darwin, while the northernmost plutons of the MPF are slightly younger.

Thus, a progressive cratonward migration of the volcanic front is registered in the southernmost Andes, which started in the mid-Cretaceous associated with the closure of the marginal basin (Stern et al. 1991) and culminated in the latest Cretaceous (Fig. 15).

During emplacement of the 60–34 Ma Seno Año Nuevo suite and the Miocene Packsaddle volcanics, the calc-alkaline volcanic front in Tierra del Fuego shifted back to the Pacific side (Fig. 15) and remained there up to its culmination in the Miocene, except for the small Recent Cook volcano eruptions (Hervé et al. 1984; Puig et al. 1984) (Fig. 1).

Contrarily, in the Southern Patagonian Andes, the Late Cretaceous to Miocene volcanic front remained in a steady position (Fig. 15) (Hervé et al. 2007a). After a period of magmatic quiescence between 40 and 25 Ma, Miocene magmatism was abundant in the core of the SPB (Hervé et al. 2007a), together with rear-arc satellite bodies up to 70 km to the east (Fig. 15) (Michael 1991; Sánchez et al. 2006; Ramos et al. 2004; Hervé et al. 2007a; Michel et al. 2008; Ramírez de Arellano et al. 2012).

Concerning the Upper Cretaceous suites, the following question arises: Why is the potassic magmatism (both UPA and MPF) absent in Cordillera Darwin, while it reappears at 52°–53°S? The fact that potassic magmatism took place cratonward of the marginal basin at its northern and southern ends, associated with its closure, leaves the possibility of occurrence all along its length. Coeval plutons of the Beagle suite in Cordillera Darwin were intruded at pressures close to 6 kbar (Maloney et al. 2011), while the MPF plutons were emplaced at 2–4 kbar (González Guillot et al. 2009). Thus, one possibility for the absence of potassic plutons at Cordillera Darwin is that they should have emplaced at higher levels and then eroded away. Figure 15 shows that the axis of the rear-arc actually passes through the northern flank of Cordillera Darwin, where low-grade metamorphic sequences crop out, which are equivalent to those to the east where the FPM is emplaced. This implies, alternatively, that potassic plutons may be present in that northern flank awaiting to be discovered.

4.1 Across-Arc Compositional Variations

There follows below a more detailed comparison of the plutonic suites that helps to elucidate the geometry of the subduction zone for the Cretaceous. Nevertheless, it is not easy to compare trace elements for the FB and rear-arc suites because there is a mixture of 40-year-old XRF and recent ICP-ES and ICP-MS results, and the complete spectrum of trace elements and REE for the former unit was not obtained.

The landward shift in the volcanic front in the mid-Cretaceous encompassed (i) the widening of the arc, broadly from 50 to 90 km in the Early Cretaceous to 80–120 km in the Late Cretaceous (Fig. 15), and (ii) progressive changes of magma composition away from the trench (Figs. 1, 2, 3, 4, 5, 6, 7, 8, 9, 10, 11, 12, 13, 14 and 15).

Both potassic suites of the rear-arc appear to represent the plutonic equivalents of a second volcanic front in what are known in modern arcs as dual volcanic chains, located above relatively low angle-subducting slabs (Marsh 1979; Tatsumi and Eggins 1995). Secondary volcanic chains are characterized by small magma volume, lower normative quartz, greater amounts of hydrous phases, and higher concentrations of incompatible elements, compared to trench-side chains (Tatsumi and Eggins 1995; Tatsumi 2003). This is related to smaller fluid flux into the magma source region (Tatsumi and Eggins 1995; Tatsumi and Kogiso 1997). Together with magma generation and segregation at greater depths beneath the rear-arc (Tatsumi et al. 1983), the less H_2O supply is translated as less percentage of partial melting (Gill 1981; Sakuyama and Nesbitt 1986). Alternatively, different metasomatic agents to the mantle wedge have been proposed for trench-side arc and rear-arc magmas, low-temperature hydrous fluids versus slab (sediment and/or basalt) melts, or supercritical fluids, respectively (Stern 1991; Kay and Mpodozis 2002; Morris and Ryan 2003; Kimura et al. 2010; Hanyu et al. 2012). Another variable in continental settings is crustal and/or subcontinental mantle contamination/assimilation of rear-arc magmas as they traverse thicker lithosphere.

Further inspection of the Cretaceous suites in the Fuegian Andes reveals not only differences between trench-side and rear-arc plutons, but also systematic changes within the rear-arc with distance to the paleo-trench (González Guillot 2009; González Guillot et al. 2012).

There is an increase in the relative abundance of felsic rocks (monzonite, syenite, and K-feldspar syenite) relative to mafic (monzodiorite, monzogabbro, diorite, and gabbro) and ultramafic rocks (hornblendite and pyroxenite). The ultramafic:mafic: felsic ratio varies from 49:51:0 in the Ushuaia pluton, through 6:72:22 in the Moat pluton to 0:27:73 in the Kranck pluton (Fig. 16), which is similar to that in the Jeu-Jepén pluton. This do not correspond to different level of exposure of the plutons, since two of them, one in the south (Ushuaia) and other in the north

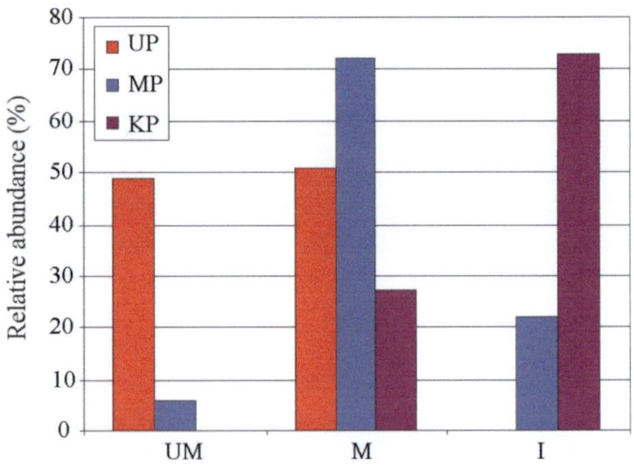

Fig. 16 Relative abundance of ultramafic (UM), mafic (M), and intermediate (I) lithologies in three FPM plutons, based on the maps of Fig. 8. Abbreviations as in Fig. 2. See text for more detail

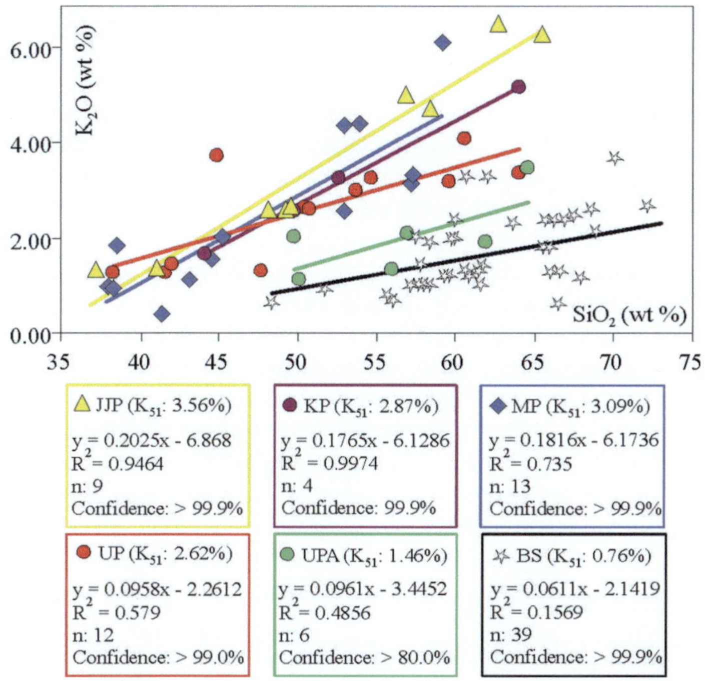

Fig. 17 K_2O versus. SiO_2 diagram with linear regression curves for FPM (JJP, KP, MP, UP), *UPA* and *FB* (Beagle suite). Same references as in Fig. 2 and Table 1. Equations and statistical parameters are provided, along with the K_{51} values for each suite. See text for further detail

(Jeu-Jepén), are exposed at or near roof level. Additionally, there is a concomitant progressive increase in K-feldspar in the most differentiated facies, as late-stage veins consist of monzonite in the Ushuaia pluton, syenite in the Moat pluton, and K-feldspar syenite in Kranck and Jeu-Jepén plutons (Fig. 4b).

The K_2O content for a given silica value also increases with distance to the paleo-trench (Fig. 17). The K_{51} values (K_2O wt% at 51 wt% SiO_2 calculated from linear regressions in Fig. 17, Tatsumi and Eggins 1995) for both rear-arc suites are greater than 1.46 %, above the lower limit of most back-arc-side volcanoes worldwide (1.2 %, Tatsumi and Eggins 1995), whereas the K_{51} value for the coeval Beagle suite is 0.76 % (Fig. 17; Table 2). The rate of increasing K_2O with silica also increases northward, as the slopes of the regression lines are progressively steeper. Additionally, LREE enrichment and HREE depletion, among many other chemical parameters, have been documented in rear-arc magmas in numerous arcs (see review in Tatsumi and Eggins 1995). Since no HREE are available for the FB, Y is used instead, which has similar behavior in magmatic systems (Rollinson 1993). The La_N/Y_N ratio (normalized to CI chondrite of Taylor and McLennan 1985) for a given silica content (51 %) varies from 4.92 in the FB (Beagle suite), to 6.03 in the Ushuaia pluton up to 9.29 in the Jeu-Jepén pluton (Table 2). The La_N/Y_{N51} in the UPA is 6.93, similar to the average in the MPF, although the ratios for greater silica values are the highest, consistent with an addition of subducted sediment partial melts to the source region (González Guillot et al. 2011, see below).

Higher Ba/Th, Pb/Th, and Pb/Ce ratios and lower Th contents and Th/La ratios in the volcanic front relative to the rear-arc are believed to be indicators of addition of slab-derived fluid to the mantle source because Ba and Pb are more fluid mobile, whereas Th, La, and Ce are relatively compatible in melts but not in hydrous fluids (Stern and Kilian 1996; Hawkesworth et al. 1997; Elliott et al. 1997; Johnson and Plank 1999; Pearce et al. 2005; Kimura et al. 2010). The variation in these values from volcanic front to rear-arc is continuous rather than abrupt and reflects a change in the composition of the slab component as the slab depth increases.

In Tierra del Fuego, from the volcanic front (Beagle suite and Hardy Formation) toward the rear-arc (UPA and FPM), there is a general increase in Th content and in Ce/Pb and Th/La ratios (Fig. 18), whereas Pb/Th ratios are similar to higher in the volcanic front with respect to the rear-arc suites (Fig. 18b). On the other hand, Ba/Th ratios are roughly similar in the Fuegian and Patagonian batholiths and the rear-arc suites, but some Hardy Formation samples show the highest ratios.

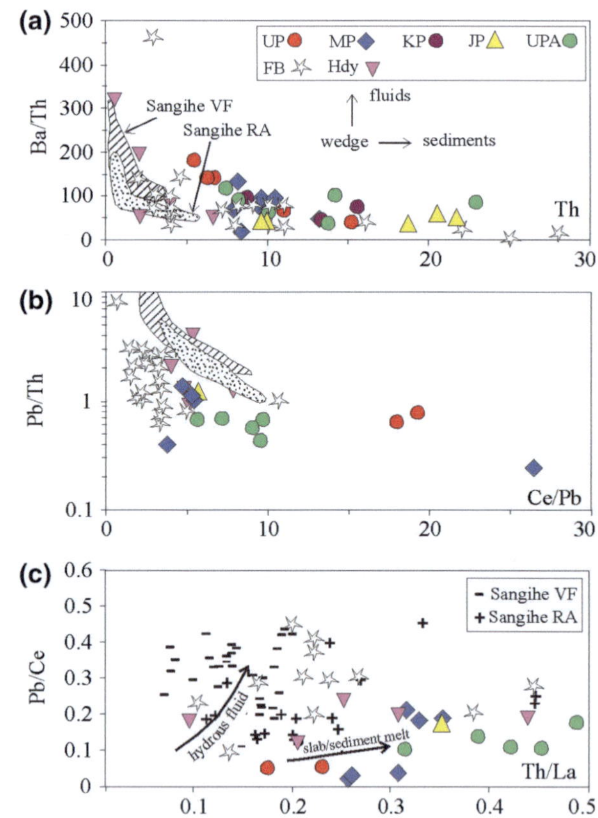

Fig. 18 Th versus Ba/Th (**a**), Ce/Pb versus Pb/Th (**b**), and Th/La versus Pb/Ce (**c**) of trench-side arc (FB and Hdy) and rear-arc suites (FPM and UPA) in Tierra del Fuego. Data from Suárez (1977a), Miller et al. (1994), González Guillot (2009), and González Guillot et al. (2011, 2012). For comparison, data of the Sangihe volcanic front and rear-arc (Indonesia) are presented (Hanyu et al. 2012)

Comparing the two rear-arc units, the UPA have greater Th contents and Th/La ratios, and lower Pb/Th ratios than the FPM (Fig. 18; Table 2).

The geochemical variability shown above is consistent with a mantle wedge metasomatized by fluids beneath the volcanic front and by subducted sediment (or basalt) melts, generated in the garnet stability field, in the zone where the UPA generated (González Guillot et al. 2011). The farther-from-the-trench suite (FPM), however, has a somewhat intermediate composition in these variables between the UPA and the FB (Fig. 18), although it shows the largest K and other LILE enrichments (Table 2). It appears then that the FPM is formed in a mantle source metasomatized by fluids, but by lower degrees of partial melting.

5 Tectonic Setting During Arc Construction

The oceanward shift of the volcanic front after emplacement of the Darwin suite (Fig. 15) can be related to the opening and widening of the Rocas Verdes basin that initiated in the latest Jurassic, which split the magmatic arc during steepening of the subducting slab (Fig. 19a–b) (e.g., Miller et al. 1994; Hervé et al. 2007a).

The geochemical signature of UPA, FPM and the Barros Arana and La Pera shoshonites is typical of rear-arc magmatism, consistent with north–northeastward subduction of the proto-Pacific plate (González Guillot et al. 2012). This is in contrast with another interpretation that assumes that the Late Cretaceous magmatism was related to southward subduction of the marginal basin oceanic floor (Mpodozis and Rojas 2006; Klepeis et al. 2010).

Thus, in the mid- and Late Cretaceous, a subduction shallowing and ensuing cratonward shift in the volcanic front took place during marginal basin closure (Fig. 19c) (Stern et al. 1991; Stern and De Wit 2003; González Guillot et al. 2011). The resulting compression produced shortening by means of continental underthrusting and obduction of the oceanic floor (e.g., Cunningham 1995; Kraemer 2003), enhancing the migration of the volcanic front. The process did not reach a mature state as in other segments of the Andes (e.g., the flat Pampean subduction zone, Barazangi and Isacks 1976); however, a brief magmatic gap between 68 and 60 Ma existed in Tierra del Fuego (Fig. 15). A later shift of the volcanic front toward the Pacific at 60 Ma in Tierra del Fuego (Fig. 15), plus the preservation of remnants of the Jurassic accretionary prism (Davidson et al. 1989), suggests that subduction erosion was not significant during the Cretaceous. However, the increase in relative convergence and plate coupling inherent to slab shallowing may have facilitated sediment supply to the magma source of the UPA (González Guillot et al. 2011).

Rear-arc potassic magmatism was diachronous, around 104 Ma (pretectonic) at 52°–53°S and 83–68 Ma (post-tectonic) at 55°S. Additionally, rear-arc magmas extruded ~35–40 km landward of the contemporaneous trench-side plutons in the Southern Patagonian Andes, whereas there is not such a spatial hiatus in Tierra del Fuego. Instead, there is a progression in age and composition, from medium-K,

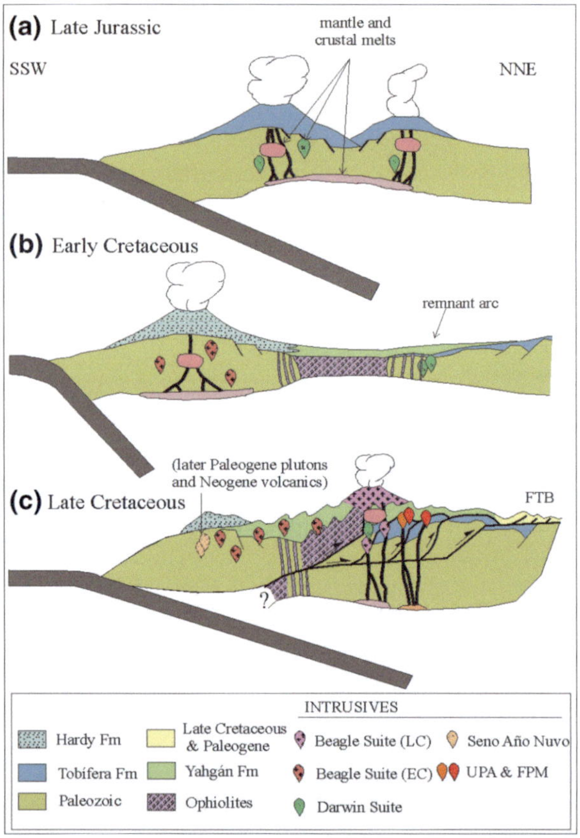

Fig. 19 Sketches showing models for the tectonomagmatic evolution of the Pacific margin of Tierra del Fuego from the latest Jurassic to the Late Cretaceous. **a** Subduction-related magmatism in an extensional arc. Generation of mantle and crustal melts to form the Darwin suite intrusions and Tobífera volcanics. **b** Slab steepening derived in slab rollback and full extension that splits the arc and develops the Rocas Verdes marginal basin. Most Darwin suite plutons remain in the remnant arc. Calc-alkaline magmatism continues in the frontal arc. **c** A slab flattening event produces inversion of the marginal basin and orogenic shortening. Magmatism shifts landward, and rear-arc potassic magmatism takes place. The position of the subsequent Paleogene and Neogene magmatism is indicated. *Sources* Stern et al. (1991), Miller et al. (1994), Hervé et al. (2007a), González Guillot et al. (2009, 2011), Klepeis et al. (2010) and Torres Carbonell and Dimieri (2013)

through high-K, to shoshonitic (Figs. 6 and 15), from the trench-side arc to the rear-arc. These constraints indicate that rear-arc magmatism in the Southern Patagonian Andes occurred shortly after the initiation of flattening, whereas it occurred long afterward at 55°S. A possible cause of this delay may be connected to a wider marginal basin in Tierra del Fuego with respect to Southern Patagonia (e.g., Stern and De Wit 2003; Kraemer 2003).

Based on the lack of radiogenic isotopic compositions, Stern et al. (1991) proposed that the Barros Arana basalts did not interact with crustal basement, as was the case of coeval plutons in the PB. These authors assumed that the rear-arc was still under extension during lava extrusion, while compressional deformation may have already begun to the west. On the other hand, the FPM was emplaced under a state of stress neutral to compressional, after the metamorphic climax. Extreme differentiation as occurred in this suite is possible in closed systems, isolated from continuous injections of fresh magma and from tapping of differentiated products through eruptive vents (Stern 1979). The slab flattening event initiated tens of million years before provides the conditions for a locked subduction system, consistent with a compressional setting.

The evidence indicates that both rear-arc shoshonitic suites were emplaced in different tectonic settings, leading to the opposite mode of occurrence: extrusive versus intrusive, and chemical and lithological compositions. Morrison (1980) concluded that shoshonitic rocks are associated with arc deformation during the termination of subduction or during flipping of the subduction zone. It seems that the Barros Arana and La Pera absarokites represent an early manifestation of subduction flattening (Stern et al. 1991), whereas the FPM and UPA represent the culmination of that phase.

From the latest Cretaceous on, a more obvious decoupling in the position of the volcanic front occurs between the FB and SPB. After the culmination of rear-arc magmatism, an important plate reconfiguration gave way to voluminous Paleogene plutonism toward the trench in Tierra del Fuego (Seno Año Nuevo, Fig. 15). The process took ~8 Myr from 68 Ma (the youngest rear-arc pluton) to 60 Ma (the oldest Paleogene pluton). However, Paleogene plutonism is scarce and occurs roughly at the central axis of the batholith in the Southern Patagonian Andes (Fig. 15) (Hervé et al. 2007a), implying a less dramatic change in the subduction parameters. The magmatic gap there occurred between 75 and 67 Ma (Hervé et al. 2007a).

Additionally, while Paleogene and Neogene magmatism remained at a stable position in the Fuegian Andes (Puig et al. 1984), Neogene plutons at 52°S–53°S are more widely distributed and extended to the east (Fig. 15). This manifests a renewed flattening and cratonward migration of magmatic activity in the Southern Patagonian Andes (e.g., Espinoza et al. 2007; Folguera and Ramos 2011; Ramírez de Arellano et al. 2012).

The Paleogene retreat of the volcanic front in Tierra del Fuego may be related to highly oblique convergence—opposed to a quasi-orthogonal convergence to the north—(Eagles and Scott 2014; Eagles 2016), rather than to steepening of the subducted plate. These conditions also controlled the Neogene contrasting evolution of magmatism on both segments. This should be taken into account for further research concerning whether or not the bend in the southernmost Andes is a true orocline, possibly initiated after ~70 Ma (cf. Maffione 2016; Rapalini et al. 2015).

According to Eagles and Scott (2014), the absence of calc-alkaline magmatism after the Seno Año Nuevo suite might be related to the approach of the Farallon/Nazca–Phoenix ridge to the trench at ~30 Ma and the subduction of young, volatile-poor, buoyant oceanic lithosphere (Breitsprecher and Thorkelson 2009). After a hiatus, calc-alkaline volcanism resumed briefly at Peninsula Hardy (Packsaddle Fm) in earliest Miocene times, before ceasing again until recently (Puig et al. 1984). Subduction of the Nazca–Phoenix ridge at ~20 Ma led to the alkaline term of the Packsaddle Fm associated with slab window processes (Breitsprecher and Thorkelson 2009; Eagles et al. 2009).

6 Conclusions

The magmatic arc in Tierra del Fuego occurred as a response of northward subduction (present coordinates) of the proto-Pacific plate under South America. It grew episodically from the Late Jurassic to the Miocene. The FB is the most voluminous constituent of the arc, followed by volcanic and volcaniclastic components and minor rear-arc intrusions. The volcanic front shifted, and the composition of magmas changed several times during arc construction. The oldest components of the FB are Upper Jurassic S-like granitoids and hornblende gabbros. They occur at the northern margin of the batholith. In the latest Jurassic–Early Cretaceous, magmatism shifted to the Pacific side of the archipelago as a consequence of the opening and widening of the Rocas Verdes marginal basin, according to some authors linked to a steepening of the slab and trench rollback (e.g., Miller et al. 1994).

A slab flattening event in the mid-Cretaceous derived in the inversion of the marginal basin, peak metamorphism, and orogenesis. This caused a cratonward migration of the volcanic front, the widening of the arc, and inception of potassic magmatism in the rear-arc (Stern et al. 1991; González Guillot et al. 2011). There is a progressive change in age and composition of Cretaceous plutons toward the continent in Tierra del Fuego, from medium-K, calc-alkaline, through high-K, calc-alkaline to shoshonitic. Shoshonitic magmatism in the rear-arc in Tierra del Fuego was partly coeval with calc-alkaline magmatism in the trench-side arc, although the former includes the youngest components of the Late Cretaceous arc. Besides, there is no spatial hiatus between coeval trench-side arc and rear-arc suites. Contrarily, in the Southern Patagonian Andes shoshonitic rear-arc magmatism occurred ~20 Myr before than in the Fuegian Andes and ~40 km landward of the contemporaneous calc-alkaline plutons. Therefore, shoshonitic magmatism represents the initial stages of flattening in Patagonia and its culmination in Tierra del Fuego.

After ~8 Myr of no magmatic activity in Tierra del Fuego, between 68 and 60 Ma, magmatism shifted back to the Pacific side of the archipelago and remained there until its culmination.

Decoupling in the locus, timing and composition of magmatism between the Fuegian and Southern Patagonian Andes from the mid-Cretaceous, but especially from the Paleogene onward, may be related to differential development of the marginal basin and probably to oroclinal bending in the latest Cretaceous–Paleogene. This conclusion agrees with structural models that consider the oroclinal anticlockwise rotation of the Fuegian Andes with respect to the Patagonian Andes, in a context where the Rocas Verdes basin was significantly wider to the south (Kraemer 2003) and is consistent with highly oblique subduction since the Paleogene (Eagles and Scott 2014).

Acknowledgments My special thanks to Márcio Martins Pimentel (Brasilia University) for providing the geochronological data commented in this review. I also appreciate Mónica Escayola and Daniel Acevedo's comments and field assistance during my Ph.D. research; part of the latter has been included in this chapter. Mara Uría helped with translation into English.

References

Acevedo RD, Roig CE, Linares E, Ostera HA, Valín-Alberdi ML, Queiroga-Mafra JM (2000) La intrusión plutónica del Cerro Jeu-Jepén. Isla Grande de Tierra del Fuego, República Argentina. Cadernos do Laboratorio Xeolóxico de Laxe 25:357–359

Acevedo RD, Linares E, Ostera HA, Valín-Alberdi ML (2002) La Hornblendita Ushuaia (Tierra del Fuego): Geoquímica y Geocronología. Rev Asoc Geol Argentina 57(2):133–142

Alabaster T, Storey BC (1990) Modified Gulf of California model for South Georgia, north Scotia Ridge, and implications for the Rocas Verdes back-arc basin, Southern Andes. Geology 18:497–500

Anguita N (2010) Petrogénesis de diques lamprofídicos postectónicos y su relación con el magmatismo shoshonítico cretácico de la Formación Barros Arana y el Complejo La Pera, Magallanes, XII Región, Chile. Graduate Thesis, Univ. Chile, p 237

Barazangi M, Isacks B (1976) Spatial distribution of earthquakes and subduction of the Nazca plate beneath South America. Geology 4:686–692

Barbeau DL, Gombosi DJ, Zahid KM, Bizimis M, Swanson-Hysell N, Valencia V, Gehrels GE (2009) U-Pb zircon constraints on the age and provenance of the Rocas Verdes basin-fill, Tierra del Fuego. Argentina. Geochem Geophys Geosys 10:Q12001

Barker PF (2001) Scotia Sea regional tectonic evolution: implications for mantle flow and palaeocirculation. Earth Sci Rev 55:1–39

Breitsprecher K, Thorkelson DJ (2009) Neogene kinematic history of Nazca-Antarctic-Phoenix slab windows beneath Patagonia and the Antarctic Peninsula. Tectonophysics 464:10–20

Bruce RM, Nelson EP, Weaver SG, Lux DR (1991) Temporal and spatial variations in the Southern Patagonian Batholith; constraints on magmatic arc development. In: Harmon RS, Rapela CW (eds) Andean Magmatism and its tectonic setting. Geol S Am S 265:1–12

Bruhn RL (1979) Rock structures formed during back-arc basin deformation in the Andes of Tierra del Fuego. Geol Soc Am Bull 90:998–1012

Bruhn RL, Stern CR, de Wit MJ (1978) Field and geochemical data bearing on the development of a Mesozoic volcano-tectonic rift zone and back-arc basin in southernmost South America. Earth Planet Sc Lett 41(1):32–46

Calderón M, Fildani A, Hervé F, Fanning CM, Weislogel A, Cordani U (2007) Late Jurassic bimodal magmatism in the northern sea-floor remnant of the Rocas Verdes basin, Southern Patagonian Andes. J Geol Soc Lond 164:1011–1022

Calderón M, Fosdick JC, Warren C, Massonne HJ, Fanning CM, Fadel Cury L, Schwanethal J, Fonseca PE, Galaz G, Gaytán D, Hervé F (2012) The low-grade Canal de las Montañas Shear Zone and its role in the tectonic emplacement of the Sarmiento Ophiolitic complex and Late Cretaceous Patagonian Andes orogeny, Chile. Tectonophysics 525:165–185

Calderón M, Hervé F, Fuentes F, Fosdick JC, Sepúlveda F, Galaz G (2016) Tectonic evolution of Paleozoic and Mesozoic andean metamorphic complexes and the Rocas Verdes ophiolites in southern Patagonia. In: Ghiglione MC (ed) Geodynamic Evolution of the Southernmost Andes. Springer Earth System Sciences, pp 7–36

Carey SW (1955) The orocline concept in geotectonics. P Roy Soc of Tasmania 89:255–288

Castelli J, Robertson R, Harambour S (1992) Evaluación Geológica y Petrolera, bloques Última Esperanza Sur e Isla Riesco. Unpubl. report. Empresa Nacional del Petróleo, Chile, p 160

Cerredo ME, Tassone A, Rapalini A, Hervé F, Fanning M (2011a) Campanian magmatism in the Fuegian Andes: new SHRIMP age of JeuJepen pluton. Argentina. XVIII Argentinian Geological Congress, Neuquén, pp 714–715

Cerredo ME, Tassone A, Peroni J, Menichetti M (2011b) Late Cretaceous magmatism in central Tierra del Fuego: Kranck Pluton, Fuegian Andes, Argentina. I Symposium on Igneous Petrology and Associated Metallogenesis, San Miguel de Tucumán, p 1

Cunningham WD (1993) Strike-slip faults in the Southernmost Andes and the development of the Patagonian Orocline. Tectonics 12:169–186

Cunningham WD (1995) Orogenesis at the southern tip of the Americas: the structural evolution of the Cordillera Darwin metamorphic complex, Southernmost Chile. Tectonophysics 244:197–229

Cunningham WD, Klepeis KA, Gose WA, Dalziel IW (1991) The Patagonian Orocline: new paleomagnetic data from the Andean magmatic arc in Tierra del Fuego, Chile. J Geophys Res 96:16061–16069

Dalziel IWD (1981) Backarc extension in the Southern Andes: a review and critical reappraisal. Philos Trans R Soc Lond 300:319–335

Dalziel IWD (1986) Collision and Cordilleran orogenesis: an Andean perspective. In: Coward MP, Riers AC (eds) Collision Tectonics, Geol Soc Spec Publ 19:389–404

Dalziel IWD, de Wit MF, Palmer KF (1974) Fossil marginal basin in the Southern Andes. Nature 250:291–294

Davidson J, Mpodozis C, Godoy E, Hervé F, Muñoz N (1989) Jurassic accretion of a high buoyancy guyot in the southernmost South America: the Diego Ramírez Islands. Rev Geol Chile 16:247–251

de Wit MJ (1977) The evolution of the Scotia Arc as a key to the reconstruction of Southwestern Gondwanaland. Tectonophysics 37:53–81

Diraison M, Cobbold PR, Gapais D, Rossello EA, Le Corre C (2000) Cenozoic crustal thickening, wrenching and rifting in the foothills of the Southernmost Andes. Tectonophysics 316:91–119

Dott R, Winn R, de Wit M, Bruhn L (1977) Tectonic and sedimentary significance of Cretaceous Tekenika Beds of Tierra del Fuego. Nature 266:620–622

Eagles G, Gohl K, Larter RD (2009) Animated tectonic reconstruction of the Southern Pacific and alkaline volcanism at its convergent margins since Eocene times. Tectonophysics 464:21–29

Eagles G, Scott B (2014) Plate convergence west of Patagonia and the Antarctic Peninsula since 61 Ma. In: Maldonado A, Dalziel IWD, Leat PT (eds) Scotia arc evolution: global implications. Glob Planet Change 123:189–198

Eagles G (2016) Tectonic reconstructions of the Southernmost Andes and the Scotia Sea during the opening of the Drake Passage. In: Ghiglione MC (ed) Geodynamic Evolution of the Southernmost Andes. Springer Earth System Sciences, pp 75–108

Elsztein C (2004) Geología y evolución del Complejo Intrusivo de la Península Ushuaia, Tierra del Fuego. Graduate Thesis (unpubl.), Univ. Buenos Aires, p 103

Elliott T, Plank T, Zindler A, White W, Bourdon B (1997) Element transport from slab to volcanic front at the Mariana Arc. J Geophys Res 102:14991–15019

Espinoza F, Morata D, Lagabrielle Y (2007) Chemical similarities between Miocene Patagonia Calc-alkaline magmatism and some "flat-slab" volcanics: a first approach. International Congress on Geology and Geophysics of the Southern Hemisphere (GeoSur), Santiago de Chile, p 55

Fildani A, Hessler AM (2005) Stratigraphic record across a retroarc basin inversion: Rocas Verdes Magallanes Basin, Patagonian Andes. Geol Soc Am Bull 117:1596–1614

Folguera A, Ramos V (2011) Repeated eastward shifts of arc magmatism in the Southern Andes: a revision to the long-term pattern of Andean uplift and Magmatism. J S Am Earth Sci 32(4):531–546

Fuenzalida R, Covacevich V (1988) Volcanismo y biosestratigrafía del Jurásico Superior y Cretácico Inferior en la Cordillera Patagónica, Región de Magallanes, Chile. V Chilean Geological Congress, Santiago, Actas 3(H):159–183

Galaz G, Hervé F, Calderón M (2005) Metamorfismo y deformación de la Formación Tobífera en la Cordillera Riesco, Región de Magallanes, Chile. Rev Asoc Geol Argent 60:762–774

Gealey WK (1980) Ophiolite obduction mechanisms. In: Panatioyou A (ed) Ophiolites, Proceedings International Ophiolite Symposium, Cyprus. Cyprus Geological Survey, Nicosia, pp 228–243

Ghiglione MC, Ramos VA (2005) Chronology of deformation in the Southernmost Andes of Tierra del Fuego. Tectonophysics 405:25–46

Ghiglione MC, Cristallini E (2007) Have the Southernmost Andes been curved since Late Cretaceous time? An analog test for the Patagonian Orocline. Geology 35:13–16

Gill JB (1981) Orogenic andesite and plate tectonics. Springer, New York

Gombosi DJ, Barbeau DL, Garver J (2009) New thermochronometric constraints on the rapid Paleogene exhumation of the Cordillera Darwin complex and related thrust sheets in the Fuegian Andes. Terra Nova 21:507–515

González Guillot M (2009) Estudio petrogenético de plutones de la Cordillera Fueguina entre el lago Fagnano y el canal Beagle y algunas consideraciones sobre las mineralizaciones asociadas. Ph.D. thesis (unpubl.), Univ. Nac. La Plata, p 327

González Guillot M, Acevedo D (2009) Facies con biotita y granate en plutones de los Andes Fueguinos de Argentina. XII Chil Geol Congr, Santiago de Chile S8–014:4

González Guillot M, Schalamuk I (2009) Mineralizaciones metalíferas en rocas ultramáficas de Tierra del Fuego. Argentina. IX Argentinian Congress of Economic Geology, Catamarca, pp 19–24

González Guillot M, Escayola M, Acevedo R, Pimentel M, Seraphim G, Proenza J, Schalamuk I (2009) The Plutón Diorítico Moat: Mildly alkaline monzonitic magmatism in the Fuegian Andes of Argentina. J S Am Earth Sci 28:345–359

González Guillot M, Escayola M, Acevedo R (2011) Calc-alkaline rear-arc magmatism in the Fuegian Andes: implications for the mid-Cretaceous tectonomagmatic evolution of southernmost South America. J S Am Earth Sci 31:1–16

González Guillot M, Prezzi C, Acevedo RD, Escayola M (2012) A comparative study of two rear-arc plutons and implications for the Fuegian Andes tectonic evolution: mount Kranck Pluton and Jeu-Jepén Monzonite, Argentina. J S Am Earth Sci 38:71–88

González Guillot M, Urraza I, Acevedo RD, Escayola M (2016) Magmatismo básico Jurásico-Cretácico en los Andes Fueguinos y su relación con la cuenca marginal Rocas Verdes. Rev Asoc Geol Argentina (in press)

Grunow AM (1993) Creation and destruction of Weddell Sea floor in the Jurassic. Geology 21:647–650

Gust DA, Biddle KT, Phelps DW, Uliana MA (1985) Associated middle to late Jurassic volcanism and extension in southern South America. Tectonophysics 116:223–253

Halpern M (1973) Regional geochronology of Chile South of 50° Latitude. Geol Soc Am Bull 84(7):2407–2422

Halpern M, Rex DC (1972) Time of folding of the Yahgan Formation and age of the Tekenika beds, Southern Chile, South America. Geol Soc Am Bull 83:1881–1886

Hanson R, Wilson T (1991) Submarine rhyolitic volcanism in a Jurassic proto-marginal basin; Southern Andes, Chile and Argentina. In: Harmon R, Rapela C (eds) Andean Magmatism and its tectonic setting. Geol S Am S 265:13–27

Hanyu T, Gill J, Tatsumi Y, Kimura JI, Sato K, Chang Q, Senda R, Miyazaki T, Hirahara Y, Takahashi T, Zulkarnain I (2012) Across- and along-arc geochemical variations of lava chemistry in the Sangihe arc: various fluid and melt slab fluxes in response to slab temperature. Geochem Geophys Geosys 13(10):Q10021

Hawkesworth C, Turner S, Peate D, McDermott F, van Calsteren P (1997) Elemental U and Th variations in island arc rocks: implications for U-series isotopes. Chem Geol 139:207–221

Hervé F, Nelson E, Kawashita K, Suárez M (1981) New isotopic ages and the timing of orogenic events in the Cordillera Darwin, Southernmost Chilean Andes. Earth Planet Sc Lett 55(2): 257–265

Hervé F, Pankhurst R, Fanning C, Calderón M, Yaxley G (2007a) The South Patagonian Batholith: 150 my of granite magmatism on a plate margin. Lithos 97:373–394

Hervé F, Massonne HJ, Calderón M, Theye T (2007b) Metamorphic P-T conditions of Late Jurassic rhyolites in the Magallanes fold and thrust belt, Patagonian Andes, Chile. J Iberian Geol 33(1):5–16

Hervé M, Suárez M, Puig A (1984) The Patagonian Batholith south of Tierra del Fuego, Chile. Timing and tectonic implications. J Geol Soc Lond 141(5):909–917

Johnson M, Plank, T (1999) Dehydration and melting experiments constrain the fate of subducted sediments. Geochem Geophys Geosys 1 GC000014

Katz HR, Watters WA (1966) Geological investigations of the Yahgán Formation (Upper Mesozoic) and associated igneous of Navarino island, southern Chile. N Z J Geol Geophys 9(3):323–359

Kay SM, Mpodozis C (2002) Magmatism as a prove to the Neogene shallowing of the Nazca plate beneath the modern Chilean flat-slab. J S Am Earth Sci 15:39–57

Kay SM, Ramos VA, Mpodozis C, Sruoga P (1989) Late Paleozoic to Jurassic silicic magmatism at the Gondwana margin: analogy to the middle Proterozoic in North America. Geology 17:324–328

Kimura J-I, Kent A, Rowe M, Katakuse M, Nakano F, Hacker B, van Keken P, Kawabata H, Stern R (2010) Origin of cross-chain geochemical variation in Quaternary lavas from the northern Izu arc: Using a quantitative mass balance approach to identify mantle sources and mantle wedge processes. Geochem Geophys Geosys 11(10):Q10011

Klepeis KA (1994a) Relationship between uplift of the metamorphic core of the Southernmost Andes and shortening in the Magallanes foreland fold and thrust belt, Tierra del Fuego, Chile. Tectonics 13:882–904

Klepeis KA (1994b) The Magallanes and Deseado fault zones: major segments of the South American-Scotia transform plate boundary in Southernmost South America, Tierra del Fuego. J Geophys Res 99(B11):22,001–22,014

Klepeis KA, Austin JA (1997) Contrasting styles of superposed deformation in the Southernmost Andes. Tectonics 16(5):755–776

Klepeis K, Betka P, Clarke G, Fanning M, Hervé F, Rojas L, Mpodozis C, Thomson S (2010) Continental underthrusting and obduction during the Cretaceous closure of the Rocas Verdes rift basin, Cordillera Darwin, Patagonian Andes. Tectonics 29 TC3014

Kohn MJ, Spear FS, Harrison TM, Dalziel IWD (1995) 40Ar/39Ar geochronology and P-T–t paths from the Cordillera Darwin metamorphic complex, Tierra del Fuego, Chile. J Met Geol 13:251–270

Kraemer PE (2003) Orogenic shortening and the origin of the Patagonian orocline (56 S Lat). J S Am Earth Sci 15:731–748

Kranck EH (1932) Geological investigations in the Cordillera of Tierra del Fuego. Acta Geogr Helsinki 4(2):1–231

Lameyre J (1987) Granites and evolution of the crust. Rev Bras Geociências 17(4):349–359

Lameyre J, Bowden P (1982) Plutonic rock type series: discrimination of various granitoids series and related rocks. J Volcanol Geoth Res 14:169–186

Lameyre J, Bonin B (1991) Granites in the main plutonic series. In: Didier J, Barbarin B (eds) Enclaves and granite petrology. Dev Petrol 13:3–17. Elsevier

Likerman L, Burlando JF, Cristallini EO, Ghiglione MC (2013) Along-strike structural variations in the Southern Patagonian Andes: insights from physical modeling. Tectonophysics 590:106–120

Lodolo E, Menichetti M, Bartole R, Ben-Avraham Z, Tassone A, Lippai H (2003) Magallanes-Fagnano continental transform fault (Tierra del Fuego, southernmost South America). Tectonics 22:1076

Maffione M (2016) Kinematic evolution of the Southern Andean orogenic arc. In: Ghiglione MC (ed) Geodynamic Evolution of the Southernmost Andes. Springer Earth System Sciences, pp 173–200

Maffione M, Speranza F, Faccenna C, Rossello E (2010) Paleomagnetic evidence for a pre-early Eocene ($\sim$50 Ma) bending of the Patagonian orocline (Tierra del Fuego, Argentina): Paleogeographic and tectonic implications. Earth Pla-net Sc Lett 289:273–286

Maffione M, Fernandez-Moreno C, Ghiglione M, Speranza F, van Hinsbergen DJJ, Lodolo E (2015) Constraints on deformation of the Southern Andes since the Cretaceous from anisotropy of magnetic susceptibility. Tectonophysics 665:236–250

Maloney KT, Clarke GL, Klepeis KA, Fanning CM, Wang W (2011) Crustal growth during back-arc closure: Cretaceous exhumation history of Cordillera Darwin, Southern Patagonia. J Metamorph Geol 29(6):649–672

Marsh BD (1979) Island arc development: some observations, experiments, and speculations. J Geol 87:687–713

Martin M, Pankhurst R, Fanning C, Thompson S, Calderón M, Hervé F (2001) Age and distribution of plutons across the Southern Patagonian Batholith: new U-Pb data on zircons. South American symposium on Isotope Geology. Sociedad Geológica de Chile, Santiago. Extended Abstracts (CD): 585–588

Martinioni DR, Linares E, Acevedo RD (1999) Significado de la edad isotópica de diques básicos intruidos en la Formación Beauvoir (Cretácico temprano) Tierra del Fuego. Argentina. Rev Asoc Geol Argentina 54(1):88–91

McAtamney J, Klepeis K, Mehrtens C, Thomson S, Betka P, Rojas L, Snyder S (2011) Along-strike variability of back-arc basin collapse and the initiation of sedimentation in the Magallanes foreland basin, Southernmost Andes (53–54.5°S). Tectonics 30 TC5001

Menichetti M, Lodolo E, Tassone A (2008) Structural geology of the Fuegian Andes and Magallanes fold and thrust belt-Tierra del Fuego island. Geologica Acta 6:19–42

Michael PJ (1991) Intrusion of basaltic magmas into a crystallizing granitic magma chamber; the Cordillera del Paine Pluton in southern Chile. Contrib Miner Petr 108:396–418

Michel J, Baumgartner L, Putlitz B, Schaltegger U, Ovtcharova M (2008) Incremental growth of the Patagonian Torres del Paine laccolith over 90 ky. Geology 36:459–462

Middlemost EAK (1991) Towards a comprehensive classification of igneous rocks and magmas. Earth Sc Rev 31:73–87

Miller CA, Barton M, Hanson RE, Fleming TH (1994) An Early Cretaceous volcanic arc/marginal basin transition zone, Peninsula Hardy, southernmost Chile. J Volcanol Geoth Res 63(1–2): 33–58

Miyashiro A (1978) Nature of alkalic volcanic rock series. Contrib Miner Petr 66:91–104

Morris JD, Ryan JG (2003) Subduction zone processes and implications for changing composition of the upper and lower mantle. In: Carlson RW (ed) The mantle and core, vol 2. Treatise on Geochemistry, Elsevier-Pergamon-Oxford, pp 451–470

Morrison GW (1980) Characteristics and tectonics of the shoshonite rock association. Lithos 13:97–108

Mpodozis C (2006) Revisión de la estratigrafía y facies de la cuenca de antepaís cretácica entre Lago Argentino y Península Brunswick. Unpubl. report. Empresa Nacional del Petróleo, Chile, p 161

Mpodozis C, Rojas L (2006) Orogénesis en los Andes Patagónicos Australes de Tierra del Fuego: cierre de una "cuenca marginal" o colisión intracontitnetal? XI Chil Geol Congr, Antofagasta, Chile 2:283–286

Mpodozis C, Alvarez P, Elgueta S, Mella P, Hervé F, Fanning M (2007) Revised Cretaceous stratigraphy of the Magallanes Foreland Basin at Seno skyring: regional implications of new SHRIMP age data on detrital zircon populations. International Congress on the Geology and Geophysics of the Southern Hemisphere (GeoSur), Santiago de Chile, p 106

Mukasa SB, Dalziel IWD (1996) Southernmost Andes and South Georgia island, north Scotia Ridge: zircon U-Pb and muscovite Ar/Ar age constraints on tectonic evolution of southwestern Gondwanaland. J S Am Earth Sci 9:349–365

Mukasa SB, Brueckner HK, Dalziel IWD (1988) Zircon U-Pb constraints on the kinematic evolution of the Northern Scotia Arc. Geol Soc Am Abstr Programs 20:A12

Nelson EP (1982) Postectonic uplift of the Cordillera Darwin orogenic core complex: evidence for fission track geochronology and closing temperature time relationships. J Geol Soc Lond 139:755–761

Nelson EP, Dalziel IWD, Milnes AG (1980) Structural geology of the Cordillera Darwin-collision style orogenesis in the Southernmost Andes. Eclogae Geol Helv 73:727–751

Nelson E, Bruce B, Elthon D, Kammer D, Weaver S (1988) Regional lithologic variations in the Patagonian Batholith. J S Am Earth Sci 1(3):239–247

Olivero E, Martinioni D (2001) A review of the geology of the Argentinian Fuegian Andes. J S Am Earth Sci 14:175–188

Olivero E, Malumián N (2007) Mesozoic-Cenozoic stratigraphy of the Fuegian Andes Argentina. Geologica Acta 6(1):5–18

Pagel M, Leterrier J (1980) The subalkaline potassic magmatism of the Ballons massif (Southern Vosges, France): shoshonitic affinity. Lithos 13:1–10

Pankhurst RJ, Leat PT, Sruoga P, Rapela CW, Márquez M, Storey BC, Riley TR (1998) The Chon Aike province of Patagonia and related rocks in West Antarctica: a silicic large igneous province. J Volcanol Geoth Res 81:113–136

Pearce J, Stern R, Bloomer S, Fryer P (2005) Geochemical mapping of the Mariana Arc-Basin System: implications for the nature and distribution of subduction components. Geochem Geophys Geosys 6 2004GC000895

Pelayo A, Wiens D (1989) Seismotectonics and relative plate motions in the Scotia Sea region. J Geophys Res 94:7293–7320

Peroni J, Tassone A, Menichetti M, Cerredo M (2009) Geophysical modeling and structure of Ushuaia Pluton, Fuegian Andes, Argentina. Tectonophysics 476:436–449

Poblete F, Roperch P, Hervé F, Diraison M, Espinoza M, Arriagada C (2014) The curved Magallanes fold and thrust belt: tectonic insights from a paleomagnetic and anisotropy of magnetic susceptibility study. Tectonics 33(12):2526–2551

Prades CF (2008) Petrología y metamorfismo de las rocas basálticas en Isla Capitán Aracena, Isla Carlos III y Estero La Pera, Región de Magallanes, Chile. Graduate Thesis, Univ. Chile, p 145

Puig A, Hervé M, Suárez M, Saunders A (1984) Calc-alkaline and alkaline Miocene and calc-alkaline recent volcanism in the Southernmost Patagonia Cordillera, Chile. J Volcanol Geoth Res 20:149–163

Ramírez de Arellano C, Putlitz B, Müntener O, Ovtcharova M (2012) High precision U/Pb zircon dating of the Chaltén Plutonic Complex (Cerro Fitz Roy, Patagonia) and its relationship to arc migration in the Southernmost Andes. Tectonics 31 TC4009

Ramos VA, Haller MJ, Butrón F (1986) Geología y Evolución tectónica de las Islas Barnevelt. Rev Asoc Geol Argentina 15:137–154

Ramos VA, Kay SM, Singer B (2004) Las adakitas de la Cordillera Patagónica: Nuevas evidencias geoquímicas y geocronológicas. Rev Asoc Geol Argentina 59(4):693–706

Rapalini A, Peroni J, Luppo T, Tassone A, Cerredo ME, Esteban F, Lippai H, Vilas F (2015) Palaeomagnetism of Mesozoic magmatic bodies of the Fuegian Cordillera: implications for the formation of the Patagonian Orocline. In: Pueyo EL, Cifelli F, Sussman AJ, Oliva-Urcia B (eds) Palaeomagnetism in Fold and Thrust Belts: New Perspectives. Geol Soc Spec Publ 425 http://doi.org/10.1144/SP425.3

Rickwood PC (1989) Boundary lines between petrologic diagrams which use oxides of major and minor elements. Lithos 22:247–263

Rollinson HR (1993) Using geochemical data: evaluation presentation, interpretation. Longman Scientific & Technical, England, p 352

Rossello EA (2005) Kinematics of the Andean sinistral wrenching along the Fagnano-Magallanes fault zone (Argentina-Chile Fuegian foothills). 6th International Symposium on Andean Geodynamics (ISAG), Barcelona. Extended Abstracts, pp 623–626

Rubio E, Torné M, Vera E, Díaz A (2000) Crustal structure of the Southernmost Chilean margin from seismic and gravity data. Tectonophysics 323:39–60

Sakuyama N, Nesbitt RW (1986) Geochemistry of the Quaternary volcanic rocks of the Northeast Japan arc. J Volcanol Geoth Res 29:413–450
Sánchez A, Saint-Blanquat M, Hervé F, Pankhurst RJ, Fanning CM (2006) A Shrimp U-Pb zircon late Miocene crystallization age for the Torres Del Paine Pluton, Chile. In: Proceedings of V South American Symposium on Isotope Geology, Punta del Este, Uruguay, 24–27 April, 2006
Saunders AD, Tarney J, Stern CR, Dalziel IWD (1979) Geochemistry of Mesozoic marginal basin floor igneous rocks from southern Chile. Geol Soc Am Bull 90:237–258
Sernageomin (2003) Mapa Geológico de Chile: versión digital, escala 1:1,000,000. Servicio Nacional de Geología y Minería, Santiago, Chile
Soffia JM (1988) Evaluación geológica y petrolera del sector sur de la provincia de Última Esperanza. Unpubl. report. Empresa Nac del Petróleo, Chile, p 89
Stern CR (1979) Open and closed system igneous fractionation within two Chilean ophiolites and their tectonic implications. Contrib Miner Petr 68:243–258
Stern CR (1980) Geochemistry of Chilean ophiolites, evidence for the compositional evolution of the mantle source of back-arc basin basalts. J Geophys Res 85(B2):955–966
Stern CR (1991) Role of subduction erosion in the generation of Andean magmas. Geology 19: 78–81
Stern CR, De Wit MJ (2003) Rocas Verdes ophiolites, Southernmost South America: remnants of progressive stages of development on oceanic-type crust in a continental margin back-arc basin. In Dilek Y, Robinson PT (eds) Ophiolites in earth history. Geol Soc Spec Publ 218:1–19
Stern CR, Kilian R (1996) Role of the subducted slab, mantle wedge and continental crust in the generation of adakites from the Andean Austral Volcanic Zone. Contrib Miner Petr 123: 263–281
Stern CR, Mohseni PP, Fuenzalida R (1991) Petrochemistry and tectonic significance of Lower Cretaceous Barros Arana Formation basalts, Southernmost Chilean Andes. J S Am Earth Sci 4:331–342
Storey BC, Alabaster T (1991) Tectonomagmatic controls on Gondwana break-up models: Evidence from the proto-Pacific margin of Antarctica. Tectonics 10:1274–1288
Suárez M (1977a) Notas geoquímicas preliminares del Batolito Patagónico al sur de Tierra del Fuego, Chile. Rev Geol Chile 4:15–33
Suárez M (1977b) Aspectos geoquímicos del Complejo Ofiolítico Tortuga en la Cordillera Patagónica del sur de Chile. Rev Geol Chile 4:3–14
Suárez M, Pettigrew TH (1976) An upper Mesozoic island-arc back-arc system in the Southern Andes and South Georgia. Geol Mag 113:305–400
Suárez M, Hervé M, Puig A (1985) Hoja Isla Hoste e islas adyacentes, XII Región. Carta Geológica de Chile 1:250.000, no. 65. Servicio Nacional de Geología y Minería, Chile, p 113
Suárez M, Puig A, Hervé M (1986) K-Ar dates on granitoids from Archipiélago Cabo de Hornos, Southernmost Chile. Geol Mag 123(5):581–584
Suárez M, Naranjo JA, Puig A (1990) Mesozoic "S-like" granites of the central and Southern Andes: a review. In: Kay SM, Rapela CW (eds) Plutonism from Antarctica to Alaska. Geol S Am S 241:27–32
Sun SS, McDonough WF (1989) Chemical and isotopic systematics of oceanic basalts: implications for mantle composition and processes. In: Saunders AD, Norry MJ (eds) Magmatism in Ocean Basins. Geol Soc Spec Publ 42:313–345
Tatsumi Y (2003) Some constraints on arc magma genesis. In: Eiler J (ed) Inside the subduction factory. AGU, Geophys Monogr 138:277–292
Tatsumi Y, Eggins S (1995) Subduction zone magmatism. Blackwell, Boston
Tatsumi Y, Kogiso T (1997) Trace element transport during dehydration processes in the subducted oceanic crust: 2. Origin of chemical characteristics in arc magmas. Earth Planet Sc Lett 148:207–222
Tatsumi Y, Sakuyama M, Fukuyama H, Kushiro I (1983) Generation of arc basalt magmas and thermal structure of the mantle wedge in subduction zones. J Geophys Res 88:5815–5825
Taylor SR, McLennan SM (1985) The Continental crust: its composition and evolution. Blackwell, Boston, MA

Torres Carbonell P, Dimieri L (2013) Cenozoic contractional tectonics in the Fuegian Andes, Southernmost South America: a model for the transference of orogenic shortening to the foreland. Geol Acta 11:331–357

Torres Carbonell P, Dimieri L, Olivero E, Bohoyo F, Galindo-Zaldívar J (2014) Structure and tectonic evolution of the Fuegian Andes (Southernmost South America) in the framework of the Scotia Arc development. Glob Planet Change 123:174–188

Yogodzinski G, Volynets O, Koloskiv A, Seliverstov N, Matvenkov V (1994) Magnesian andesites and the subduction component in a strongly calc-alkaline series at Piip volcano, Far Western Aleutians. J Petrol 35:163–204

Yogodzinski G, Kay R, Volynets O, Koloskov A, Kay S (1995) Magnesian andesite in the western Aleutian Komandorsky region: implications for slab melting and processes in the mantle wedge. Geol Soc Am Bull 107:505–519

Tectonic Reconstructions of the Southernmost Andes and the Scotia Sea During the Opening of the Drake Passage

Graeme Eagles

Abstract Study of the tectonic development of the Scotia Sea region started with basic lithological and structural studies of outcrop geology in Tierra del Fuego and the Antarctic Peninsula. To nineteenth- and early twentieth-century geologists, the results of these studies suggested the presence of a submerged orocline running around the margins of the Scotia Sea. Subsequent increases in detailed knowledge about the fragmentary outcrop geology from islands distributed around the margins of the Scotia Sea, and later their interpretation in the light of the plate tectonic paradigm led to large modifications in the hypothesis such that by the present day the concept of oroclinal bending in the region persists only in vestigial form. Of the early comparative lithostratigraphic work in the region, only the likenesses between Jurassic–Cretaceous basin floor and fill sequences in South Georgia and Tierra del Fuego are regarded as strong enough to be useful in plate kinematic reconstruction by permitting the interpretation of those regions' contiguity in mid-Mesozoic times. Marine and satellite geophysical data sets reveal features of the remaining, submerged, 98 % of the Scotia Sea region between the outcrops. These data enable a more detailed and quantitative approach to the region's plate kinematics. In contrast to long-used interpretations of the outcrop geology, these data do not prescribe the proximity of South Georgia to Tierra del Fuego in any past period. It is, however, possible to reinterpret the geology of those two regions in terms of the plate kinematic history that the seafloor has preserved.

Keywords Southernmost Andes · Scotia Sea · Scotia Ridge · Antarctica · Antarctic Peninsula · Drake Passage

G. Eagles (✉)
Helmholtz Centre for Marine and Polar Research, Alfred Wegener Institute, Bremerhaven, Germany
e-mail: geagles@awi.de

© Springer International Publishing Switzerland 2016

M.C. Ghiglione (ed.), *Geodynamic Evolution of the Southernmost Andes*, Springer Earth System Sciences, DOI 10.1007/978-3-319-39727-6_4

1 Introduction

The Scotia Sea is the name given to the area of the Southern Ocean lying to the east of Drake Passage, the strait that separates Tierra del Fuego from the Antarctic Peninsula (Fig. 1). It is named after the ship *S.Y. Scotia* that brought W.S. Bruce's Scottish National Antarctic Expedition to the region in 1902–1904, but has been known to Western scientists, including the young Charles Darwin, since the early nineteenth century. Since that time, Drake Passage and the Scotia Sea have been suggested to play important roles for processes as diverse as migration, colonisation and evolution of the land fauna of South America and Australia (e.g. Reguero and Marenssi 2010), evolution and migration of volcanic vent-specific fauna between the Pacific and Atlantic oceans (Roterman et al. 2013), whale and penguin evolution (Fordyce 2003; Hospitaleche et al. 2013) and the flow of asthenospheric rocks from beneath the Pacific plates to beneath those of the Atlantic (Helffrich et al. 2002; Müller 2001; Nerlich et al. 2013). Most famously, however, an open Drake Passage is frequently referred to as a necessary prerequisite for, or even a driver of, the onset of the Antarctic Circumpolar Current and the attendant permanent glaciation of Antarctica in Neogene times (Kennett 1977; Toggweiler and Bjornsson 2000).

All these processes are active and changeable on geological timescales, making the question of the first completion of Drake Passage pertinent to understanding them better. The passage exists as a result of oblique separation of the South American and Antarctic plates, which in turn caused the local development and relative motion of a number of smaller, shorter-lived plates, whose remnants constitute the floor of the Scotia Sea and its margins. The main archives of the motions of these plates are the exposed rocks dotted around the Scotia Sea margins, and the patterns of fracture zones and magnetic reversal isochrons in the floor of the Scotia Sea itself. This chapter reviews what can be interpreted of the plate tectonic history of the region by using these resources.

2 Comparative Tectonostratigraphy

Similarities between the Southernmost Andes and the mountain ranges of the Antarctic Peninsula have been sought and presented since the early nineteenth century. Arctowski (1895) first proposed that the similarities are the result of oroclinal folding on a grand scale and that the rocks of the Andes continue into the peninsula via Shag Rocks and South Georgia on the North Scotia Ridge, the South Sandwich Islands, the South Orkney Islands on the South Scotia Ridge and the South Shetland Islands. These rocks and islands are all remote and their emergent parts encompass less than 2 % of the Scotia Sea (Fig. 1). Less still is accessible outcrop.

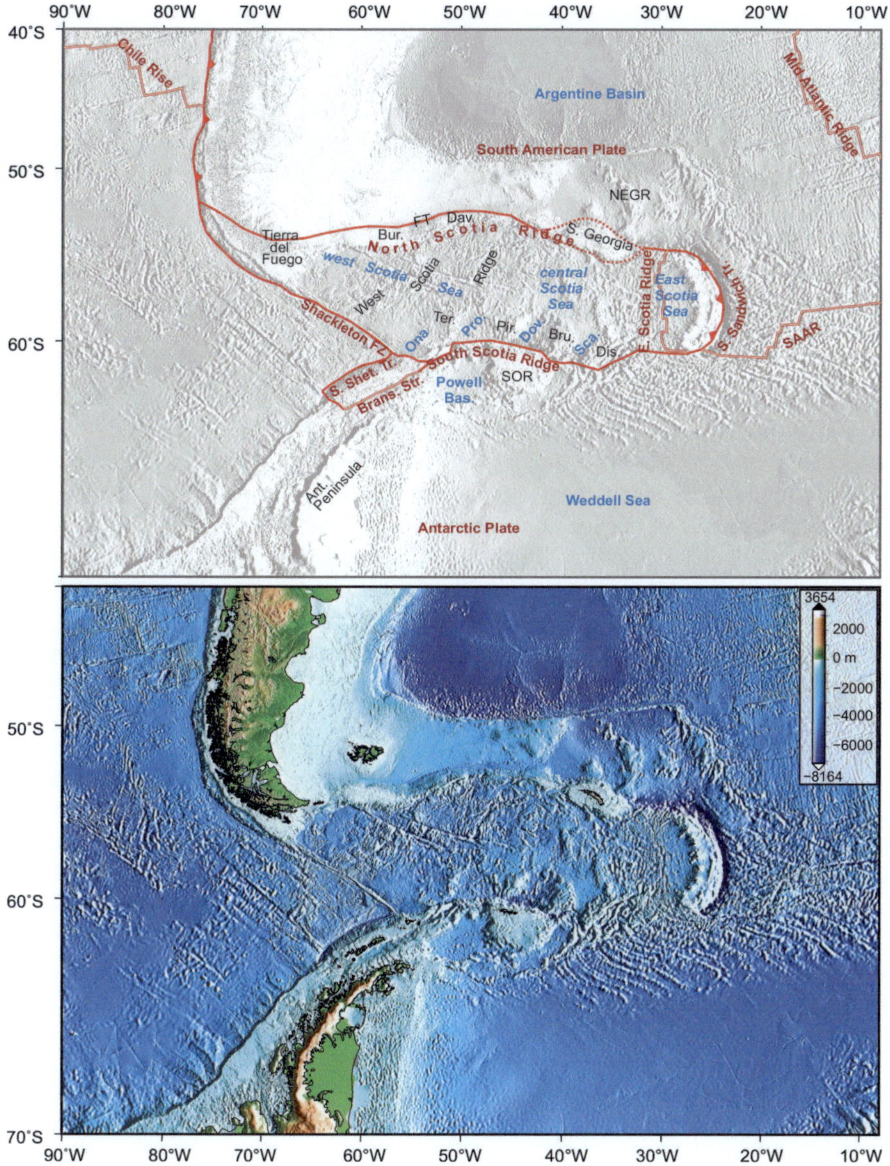

Fig. 1 Overview of the Scotia Sea region in bathymetric data (SRTM 15 PLUS; Becker et al. 2009). *Top Red lines*—present-day plate boundaries (modified from Bird 2003). *Red labels*: presently active tectonic features. *Blue labels*: basins and basin regions. *Black labels*: extinct tectonic features and bathymetric highs. Brans. Str: Bransfield Strait; *Bru* Bruce Bank; *Bur* Burdwood Bank; *Dav* Davis Bank; *Dis* Discovery Bank; *Dov* Dove Basin; *FT* Falkland Trough; *NEGR* North-east Georgia Rise; *Pir* Pirie Bank; *Pro* Protector Basin; *Sca* Scan Basin; *SOR* South Orkney Microcontinent; *S.* Sandwich *Tr.* South Sandwich Trench; *S.* Shet. *Tr.* South Shetland Trench; *Ter* Terror Rise. *Bottom colour* background image without references

The idea of a tremendous Scotian Orocline remained popular through the first part of the twentieth century as visits to these outcrops confirmed the presence of Mesozoic rock assemblages like those of the Southernmost Andes and Antarctic Peninsula (Pirie 1905; Suess 1909; Tilley 1935; Trendall 1953, 1959). The topic of oroclinal bending in the region has remained the subject of intense debate amongst structural geologists and palaeomagnetists to this day (e.g. Cunningham 1995; Maffione et al. 2010, 2015; Maffione 2016).

By the second half of the twentieth century, the explanatory power of plate tectonics, along with improved sampling and dating techniques, brought the opportunity to assess details of the similarities of exposed rocks around the Scotia Sea's margins. The tectonostratigraphy of the Antarctic Peninsula and Tierra del Fuego came to be differentiated on the basis of differences in lithology, rock age and metamorphism. By then, close comparisons between the rocks of the South Orkney Islands and those of South Georgia or the South Shetland Islands were no longer supported by observations. Where correlations for the South Orkney Islands are still suggested, these are with upper Palaeozoic and lower Mesozoic rocks of the Antarctic Peninsula and the Southern Patagonian Andes north of Tierra del Fuego. These correlations are interpreted in terms of a shared history of subduction during that period, prompting suggestions of geographical proximity (Barron et al. 1978; Calderón et al. 2007, 2016; Hervé et al. 2003, 2005).

The most celebrated set of affinities in the region, however, are those between the Upper Jurassic and Lower Cretaceous rocks of Tierra del Fuego and South Georgia, which today occupy locations 1600 km apart at opposing ends of the boundary between the Scotia and South American plates (Fig. 1). These affinities are seen in the age and source of deposition of turbidites and interpreted for the timing and setting of their deformation. The inset in Fig. 2 summarises the often-reported 'grain for grain' comparability of the Aptian–Albian, Sandebugten and Cumberland Bay formations of South Georgia to the Yahgán Formation of central Tierra del Fuego (Dalziel et al. 1975; Winn 1978). Both sequences consist of turbidites of andesitic debris overlain by more quartz-rich turbidites sourced by erosion of rhyolitic rocks to the north. Carter et al. (2014) presented U–Pb dating work on zircon grains that confirm how the three formations share similar Mesozoic sources. Furthermore, both turbiditic sequences are bounded to the south by ophiolitic complexes of Early Cretaceous age. The rocks are folded on similar wavelengths and the axes have similar orientations, which have been interpreted to indicate a common deformational setting at some period. This period has usually been put in mid-Cretaceous times, because various lines of evidence from the Yahgán rocks and elsewhere in Tierra del Fuego suggest that a protracted period of basin closure started in the period before 86 Ma (Klepeis et al. 2010). Less prescriptively, Rb–Sr analyses on minerals from a biotite schist (84 Ma; Curtis et al. 2010) and K–Ar whole-rock dating (135–51 Ma with best fits for 92–81 Ma; Thomson et al. 1982) record the timing of greenschist facies metamorphism around the same time near the Cooper Bay Shear Zone in south-eastern South Georgia. At other locations on the island, widespread prehnite occurrence and zircon- and apatite-based thermal histories show that granites were intruded at $\sim$100 Ma, that

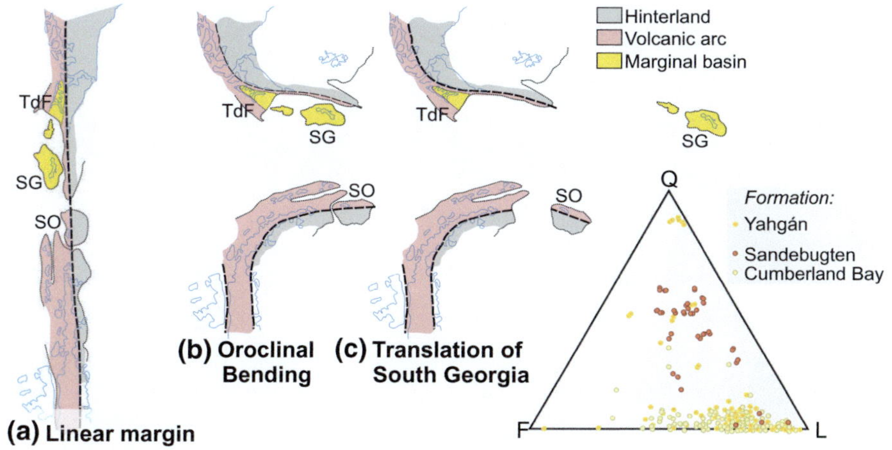

Fig. 2 Survival of the late-nineteenth-century orocline concept into the plate tectonic age and its modification by **a** segregation of the limbs into three elements related to subduction (volcanic arc, back-arc marginal basin and hinterland) and **b** bending and **c** post-bending translation of continental fragments (South Georgia; SG) and the South Orkney microcontinent (SO) away from Tierra del Fuego (TdF) (redrawn from Dalziel and Elliot 1971). *Bottom right*: Quartz–feldspar–lithics plots demonstrate the comparability of sedimentary rocks on South Georgia (Cumberland Bay and Sandebugten Formations) to those of similar age on Tierra del Fuego (Yahgán Formation). From Dalziel et al. (1975) and Winn (1978)

the neighbouring Cumberland Bay and Sandebugten formations were subsequently buried at shallow-to-moderate depths and that rapid uplift occurred starting in Eocene times (Clayton 1982; Carter et al. 2014).

These correlations are strong enough to support the notion of South Georgia having occupied a location immediately to the east of Isla Navarino and Tierra del Fuego in Early Cretaceous times (Dalziel and Elliot 1971). According to this idea, the Yahgán, Cumberland Bay and Sandebugten formations were deposited at that time side-by-side in a basin floored by oceanic crust, the so-called Rocas Verdes Marginal Basin (Fig. 3) (see Calderón et al. 2016; Guillot 2016). Later, both sets of rocks experienced orogenesis as the basin closed, which is usually attributed to an acceleration in the rate of seafloor spreading in the South Atlantic that increased the vigour of collision between the plates of the palaeo-Pacific ocean and South America (Dalziel 1986; Curtis et al. 2010). Since significant lateral translation is required for South Georgia to have subsequently reached its present-day position on the North Scotia Ridge, the orocline concept had to be modified to accommodate the idea of its limbs' fragmentation (Dalziel and Elliot 1971; Fig. 2a–c).

Figure 3 summarises the power of tectonostratigraphic correlations for regional reconstructions of the Scotia Sea. Available correlations can clearly support a wide range of such reconstructions. The similarities of South Georgia and Tierra del Fuego constitute essentially a single correlation point, permitting a wide angular range of possible locations for South America with respect to the Antarctic peninsula. Dalziel (1983) notes that, without additional information from plate kinematic

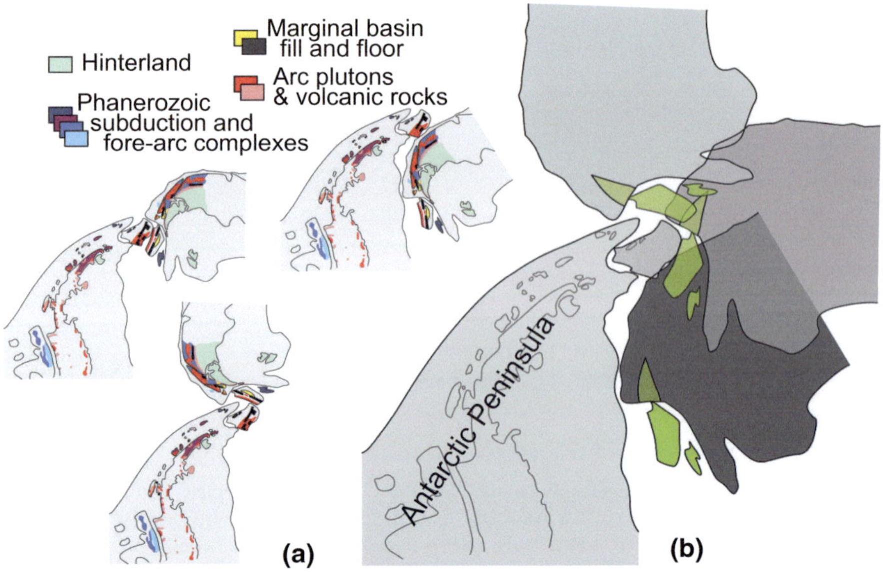

Fig. 3 **a** Three regional reconstructions that can be built on the basis of tectonostratigraphic correlations between outcrop geology in and around the Scotia Sea. Redrawn from Dalziel (1983). **b** Without other sources of constraint than the sedimentological similarities between the Cumberland Bay, Sandebugten and Yahgán formations of South Georgia and the Rocas Verdes basin (*green*), the three reconstructions must be seen as equally likely and thus of limited value for reconstructing the opening of the Scotia Sea

studies of the basins forming the Scotia Sea floor, there would be little reason to prefer any of these possibilities over the others. The next section reviews studies of this kind and their results, which have been developed since the late 1960s.

3 Basin Restoration

The recognition of magnetic anomaly isochrons over the seafloor in the Scotia Sea accompanied publication of the first magnetic reversal timescales in the late 1960s and early 1970s (Barker 1970, 1972). Using those timescales, a number of oceanic crustal provinces were identified on the basis of variable anomaly orientations and tentatively dated to periods over the last 40 Ma (Fig. 4). These provinces were identified as the products of episodes of plate divergence that had given rise to the creation of oceanic crust by volcanism. Taking advantage of the tenets of plate tectonics, that the interiors of plates are rigid and that deformation occurs only at the boundaries between pairs of plates, it is possible to reconstruct the development of the seafloor by reuniting the magnetic isochron pairs. The newly identified isochrons showed that the majority of the Scotia Sea floor had developed by east–west

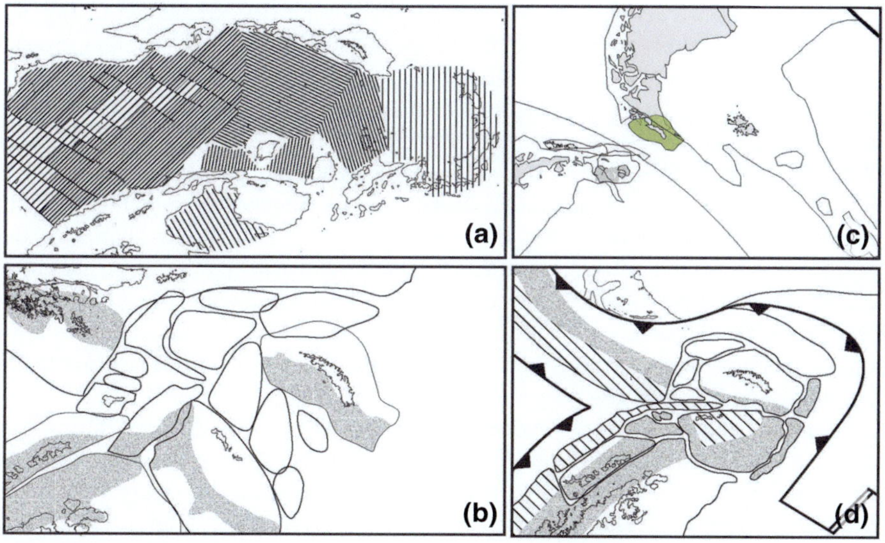

Fig. 4 **a** Magnetic anomaly provinces as interpreted from magnetic anomalies over the floor of the Scotia Sea by Barker (1970, 1972). The orientation of the hatching indicates the strike of the magnetic anomalies. Close hatching dated as 'post-Cretaceous' and wide hatching as <10 Myr old seafloor. **b** Plate kinematic reconstruction produced on a flat map by removing the oceanic crustal provinces between shallower, presumed continental, areas of crust. Stipple: areas of intense magnetic anomalies. Note that a close placement of South Georgia to Tierra del Fuego, as prescribed by sedimentological comparisons, is not a simple result of closing the oceanic basins of the Scotia Sea. **a** and **b** both redrawn from Barker and Griffiths (1972). **c** 50 Ma sphere-based plate reconstruction adapted from Lawver and Gahagan (2003), with emphasis on fidelity to tectonostratigraphic comparisons between South Georgia (*green*) and Tierra del Fuego. Note that this reconstruction, despite its numerical precision, ignores much of the detail addressed by Barker and Griffiths (1972). **d** Flat map reconstruction with all basins closed, generating a cuspate connection and placing South Georgia and Tierra del Fuego in closer proximity than (**b**). Stipple: areas with strong magnetic anomalies. Hatching: areas with weak magnetic anomalies. Redrawn with modifications from King and Barker (1988)

or ESE–WNW-directed plate divergence episodes. Restoring the plate divergence that had led to their formation allowed for a compact cusp-like restoration of the basins' bounding blocks (Fig. 4b, c). The main exception to this pattern was an area in the central Scotia Sea that appeared to have opened by north–south-directed divergence. This province made it conceivable to restore South Georgia via two phases of plate motion, one oriented north–south and one ESE–WNW, to a position close, but not directly adjacent, to the continental margin SE of Tierra del Fuego (Barker and Griffiths 1972). A variety of smaller continental blocks were depicted, from the margins of the closed basins, tightly packed into the spaces around and between South Georgia and Tierra del Fuego (Fig. 4b).

Figure 4 gives a feeling for the sources of complexity in reconstructing the numerous continental and arc blocks of the Scotia Sea to one another by the removal of intervening oceanic basins. This complexity became clear after much effort was invested in characterising the floor of the Scotia Sea in detail both magnetically and bathymetrically (e.g. Barker and Burrell 1977; Hill and Barker 1980; Barker et al. 1982, 1984). Hence, whilst conclusions about the locations and opening directions of the individual basins in the Scotia Sea remained largely unchanged after the early 1970s, ideas about the number and the shapes of the continental blocks bounding the numerous individual basins evolved considerably. Knowledge about the basins thus enabled the relative locations of the blocks to be reconstructed with confidence, although the short lengths of palaeoplate boundaries that the basins represent, mean inevitably that the reconstructions could not be of particularly high resolution.

Seismic reflection profiles were also added to the regional data set during this second phase of marine geophysical discovery, allowing detailed interpretations of the Scotia Sea's active and fossil plate boundaries to be made (e.g. King and Barker 1988; Maldonado et al. 2003; Kavoun and Vinnikovskaya 1994; Coren et al. 1997). A lack of well-tie locations has consistently made these data difficult to use for correlating the basin's fills and therefore their stratigraphic age. Despite this, seismic data collection allowed for numerous refinements to Barker et al. (1982) reconstruction (Fig. 4b) and its placement in a plausible geodynamic context dominated by back-arc extension behind a migrating west-directed subduction zone that is today expressed by the South Sandwich Trench (Fig. 1).

Until this point, Scotia Sea reconstructions had all been built like jigsaw puzzles using paper and scissors, or simply sketched directly onto paper. Both of these methods fail to account for the distortions inherent in projecting features of a spherical surface onto a flat sheet of paper. Some of the first computer-based map-projected reconstructions of the region were presented by the PLATES group at the University of Texas (Lawver et al. 1992; Lawver and Gahagan 1998). Although the PLATES treatment at this time did not explicitly account for all of the basins and bounding blocks of the Scotia Sea floor, the South Georgia microcontinent was shown moving in ways consistent with ideas developed, using marine geophysical data, on the basis of Barker and Griffiths (1972) interpretation of plate motions in the west and central Scotia Sea.

By the mid-1990s, therefore, near-consensus had been reached that both the tectonostratigraphic and basin restoration approaches advocate pre-Scotia reconstructions featuring a compact Andean cusp at $\sim$40 Ma built around a close placement of South Georgia to Tierra del Fuego (Fig. 4d). Around this time, processing of satellite altimeter data to give an idea of sea surface slope, and with it free-air gravity over the oceans, greatly increased detail in knowledge about the shapes of the basin floors (Livermore et al. 1994). This data set paved the way for plate kinematic reconstruction techniques whose aims were to minimise and quantify uncertainties in basin restorations and allow their full integration into map-projected plate kinematic reconstructions (Eagles 2000, 2004; Eagles et al. 2005). Using the results of these techniques to build reconstructions of gridded data sets allowed for new constraints, implicit in the rules of plate tectonics, to play a role in reconstructing the Scotia Sea. The most important of these constraints are those that concern plate rigidity and plate boundary continuity. The most relevant consequence was the recognition that the central Scotia Sea was unlikely to house both halves of a Cenozoic-aged basin. Multiple lines of evidence instead suggested that the basin formed during Cretaceous north–south-directed plate divergence (Eagles 2000, 2010b; Dalziel et al. 2013a). This, in turn, removed the principal supporting interpretation of the mid-1990s consensus, as it implied South Georgia could not have reached its present location by a combination of ESE–WNW plate divergence in the southern Scotia Sea and north–south-directed divergence in the central Scotia Sea.

By today, therefore the largest remaining controversy in reconstruction of the Scotia Sea concerns whether or not the tectonostratigraphic correlations between South Georgia and Tierra del Fuego really should prescribe their geographical contiguity for any time in the past. The PLATES group continues to generate reconstructions that assume it should (e.g. Dalziel et al. 2013b), whilst acknowledging the lack of evidence for eastwards translation of South Georgia that is required of doing so. In contrast to this, the following pages present a plate tectonic history of the Scotia Sea in reverse stratigraphic order from 6 Ma until 50 Ma that has been derived with stricter adherence to the seafloor tectonic record. In detail, it is based on the plate kinematic model of Eagles and Jokat (2014) and illustrated by a set of reconstructions of Sandwell et al. (2014) gravity anomaly grid. A later section will reconsider the strength of testimony from South Georgia's tectonostratigraphy in the context of some pre-50 Ma reconstructions of the region.

4 Development of the Scotia Sea

4.1 6 Ma

The Scotia Sea has probably existed in its current plate tectonic form since around 6 Ma, when the mid-ocean ridge in the western part of the Scotia Sea ceased to operate. The two plates that had been diverging on its north-western and south-eastern flanks amalgamated to form the modern Scotia Plate (Fig. 5). This instituted a long and complex transcurrent plate boundary between the Scotia and South American plates along the North Scotia Ridge, and a similarly long and complex transcurrent plate boundary along the South Scotia Ridge that separates the Scotia and Antarctic plates. The South Scotia Ridge is characterised by a handful of deep basins that seem to form at releasing bends on the plate boundary (Galindo-Zaldívar et al. 1996). At the eastern end of the Scotia Plate, a divergent plate boundary is working by seafloor spreading to generate the oceanic floor of a small back-arc basin called the East Scotia Sea (Fig. 1). The eastern half of the basin is floored by the small Sandwich plate, whose eastern margin is a convergent margin at which part of the South American plate is being subducted towards the

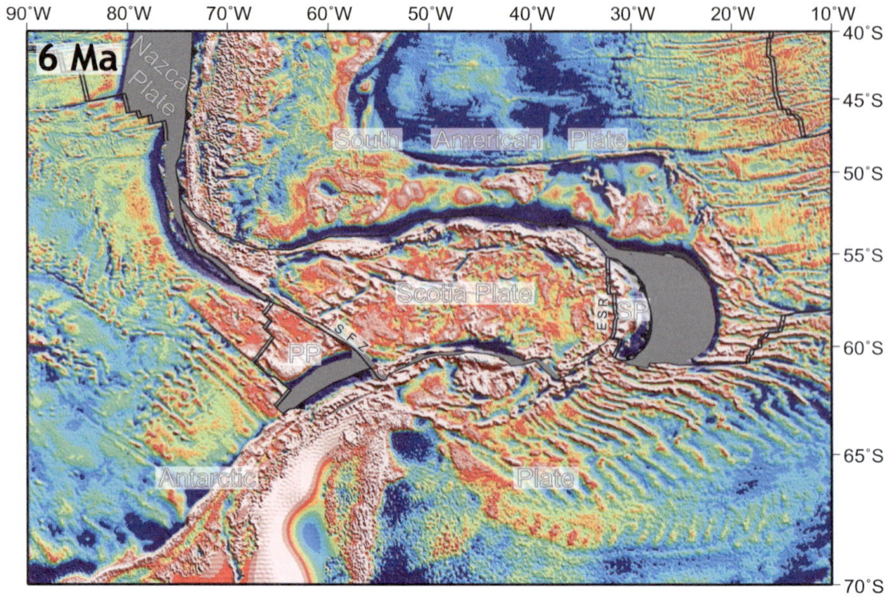

Fig. 5 Reconstruction of gridded free-air gravity anomalies (Sandwell et al. 2014) at 6 Ma, using rotations from Eagles and Jokat (2014). *ESR* East Scotia Ridge; *PP* Phoenix Plate; *SFZ* Shackleton Fracture Zone; *SP* Sandwich Plate

west beneath the Scotia Sea (Figs. 1 and 5). At the western end of the Scotia Sea, a SE-trending segment of the Scotia Plate's margin runs along the Shackleton Fracture Zone, separating it from the small oceanic Phoenix plate. At the Shackleton Fracture Zone, part of this plate has experienced flexural uplift owing to a very slight component of plate convergence at its margin with the plates in the West Scotia Sea (Livermore et al. 2004). Other parts of the Phoenix Plate at this time are undergoing subduction at a short segment of its margin, in the south-east, which sees it thrust beneath the Antarctic plate at the South Shetland Islands. The opposing segments of the Phoenix Plate's margin, also shared with the Antarctic plate, are divergent and experiencing seafloor spreading. This unusually simple plate kinematic setting has seen the Phoenix Plate used as a laboratory for studying plate driving forces. These studies have repeatedly shown that slab pull seems to have played an important role in its motion with respect to the Antarctic plate (Larter and Barker 1991; Eagles 2004; Eagles and Scott 2014).

4.2 10 Ma

The reconstruction in Fig. 6 illustrates the situation shortly before the extinction of the West Scotia Ridge, when the floor of the West Scotia Sea was subdivided into

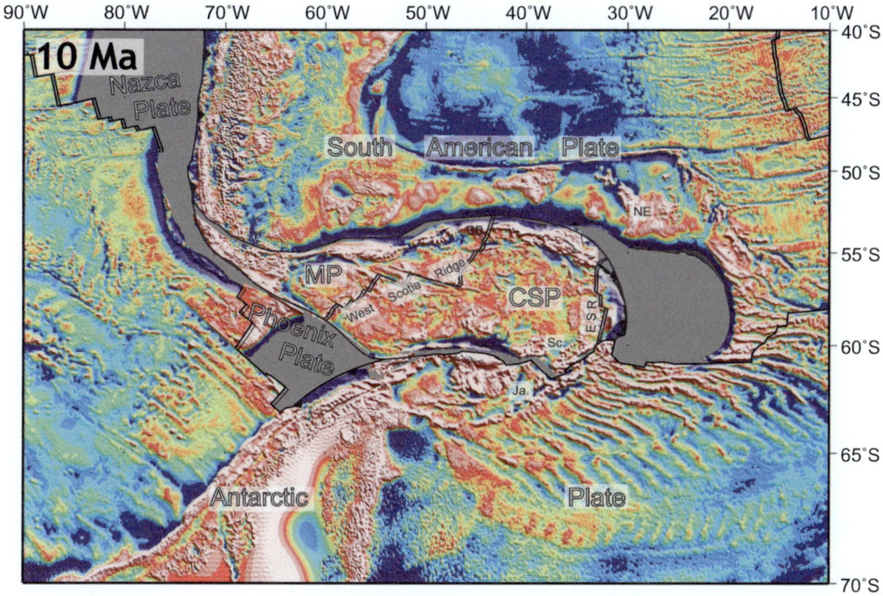

Fig. 6 Reconstruction of gridded free-air gravity anomalies (Sandwell et al. 2014) at 10 Ma, using rotations from Eagles and Jokat (2014). *BB* Aurora (Barker) Bank; *CSP* Central Scotia Plate; *MP* Magallanes Plate; *Sc.* Scan Basin, *Ja.* Jane Basin; *NE* NE Georgia Rise

sections carried on the Magallanes and Central Scotia Plates (Fig. 6) (Eagles et al. (2005). A small 15-km underlap suggests that slow sinistral transpression has been accommodated along the North Scotia Ridge since the time of this reconstruction. The spreading rate on the West Scotia Ridge is quite slow ($\sim$10 km/Myr half rate) at this time (Eagles et al. 2005). This is reflected in the rough seafloor fabric close to the flanks of the present-day extinct ridge (Fig. 1). Dalziel et al. (2013a) and Pearce et al. (2014) relate intraplate volcanism in the Central Scotia Plate to low-angle subduction of part of the North-east Georgia Rise large igneous province beneath the Scotia Sea. This subduction occurs at the South Sandwich Trench, the eastern boundary of a young and small Sandwich plate (Fig. 6).

4.3 17 Ma

The 17 Ma period (Fig. 7) marks two important aspects of the evolution of the West Scotia Sea, which developed shortly before it reached almost its full volume. The first is the onset of a period of slow spreading rates in the west Scotia Sea. The second is that the West Scotia Ridge has propagated northwards to reach the Barker/Aurora Bank, reaching its full length. Eagles and Jokat (2014) estimated that this occurred close to 18.5 Ma, based on magnetic anomalies crossing the bank, which may

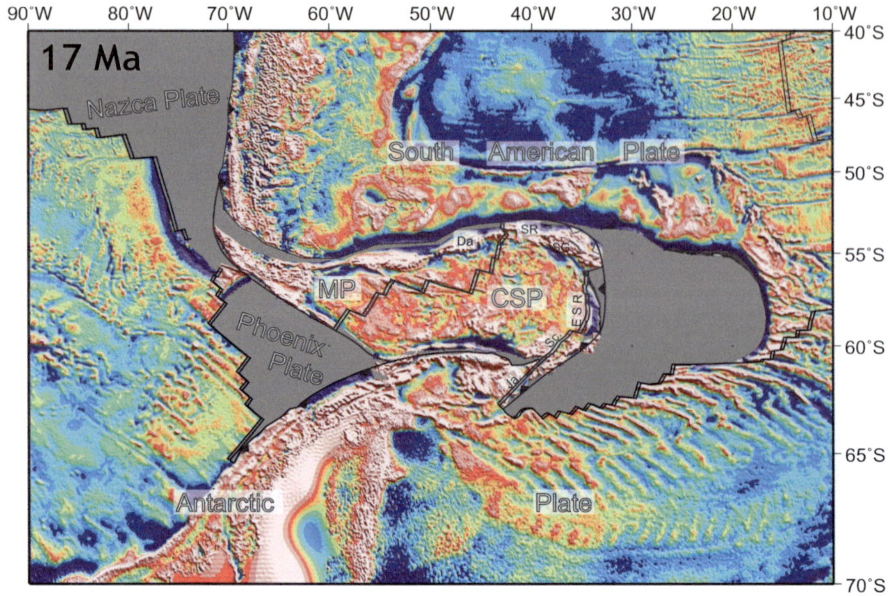

Fig. 7 Reconstruction of gridded free-air gravity anomalies (Sandwell et al. 2014) at 17 Ma, using rotations from Eagles and Jokat (2014). *Da* Davis Bank; *SG* South Georgia; *SR* Shag Rocks

therefore consist largely of sediments resting on oceanic crust. The subsequent plate divergence at this latitude led to development of a gap in the North Scotia Ridge that today allows deep water of the Antarctic Circumpolar Current to cross out of the Scotia Sea and into the South Atlantic. Tectonically, the rest of the North Scotia Ridge acts in much the same way as in the 10-Ma reconstruction. A narrow underlap depicted across the South Scotia Ridge is consistent with interpretations of light transpressional strain made from seismic reflection profiles across it (Galindo-Zaldívar et al. 1996; Kavoun and Vinnikovskaya 1994; Lodolo et al. 2010).

A very small Sandwich Plate is shown in the East Scotia Sea, based on Vanneste and Larter's (2002) interpretation of isochron 5C on the western flank of the East Scotia Ridge. Those authors suggested that the absence of a conjugate 5C on the Sandwich Plate is a consequence of subduction erosion, the process of mechanical removal of crustal material from the leading edge of the overriding plate and its delivery into the mantle by the subducting plate. This earliest Sandwich Plate is shown with a longer north–south extent than at later times, which Eagles and Jokat (2014) attributed to opening of the Scan and Jane basins behind a longer ancestral trench. Barker et al. (1982, 1984) show how parts of this trench may have deactivated, shortening the plate, as segments of the South American–Antarctic Ridge in the Weddell Sea collided with it in Miocene times. Eagles et al. (2005) noted that the initiation of this young Sandwich Plate meant that a lesser proportion of the eastwards migration of the trench would need to be accommodated at the West Scotia Ridge, consistent with its deceleration at this time.

The description of Scotia Sea tectonics at 17 Ma given above can be challenged based on alternative interpretations of basin ages given in the literature. These interpretations concern the Protector, Scan and Jane basins and the central Scotia Sea. Lawver et al. (1991) suggested, on the basis of heat flow determinations from sediment temperature gradients, that Jane Basin may have opened far earlier than shown here, in the period 32–25 Ma. This would allow for the basin to have played an important role in early deep water circulation through Drake Passage. Although the interpretation requires the basin fill to be dominated by clay, which could be considered unusual for a back-arc basin, its adoption in Fig. 7 would not strongly alter the overall plate kinematic scenario depicted.

This is not the case for Protector Basin and the central Scotia Sea, both of which in Fig. 7 are shown as open and passive at 17 Ma. Hill and Barker (1980) suggested in contrast that both Protector Basin and the central Scotia Sea were opening by seafloor spreading at 17 Ma. Maldonado et al. (2003) and Galindo-Zaldívar et al. (2006) have reproduced the magnetic anomaly models that these young interpretations are based on and depicted them in kinematic sketches. Figure 8 portrays what these interpretations would mean for plate boundaries within the Scotia Sea at 17 Ma. The thick black lines depict those parts of the plate boundaries that have been identified and suggested as active at 17 Ma. The dashed red lines depict additional tectonic features that would be required to maintain a closed circuit of plate boundaries around rigid plates within the Scotia Sea. Dashed black lines depict the magnetic bights that would be expected at the junction of the spreading centres in the central and east Scotia Seas. Evidence for these extinct plate boundaries and magnetic bight has not yet been found.

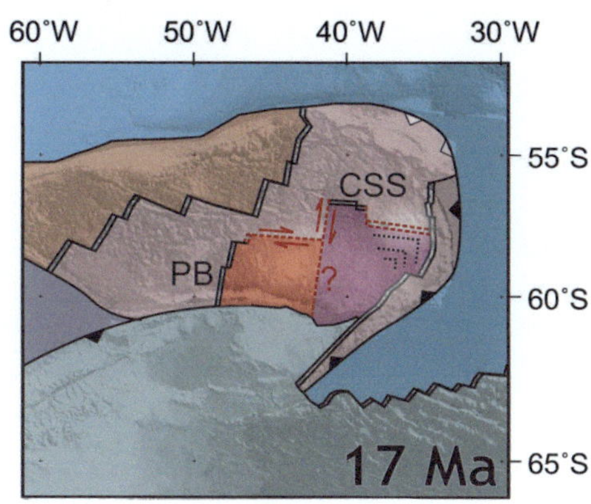

Fig. 8 Unsubstantiated plate tectonic complexity of a Scotia Sea with active mid-ocean ridges in the central Scotia Sea (CSS) and Protector Basin (PB) at 17 Ma. *Red shading*: plates within the Scotia Sea. *Blue shading*: plates in the surrounding circuits. *Red* and *black dashed lines*: implied plate boundary segments and magnetic bights for which no evidence is known. Background shows the 17 Ma bathymetric reconstruction of Eagles and Jokat (2014)

4.4 20 Ma

Figure 9 illustrates the tectonics of the Scotia Sea at 20 Ma. In contrast to the situation at 17 Ma, the back-arc basin in the eastern part of the Scotia Sea has yet to form and, concomitantly, the West Scotia Ridge is spreading more rapidly. The West Scotia Ridge is also shorter; its northern end lies somewhere east of Davis Bank. Just north of this, the reconstruction underlap on continental crust at the North Scotia Ridge is broader than that shown for 17 Ma. This implies $\sim$75 km of shortening on the Scotia–South America plate boundary in the intervening period. Eagles and Jokat (2014) showed that the magnitude of this shortening is smaller than the uncertainties involved in constructing the plate circuit that depicts it. The occurrence of this convergence, however, is more confidently implied by the presence of a seemingly young accretionary prism at the southern edges of Burdwood and Davis banks, and of Miocene transpression in Tierra del Fuego and along northern Burdwood Bank (Bry et al. 2004; Cunningham et al. 1995; Klepeis and Austin 1997; Kraemer 2003). The South Scotia Ridge, in contrast, is depicted much as in the later reconstructions, implying its history since Miocene times has been one of very slow sinistral transpression. As such, this reconstruction shows a period very early in the evolution of the Scotia Sea with northern and southern boundaries similar to today's.

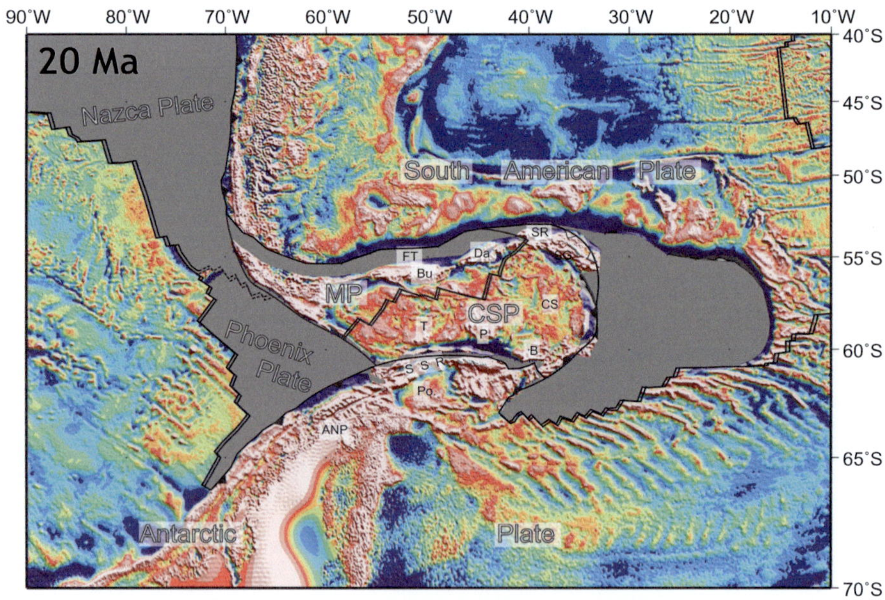

Fig. 9 Reconstruction of gridded free-air gravity anomalies (Sandwell et al. 2014) at 20 Ma, using rotations from Eagles and Jokat (2014). *ANP* Antarctic Peninsula; *B* Bruce Bank; *Bu.* Burdwood Bank; *CS* Central Scotia Sea; *FT* Falkland Trough; *P* Pirie Bank; *Po.* Powell Basin; *SSR* South Scotia Ridge; *T* Terror Rise

Further in the west, Fig. 7 shows the possibility that a segment of the Phoenix–Nazca plate boundary was subducted beneath Cape Horn. Ongoing divergence of the surface portions of the two plates would have seen a space open within the mantle between their subducted parts. Breitsprecher and Thorkelsen (2009) and Eagles et al. (2009) have suggested that decompression melting of mantle rocks rising through this so-called slab window produced melts that are represented by minor occurrences of alkali basalts in Tierra del Fuego (Puig et al. 1984). Eagles and Jokat (2014) suggested that the loss of this ridge in subduction may have presented a new barrier to the dispersal of vent-specific Kiwaid crabs, whose Pacific and Atlantic lineages are reckoned to have diverged in the period 25.9–13.4 Ma (Roterman et al. 2013).

As in the 17 Ma reconstruction, there are alternative interpretations of some of the smaller basins in the Scotia Sea that could alter the picture of plate tectonics in the Scotia Sea at 20 Ma. As well as the central Scotia Sea, whose interpretation as a Miocene back-arc basin (Hill and Barker 1980) would affect the 20 Ma reconstruction in much the same way as the 17 Ma reconstruction, there is also the suggestion that the Scan and Dove basins may have been active at 20 Ma (Galindo-Zaldívar et al. 2014). Figure 10 shows the implication of these ages for the plate tectonic setting in the Scotia Sea at 20 Ma. The red dashed lines show plate boundaries for which good evidence is missing. Most conspicuous amongst these is a mid-ocean ridge in Scan Basin. The sketch also shows one of the best-known puzzles concerning Hill and Barker's (1980) back-arc interpretation for the central Scotia Sea: why should a back-arc basin behind an east-migrating trench have opened by north–south extension?

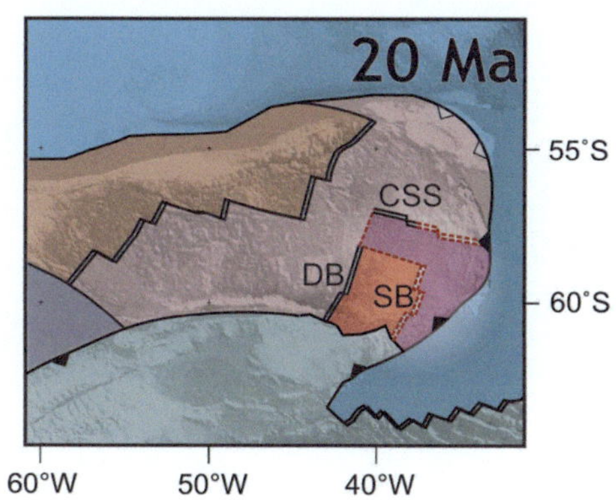

Fig. 10 Unsubstantiated plate tectonic complexity of a Scotia Sea with active mid-ocean ridges in the central Scotia Sea (CSS), Dove Basin (DB) and Scan Basin (SB) at 20 Ma. *Red shading*: plates within the Scotia Sea. *Blue shading*: plates in the surrounding circuits. *Red black dashed lines*: implied plate boundary segments for which no evidence is known. Background shows the 17 Ma bathymetric reconstruction of Eagles and Jokat (2014)

4.5 26 Ma

26 Ma was part of a transitional period in the development of the Scotia Sea (Fig. 11). Although seafloor spreading was underway on the young West Scotia Ridge, its flanking magnetic isochrons suggested it had yet to propagate any further north than the eastern margin of Burdwood Bank. Dredging work has returned evidence for volcanism near what was the northern end of the ridge at 29 Ma (Dalziel et al. 2013a; Pearce et al. 2014). Dalziel et al. (2013a) suggest that this volcanism was related to subduction of South American lithosphere beneath the Scotia Sea, on the basis of the geochemical signature of the dredged rocks. Eagles and Jokat (2014) noted that this would require a very shallow angle (10–20°) slab, based on the distance to the trench in the reconstruction. They suggested alternatively that the source of the dredged rocks may be immature lower crust remelted by the addition of mantle melts during extension related to the northward propagation of the West Scotia Ridge, noting the analogy to the region's older Chon Aike large igneous province (Pankhurst and Rapela 1995).

Regardless of its role in this melting, it seems that the divergent Magallanes–Central Scotia plate boundary reached far enough north that parts of the North Scotia Ridge could have adopted a role as the northern margin of the Scotia Sea, acting as it did later as two separate plate boundaries between the South American

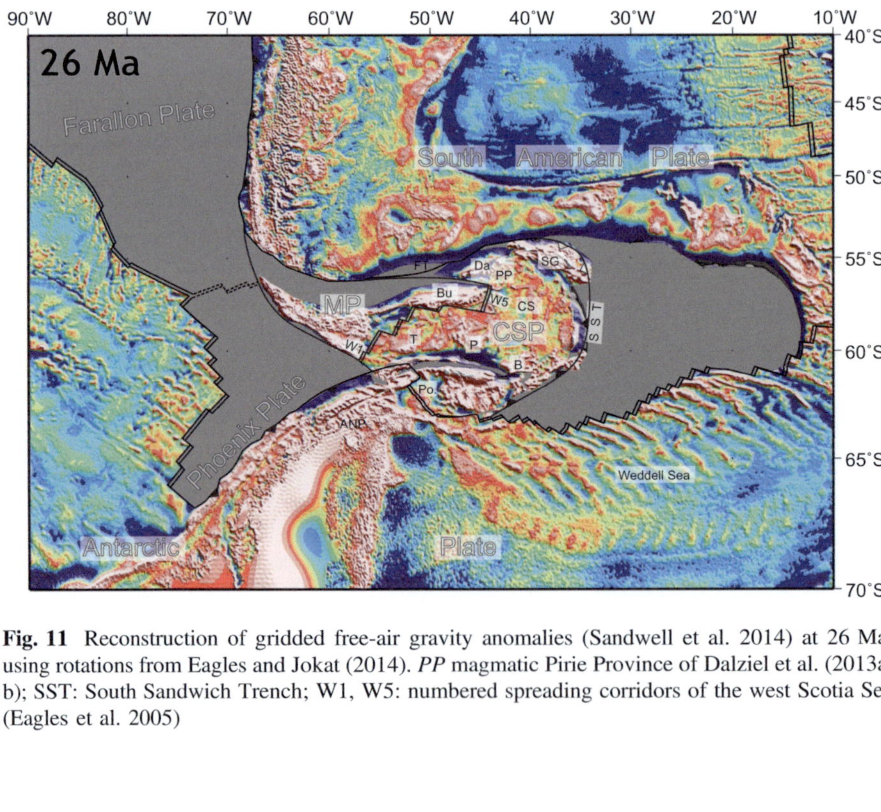

Fig. 11 Reconstruction of gridded free-air gravity anomalies (Sandwell et al. 2014) at 26 Ma, using rotations from Eagles and Jokat (2014). *PP* magmatic Pirie Province of Dalziel et al. (2013a, b); SST: South Sandwich Trench; W1, W5: numbered spreading corridors of the west Scotia Sea (Eagles et al. 2005)

plate and the Magallanes and Central Scotia plates. The reconstruction underlap on the Magallanes–South America segment implies continuous slow convergence in the period 26–20 Ma, albeit below the statistical resolution of the plate circuit Eagles and Jokat (2014) built. Further south, in contrast to later times, the southern margin of the Scotia Sea did not run along the South Scotia Ridge but instead through a small oceanic basin off the northern end of the Antarctic Peninsula, Powell Basin. There is consensus that seafloor spreading in Powell Basin was an Oligocene and earliest Miocene process (King and Barker 1988; Lawver et al. 1994; Coren et al. 1997; Eagles and Livermore 2002). The southern boundary's location may have switched in response to early Miocene collision of segments of the South American–Antarctic Ridge with the eastern convergent plate boundary of the South Orkney Microcontinent, beyond Powell Basin's eastern flank (Barker 1995; Barker et al. 1982, 1984; Hamilton 1989). The approach and eventual collision of the ridge may have reduced subduction-related driving forces that contributed to the basin's opening.

4.6 30 Ma

30 Ma (Fig. 12) marks what has been presented as an important epoch in the development of the Scotia Sea. This period has returned the oldest direct evidence for active subduction of South American lithosphere beneath the central Scotia Sea, in the form of calc-alkaline volcanic rocks dredged from a rotated fault block in the modern fore-arc (Barker 1995; Livermore et al. 1994). Eagles and Jokat (2014) suggested this could be interpreted to mean that the South Sandwich Trench had, for the first time, reached as far north as South Georgia, where it bent into an east–west orientation to take up the very earliest transcurrent plate motions on the eastern North Scotia Ridge. The reconstruction shows no back-arc basin immediately behind the trench at this time, although suggestions that Scan Basin may date from around this time, based on heat flow values modelled from sediment temperature gradient data (Barker et al. 2013), could be interpreted in terms of such a feature. Further west, the reconstruction shows the young West Scotia Ridge reaching only as far north as Isla de los Estados, where its motion is taken up on an oblique transform fault running NW into Tierra del Fuego. Here, this transform meets the east-trending transform fault that runs through or near South Georgia to the trench. Together with the young divergent plate boundary in Powell Basin, these plate boundaries enabled Eagles and Jokat (2014) to interpret this period as the one in which the central Scotia Sea first came to be encircled by plate boundaries, setting in place the core of the modern Scotia Plate.

The shape of the reconstruction misfit occupied by this plate boundary in Tierra del Fuego is largely unchanged in the 26 Ma reconstruction, suggesting at least the possibility that there were periods of no relative movement between the Magallanes and South American plates during the early development of the Scotia Sea.

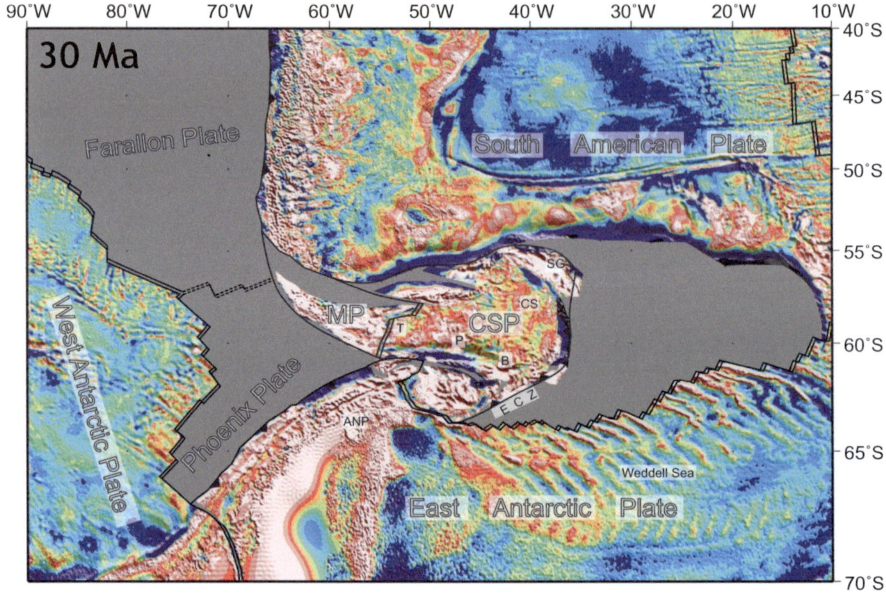

Fig. 12 Reconstruction of gridded free-air gravity anomalies (Sandwell et al. 2014) at 30 Ma, using rotations from Eagles and Jokat (2014). *ECZ* Endurance Collision Zone (Ghidella et al. 2002)

The West Scotia Ridge in this reconstruction is the very oldest that Eagles and Jokat (2014) show. Following Livermore et al. (2005), they noted that a tight fit between the eastern and western margins of the West Scotia Sea (at Terror Rise and Tierra del Fuego) dated to this time when using a spreading rate before chron 8 that was the same as immediately afterwards. They further noted that this is the same time that Eagles et al. (2006) determined for the end of seafloor spreading in Protector Basin on the opposing side of Terror Rise, suggesting simple abandonment of one mid-ocean ridge had occurred in favour of another.

Direct evidence for this earliest phase of spreading on the West Scotia Ridge is difficult to interpret, however, and is the subject of moderate controversy because the time frame lies very close to the end-Eocene build up of ice on Antarctica, which has been linked to the onset of the Antarctic Circumpolar Current that in turn has been suggested as a simple consequence the opening of the Drake Passage (Kennett 1977; De Conto and Pollard 2003). Figure 13 summarises this evidence, which comes from magnetic anomaly profiles crossing the southernmost corridors of seafloor spreading in the West Scotia Sea, directly adjacent to the Shackleton Fracture Zone.

The profiles, three for each flank in these corridors, show that anomaly isochrons younger than and including 8 are interpretable in many locations, although by no

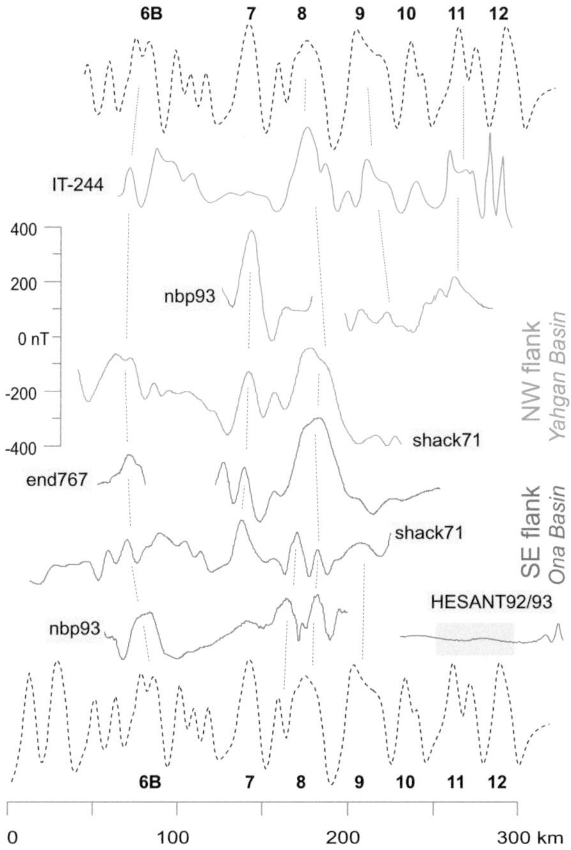

Fig. 13 Magnetic anomaly wiggles in the south-westernmost spreading corridors of the West Scotia Sea. Old end of profiles towards the *right*. *Green lines* show profiles from the NW flank, *pink lines* the SE flank. Dashed profiles are synthetic magnetic anomalies generated for a source layer 1 km thick lying at 4.5 km depth, labelled with chron number. *Pink box* shows the portion of the HESANT92/93 profiles in which reversal isochrons 11 and 12 were modelled at very small amplitudes by Maldonado et al. (2014)

means are they ubiquitously so. Three of the profiles supply hints that isochron 9 may be present in these corridors. Beyond this, two profiles confined to the north-western half of the corridors have been interpreted to reveal isochrons 10 and 11 (Lodolo et al. 2006), although it should be noted that the waveforms are far from coherent between them. Chron 12 has recently been interpreted from the south-eastern flank in magnetic profiles crossing the Ona Basin, close to the continental shelf north of Elephant Island (Maldonado et al. 2014). Plotted at the same scale as the profiles in which the younger magnetic isochrons were identified, it is clear, however, that this interpretation requires oceanic crust with an order of magnitude smaller magnetic susceptibility than anywhere else in the West Scotia Sea. It seems then that the oldest likely oceanic crust preserved in the West Scotia Sea dates from chron 11 ($\sim$30 Ma; Gradstein et al. 2012), as in the reconstruction. Whether this crust was developed on an organised linear mid-ocean ridge, as depicted, or in a more transitional setting like that described as 'chaotic seafloor spreading' by Barker and Burrell (1977), remains unclear with current data.

4.7 33 Ma

All of the Oligocene or younger reconstructions shown so far have featured one or more plate boundary segments along the North Scotia Ridge. The 33 Ma reconstruction in Fig. 14 represents a step further back towards an embryonic Scotia Sea by being the youngest to lack such a plate boundary. At this time, the reconstruction shows the young South Sandwich Trench reaching only as far north as Discovery Bank, although dating of available samples only confirm the bank as a site of subduction-related volcanism as far back as 20–12 Ma (Barker et al. 1982).

At the northern edge of Discovery Bank, the reconstruction shows the palaeo-subduction zone bending abruptly into an east–west orientation much like its present-day manifestation does to run along the North Scotia Ridge near South Georgia. Both of these features are examples of so-called tear faults (Fig. 15). Eagles and Jokat (2014) termed the 33 Ma tear fault the Burdwood transform fault. Other than the gravity anomalies along the northern edges of Bruce and Pirie banks, and the southern edge of Burdwood Bank, there is little direct evidence for the remnants of this fault, although it should be noted that the area is buried by thick sediments (Perez et al. 2014) and the plate circuit, as constructed, suggests the cross-axial component of the movements it accommodated may have been slight. The 30.2 Ma basement heat flow estimate at the northern edge of Scan Basin (Barker et al., 2013) might be interpreted, instead of in terms of basin age, as the time

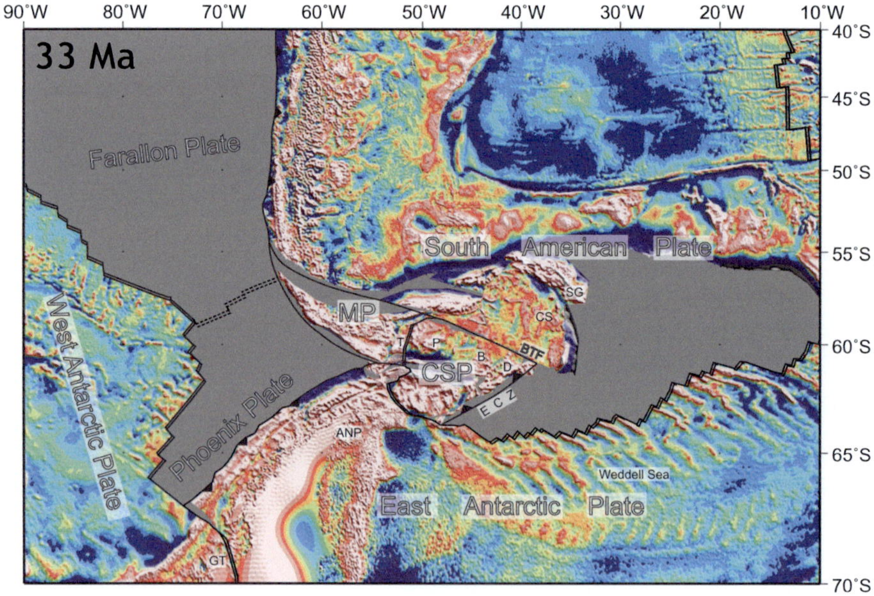

Fig. 14 Reconstruction of gridded free-air gravity anomalies (Sandwell et al. 2014) at 33 Ma, using rotations from Eagles and Jokat (2014). *D* Discovery Bank

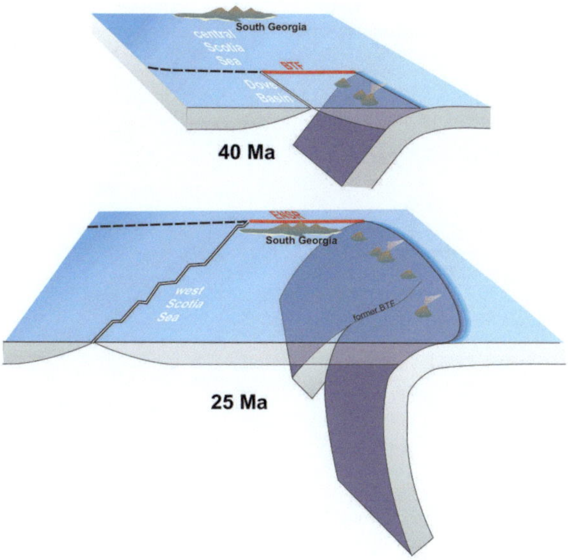

Fig. 15 Successive locations of tear faults in the Scotia Sea following northwards propagation of the ancestral South Sandwich Trench. *BTF* Burdwood tear fault; *ENSR* eastern North Scotia Ridge

of cessation of the fault's movement. This aside, there is stronger evidence for other plate boundaries further south in the form of the continental extensional basins bordering Powell Basin (King and Barker 1988) and the short mid-ocean ridge in Protector Basin (Eagles et al. 2006). The reconstruction shows these boundaries, along with the Burdwood transform fault, as defining the body of an independently moving small overriding arc plate at the ancestral South Sandwich Trench.

With the Burdwood transform fault running along part of its southern edge, the central Scotia Sea and South Georgia are shown as features embedded within the body of the South American Plate. To the west, the length of plate boundary running through continental South America is much shorter than in all the younger reconstructions; it is confined to Tierra del Fuego where a smaller underlap than in the next-younger (30 Ma) reconstruction can be interpreted in terms of a phase of plate divergence. As for Eagles and Jokat's (2014) reconstructions at all other times, the amount of divergence shown on this particular boundary is unlikely to be greater than the uncertainties in the various elements of the plate circuit might combine to show, whereas the evidence for continental extension in the Sloggett Basin of south-eastern Tierra del Fuego (Ghiglione et al. 2008) is not.

4.8 41 Ma

The general configuration at 41 Ma (Fig. 16) is in most aspects a shortened version of that at 33 Ma, suggesting a simple evolution from one to the other by ongoing action of the plate boundaries. The only large difference is in the location of the

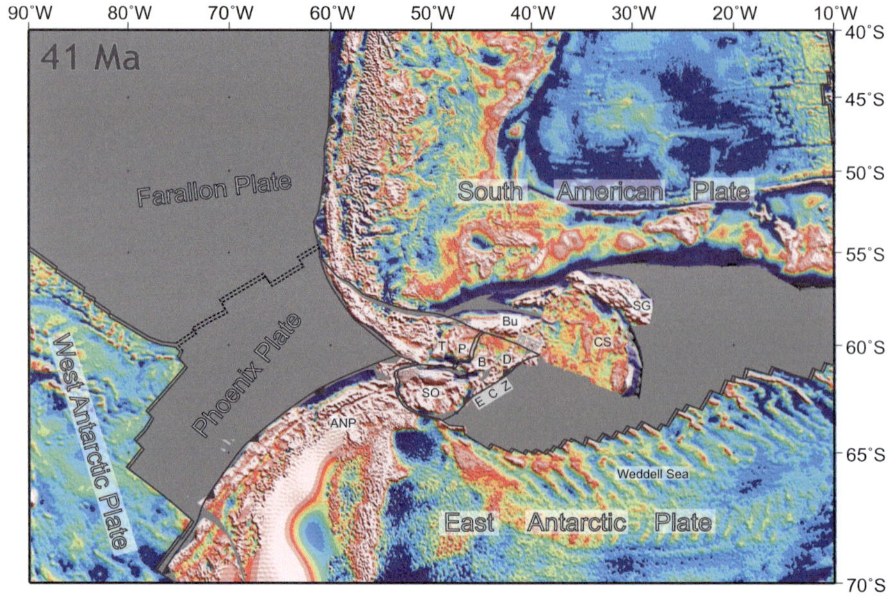

Fig. 16 Reconstruction of gridded free-air gravity anomalies (Sandwell et al. 2014) at 41 Ma, using rotations from Eagles and Jokat (2014). *SO* South Orkney microcontinent

western, divergent boundary of the arc plate, whose northern reaches pass along a mid-ocean ridge in Dove Basin, rather than Protector Basin. The Burdwood transform fault is somewhat shorter because the Protector Basin was yet to open. These configurations for Protector and Dove Basin allow the Scotia Sea to be depicted as a relatively simple system of two plates behind the ancestral South Sandwich Trench at any time between 42 and 30 Ma. One of these plates is the overriding plate at the trench and probably featured a volcanic arc at its eastern edge. The plate behind the arc plate bears Tierra del Fuego and moves only slowly in a collisional sense, if at all, with respect to the South American plate. As before, the resolution of the plate circuit is too coarse to be able to define this motion, or the lack of it, as statistically relevant, but interpretations of active shortening in Tierra del Fuego in middle Eocene times seem to confirm that convergence did occur (Ghiglione and Ramos 2005; Gombosi et al. 2009; see Ghiglione 2016 for a review).

Eagles and Jokat (2014) calculated that the arc plate was moving rather more slowly with respect to Antarctica than with respect to South America in this period and noted that this slow motion did not give rise to oceanic growth, but rather to continental extension, as expressed in the eruption of alkali basalts and the formation of grabens around and across the South Orkney Microcontinent (Barber et al. 1991; King and Barker 1988).

4.9 50 Ma

The onset of independent motions of small plates in the Scotia Sea began with continental crustal extension over the South Orkney Microcontinent and of Pirie and Bruce banks, which gave way eventually to seafloor spreading in the Powell and Dove basins. Figure 17 shows the region before any of these motions had started, at around 50 Ma but neither it, nor that in Eagles and Jokat (2014), attempts to reconstruct any of the continental extension. Eagles and Livermore's (2002) reconstruction of Powell Basin gives an idea of how such a reconstruction might appear around the South Scotia Ridge. Beyond this, the reconstruction can be directly compared to the tectonostratigraphic and flat map reconstructions of Figs. 3 and 4. The clearest difference between them is in the area south of Burdwood Bank, where Pirie and Bruce banks and Terror Rise lie as a result of reconstructing the growth of the West Scotia Sea, Protector and Dove basins. The dredged stratigraphy and magnetic anomalies of these blocks match comfortably those of Tierra del Fuego (Eagles et al. 2006; Schenke and Udintsev 2009; Udintsev et al. 2012). South Georgia, on the other hand, lies south of Maurice Ewing Bank at the eastern end of the Falkland Plateau, a few hundred kilometres west of its current location. As the previous reconstructions have shown, there is no preserved evidence anywhere in the seafloor of the Scotia Sea for its translation from a location further west, in spite of the widely reported tectonostratigraphic arguments for such.

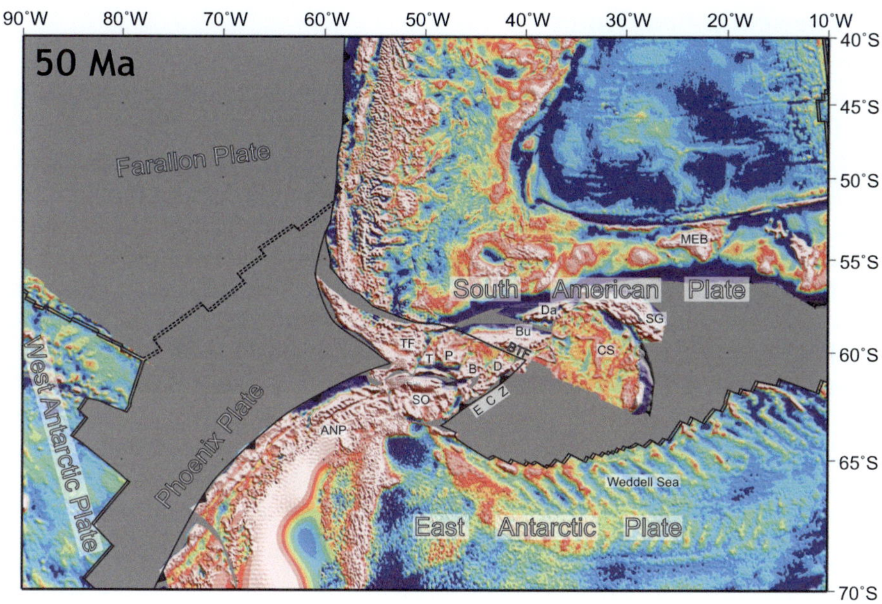

Fig. 17 Reconstruction of gridded free-air gravity anomalies (Sandwell et al. 2014) at 6 Ma, using rotations from Eagles and Jokat (2014). *MEB* Maurice Ewing Bank; *TF* Tierra del Fuego

Ghiglione et al. (2008) detail observations in geophysical and outcrop data for South American–Antarctic early Eocene plate divergence in Tierra del Fuego further north, which Livermore et al. (2005) had modelled as being slow, north–south-directed, and coming to an end soon after 50 Ma. Prior to this, the eastern margin of Omond Land, the continental mass formed of contiguous South Orkney Microcontinent, Terror Rise, Pirie, Bruce and Discovery banks had evolved from a transform fault into a continent–ocean collision zone as the azimuth of South American–Antarctic plate motions rotated from NE–SW to NW–SE in Maastrichtian through Palaeogene times (Barker et al. 1991; Eagles 2010b). It was this so-called Endurance Collision Zone (Ghidella et al. 2002) that evolved to become the ancestral South Sandwich subduction zone, which slowly propagated northwards and migrated eastwards, as seen from Antarctica, throughout Cenozoic times. The 50 Ma reconstruction shows the tear fault, the Burdwood transform fault, at the northern end of this collision zone. At this time, the fault's shortness and orientation allow it to act as a simple transcurrent fault between the South American and Antarctic plates. In this way, relative motions of the Endurance Collision Zone/South Sandwich Trench, and the South American and Antarctic plates at 50 Ma are accommodated by plate boundary-scale processes on a single fault. As previous sections have shown, the subsequent plate tectonic history of the Scotia Sea can be seen as the history of this fault's lengthening and elaboration to accommodate changes in those relative motions by plate-scale processes instead.

5 Pre-scotia Sea Times

Unlike the reconstructions shown so far, Sandwell et al.'s (2014) gravity data for the 125 Ma reconstruction in Fig. 18 have been filtered to remove long wavelengths, probably related to very large-scale convection-related structures in the mantle, so that the remaining anomaly field matches across the mid-ocean ridge in the South Atlantic. This eases visualisation of the tectonic fabric, which tends to occur on shorter wavelengths. The reconstruction is developed by rotating the 50 Ma configuration of the Scotia Sea as if it had been embedded within the South American plate in the period 125–50 Ma. Doing this requires the assumption that the Burdwood transform fault had been newly inaugurated at 50 Ma and that at earlier times the South American–Antarctic plate boundary lay further south, in the Weddell Sea, as suggested by magnetic seafloor spreading anomalies there (Livermore and Woollett 1993; Livermore and Hunter 1996; Livermore et al. 2005). The reconstruction shows a pair of extended continental margins that formed in the divergence of west and east Gondwana, and areas of Mesozoic oceanic crust that accreted to them in the central Scotia Sea (Eagles 2010a) and Weddell Sea (Livermore and Hunter 1996).

Further west, the reconstruction shows Graham Land and other parts of the Antarctic Peninsula as if they had been embedded within the South American plate for a period ending at 103 Ma, the time of the Palmer Land Event, a Cretaceous collisional episode interpreted from sparse outcrop data in southern parts of the peninsula (Vaughan et al. 2012). The placement should be viewed as speculative; not only does it rely on disputed tectonostratigraphic correlations between the South Orkney Islands and rocks further north in the Patagonian Andes, but also the very occurrence of a Palmer Land Event has recently been called into question (Burton-Johnson and Riley 2015).

Figure 18 also shows where, at 125 Ma, the Yahgán, Cumberland Bay and Sandebugten formations would have been deposited at widely separated sites in the Rocas Verdes Basin of Tierra del Fuego and in South Georgia. By this, it is evident that the plate kinematic history given so far fails to reconcile with the long-suspected geographical proximity of South Georgia and Tierra del Fuego in Aptian–Albian times. Faced with this, Eagles (2010b) attempted to reinterpret the similarities between South Georgia and Tierra del Fuego in terms of an Early Cretaceous plate boundary configuration from which the 125 Ma setting shown

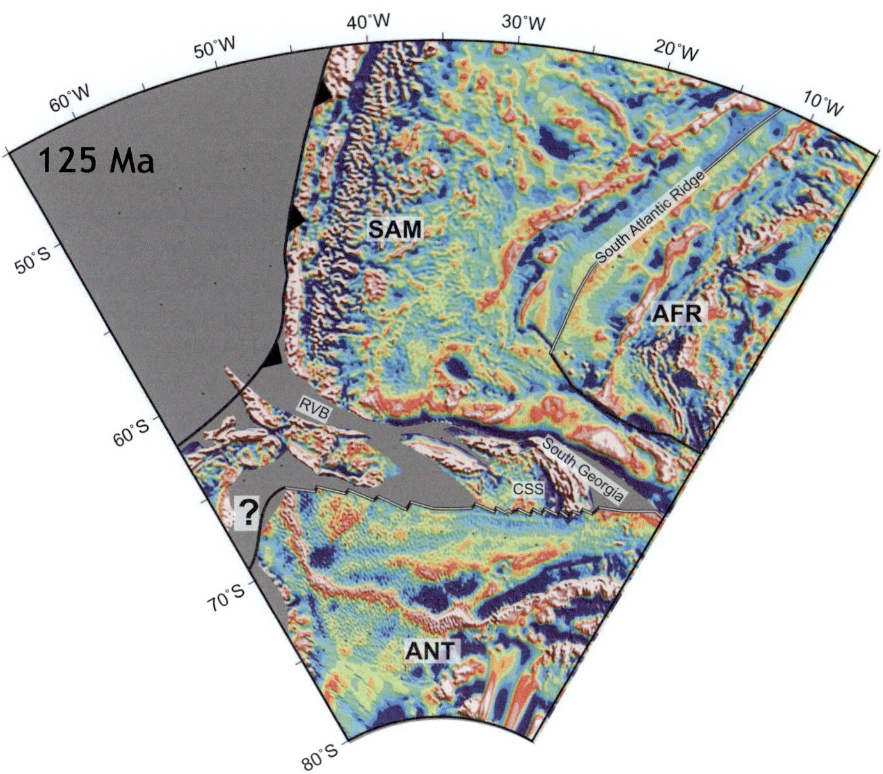

Fig. 18 125 Ma reconstruction. *AFR* African plate; *ANT* Antarctic plate; *SAM* South American plate '?': Precursor plate boundary to Palmer Land Shear Zone of Vaughan et al. (2012); *CSS* central Scotia Sea; *RVB* inactive Rocas Verdes Basin

above might have evolved. His scheme presented South Georgia as part of the South American conjugate to the extended continental margin in the southern Weddell Sea. He suggested that breakup-related volcanism at this margin was distinguished geochemically because of its occurrence in lithosphere that had previously experienced an unusual combination of processes including metasomatism during flat-slab subduction and the arrival of a mantle plume (Pankhurst and Rapela 1995; Pankhurst et al. 2000). The Chon Aike large igneous province represents this volcanism in outcrop and subcrop at sites spread throughout Patagonia and the Antarctic Peninsula (Pankhurst et al. 2000). This volcanic province probably extends from the Deseado Massif in Patagonia offshore towards the Rio Chico high, and from there to the North Scotia Ridge and Falkland Plateau (Baristeas et al. 2013). It occurs in association with sites of crustal extension and is characterised by bimodal andesitic–rhyolitic volcanism, as reflected in the petrographic detrital composition of Yahgán, Cumberland Bay and Sandebugten formations. Detrital zircon U–Pb ages from Sandebugten Formation of South Georgia Island includes a strong 171 Ma peak (Carter et al. 2014) corresponding to the V2 Jurassic volcanic stage of the Chon Aike province where it crops out in the Deseado Massif (Pankhurst et al. 2000).

Figure 19 summarises this new tectonostratigraphic interpretation and compares it to that of Dalziel and Elliot (1971). Both have been annotated with proposed locations for the sources of the South Georgia´s sequences. Neither scenario requires any of the Cumberland Bay or Sandebugten formations' sediments to have been sourced from the Falkland Islands, despite Carter et al.'s (2014) attempt to distinguish between them on the basis of the absence of Proterozoic 'Gondwanan interior' zircons from South Georgia. If anything, it is the active margin location that is more difficult to reconcile with this absence, given the palaeocurrent directions shown. Seen in this context, the sedimentological similarities between the Yahgán, Cumberland Bay and Sandebugten formations need not be prescriptive of geographical proximity.

The other claimed strong correlation between South Georgia and Tierra del Fuego is in the timing of tectonism that occurred to close the Rocas Verdes and 'Cumberland Bay' basins. In both locations, this event is dated to 86–84 Ma, or beforehand. This requires a plate boundary location for both basins at that time. Figure 20 shows a reconstruction for 84 Ma that enables interpretation of the locations of plate boundaries in the region. This reconstruction is well constrained because of the ease of identifying magnetic reversal isochron 34y; statistical reconstructions for this isochron using three pairs of plates or three plates simultaneously have repeatedly demonstrated that a fourth plate did not exist in the Weddell Sea at this time (Livermore and Hunter 1996; Nankivell 1997; Eagles and Vaughan 2009). There are no indications from structural geology or palaeomagnetic work that the Antarctic Peninsula at this time was anything other than fixed with respect to the interior of Antarctica, just as it is today (Vaughan et al. 2012; Grunow 1993). The plate boundaries shown are thus necessary and sufficient to explain all known plate kinematic and palaeomagnetic constraints. Within the constraints of these boundaries, the N–S trending path sketched through the Rocas Verdes Basin

(a)
(b)

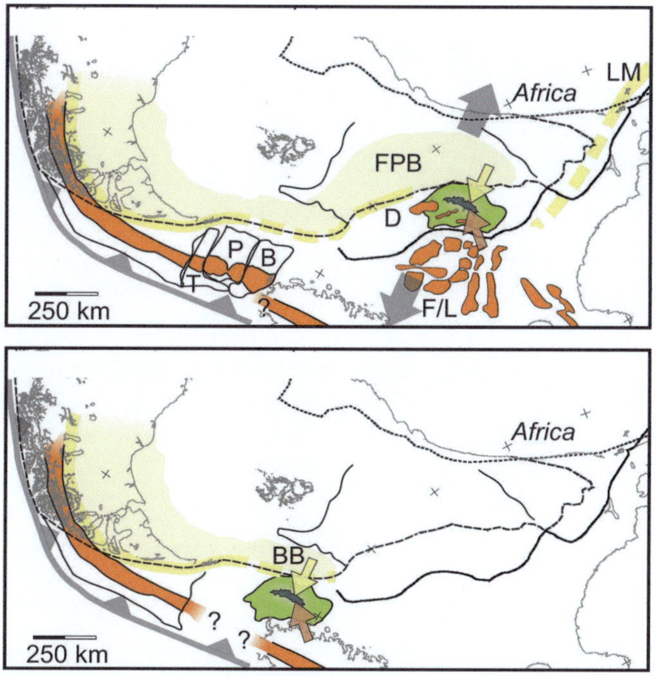

Fig. 19 Alternative contexts for deposition of the Yahgán, Cumberland Bay and Sandebugten formations. The basemaps, redrawn from Dalziel et al. (1975), depict a time early in the Cretaceous. **a** Gondwana breakup over metasomatised mantle, **b** subduction and back arc extension *Bright yellow lines*: outcrop or suspected outcrop (*dashed*) of silicic volcanic rocks of the Chon Aike large igneous province. *Orange*: interpreted locations of andesitic volcanic rocks. *BB* Burdwood Bank; *TPB* 'batholith' magnetic anomalies of Terror Rise, Pirie Bank and Bruce Bank. *D* Davis Bank dredge sample (Pandey et al. 2010); *LM* Lebombo monocline rhyolites; *F/L* Filchner and 'Lozenge' magnetic anomalies of Ferris et al. (2000); *FPB* Falkland Plateau Basin. *Yellow* and *orange arrows*: palaeocurrent directions determined for later deposition of Sandebugten and Cumberland Bay formations rocks. *Grey arrows*: orientation of Gondwana breakup-related plate divergence

can explain the basin's shortening along a segment of the Antarctic–South American plate boundary. The main faults and folds of South Georgia, in contrast, suggest a plate boundary segment oriented ESE. This orientation could not have acted in a convergent sense in either of the locations suggested for South Georgia. The evidence for Rb–Sr system closure in South Georgia's schists and the K–Ar dating of its biotites' growth are therefore very difficult to relate to Andean orogenesis at $\sim$84 Ma without invoking extra plates whose motion would be otherwise unattested. An alternative setting, consistent with the thermal history presented by Carter et al. (2014) and with the location of the highest metamorphic grades in the closest proximity to the exposed basement and oceanic complexes in the SE of the island, may simply be deep burial of sediments in a basin situated on a young, warm, continental margin.

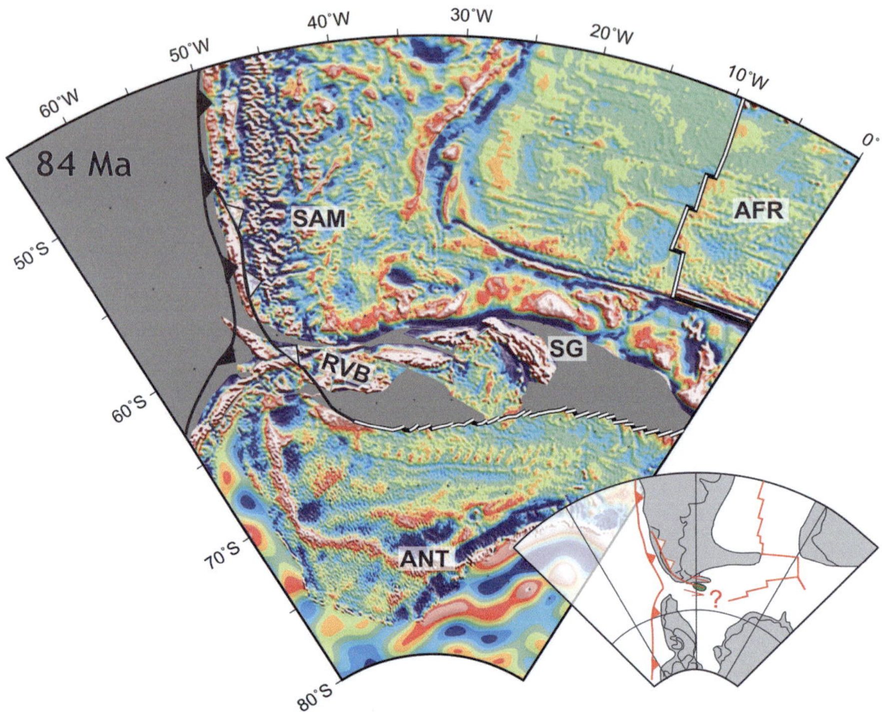

Fig. 20 The South Atlantic at 84 Ma, around the time of inversion of the Rocas Verdes Basin (RVB) and of the 'Cumberland Bay' basin in South Georgia (SG). At this time, the region was evolving by divergence of three large plates, South America (SAM), Africa (AFR) and Antarctica (ANT) beneath which plates of the palaeo-Pacific margin were subducting from the west. The Antarctic Peninsula was not moving with respect to the interior of Antarctica. Neither of the locations implied by reconstruction of the seafloor in the Scotia Sea (shown) or by comparative tectonostratigraphy (in the *inset*) places South Georgia near the Pacific margin, or indeed any plate margin, that might be invoked to explain the deformation proposed from dating of Rb–Sr closure in biotite and titanite minerals at this time. *Inset* non-closure (?) of plate boundaries with 'Rocas Verdes' location of South Georgia, redrawn and adapted from Curtis et al. (2010)

6 Summary

This chapter has revisited and summarised nearly 200 years of thinking on the tectonic history of the Scotia Sea region, which started from basic and fragmentary information on outcrop geology in Tierra del Fuego and the Antarctic Peninsula. The first century and a half of this work was dominated by increasing knowledge and understanding of the intervening outcrops, dotted amongst islands around the periphery of the Scotia Sea. By the time of the plate tectonic revolution, much was known in detail about contrasts and correlations between these outcrops, and little of the early conviction that an orocline of uniform Andean composition snaked continuously around the Scotia Sea could be sustained. Only a set of strong likenesses between the Jurassic oceanic basement rocks, and Aptian–Albian basin fills of South Georgia and Tierra del Fuego, and interpretations of their shared experience of orogenesis later in the Cretaceous remained. Since the second half of the twentieth century, new data sets offering new chances to consider the Scotia Sea's growth became available from marine and satellite geophysical techniques that revealed the bathymetric, gravimetric and magnetic structure of the 98 % of the Scotia Sea that had previously lain hidden beneath the sea surface between its islands. Until now, these data have proved impossible to interpret in any way that enables the likenesses between the late Jurassic and Cretaceous rocks of South Georgia and Tierra del Fuego to be explained in terms of their contiguity at any time in the past. An alternative tectonostratigraphy that is consistent with the plate kinematic history interpretable from the seafloor has been suggested and is beginning to be tested. This task is of importance in more than just its own right, because it stands in the foreground of a set of interdisciplinary studies that all stand to benefit from a more precise and confident understanding of the tectonic and bathymetric development of the Scotia Sea.

References

Arctowski H (1895) Observations sur l'interet qui presente l'exploration des Terres Australes. Bull Soc Geol Fr (ser 3) 23:589–591

Barber PL, Barker PF, Pankhurst RJ (1991) Dredged rocks from Powell Basin and the South Orkney microcontinent. In: Thomson MRA, Crame JA, Thomson JW (eds) Geological evolution of Antarctica. Cambridge University Press, Cambridge, pp 361–367

Baristeas N, Anka Z, di Primio R, Rodriguez JF, Marchal D, Dominguez F (2013) New insights into the tectono-stratigraphic evolution of the Malvinas Basin, offshore of the southernmost Argentinean continental margin. Tectonophysics. doi:10.1016/j.tecto.2013.06.009

Barker PF (1970) Plate tectonics of the Scotia Sea region. Nature 228:1293–1296

Barker PF (1972) Magnetic lineations in the Scotia Sea. In: Adie RJ (ed) Antarctic geology and geophysics. Universitetsforlaget, Oslo, pp 3–11

Barker PF (1995) Tectonic framework of the east Scotia Sea. In: Taylor B (ed) Backarc Basins: tectonics and magmatism plenum. New York, pp 281–314

Barker PF, Burrell J (1977) The opening of Drake Passage. Mar Geol 25:15–34

Barker PF, Griffiths DH (1972) The evolution of the Scotia Ridge and Scotia Sea. Philos T Roy Soc A 271:151–183

Barker PF, Hill IA, Weaver SD, Pankhurst RJ (1982) The origin of the eastern South Scotia ridge as an intra-oceanic island arc In: Craddock C (ed) Antarctic Geosciences Univ Wisc Press, Madison, pp 203–211

Barker PF, Barber PL, King EC (1984) An early miocene ridge crest–trench collision on the South Scotia Ridge near 36W. Tectonophysics 102:315–332

Barker PF, Lawver LA, Larter RD (2013) Heat-flow determinations of basement age in small oceanic basins of the southern central Scotia Sea. Geol Soc Spec Pub, 381:139–150

Barron EJ, Harrison CGA, Hay WW (1978) A revised reconstruction of the southern continents. Eos T Am Geophys U 59:436–449

Becker JJ, Sandwell DT, Smith WHF, Braud J, Binder B, Depner J, Weatherall P (2009) Global bathymetry and elevation data at 30 arc seconds resolution: SRTM30_PLUS. Mar Geod 32 (4):355–371

Bird P (2003) An updated digital model of plate boundaries. Geochem Geophy Geosy 4(3)
Breitsprecher K, Thorkelson DJ (2009) Neogene kinematic history of Nazca-Antarctic-Phoenix slab windows beneath Patagonia and the Antarctic Peninsula. Tectonophysics 464:10–20
Bry M, White N, Singh S, England R, Trowell C (2004) Anatomy and formation of oblique continental collision: South Falkland basin. Tectonics 23(4):TC401
Burton-Johnson A, Riley TR (2015) Autochthonous vs accreted terrane development of continental margins: a revised in situ tectonic history of the Antarctic Peninsula. J Geol Soc London 172:822–835
Calderón M, Fildani A, Hervé F, Fanning CM, Weislogel A, Cordani U (2007) Late Jurassic bimodal magmatism in the northern sea-floor remnant of the Rocas Verdes basin, Southern Patagonian Andes. J Geol Soc London 164:1011–1022
Calderón M, Hervé F, Fuentes F, Fosdick JC, Sepúlveda F, Galaz G (2016) Tectonic evolution of Paleozoic and Mesozoic andean metamorphic complexes and the Rocas Verdes ophiolites in southern Patagonia. In: Ghiglione MC (ed) Geodynamic Evolution of the Southernmost Andes. Springer Earth System Sciences, pp 7–36
Carter A, Curtis M, Schwanethal J (2014) Cenozoic tectonic history of the South Georgia microcontinent and potential as a barrier to Pacific-Atlantic through flow. Geology 42 (4):299–302
Clayton (1982) The Geology of North-Western South Georgia: III petrology of the Cumberland Bay formation. Br Antarct Surv Bull 51:79–88
Coren F, Ceccone G, Lodolo E, Zanolla C, Zitellini N, Bonazzi C, Centonze J (1997) Morphology, seismic structure and tectonic development of the Powell Basin, Antarctica. J Geol Soc London 154:849–862
Cunningham WD (1995) Orogenesis at the southern tip of the Americas: the structural evolution of the Cordillera Darwin metamorphic complex, southernmost Chile. Tectonophysics 244:197–229
Cunningham WD, Dalziel IW, Lee TY, Lawver LA (1995) Southernmost South America-Antarctic Peninsula relative plate motions since 84 Ma: Implications for the tectonic evolution of the Scotia Arc region. J Geophys Res-Sol EA 100(B5):8257–8266
Curtis ML, Flowerdew MJ, Riley TR, Whitehouse MJ, Daly JS (2010) Andean sinistral transpression and kinematic partitioning in South Georgia. J Struct Geol 32:464–477
Dalziel IWD (1983) The evolution of the Scotia Arc: a review. In: Oliver RL, James PR, Jago JB (eds) Antarctic earth science. Cambridge University Press, Cambridge, pp 283–288
Dalziel IWD (1986) Collision and cordilleran orogenesis: an Andean perspective. Geol Soc Spec Pub, 19:389-404
Dalziel IWD, Elliot DH (1971) Evolution of the Scotia Arc. Nature 233:246–251
Dalziel IWD, Dott Jr RH, Winn Jr RD, Bruhn RL (1975) Tectonic relations of South Georgia Island to the Southernmost Andes. Geol Soc Am Bull 86:1034–1040
Dalziel IWD, Lawver LA, Pearce JA, Barker PF, Hastie AR, Barfod DN, Schenke H-W, Davis MB (2013a) A potential barrier to deep Antarctic circumpolar flow until the late Miocene?. Geology 41(9):947–950
Dalziel IWD, Lawver LA, Norton IO, Gahagan LM (2013b) The Scotia Arc: genesis, evolution, global significance Annu Rev Earth Pl Sci 41, 767–793
De Conto RM, Pollard D (2003) Rapid Cenozoic glaciations of Antarctica induced by declining atmospheric CO_2. Nature 421:245–249
Eagles G (2000) Modelling plate kinematics in the Scotia Sea. Unpubl. Ph.D. Thesis, University of Leeds, UK
Eagles G (2004) Tectonic evolution of the Antarctic-Phoenix plate system since 15 Ma. Earth Planet Sci Lett 217:97–109
Eagles G (2010a) South Georgia and Gondwana's Pacific margin: lost in translation? J S Am Earth Sci 30:65–70
Eagles G (2010b) Age and origin of the central Scotia Sea. Geophys J Int 183:587–600
Eagles G, Jokat W (2014) Tectonic reconstructions for paleobathymetry in Drake Passage. Tectonophysics 611:28–50

Eagles G, Livermore RA (2002) Opening history of Powell Basin, Antarctic Peninsula. Mar Geol 185:195–205

Eagles G, Scott B (2014) Plate convergence west of Patagonia and the Antarctic Peninsula since 61 Ma. Global Planet Change 123:189–198

Eagles G, Vaughan APM (2009) Gondwana breakup and plate kinematics: business as usual. Geophys Res Lett 36:L10302

Eagles, G, Livermore, RA, Fairhead, JD, Morris, P, (2005) Tectonic evolution of the west Scotia Sea J Geophys Res 110:B02401

Eagles G, Livermore RA, Morris P (2006) Small basins in the Scotia Sea: the Eocene drake passage gateway. Earth Planet Sci Lett 242:343–353

Eagles G, Gohl K, Larter RD (2009) Animated tectonic reconstruction of the Southern Pacific and alkaline volcanism at its convergent margins since Eocene times. Tectonophysics 464:21–29

Ferris JK, Vaughan APM, Storey BC (2000) Relics of a complex triple junction in the Weddell Sea embayment, Antarctica. Earth Planet Sci Lett 178(3):215–230

Fordyce RE (2003) Cetacea evolution and Eocene-Oligocene oceans revisited. In: Prothero DR, Ivany LC, Nesbitt E (eds) From greenhouse to icehouse: the marine eocene-oligocene transition. Columbia University Press, New York, pp 154–170

Galindo-Zaldívar J, Jabaloy A, Maldonado A, Sanz de Galdeano C (1996) Continental fragmentation along the South Scotia Ridge transcurrent plate boundary (NE Antarctic Peninsula). Tectonophysics 258:275–301

Galindo-Zaldívar J, Bohoyo F, Maldonado A, Schreider A, Suriñach E, Vázquez JT (2006) Propagating rift during the opening of a small oceanic basin: the Protector Basin (Scotia Arc, Antarctica). Earth Planet Sc Lett 241:398–412

Galindo-Zaldívar J, Puga E, Bohoyo F, González FJ, Maldonado A, Martos YM, de Federico Antonio D (2014) Magmatism, structure and age of Dove Basin (Antarctica): A key to understanding South Scotia Arc development. Glob Planet Change 122:50–69

Ghidella ME, Yáñez G, LaBrecque JL (2002) Revised tectonic implications for the magnetic anomalies of the western Weddell Sea. Tectonophysics 347:65–86

Ghiglione MC (2016) Orogenic growth of the Fuegian Andes (52–56°) and its relation with tectonics of the Scotia Arc. In: Folguera A, Naipauer M, Sagripanty L, Ghiglione MC, Orts D, Giambiagi LB (eds) Growth of the Southern Andes. Springer Earth System Sciences, pp 241–267

Ghiglione MC, Ramos VA (2005) Chronology of deformation in the Southernmost Andes of Tierra del Fuego. Tectonophysics 405:25–46

Ghiglione MC, Yagupsky D, Ghidella M, Ramos VA (2008) Continental stretching preceding the opening of the drake passage: evidence from Tierra del Fuego. Geology 36:643–646

Gombosi DJ, Barbeau DL, Garver J (2009) New thermochronometric constraints on the rapid Paleogene exhumation of the Cordillera Darwin complex and related thrust sheets in the Fuegian Andes. Terra Nova 21:507–515

Guillot MG (2016) Magmatic evolution of the Southernmost Andes and its relation with subduction processes. In: Ghiglione MC (ed) Geodynamic Evolution of the Southernmost Andes. Springer Earth System Sciences, pp 37–74

Gradstein FM, Ogg JG, Schmitz M, Ogg G (2012) The geologic time scale 2012, 2-volume set, Elsevier

Grunow AM (1993) New paleomagnetic data from the Antarctic Peninsula and their tectonic implications. J Geophys Res 98:13815–13833

Hamilton IW (1989) Geophysical investigations of subduction related processes in the Scotia Sea. Unpubl. Ph.D. thesis, University of Birmingham, Birmingham, UK

Helffrich G, Wiens DA, Vera E, Barrientos S, Shore P, Robertson S, Adaros R (2002) A teleseismic shear-wave splitting study to investigate mantle flow around South America and implications for plate-driving forces. Geophys J Int 149(1):F1–F7

Hervé F, Fanning CM, Pankhurst RJ (2003) Detrital zircon age patterns and provenance of the metamorphic complexes of southern Chile. J South Am Earth Sci 16:107–123

Hervé F, Miller H, Pimpirev C (2005) Patagonia-Antarctica connections before Gondwana Break-Up. In: Fütterer DK, Damaske D, Kleinschmidt G, Miller H, Tessensohn F (eds) Antarctica: contributions to global earth sciences. Springer, Berlin, Heidelberg & New York, pp 215–226
Hill IA, Barker PF (1980) Evidence for Miocene back-arc spreading in the central Scotia Sea. Geophys J Roy Astr S 63:427–440
Hospitaleche CA, Reguero M, Scarano A (2013) Main pathways in the evolution of the Paleogene Antarctic Sphenisciformes. J S Am Earth Sci 43:101–111
Kavoun M, Vinnikovskaya O (1994) Seismic stratigraphy and tectonics of the north-western Weddell Sea (Antarctica) inferred from marine geophysical surveys. Tectonophysics 240:299–323
Kennett JP (1977) Cenozoic evolution of Antarctic glaciation, the circum-Antarctic Ocean, and their impact on global paleoceanography. J Geophys Res 82:3843–3860
King EC, Barker PF (1988) The margins of the South Orkney microcontinent. J Geol Soc London 145:317–331
Klepeis KA, Austin JA Jr (1997) Contrasting styles of superposed deformation in the Southernmost Andes. Tectonics 16:755–776
Klepeis K, Betka P, Clarke G, Fanning M, Hervé F, Rojas L, Mpodozis C, Thomson S (2010) Continental underthrusting and obduction during the Cretaceous closure of the Rocas Verdes rift basin, Cordillera Darwin, Patagonian Andes. Tectonics 29(3):TC3014
Kraemer PE (2003) Orogenic shortening and the origin of the Patagonian orocline (56 S Lat). J S Am Earth Sci 15:731–748
Larter RD, Barker PF (1991) Effects of ridge crest–trench interaction on Antarctic-Phoenix spreading: forces on a young subducting plate. J Geophys Res 96:19583–19607
Lawver LA, Gahagan LM (1998) Opening of drake passage and its impact on Cenozoic ocean circulation. Oxf Monogr Biogeogr 39:212–226
Lawver LA, Gahagan LM (2003) Evolution of Cenozoic seaways in the circum-Antarctic region. Palaeogeogr Palaeoclimatol Palaeoecol 198(1):11–37
Lawver LA, Della Vedova B, Von Herzen RP (1991) Heat flow in Jane Basin, northwest Weddell Sea. J Geophys Res Sol EA 96(B2):2019–2038
Lawver LA, Gahagan LM, Coffin MF (1992) The development of paleoseaways around Antarctica. In: Kennett JP, Warkne DA (eds) The Antarctic Paleoenvironment: a perspective on global change: part one. Antarct Res Ser, vol 56 AGU, Washington DC, pp 7–30
Lawver LA, Williams T, Sloan B (1994) Seismic stratigraphy and heat flow of Powell Basin. Terra Ant Reports 1(2):309–310
Livermore RA, Hunter R (1996) Mesozoic seafloor spreading in the southern Weddell Sea. Geol Soc Spec Publ 108:227–241
Livermore RA, Woollett RW (1993) Seafloor spreading in the Weddell Sea and southwest Atlantic since the Late Cretaceous. Earth Planet Sci Lett 117:475–495
Livermore RA, McAdoo D, Marks K (1994) Scotia Sea tectonics from high-resolution satellite gravity Earth. Planet Sci Lett 123:255–268
Livermore RA, Eagles G, Morris P, Maldonado A (2004) Shackleton Fracture Zone: no barrier to early circumpolar ocean circulation. Geology 32:797–800
Livermore RA, Nankivell AP, Eagles G, Morris P (2005) Paleogene opening of Drake Passage. Earth Planet Sci Lett 236:459–470
Lodolo E, Donda F, Tassone A (2006) Western Scotia Sea margins: improved constraints on the opening of the Drake Passage. J Geophys Res 111:B06101
Lodolo E, Civile D, Vuan A, Tassone A, Geletti R (2010) The Scotia-Antarctica plate boundary from 35°W to 45°W. Earth Planet Sci Lett 293:200–215
Maffione M (2016) Kinematic evolution of the Southern Andean orogenic arc. In: Ghiglione MC (ed) Geodynamic Evolution of the Southernmost Andes. Springer Earth System Sciences, pp 173–200

Maffione M, Speranza F, Faccenna C, Rossello E (2010) Paleomagnetic evidence for a pre-early Eocene (~50 Ma) bending of the Patagonian orocline (Tierra del Fuego, Argentina): Paleogeographic and tectonic implications. Earth Planet Sci Lett 289:273–286

Maffione M, Fernandez-Moreno C, Ghiglione M, Speranza F, van Hinsbergen DJJ, Lodolo E (2015) Constraints on deformation of the Southern Andes since the Cretaceous from anisotropy of magnetic susceptibility. Tectonophysics 665:236–250

Maldonado A, Barnolas A, Bohoyo F, Galindo-Zaldívar J, Hernández-Molina J, Lobo F, Rodríguez-Fernández J, Somoza L, Vázquez JT (2003) Contourite deposits in the central Scotia Sea: the importance of the Antarctic Circumpolar Current and the Weddell Gyre flows. Palaeogeogr Palaeoclimatol Palaeoecol, 198:187–221

Maldonado A, Bohoyo F, Galindo-Zaldívar J, Hernández-Molina FJ, Lobo FJ, Lodolo E, Somoza L (2014) A model of oceanic development by ridge jumping: opening of the Scotia Sea. Global Planet Change 123:152–173

Müller C (2001) Upper mantle seismic anisotropy beneath Antarctica and the Scotia Sea region. Geophys J Int 147:105–122

Nankivell AP (1997) Tectonic evolution of the Southern Ocean between Antarctica, South America and Africa over the last 84 Ma. Ph.D. thesis Univ of Oxford, Oxford, UK

Nerlich R, Clark SR, Bunge HP (2013) The Scotia Sea gateway: no outlet for Pacific mantle. Tectonophysics 604:41–50

Pandey A, Parson L, Milton A (2010) Geochemistry of the Davis and Aurora Banks: possible implications on evolution of the North Scotia Ridge. Mar Geol 268:106–114

Pankhurst RJ, Rapela CW (1995) Production of Jurassic rhyolite by anatexis of the lower crust of Patagonia. Earth Planet Sci Lett 134:23–36

Pankhurst RJ, Riley TR, Fanning CM, Kelley SP (2000) Episodic silicic volcanism in Patagonia and the Antarctic Peninsula: chronology of magmatism associated with the break-up of Gondwana. J Petrol 41:605–625

Pearce JA, Hastie AR, Leat PT, Dalziel IWD, Lawver LA, Barker PF, Bevins RE (2014) Composition and evolution of the Ancestral South Sandwich Arc: implications for the flow of deep ocean water and mantle through the Drake Passage gateway. Glob Planet Change 123:298–322

Pérez LF, Lodolo E, Maldonado A, Hernández-Molina FJ, Bohoyo F, Galindo-Zaldívar J, Lobo FJ, Burca M (2014) Tectonic development, sedimentation and paleoceanography of the Scan Basin (southern Scotia Sea, Antarctica). Global Planet Change, 123:344-358

Pirie JHH (1905) On the graptolite-bearing rocks of the South Orkneys. Proc R Soc Edinb 25:463–470

Puig A, Herve M, Suarez M, Saunders AD (1984) Calc-alkaline and alkaline Miocene and calc-alkaline recent volcanism in the southernmost Patagonian cordillera, Chile. J Volcanol Geotherm Res 21:149–163

Reguero MA, Marenssi SA (2010) Paleogene climatic and biotic events in the terrestrial record of the Antarctic Peninsula: an overview. In: Madden RH, Carlini AA, Vucetich MG, Kay RF (eds) The paleontology of Gran Barranca—evolution and environmental change through the Middle Cenozoic of Patagonia. Cambridge University Press, Cambridge, UK, pp 383–397

Roterman CN, Copley JT, Linse KT, Tyler PA, Rogers AD (2013) The biogeography of the yeti crabs (Kiwaidae) with notes on the phylogeny of the Chirostyloidea (Decapoda: Anomura). Proc R Soc Biol Sci Ser B 280:20130718

Sandwell DT, Müller RD, Smith WH, Garcia E, Francis R (2014) New global marine gravity model from CryoSat-2 and Jason-1 reveals buried tectonic structure. Science 346:65–67

Schenke HW, Udintsev GB (2009) The Central Scotia Sea floor—is it an paleo-oceanic plate, an young rifted plate or an paleo-land Scotia? Ukrainian Antarct J 8:36–45

Suess E (1909) Das Antlitz der Erde Freytag. Leipzig 3 volumes

Thomson MRA, Tanner PWG, Rex DC (1982) Fossil and radiometric evidence for ages of deposition and metamorphism of the sedimentary sequences on South Georgia. In: Craddock C (ed) Antarctic Geoscience. University of Wisconsin Press, Madison, pp 177–184

Tilley CE (1935) Report on rocks from the South Orkney Islands. Discov Rep 10:383–390

Toggweiler JR, Bjornsson H (2000) Drake Passage and paleoclimate. J Quat Sci 15:319–328
Trendall AF (1953) The geology of South Georgia: I Falkland Islands. Dependencies survey scientific reports 7, 26 p
Trendall AF (1959) The geology of South Georgia: II Falkland Islands. Dependencies survey scientific reports 19, 48 p
Udintsev GB, Kurentsova NA, Beresnev AF, Koltsova AV, Domoratskaya LG, Schenke GV, Bakhmutov VG, Solovev VD (2012) Tectonics of the drake Passage-Scotia Sea zone in the southern ocean. Dokl Earth Sci 445:1029–1035
Vanneste LE, Larter RD (2002) Sediment subduction, subduction erosion and strain regime in the northern South Sandwich forearc. J Geophys Res Sol EA 107:EPM 5
Vaughan APM, Eagles G, Flowerdew MJ (2012) Evidence for a two-phase Palmer Land event from crosscutting structural relationships and emplacement timing of the Lassiter Coast Intrusive Suite, Antarctic Peninsula: implications for mid-Cretaceous Southern Ocean plate configuration. Tectonics 31:TC1010
Winn RD (1978) Upper Mesozoic flysch of Tierra del Fuego and South Georgia Island: a sedimentologic approach to lithosphere plate restoration. Geol Soc Am Bull 89(4):533–547

The Relation Between Neogene Denudation of the Southernmost Andes and Sedimentation in the Offshore Argentine and Malvinas Basins During the Opening of the Drake Passage

Matías C. Ghiglione, Christian Sue, Miguel E. Ramos, Jonathan E. Tobal and Rocío E. Gallardo

Abstract The Neogene orogenic growth of the Southern Patagonian Andes has been related to the approximation and collision of a series of segments of the Chile seismic ridge, which separates the Antarctic and Nazca plates, against South America. The compiled thermochronological data consistently indicates an eastward moving trend of exhumation, were uplift of the western basement domain occurred from ~34 to 15 Ma, and was followed by denudation of the basement front and the fold and thrust belt between ~20 and 5 Ma. There has been an assumption that tectonic growth in southern Patagonia ended in late Miocene times, largely based on the top age of the molasse deposits of the Santa Cruz Formation, spanning from ~22–19 to 14 Ma. There is, however, multiple thermochronological evidence that exhumation in the hinterland continued profusely, with large volumes of rock denudated rapidly between ~15 and 5 Ma, and steadily since ~7 Ma. However, continental sedimentation rate was very low in the Magallanes–Austral Basin of the Southernmost Andes after 14 Ma, an effect produced by the dynamic uplift of Patagonia. Contrastingly, the upper Miocene–lower Pliocene constitutes an aggradational period very well developed in the offshore Argentine continental margin. We propose that the great volumes of sediments produced by Miocene–Pliocene denudation of the Southernmost Andes bypassed Patagonia and reached the Argentine and Malvinas basins, where they were accommodated in thick sequences with high sedimentation rates. Those sediments were distributed along the Southern Atlantic margin by sub-Antarctic currents, which propagated into the Argentine continental margin during the deepening of the Drake Passage. The sediments were

M.C. Ghiglione (✉) · M.E. Ramos · J.E. Tobal · R.E. Gallardo
Departamento de Ciencias Geológicas, Instituto de Estudios Andinos "Don Pablo Groeber" CONICET, FCEN—Universidad de Buenos Aires, Buenos Aires, Argentina
e-mail: matias@gl.fcen.uba.ar

C. Sue
Bourgogne Franche-Comté University, Laboratoire Chrono-environnement, Besançon, France

© Springer International Publishing Switzerland 2016
M.C. Ghiglione (ed.), *Geodynamic Evolution of the Southernmost Andes*, Springer Earth System Sciences, DOI 10.1007/978-3-319-39727-6_5

probably funneled through gargantuan fluvial and glacifluvial W–E systems, similar to those preserved in Patagonia from the last glaciation, and axially through the Fuegian Andes foothills toward the offshore basins.

Keywords Atlantic offshore basins · Argentine Basin · Malvinas Basin · Magallanes–Austral Basin · Patagonian Andes · Fuegian Andes · Drake Passage · Malvinas Current · Sub-Antarctic Current

1 Introduction

The beginning of the Oligocene–Neogene orogenic cycle of the Southernmost Andes is marked by the exhumation of its western crystalline domain (Fig. 1) (Nelson 1982; Thomson et al. 2001; Gombosi et al. 2009) and the onset of a widespread clastic wedge with conglomerates resting on a lower Oligocene regional unconformity (Malumián et al. 2000; Marenssi et al. 2005; Barreda et al. 2009; Malumián and Jannou 2010). The kinematics involved in each Andean segment during the Oligocene–Neogene deformation was markedly different, from compression linked to ridge collision in the Southern Patagonian Andes (Ramos 1989; see Ghiglione et al. 2016 for a review), to wrench kinematics associated to the opening of the Scotia Sea in Tierra del Fuego (Fig. 1) (Ghiglione 2002, 2016; Lodolo et al. 2003; Eagles et al. 2005; Eagles 2016; Lagabrielle et al. 2009; Ghiglione et al. 2010; Sue and Ghiglione 2016). These geodynamic processes sustained deformation and continuous orogenic denudation during the Neogene in the Southernmost Andes (Nelson 1982; Ghiglione and Ramos 2005; Thomson et al. 2010; Gombosi et al. 2009; Maffione et al. 2010, 2015).

The late Eocene–Oligocene was characterized by a generalized marine transgression that covered all of Patagonia and Tierra del Fuego. These conditions contrast with coeval global high $\delta^{18}O$ values and a late Eocene–early Oligocene regression in the Patagonian platform, suggesting that the water deepening recorded in the Fuegian (and Southern Patagonian) Andes was due mainly to tectonic loading that facilitated the incursion of Antarctic waters into the Magallanes–Austral basin (Malumián and Jannou 2010). The onset of this transgression appeared first in Tierra del Fuego, during the middle Eocene (Mella 2001; Malumián 2002; Gallardo 2014). In the Southern Patagonian Andes, the flooding of the shallow marine

Fig. 1 Location of main morphostructural units and tectonic features discussed in the text and ► location of figures in the following of this chapter, after Ghiglione et al. (2010) and references therein. *Yellow* Myr ages in the Pacific trench indicate the time of collision of each ridge segment between oceanic fault zones (refer to Fig. 6). *CaB* Canal Beagle; *CoD* Cordillera Darwin; *FZ* fault zone; *IrB* Irigoyen Basin; *LAr* Lago Argentino; *LBA* Lago Buenos Aires; *LBe* Lago Belgrano; *LF* Lago Fagnano; *LPo* Lago Posadas; *LSM* Lago San Martín; *LVi* Lago Viedma; *MEs* Magallanes Strait; *NSR* North Scotia Ridge; *RTu* Río Turbio; *SAz* Seno Almirantazgo; *UEs* Última Esperanza. Contours indicate sediment thickness in kilometers

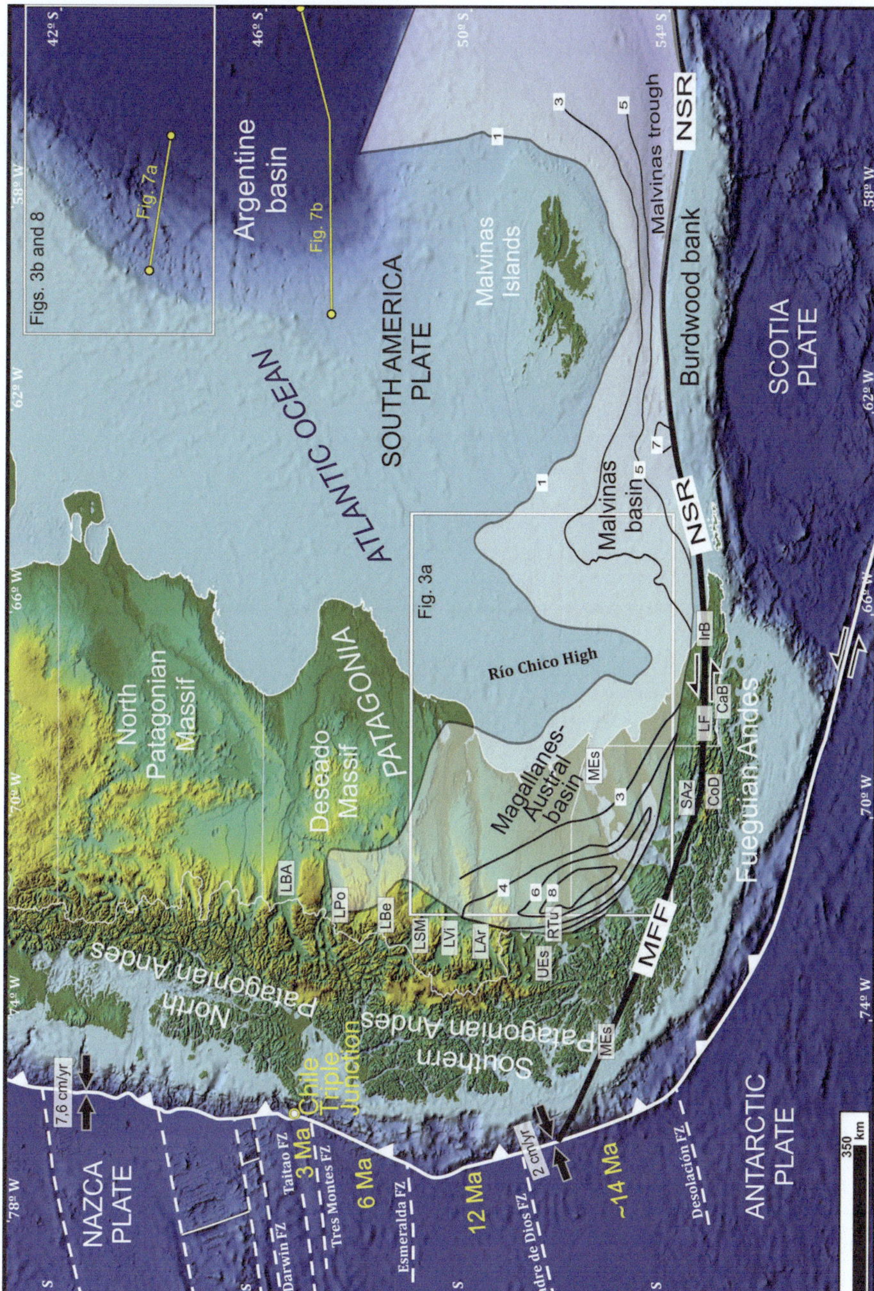
NAZCA PLATE
ANTARCTIC PLATE
SCOTIA PLATE
SOUTH AMERICA PLATE
ATLANTIC OCEAN
Argentine basin
Malvinas Islands
Malvinas trough
Malvinas basin
Burdwood bank
NSR
MFF
Río Chico High
Magallanes-Austral basin
North Patagonian Massif
Deseado Massif
PATAGONIA
North Patagonian Andes
Southern Patagonian Andes
Fueguian Andes
Chile Triple Junction
7,6 cm/yr
2 cm/yr
3 Ma
6 Ma
12 Ma
~14 Ma
Darwin FZ
Taitao FZ
Tres Montes FZ
Esmeralda FZ
Madre de Dios FZ
Desolación FZ
Figs. 3b and 8
Fig. 7a
Fig. 7b
Fig. 3a
LBA
LPo
LBe
LSM
LVi
LAr
UEs
RTu
MEs
SAz
CoD
LF
CaB
IrB
42º S
46º S
50º S
54º S
78º W
74º W
70º W
66º W
62º W
58º W
0
350
km

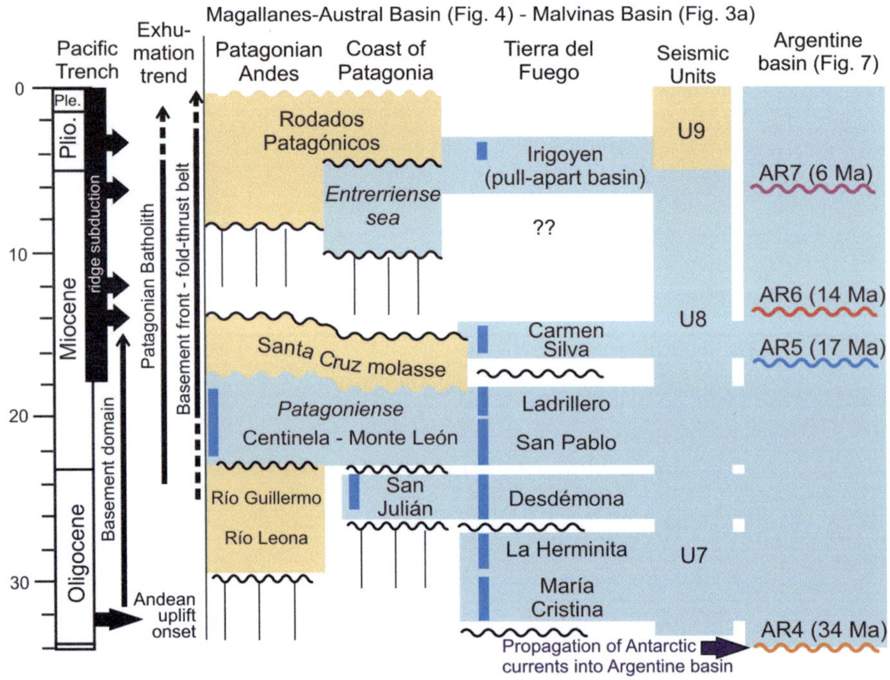

Fig. 2 Correlation of most relevant formations and transgressions in the Magallanes–Austral, Malvinas and Argentine basins. After Hinz et al. (1999), Ghiglione and Ramos (2005), Parras et al. (2008), Guillaume et al. (2009), Malumián and Jannou (2010), del Rio et al. (2013), Gruetzner et al. (2012), and Sachse et al. (2015). *Light blue bands* correspond to marine sequences and transgression periods; *brown background* to continental units. Tectonic events and exhumation trends: see discussion in Sect. 3.2 and references therein

"Patagoniense" transgression occurred later, during the latest Oligocene—early Miocene (Fig. 2) (Malumián 2002; Cuitiño and Scasso 2010; Parras et al. 2012; Cuitiño et al. 2012, 2013).

The transgressive marine phase was followed by a middle Miocene regressive event that marks the onset of continental conditions in Patagonia (Malumián and Jannou 2010), represented by a coarsening upward succession of fluvial deposits from the Santa Cruz Formation (Argentina) (Fig. 2) or the equivalent Zeballos Group (Chile), thinning eastward from a maximum 600 m in the foothills to 250 m on the Atlantic coast (Flynn et al. 2002; Malumián 2002). These continental deposits are considered the synorogenic molasse (Ramos 1989; Lagabrielle et al. 2009) and the first-order indicator for the building up of the necessary topography to produce an effective rain shadow on the leeside of the mountains (Blisniuk et al. 2005, 2006).

In Tierra del Fuego, an Oligocene prograding wedge visualized in seismic sections (Foreland III from Gallardo 2014), represented by outcrops of the shallow marine to deltaic Carmen Silva Formation (Fig. 2) (Malumián and Olivero 2006)

progressively advanced toward the east until the depositional shelf break reached the Atlantic platform during the Miocene (Galeazzi 1998).

Continental sedimentation rate was very low in the Magallanes–Austral Basin after the middle Miocene. The Santa Cruz Formation is overlie by a thin carpet of upper Cenozoic piedmont and glacifluvial gravel deposits (Panza 2002; Martínez et al. 2009) known as Rodados Patagónicos (Feruglio 1949) covering most of extra-Andean Patagonia (Fig. 2). Close to the foothills, widespread upper Miocene basaltic Mesetas and some thin sequences cover the Santa Cruz Formation.

Contrastingly, the upper Miocene–lower Pliocene constitutes a transgressive cycle very well developed in the offshore Argentine (Hinz et al. 1999; Hernández-Molina et al. 2009, 2010; Gruetzner et al. 2012, 2014) and Malvinas basins (Galeazzi 1998; Baristeas et al. 2013) that partially filled the Malvinas Trough during the Pliocene (Fig. 1) (Fish 2005). The Miocene transgression reached only the present coast of Patagonia and is represented by thin marine sequences from the Entrerriense sea with abundant pectinid fossils of late Miocene age (Fig. 2) (Scasso et al. 2001; del Rio et al. 2013).

There has been an assumption that tectonic growth of the Southern Patagonian Andes ended in late Miocene times, largely based on the top age of the Santa Cruz Formation (Ramos 1989; Blisniuk et al. 2005; Lagabrielle et al. 2009) and the paucity of late Miocene–Pliocene or younger structures in the fold and thrust belt (Fosdick et al. 2013). There is, however, multiple thermochronological evidence that exhumation in the hinterland continued profusely, with large volumes of rock denudated rapidly between ~ 15 and 5 Ma, and steadily since ~ 7 Ma (Thomson et al. 2010; Guillaume et al. 2013; Fosdick et al. 2013).

The tectonic scenario was quite different in Tierra del Fuego and in the offshore Malvinas basin, where sub-horizontal sequences covering the fold and thrust belt indicate that orogenic growth stalled after the Oligocene (Ghiglione 2002; Malumián and Olivero 2006; Ghiglione et al. 2010; Baristeas et al. 2012, 2013). Strike-slip dominated south of 53°, although slow, post-tectonic uplift and denudation of the Fuegian Andes basement domain was continuous and significant during the late Oligocene and Neogene (Nelson 1982; Gombosi et al. 2009), contemporary with the transpressional uplift of the fold and thrust belt (Ghiglione 2002; Punta Gruesa thrusting from Ghiglione and Ramos 2005).

Late Miocene to Pliocene uplift during denudation has been assigned to isostatic rebound (Fosdick et al. 2013; Nelson 1982) and dynamic topography due to an asthenospheric window (Guillaume et al. 2013). The Patagonian glaciations that started sometime between ~ 7 and 5 Ma as indicated by basalt intercalated between the oldest till deposits (Mercer and Sutter 1982; Ton-That et al. 1999; Lagabrielle et al. 2007), unquestionably played also an important role during the last phase of Andean denudation.

That leaves the open questions of which factors controlled the present distribution of main faults and structural domains delineating the orogen? And, at which time this configuration was consolidated? Thomson et al. (2010) argue that the timing of deformation and orogen dynamics was largely influenced by late

Cenozoic glaciations. On the other hand, Fosdick et al. (2013) propose a 22–18 Ma main phase of tectonic uplift and exhumation and attribute post-10 Ma denudation solely to fluvio-glacial processes without significant shortening being involved. Concomitant with orogen exhumation, in extra-Andean Patagonia a dynamic response to active spreading ridge subduction produced a late Miocene transition from subsidence to generalized uplift (Guillaume et al. 2009).

Regardless of the discussion about the mechanisms producing Neogene orogenic exhumation, that is the subject of another work in progress, we here question what happened to those large volumes of denudated rocks from the Andes. They were not deposited in Patagonia as shown by the stratigraphical record (Fig. 2), and although surely some volumes went westward to fill the forearc basins in the Pacific trench (cf. Melnick and Echtler 2006), we propose that a larger portion bypassed the continent toward the offshore Argentine and Malvinas basins where thick Neogene sequences were distributed by Antarctic waters and deposited along the Southern Atlantic margin (Fig. 1).

2 Geological and Tectonic Setting

The Southern Patagonian and Fuegian Andes are characterized by well-defined and continuous tectonomorphic zones. These zones are differentiated by their structural style and stratigraphy (Fig. 1) (following Kraemer 1998, 2003; Ghiglione et al. 2009, 2010) in a (i) basement domain encompassing the biggest volume of the Cordillera to the west, and an eastern, (ii) fold and thrust belt domain divided in an internal and external belt.

The basement domain presents internal ductile deformation and is characterized by a regional thick-skinned style, producing the uplift of the metamorphic and magmatic basement of the region. The core and backbone of this domain are the Patagonian and Fuegian batholiths encompassing Jurassic to Neogene calc-alkaline magmatic arc rocks (González Guillot et al. 2011, 2012; Guillot 2016). The Western Andes Metamorphic Complex outcropping along the Pacific archipelago west of the Patagonian Batholith represents the upper Paleozoic–Upper Triassic accretionary prism including exotic amalgamated blocks, such as the Madre de Dios platform (Hervé et al. 2008; cf. Calderón et al. 2016). The late Paleozoic geological history during Gondwana times is represented by widespread low-grade metamorphic rocks (Riccardi and Rolleri 1980; Giacosa and Márquez 2002) cropping out along the eastern flank of the Patagonian Batolith. These polydeformed turbiditic successions reaching up to greenschist facies interspersed with minor limestone and metabasites are grouped in the Eastern Andes Metamorphic Complex (Hervé et al. 2008; Calderón et al. 2016). In Tierra del Fuego, basement rocks are concentrated in Cordillera Darwin (Fig. 1), a high-grade metamorphic complex exposing kyanite–staurolite schist (Dalziel et al. 1974; Nelson 1982; Kohn et al. 1995). It is not known at present if these metamorphic rocks of Cordillera Darwin were originally part of the Eastern Andes Metamorphic Complex, or if they were part of the Coastal Accretionary Complexes of the Southern Patagonian Andes (cf. Calderón et al. 2016). From a mechanical point of view, the eastern basement domain in the Patagonian Andes can be correlated with Cordillera Darwin in Tierra del Fuego, since they both transferred shortening to the fold and thrust belt during the Late Cretaceous and Cenozoic (Ghiglione and Cristallini 2007, Ghiglione et al. 2014; Maffione et al. 2015; Maffione 2016). The units outcropping along the concave edge of the basement domain in the Patagonian–Fuegian Andes are fault-controlled Jurassic synrift deposits.

The Early to Middle Jurassic was characterized by regional extension and the synrift deposition of basal conglomerates cover by thick volcanics of El Quemado Complex or Tobífera/Lemaire Formation (Pankhurst et al. 2000). The Jurassic fault array consists of N–NW-oriented extensional faults bounding grabens, and E–W-oriented transfer faults separating depocenters with different degrees of extension (Uliana et al. 1989; cf. Ghiglione et al. 2013). The Jurassic extensional faults and grabens were inverted during the Andean cycle, influencing style and distribution of compressional structural domain in the Southern Patagonian Andes (Kraemer 1998; Likerman et al. 2013; Ghiglione et al. 2014). The Lower Cretaceous Springhill and Río Mayer formations are interspersed in inversion structures along the basement front (Kraemer 1998; Ghiglione et al. 2009, 2014; Giacosa et al. 2012).

The Andean compressive cycle started during the Late Cretaceous in the Southernmost Andes of Patagonia and Tierra del Fuego, producing the onset of the foreland stage in the Magallanes–Austral Basin and offshore basins of southernmost South America (Ghiglione et al. 2010). Late Cretaceous metamorphism, uplift, and exhumation of the basement domain (Cunningham 1995), were followed by the onset of clastic foredeep sequences (Wilson 1991; Arbe and Hechem 1984; Kraemer and Riccardi 1997). The Upper Cretaceous sequences are well developed south of 49°S, reaching a thickness of ~3800 m between Lagos Viedma and Argentino and of ~4300 m in Última Esperanza (Fig. 1) (cf. Ghiglione et al. 2014, 2016).

The following cycle of synorogenic deposition and propagation of thrusts occurred during the latest Cretaceous–Paleogene (Biddle et al. 1986; Marenssi et al. 2002) and was accompanied by strong sedimentary rates, as indicated by growth strata in Río Turbio (51°30′S, Fig. 1) (Malumián et al. 2000).

Tierra del Fuego was affected by middle Eocene–Oligocene coeval strike-slip and compressional deformation due, respectively, to the fast (2.4 cm/year) left-lateral motion between South America and Antarctica (Livermore et al. 2005), and the synergetic combination of fast spreading rates in the Drake Passage (Ghiglione et al. 2008; Barbeau et al. 2009) with rapid convergence and northward-directed subduction of the Farallón plate against South America (Ghiglione and Cristallini 2007). The progression of exhumation and sedimentation of the Oligocene–Neogene tectonic cycle is the topic of the current review and would be explained in detail below.

3 Oligocene–Neogene Tectonic Cycle

3.1 Geodynamic Setting and Tectonic Style

3.1.1 Southern Patagonian Andes

The last Andean tectonic cycle is very well developed in the Southern Patagonian Andes and has been related to the approximation and collision of a series of segments of the Chile seismic ridge, which separates the Antarctic and Nazca plates, below South America (Fig. 1) (Ramos 1989; Folguera and Ramos 2002; Lagabrielle et al. 2004; Scalabrino et al. 2009). The first segment of the Chile ridge collided nearly parallel to the Pacific trench around 18 Ma at ~54°S as calculated by kinematic reconstructions (Breitsprecher and Thorkelson 2008). The resulting Antarctica-Nazca-South America triple junction migrated progressively northward with each segment collision, to its present position at ~46°S (Fig. 1) (Forsythe and Nelson 1985; Cande and Leslie 1986). A speed difference between the Nazca (~10 cm/year) and Antarctica (~2 cm/year) plates produced an increasingly open asthenospheric slab window, giving way to widespread alkali Patagonian plateau lavas (Gorring et al. 1997; Bourgois and Michaud 2002).

Although there is a time lag between the ~18 Ma first ridge collision, and the 30 Ma onset of exhumation in the basement domain (Fig. 2) (Thomson 2002), a cause to effect relationship has been proposed (Ramos 1989). Compression preceding ridge subduction could have been the result of increased coupling and shallow subduction due to the approach of progressively younger and more buoyant oceanic crust to the trench (Thomson et al. 2001; Folguera and Ramos 2002). Another timing discrepancy exists between an important 10 to <5 Ma denudation of the cordillera (Thomson et al. 2001), and the ~22–19 to 14 Ma deposition of the Santa Cruz Formation synorogenic deposits (Fig. 2) (Marenssi et al. 2005; Blisniuk et al. 2005; Cuitiño et al. 2012, 2015) that would be discussed below (Sect. 4).

3.1.2 Fuegian Andes, Antarctica, and the Scotia Sea

After the middle Eocene–Oligocene compressional event in Tierra del Fuego (Ghiglione and Ramos 2005), the tectonic plate scenario changed drastically for the Southernmost Andes–Antarctica region (see Ghiglione 2016 for a review). Relative movement between South America and Antarctica was accommodated by spreading in the West Scotia Sea from about 28 Ma (Lodolo et al. 2006) until ~16 Ma when the future Scotia Plate started to move independently of the South American plate (cf. Eagles 2016; Eagles et al. 2005; Lagabrielle et al. 2009). A sinistral strike-slip fault zone on the offshore North Scotia Ridge and continental Magallanes–Fagnano fault system (Fig. 1) was developed to accommodate the movement between the two plates (Thomas et al. 2003). A left-lateral strike-slip regime produced widespread deformation in Tierra del Fuego Island ever since

(Diraison et al. 2000; Lodolo et al. 2003; Sue and Ghiglione 2016). A still active rifting system perpendicular to the orogen produced the opening of the eastern Magallanes Strait (Diraison et al. 1997, 2000) and the transverse segmentation of the external fold and thrust belt (Ghiglione et al. 2013). Since spreading rates in Drake Passage were higher than required by South America–Antarctica motion, this resulted in deformation within southern South America (Livermore et al. 2005). Transpressional structures have been reported in Miocene sequences outcropping in the Atlantic coast of Tierra del Fuego (Ghiglione et al. 2002). Transpression along the proto-North Scotia Ridge–Magallanes–Fagnano fault system during the 26–14 Ma period produced a change from deep marine to very shallow conditions along the newly formed plate boundary (Lagabrielle et al. 2009). Baristeas et al. (2012, 2013) defined a transpressional foreland phase for the Malvinas Basin, during which the sediments were derived from the uplifted Burdwood bank (Fig. 1) (cf. Galeazzi 1998). Recently, deformation along the North Scotia Ridge is transpressive due to a restraining bend geometry, while along the Magallanes–Fagnano fault system a releasing bend develops (Lodolo et al. 2003).

3.2 Exhumation Age from Thermochronological Data

The samples with the oldest exhumation ages in the Southern Patagonian Andes are also the westernmost, located in outcrops along the Pacific archipelago in metamorphic and magmatic rocks of the Basement domain, around 100 km from the trench. These samples show similar apatite fission-track (AFT) central ages between ~34 and 25 Ma, indicative of rapid cooling through the apatite partial annealing zone (APAZ) from above ~110 to ~60 °C (Thomson et al. 2001, 2010). The AFT data indicate at least ~4 km of denudation in the western basement domain since 30 Ma (Thomson et al. 2001), while apatite (U–Th)/He (AHe) cooling ages within the same sector indicate similar denudation rates between 24 and 15 Ma (Thomson et al. 2010).

Along the central and eastern Patagonian Batholith, AFT ages between ~24–18 and 15 Ma are in many cases concordant with zircon fission-track ages (ZFT) (Thomson et al. 2010). The coincidence can be explained by in situ cooling, following intrusion into cool shallow crust (Thomson et al. 2001). The total amount of cooling in these rocks since 24 Ma is restricted between the APAZ (110 °C) and the closure temperature of fission tracks in zircon (~280 °C), representing between 9 and 4 km of denudation (Thomson et al. 2001). When AHe from the Patagonian Batholith is considered, ages between 20 and 11 Ma seem to be consistently distributed, although also some ages between 10 and 5 Ma are found north of 49°S.

Between the 50 and 52°S parallels, AFT and ZFT show active denudation of the basement front and the fold and thrust belt between 24 and 5 Ma (Fosdick et al. 2013), while to the north of 48°30′S ages between 14 and 5 Ma prevail (Thomson et al. 2010). Along the eastern margin of the basement domain (Eastern Andes Metamorphics), young AFT ages between 14 and 5 Ma (Fosdick et al. 2013) or

12–8 Ma (Thomson et al. 2001) indicate between 9 and 4 km of denudation. Zircon (U–Th)/He (ZHe) results from the basement front and the fold and thrust belt in Última Esperanza indicate an approximate ~6 km exhumation depth since ~22 Ma (Fosdick et al. 2013). AHe results from that region record young ages centered between 6 and 4 Ma, indicating significant late Miocene and Pliocene denudation, consistent with efficient fluvio-glacial erosion (Fosdick et al. 2013).

Haschke et al. (2006) analyzed an AFT vertical profile from the basement front at Cerro Barrancos granite, southwest of Lago Posadas at 47°30′S (Fig. 1). These authors estimated accelerated cooling between 17 and 10 Ma with a minimum of 3–4 km of exhumation. A reheating between 10 and 6 Ma followed, and final cooling commenced at ~4 Ma until the samples were exhumed to the surface (Haschke et al. 2006). Guillaume et al. (2013) reappraised the thermal history of Cerro Barrancos granite by adding AHe data and calculated slow (~70 m/Myr) denudation rates for the 25–15 Ma and 10–5 Ma periods, followed by a phase of rapid cooling, with denudation values as high as 650 m/Myr between 5 and 3 Ma.

The compiled thermochronological data consistently indicates the following, eastward moving trend of exhumation in the Southern Patagonian Andes (left column in Fig. 2): (i) western basement domain from ~34 to 15 Ma, (ii) Patagonian Batholith since 24 Ma, (iii) basement front and the fold and thrust belt between ~25 or 15 Ma and 5–3 Ma, and several kilometers of denudation since ~7–5 Ma.

Toward the south, multiple AHe data indicate sustained exhumation in the 25–3 Ma period along the southern Patagonian Batholith between 52° and 54°S (Thomson et al. 2010). Along the Canal Beagle (Fig. 1), three AFT ages reached post-Eocene exhumation ages ranging between 29 and 1 Ma (Thomson et al. 2010). Nelson (1982) using fission tracks and geochronology argued that a significant uplift and exhumation of Rocas Verdes and Cordillera Darwin occurred post-tectonically since 30 Ma. This late Oligocene–Holocene uplift of up to ~2 km, representing around two-thirds of the actual relief, occurred at constant low rates (0.05–0.20 mm/year) (Nelson 1982). Modeling of new thermochronometry data from the basement domain in the Fuegian Andes also reveals a slow but sustained post-Eocene cooling (Gombosi et al. 2009).

3.3 Synorogenic Sequences

The Oligocene–Neogene cycle of sedimentation forms a clastic wedge that thickens to the west and to the south of the Magallanes–Austral Basin (Malumián 2002). This synorogenic sequence presents very continuous exposures in the Patagonian Andes's foothills and overlaps the offshore basement Río Chico high (Fig. 3a) (Baristeas et al. 2013; Sachse et al. 2015).

The upper Cenozoic foredeep sequence starts with the Oligocene–lower Miocene fluvial systems of the Río Guillermo and Río Leona formations (Kraemer 1998; Marenssi et al. 2005) resting unconformably on a discontinuity formed

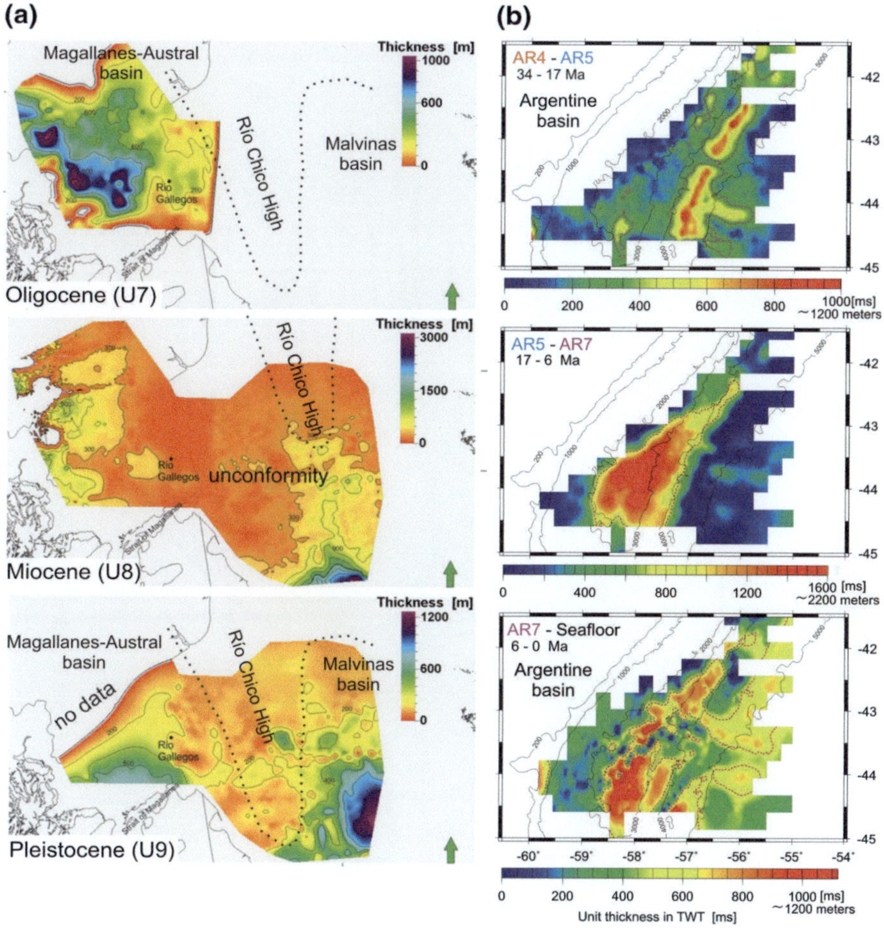

Fig. 3 Comparison between Oligocene–Neogene thickness patterns of sedimentation from seismic information, between Magallanes–Austral Basin (*to the left* in meters) and Argentine Basin (*to the right* in ms two-way traveltime, twt). See Fig. 1 for location and Fig. 2 for correlation of seismic and outcrop units: **a** thickness maps (meters) of the Austral and western Malvinas basins from Sachse et al. (2015). **b** Maps of unit thicknesses in twt for identified seismic units (approximate equivalence in meters is indicated). *Purple hatched lines* mark depocenter outlines (rms of unit thickness). Bathymetric contours are indicated as *thin black lines*. From Gruetzner et al. (2012)

during late Eocene and early Oligocene times (Fig. 2) (Malumián et al. 2000). A fining upward trend for Rio Leona Formation indicates a tectonically derived upward increase in accommodation space (Marenssi et al. 2005). This deepening trend resulted in the flooding of the shallow marine Patagoniense transgression (Fig. 4a) (Malumián 2002), represented by lower Miocene deposits of the Centinela Formation (Fig. 2) (equivalent to Estancia 25 de Mayo and El Chacay formations) outcropping along the Patagonian foothills (Cuitiño and Scasso 2010; Cuitiño et al. 2012). To the east, the Monte León, Chenque, and Gaiman Formations (Malumián 2002; Parras et al. 2008, 2012), and in Tierra del Fuego the San Pablo and

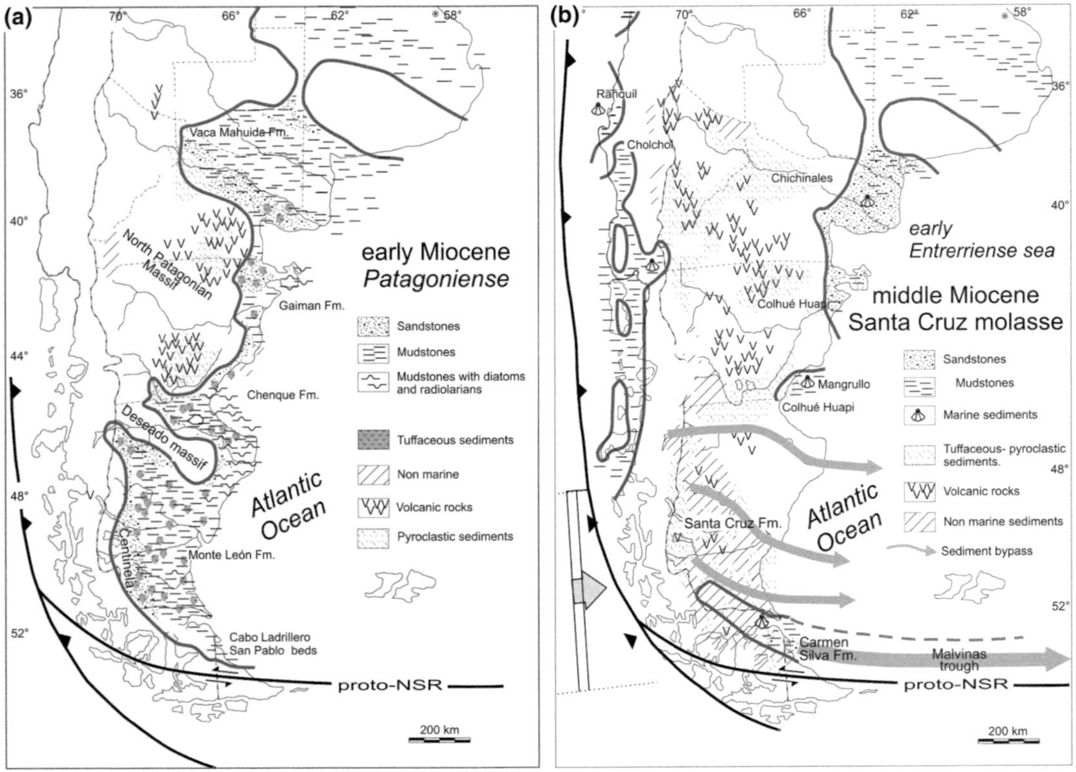

Fig. 4 Paleogeographic reconstructions from the **a** early and **b** middle Miocene. See text for discussion. *NSR* North Scotia Ridge. Modified from Malumián and Nañez (2011) and references therein

Ladrillero beds represent distal equivalents for the early Miocene transgression (Fig. 4a) (Malumián and Jannou 2010). The age of the Patagoniense has been constrained by high resolution isotopic ages between 20 and 18 Ma for the western wedge (lower Miocene; Cuitiño et al. 2012, 2015). These ages overlap with a low eustatic sea level interval, and an erosional sequence boundary, suggesting an origin caused by a strong basin floor subsidence event related to tectonic loading (Cuitiño and Scasso 2010).

The onset of the late Cenozoic transgression occurred during the middle Eocene in Tierra del Fuego, i.e. earlier than in Patagonia (Mella 2001; Malumián 2002) reaching deep water facies in the early Oligocene (María Cristina beds, Fig. 2) (Malumián and Jannou 2010). The south to north diachronism can be related to an earlier phase of deformation in Tierra del Fuego (middle to late Eocene Río Bueno thrusting; Ghiglione and Ramos 2005) that can be related to the accelerated opening of the Scotia Sea (Ghiglione et al. 2008; Barbeau et al. 2009; Gombosi et al. 2009; Eagles 2016).

The continued advance of topographic uplift produced an ensuing regression in the Southern Patagonian Andes, represented by the fluvial and lacustrine Santa Cruz Formation (Fig. 4b). There is a transitional contact with the various underlying marine units (Fig. 2), evidenced by intermediate estuarine deposits (Feruglio 1938; Cuitiño and Scasso 2010; Cuitiño et al. 2013). The age of the conglomeratic to mammal-bearing lacustrine sediments of the Santa Cruz Formation is bracketed by intercalated tuffs between $\sim$22–19 and 14 Ma (Marshall et al. 1997; Blisniuk et al. 2005; Perkins et al. 2012). Calculated average sediment accumulation rates are high, in the range of 200–100 m/Myr (Blisniuk et al. 2005; Perkins et al. 2012).

Patagonian sedimentation practically stalled after the middle Miocene. Southwest of Lago Posadas a $\sim 14 \pm 1$ Ma tuff from the Santa Cruz Formation is just 3 m below the base of a 12 Ma old basalt which is topped by a scarce succession of basalts and sediments (Blisniuk et al. 2005; Gorring et al. 1997). In extra-Andean Patagonia widespread gravel mantels known as *Rodados Patagónicos* cover the Santa Cruz Formation (Fig. 2). As expressed in a recent review by Martínez and Kutschker (2011), these conglomerates are one of the most distinctive features of the regional landscape in extra-Andean Patagonia (see also Panza 2002). Their significant roundness, along with additional sedimentary and geomorphological evidence, indicates that their origin is linked to surface drainage networks of relatively high energy (Martínez and Kutschker 2011). Such high energy was induced by powerful flow rates and/or strong water gradients, attributable to tectonism, epirogenesis, and/or lowering of sea level (Martínez and Kutschker 2011).

A mapping and morphological study of these post-middle Miocene fluvial terraces carry out by Guillaume et al. (2009) in the southern part of the North Patagonian Massif (Fig. 1) evidenced regional-scale tilting of terraces overlaid by coarse fluvial series. These authors suggest a generalized dynamic uplift of the continental plate from the Andes to the Atlantic Ocean that started in the middle–late Miocene, during the opening of the asthenospheric window (Fig. 5). This uplift probably migrated toward the northeast, following the procession of ridge segment collisions (Fig. 6).

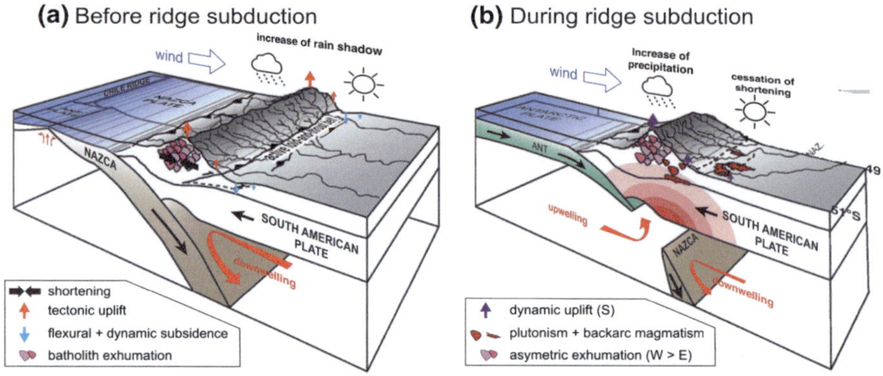

Fig. 5 Synthetic 3D sketch showing the response of the South America plate to the formation of the slab window between 51° and 47°S (modified from Guillaume et al. 2009): **a** segments of the Chile Ridge did not yet reach the trench, and convergence is accommodated by shortening in the fold and thrust belt. **b** Subduction of the ridge segments produces the opening of a slab window inducing magmatism that advect heat through the lithosphere, and a dynamic uplift associated with upwelling of sub-slab mantle

Fig. 6 Sketch showing the uplift of Patagonia resulting from the episodic subduction of ridge segments accompanying the northward migration of the triple junction. Model of ~800 m dynamic uplift in response to subduction of four Chile Ridge's segments with lengths of 400 (14 Ma), 200 (12 Ma), 200 (6 Ma), and 50 km (3 Ma). Redrawn from Guillaume et al. (2009)

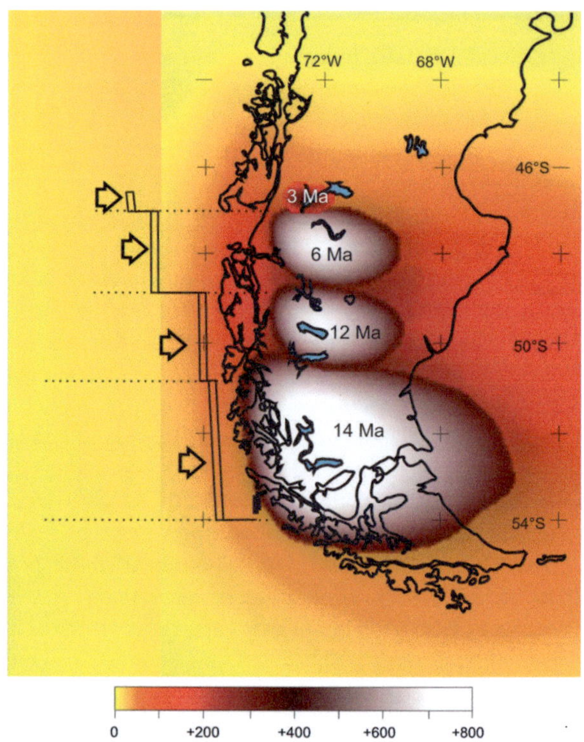

Contrasting with the situation in continental Patagonia, the middle Miocene is an aggradational phase in the Argentine continental margin (Figs. 3b and 7) (Hinz et al. 1999; Gruetzner et al. 2012) associated with major subsidence during the deepening of the Drake Passage (Hernández-Molina et al. 2010). Post-middle Miocene seismic sequences are very thick and well developed in the Argentine Basin (Figs. 3b and 7) (Hinz et al. 1999; Hernández-Molina et al. 2009, 2010; Gruetzner et al. 2014); registering an up to tenfold increase in sedimentation rate after the AR5 (17 Ma) unconformity (Fig. 8; Gruetzner et al. 2012). The middle Miocene–Pleistocene conform seismic units up to ~2000 ms (two-way traveltime) in vertical extension (Fig. 7) (around 2400 m thick, Gruetzner et al. 2012). On the other hand, the upper Miocene–lower Pliocene sequences of the Entrerriense Sea (Fig. 2) cropping out along the coast of Patagonia are less than 50 m thick. Based on $^{87}Sr/^{86}Sr$ dating of calcite shells belonging to pectinids, this sea flooding event that barely reached the Atlantic coast in the Magallanes–Austral Basin (Fig. 4b) was dated between ~10 and 5 Ma (Scasso et al. 2001; del Rio et al. 2013). At that time, a carpet of the mainly fluvial Rodados Patagónicos was covering Patagonia (Fig. 2).

In Tierra del Fuego, sedimentation was active during the middle—late Miocene transgression that completed the remaining accommodation space with shallow marine, deltaic system, and fluvial deposits of the Carmen Silva Formation (Fig. 4b) (Malumián and Jannou 2010; Gallardo 2014). However, sedimentation was funneled toward the offshore afterward, filling the Malvinas trough during the Pliocene (Fish 2005). Post-Miocene sequences show highest sediment accumulations in the Malvinas Basin (Pliocene U9, Fig. 3a). The Magallanes–Austral Basin could not accumulate significant sediments anymore; thus, it is highly likely that the sediments were further transported into the offshore basins (Sachse et al. 2015).

4 Discussion and Concluding Remarks: The Relation Between Andean Deformation, Dynamic Topography, and Sediment Bypass Toward the Offshore

Approximation and collision of the Chile ridge produced renewed tectonic uplift of the Southern Patagonia Andes since ~30 Ma, with denudation focused in the eastern sector of the Basement domain (Thomson et al. 2010), and the fold and thrust belt after ~20 Ma (Fig. 2) (Fosdick et al. 2013). The ensuing opening of an asthenospheric window produced the dynamic uplift of the retroarc, migrating from the foothills (Scalabrino et al. 2009) toward extra-Andean Patagonia (Fig. 5) (Guillaume et al. 2009). The wave of dynamic topography and a late Miocene transition from subsidence to generalized uplift could have migrated also from south to north, as different spreading segments entered the trench (Fig. 6) (Guillaume et al. 2009, 2013), and toward the east, ultimately producing the uplift of shorelines along the Atlantic coast (Pedoja et al. 2011; Isla and Angulo 2015).

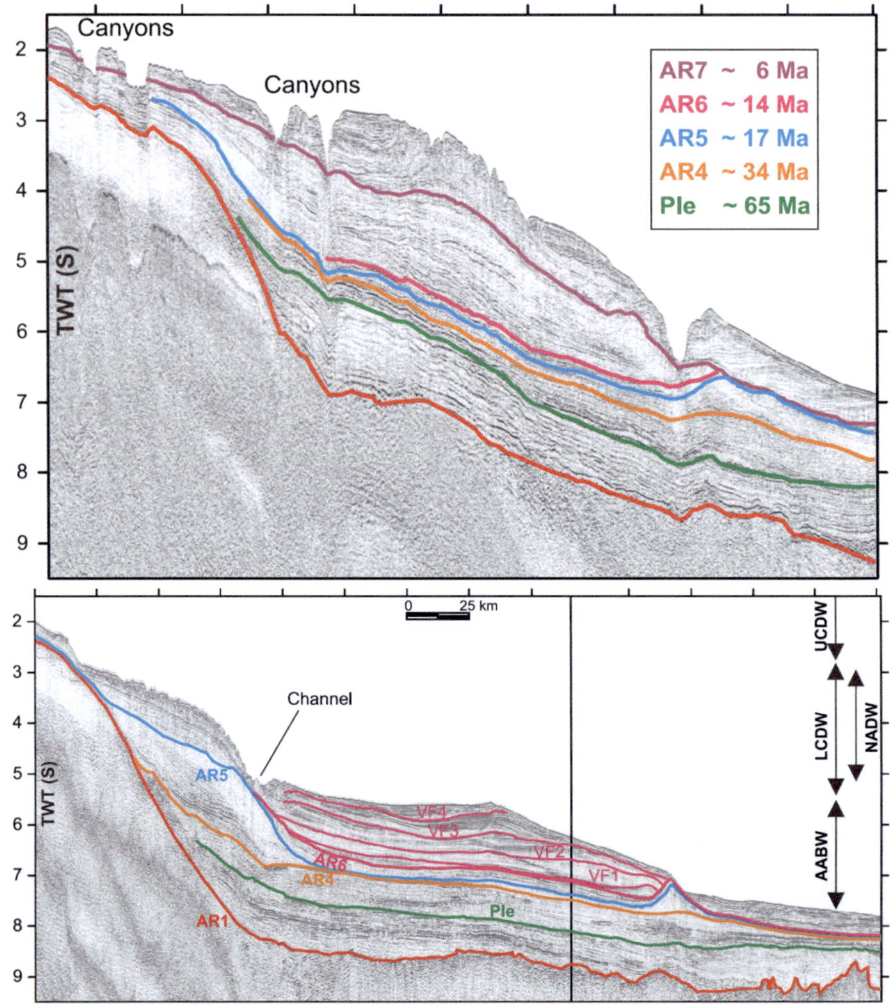

Fig. 7 Seismic profiles across the Argentine Basin. See Fig. 1 for location and Fig. 2 for correlation of seismic and outcrop units. *Arrows* indicate the extent of present-day water masses AABW (Antarctic Bottom Water), LCDW (Lower Circumpolar Deep Water), UCDW (Upper Circumpolar Deep Water), and NADW (North Atlantic Deep Water). Reflectors VF1 to VF4 indicate a refined seismostratigraphic model, where reflector VF3 is an elongated mounded drift interpreted to indicate strengthening of bottom flow during the late Miocene/early Pliocene. Modified from Gruetzner et al. (2012)

Fig. 8 Maps of sedimentation rates in the Argentine Basin from seismic data. Notice the tremendous increase in sedimentation after AR5. See Fig. 1 for location and Fig. 2 for correlation of seismic and outcrop units. Sedimentation rates were derived from reflector depths based on a velocity–depth model and age information. Bathymetric contours are indicated as *thin black lines*. From Gruetzner et al. (2012)

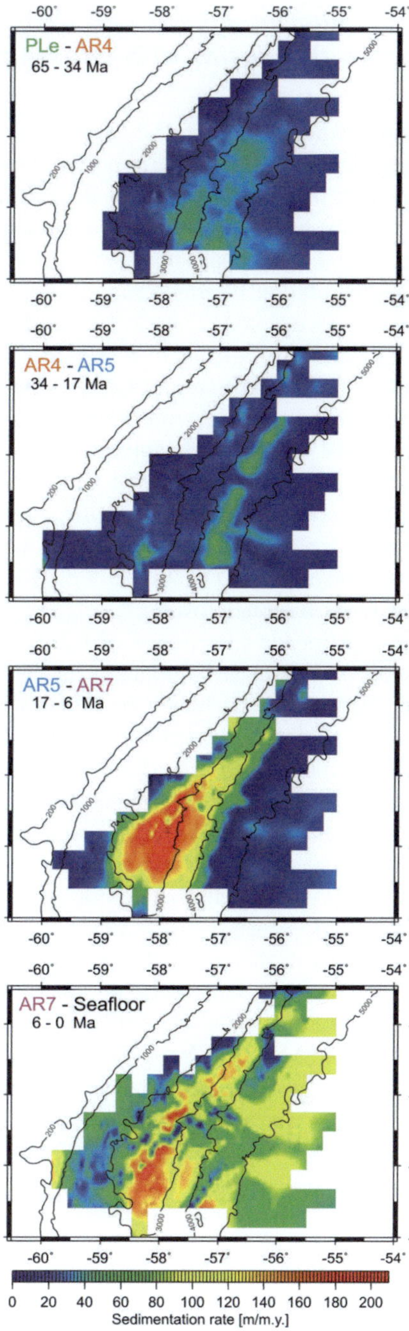

4.1 Miocene–Pleistocene Distribution of Sedimentation in the Basins of Southernmost South America

While exhumation in the Southernmost Andes was continuous during the Oligocene–Neogene, an important change in the Magallanes–Austral Basin dynamics occurred during that period, related to uplift of the retroarc. While Late Cretaceous to Eocene foreland depocenters were concentrated in the 200-km-wide strip bounding the basement thrust front (cf. Ghiglione et al. 2010, 2016), the locus of sedimentation migrated rapidly toward the offshore basins after the Oligocene (Fig. 3) (Gructzner et al. 2012; Sachse et al. 2015). Along the foothills of the Patagonian and Fuegian Andes, sedimentation almost ceased at 14 Ma (Blisniuk et al. 2005; Malumián and Olivero 2006). In contrast with the situation in continental Patagonia, major subsidence and sedimentation is registered in the Argentine continental margin and Malvinas basin since the middle Miocene (Figs. 3 and 8) (Hinz et al. 1999; Baristeas et al. 2012, 2013; Gruetzner et al. 2012; Sachse et al. 2015). It seems that since that time, Andean sediments bypassed Patagonia to reach the offshore Argentine and Malvinas basin.

Continental uplift produced a marked regression in Tierra del Fuego, where the shelf break migrated toward the east, reaching the actual Atlantic coast ($\sim$68°W) by middle Miocene times (east of Carmen Silva delta, Fig. 4b). Since the Miocene, overall sediment accumulation in the Malvinas Basin exceeded those of the Magallanes–Austral Basin (Fig. 3a) (Sachse et al. 2015). After 5.5 Ma, during the Pliocene–Pleistocene, marine sedimentation was focused east of $\sim$65°S (Galeazzi 1998), while the Río Chico high was under erosion, and an important depocenter was developed along its eastern flank (Fig. 3a) (Baristeas et al. 2013; Sachse et al. 2015).

Sediment bypass produced a strong change and long-lasting trend in sedimentation along the Argentine continental margin, marked by the A5-17 Ma unconformity (Figs. 7 and 8). Below A5, the 65–17 Ma seismic sequence has low sedimentation rates (10–46 m/Myr), similar to hemipelagic settings in the oceans (Fig. 8) (Gruetzner et al. 2012). On the other hand, the 17–0 Ma sequences exhibit 120–140 m/Myr sedimentation rates, that is, >200 m/Myr in some depocenters (Fig. 8) (Gruetzner et al. 2012), i.e. similar or higher than the calculated average 200–100 m/Myr sediment accumulation rates for the Santa Cruz Formation (Blisniuk et al. 2005; Perkins et al. 2012). At the same time, an elongated mounded drift is interpreted to indicate strengthening of bottom flow during the late Miocene/early Pliocene (Fig. 7) (Gruetzner et al. 2012).

4.2 Dynamic Uplift in Patagonia, Wrench Deformation in Tierra del Fuego and Sedimentation Bypass

Late Miocene interruption of the long-lasting Magallanes–Austral foreland depocenter (Fig. 2) seems to be the product of dynamic uplift produced by an

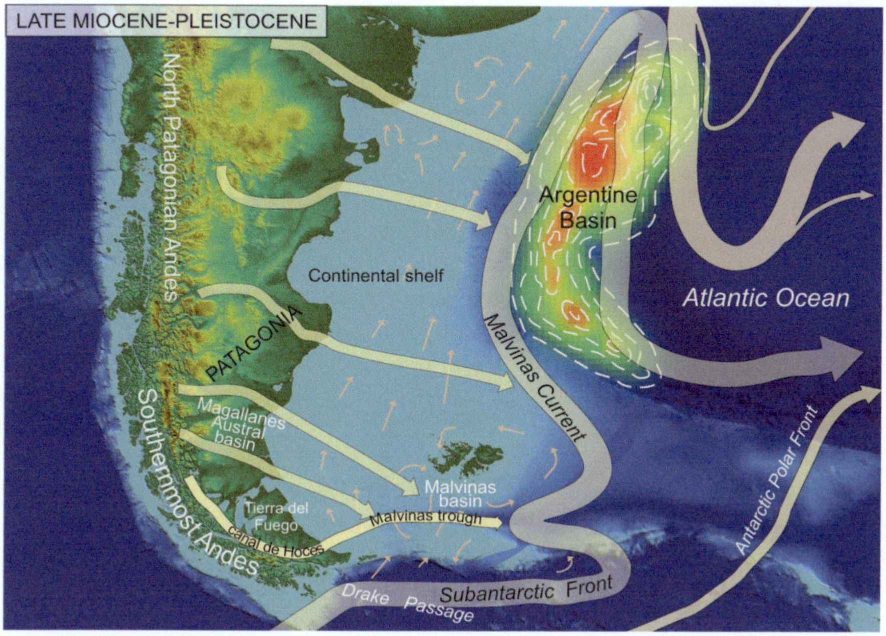

Fig. 9 Post-Miocene paleogeography, after the deepening of the Drake Passage and the onset of the sub-Antarctic Front and Malvinas Current, during the exhumation of the Patagonian–Fuegian Andes. The sedimentary systems bypassing the continent toward the depicted offshore depocenters are shown. Main ocean currents are from Piola and Matano (2001). Post-A5 (17 Ma) sediment thickness in the Argentine Basin was compiled and drawn from Hinz et al. (1999), Hernández-Molina et al. (2009, 2010) and Gruetzner et al. (2012, 2014)

asthenospheric window (Fig. 6), as proposed by Guillaume et al. (2009). To the north of 52°S, it is unclear to us by which mechanism the great volumes of rocks from the late Miocene to Pliocene exhumation and denudation of the Cordillera reached the Argentine continental shelf. Clearly some significant amount of sediment flux was diverted toward the Pacific trench (cf. Melnick and Echtler 2006). It seems probable, however, that fluvial and fluvio-glacial systems similar to those actual present in Patagonia could have acted has effective means of sediment transport toward the offshore (Fig. 9). The increase in sediment transport from the river and shelf toward the deeper part of the basin has been previously assigned to a relative sea level fall and/or tectonic uplift by Gruetzner et al. (2012).

To the south of 52°S, in the Tierra del Fuego sector affected by Neogene wrench kinematics (cf. Sue and Ghiglione 2016), the Foreland III prograding wedge (Gallardo 2014), represented by the shallow marine to deltaic Carmen Silva Formation (Fig. 2) (Malumián and Olivero 2006), formed a sedimentary system axial to the Fuegian Andes, that prograded toward the offshore Malvinas and South Malvinas basins (Fig. 4b).

There are two different deformation mechanisms to explain the existence of a Fuegian axial trough in its context of wrench kinematics, a first one acting from ~26 to 16 Ma, and a second one active after that period onwards.

26–16 Ma transpression along the proto-North Scotia Ridge fault system was proposed by Lagabrielle et al. (2009) based on tectonic models by Eagles et al. (2005) showing shortening on the Scotia—South America plate boundary in the intervening period (cf. Eagles 2016). The depicted late Oligocene–early Miocene context is also in agreement with the 24–16 Ma Punta Gruesa stage of transpressional faulting recognized along the Atlantic coast of Tierra del Fuego (Ghiglione and Ramos 2005) and in the Malvinas basin (Baristeas et al. 2012, 2013). The sediments could have been funneled along a transpressional depression in the Fuegian Andes, and toward the offshore Malvinas trough (Fig. 4b). The southern trough was probably accentuated by tectonic load due to transpressional uplift of the Basement domain and the Burdwood bank along the North Scotia Ridge that also acted as source of sediments (Galeazzi 1998; Baristeas et al. 2013).

Plate divergence since ~18 Ma (Eagles 2016) produced generalized transtensional deformation in Tierra del Fuego (Ghiglione et al. 2013), and in particular a series of pull-apart depocenters along the South America–Scotia continental plate boundary, such as the eastern Magallanes Strait, the Seno Almirantazgo, Lago Fagnano (Lodolo et al. 2003), and Irigoyen river basins (Fig. 1) (Ghiglione 2003; Ghiglione and Ramos 2005; Malumián and Jannou 2010). This concatenation of pull-apart depocenters could have acted as an effective way of funneling detritus from the continent toward the offshore (Fig. 9). The possibility that those pull-apart depocenters allowed the interchange of Pacific and Atlantic waters was already proposed by Malumián and Scarpa (2005), based on the presence of assemblages of benthic foraminifers with Pacific affinity in the Irigoyen Basin in the Atlantic coast (Fig. 1). These authors suggested a direct shallow connection between the Atlantic and Pacific Oceans by means of the Canal de Hoces along the Magallanes–Fagnano fault system (Fig. 9).

4.3 *Onset of sub-Antarctic Front–Malvinas Current During the Opening of the Drake Passage and Distribution of Sediments*

Plate divergence since 18 Ma led to development of a gap in the North Scotia Ridge that today allows deep water of the Antarctic Circumpolar Current to cross out of the Scotia Sea and into the South Atlantic (Fig. 9) (Eagles 2016). Modern hydrographic observations suggest that most of the water sowing eastward along the sub-Antarctic Front, in the northern Drake Passage, loops northward to form the Malvinas Current heading toward the Argentine Basin (Piola and Matano 2001). Accordingly, an elongated mounded drift (VF3 in Fig. 7) is interpreted to indicate strengthening of bottom flow during the late Miocene/early Pliocene (Gruetzner et al. 2012). Before the onset of these conditions, the 65–17 Ma seismic sequence in the Argentine Basin shows low bottom current activity, similar to an hemipelagic setting (Fig. 8) (Gruetzner et al. 2012).

It seems that the moment at which orogenic denudation, together with uplift in the retroarc, produced sediment bypass toward the offshore, coincided with the onset of the Southpolar Front–Malvinas Current circuit during the opening of a deep Drake Passage that helped distribute the sediments in the Argentinean continental shelf (Fig. 9).

Acknowledgements This work has been carried out with the financial support of grants projects Agencia PICT-2013-1291 and CONICET PIP 2014-2016 GI directed by M.C. Ghiglione, and the ECOS-SUD project A15U02 (Sue and Ghiglione). The authors are grateful to Parques Nacionales de Argentina for granting access to the study zone. Valuable discussions during fieldwork with José Cuitiño are greatly acknowledged.

References

Arbe HA, Hechem JJ (1984) Estratigrafía y facies de depósitos marinos profundos del Cretácico Superior, Lago argentino, Provincia de Santa Cruz. IX Congreso Geológico Argentino (Bariloche). Actas 5:7–41

Barbeau DL, Olivero E, Swanson-Hysell N, Zahid K, Murray K, Gehrels G (2009) Detrital-zircon geochronology of the eastern Magallanes foreland basin: Implications for Eocene kinematics of the northern Scotia Arc and Drake Passage. Earth Planet Sci Lett 20:23–45

Baristeas N, Ankyr Z, di Primio R, Rodriguez JF, Marchal D, Dominguez F (2012) Distribution of hydrocarbon leakyrge indicators in the Malvinas Basin, offshore Argentine continental margin. Mar Geol 332:56–74

Baristeas N, Ankyr Z, di Primio R, Rodriguez JF, Marchal D, Dominguez F (2013) New insights into the tectono-stratigraphic evolution of the Malvinas Basin, offshore of the southernmost Argentinean continental margin. Tectonophysics 604:280–295

Barreda VD, Palazzesi L, Marenssi S (2009) Palynological record of the Paleogene Río Leona Formation (southernmost South America): stratigraphical and paleoenvironmental implications. Rev Palaeobot Palynol 151:22–33

Biddle KT, Uliana MA, Mitchum RM, Fitzgerald MG, Wright RC (1986) The stratigraphy and structural evolution of the central and eastern Magallanes Basin, southern South America. In: Allen A, Homewood P (eds) Foreland basins. Blackwell Scientific Publications, London. Int Assoc Sediment, Spec Publ 8: 41–61

Blisniuk PM, Stern LA, Chamberlain CP, Idleman B, Zeitler PK (2005) Climatic and ecologic changes during Miocene surface uplift in the Southern Patagonian Andes. Earth Planet Sci Lett 230:169–186

Blisniuk PM, Stern LA, Chamberlain CP, Zeitler PK, Ramos VA, Sobel ER, Haschke M, Strecker MR, Warkus F (2006) Links between mountain uplift, climate, and surface processes in the Southern Patagonian Andes. The Andes. Springer, Berlin, pp 429–440

Bourgois J, Michaud F (2002) Comparison between the Chile and Mexico triple junction areas substantiates slab window development beneath northwestern Mexico during the past 12–10 Myr. Earth Planet Sci Lett 201:35–44

Breitsprecher K, Thorkelson DJ (2008) Neogene kinematic history of Nazca-Antarctic-Phoenix slab windows beneath Patagonia and the Antarctic Peninsula. Tectonophysics 464:10–20

Calderón M, Hervé F, Fuentes F, Fosdick JC, Sepúlveda F, Galaz G (2016) Tectonic evolution of Paleozoic and Mesozoic andean metamorphic complexes and the Rocas Verdes ophiolites in southern Patagonia. In: Ghiglione MC (ed) Geodynamic Evolution of the Southernmost Andes. Springer Earth System Sciences, pp 7–36

Cande SC, Leslie RB (1986) Late Cenozoic tectonics of the southern Chile Trench. J Geophys Res-Sol EA 91(B1):471–496

Cuitiño JI, Scasso RA (2010) Sedimentología y paleoambientes del Patagoniano y su transición a la Formación Santa Cruz al sur del Lago Argentino Patagonia Austral. Rev Asoc Geol Argentina 66:406–417

Cuitiño JI, Pimentel MM, Ventura Santos R, Scasso RA (2012) High resolution isotopic ages for the early Miocene "Patagoniense" transgression in Southwest Patagonia: stratigraphic implications. J S Am Earth Sci 38:110–122

Cuitiño JI, Ventura Santos R, Scasso RA (2013) Insights into the distribution of shallow marine/estuarine early Miocene oysters from Southwestern Patagonia: sedimentologic and stable isotope constraints. Palaios 28:583–598

Cuitiño JI, Santos RV, Muruaga PJ, Scasso RA (2015) Estratigrafía de Sr y evolución sedimentaria de los depósitos marinos del Mioceno temprano en el antepaís del norte de la Cuenca Austral (o Magallanes), Argentina. Andean Geol 42(3):364–385

Cunningham WD (1995) Orogenesis at the southern tip of the Americas: the structural evolution of the Cordillera Darwin metamorphic complex, southern Chile. Tectonophysics 244(4):197–229

Dalziel IWD, De Wit MF, Palmer KF (1974) Fossil marginal basin in the southern Andes. Nature 250:291–294

del Rio CJ, Griffin M, McArthur JM, Martinez S, Thirlwall MF (2013) Evidence for early Pliocene and late Miocene transgressions in southern Patagonia (Argentina): Sr-87/Sr-86 ages of the pectinid "Chlamys" actinodes (Sowerby). J S Am Earth Sci 47:220–229

Diraison M, Cobbold PR, Gapais D, Rossello EA (1997) Magellan strait: part of a neogene rift system. Geology 25:703–706

Diraison M, Cobbold PR, Gapais D, Rossello EA, Le Corre C (2000) Cenozoic crustal thickening, wrenching and rifting in the foothills of the Southernmost Andes. Tectonophysics 316:91–119

Eagles G (2016) Tectonic reconstructions of the Southernmost Andes and the Scotia Sea during the opening of the Drake Passage. In: Ghiglione MC (ed) Geodynamic Evolution of the Southernmost Andes. Springer Earth System Sciences, pp 75–108

Eagles G, Livermore RA, Fairhead JD, Morris P (2005) Tectonic evolution of the west Scotia Sea. J Geophys Res-Sol EA 110:1–19

Feruglio E (1938) Relaciones estratigráficas entre el Patagoniano y el Santacruciano en la Patagonia Austral. Rev Museo de La Plata, Geología 4:129–159

Feruglio E (1949) Descripción Geológica de la Patagonia. Dirección General de Yacimientos Petrolíferos Fiscales, Buenos Aires, p 334

Fish P (2005) Frontier South, East Falkland basins reveal important exploration potential. Oil Gas J 103:34–40

Flynn JJ, Novacek MJ, Dodson HE, Frassinetti D, McKenna M, Norell M, Sears K, Swisher C, Wyss A (2002) A new fossil mammal assemblage from the southern Chilean Andes: implications for geology, geochronology, and tectonics. J S Am Earth Sci 15:285–302

Folguera A, Ramos VA (2002) Partición de la deformación durante el Neógeno en los Andes Patagonicos Septentrionales (37°–46°S). Rev Soc Geol España 15:81–93

Forsythe R, Nelson E (1985) Geological manifestations of ridge collision: evidence from the Golfo de Penas-Taitao basin, southern Chile. Tectonics 4:477–495

Fosdick JC, Grove M, Hourigan JK, Calderón M (2013) Retroarc deformation and exhumation near the end of the Andes, southern Patagonia. Earth Planet Sci Lett 361:504–517

Galeazzi JS (1998) Structural and stratigraphicevolution of the Western Malvinas Basin, Argentina. Am Assoc Petrol Geol Bull 82:596–636

Gallardo RE (2014) Seismic sequence stratigraphy of a foreland unit in the Magallanes-Austral Basin, Dorado Riquelme Block, Chile: implications for deep-marine reservoirs. Lat Am J Sed Basin Anal 21(1):49–64

Ghiglione MC (2002) Diques clásticos asociados a deformación transcurrente en depósitos sinorogénicos del Mioceno inferior de la Cuenca Austral. Rev Asoc Geol Argentina 57:103–118

Ghiglione MC (2003) Estructura y evolución tectónica del Cretácico-Terciario de la costa Atlántica de Tierra del Fuego. Unpublished PhD thesis, Facultad de Ciencias Exactas y Naturales, Universidad de Buenos Aires, p 150
Ghiglione MC (2016) Orogenic growth of the Fuegian Andes (52-56°) and its relation with tectonics of the Scotia Arc. In: Folguera A, Naipauer M, Sagripanty L, Ghiglione MC, Orts D, Giambiagi LB (eds) Growth of the Southern Andes. Springer Earth System Sciences, pp 241–267
Ghiglione MC, Cristallini E (2007) Have the Southernmost Andes been curved since Late Cretaceous time? An analog test for the Patagonian Orocline. Geology 35:13–16
Ghiglione MC, Ramos VA (2005) Chronology of deformation in the Southernmost Andes of Tierra del Fuego. Tectonophysics 405:25–46
Ghiglione MC, Ramos VA, Cristallini EO (2002) Estructura y estratos de creci-miento en la faja plegada y corrida de los Andes fueguinos. Rev Geol Chile 29:17–41
Ghiglione MC, Yagupsky D, Ghidella M, Ramos VA (2008) Continental stretching preceding the opening of the drake passage: evidence from Tierra del Fuego. Geology 36:643–646
Ghiglione MC, Suarez F, Ambrosio A, Da Poian G, Cristallini EO, Pizzio MF, Reinoso RM (2009) Structure and evolution of the Austral basin fold–thrust belt, Southern Patagonian Andes. Rev Asoc Geol Argentina 65:215–226
Ghiglione MC, Quinteros J, Yagupsky D, Bonillo-Martínez P, Hlebszevtich J, Ramos VA, Vergani G, Figueroa D, Quesada S, Zapata T (2010) Structure and tectonic history of the foreland basins of southernmost South America. J S Am Earth Sci 29:262–277
Ghiglione MC, Navarrete-Rodríguez AT, González Guillot M, Bujaleski G (2013) The opening of the Magellan Strait and its geodynamic implications. Terra Nova 25(1):13–20
Ghiglione MC, Likerman J, Barberón V, Giambiagi L, Aguirre-Urreta MB, Suarez F (2014) Geodynamic context for the deposition of coarse-grained deep-water axial channel systems in the Patagonian Andes. Basin Res 26(6):726–745
Ghiglione MC, Naipauer M, Sue C, Barberón V, Valencia V, Aguirre-Urreta MB, Ramos VA (2015) U-Pb zircon ages from the northern Austral basin and their correlation with the Early Cretaceous exhumation and volcanism of Patagonia. Cretac Res 55:116–128
Ghiglione MC, Ramos V, Cuitiño J, Barberón V (2016) Growth of the Southern Patagonian Andes (46–53°S) and its relation with subduction processes. In: Folguera A, Naipauer M, Sagripanty L, Ghiglione MC, Orts D, Giambiagi LB (eds) Growth of the Southern Andes. Springer Earth System Sciences, pp 201–240
Giacosa RE, Márquez MM (2002) El basamento paleozoico de la cordillera Patagónica. In Haller MJ (ed) Geología y Recursos Naturales de Santa Cruz, Relatorio XV Congreso Geológico Argentino, pp 45–55
Giacosa R, Fracchia D, Heredia N (2012) Structure of the Southern Patagonian Andes at 49°S. Geológica Acta 10:265–282
Gombosi DJ, Barbeau DL, Garver J (2009) New thermochronometric constraints on the rapid Paleogene exhumation of the Cordillera Darwin complex and related thrust sheets in the Fuegian Andes. Terra Nova 21:507–515
Guillot MG (2016) Magmatic evolution of the Southernmost Andes and its relation with subduction processes. In: Ghiglione MC (ed) Geodynamic Evolution of the Southernmost Andes. Springer Earth System Sciences, pp 37–74
González Guillot M, Escayola M, Acevedo R (2011) Calc-alkyrline rear-arc magmatism in the Fuegian Andes: implications for the mid-Cretaceous tectonomagmatic evolution of southernmost South America. J S Am Earth Sci 31:1–16
González Guillot M, Prezzi C, Acevedo RD, Escayola M (2012) A comparative study of two rear-arc plutons and implications for the Fuegian Andes tectonic evolution: mount Kranck Pluton and Jeu-Jepén Monzonite, Argentina. J S Am Earth Sci 38:71-88
Gorring ML, Kyry SM, Zeitler PK, Ramos VA, Rubiolo D, Fernandez MI, Panza JL (1997) Neogene Patagonian plateau lavas: continental magmas associated with ridge collision at the Chile Triple Junction. Tectonics 16:1–17

Gruetzner J, Uenzelmann-Neben G, Franke D (2012) Variations in sediment transport at the central Argentine continental margin during the Cenozoic. Geoch Geophys Geosyst 13(10): Q10003

Gruetzner J, Uenzelmann-Neben G, Franke D, Arndt JE (2014) Slowdown of Circumpolar Deepwater flow during the Late Neogene: evidence from a mudwave field at the Argentine continental slope. J Geophys Res Lett 41:2070–2076

Guillaume B, Martinod J, Husson L, Roddaz M, Riquelme R (2009) Neogene uplift of central eastern Patagonia: Dynamic response to active spreading ridge subduction? Tectonics 28(2)

Guillaume B, Gautheron C, Simon-Labric T, Martinod J, Roddaz M, Douville E (2013) Dynamic topography control on Patagonian relief evolution as inferred from low temperature thermochronology. Earth Planet Sci Lett 364:157–167

Haschke M, Sobel ER, Blisniuk P, Strecker MR, Warkus F (2006) Continental response to active ridge subduction. Geophys Res Lett 33:L15315

Hernández-Molina FJ, Paterlini M, Violante R, Marshall P, de Isasi M, Somoza L, Rebesco M (2009) Contourite depositional system on the Argentine Slope: an exceptional record of the influence of Antarctic water masses. Geology 37:507–510

Hernández-Molina FJ, Paterlini M, Somoza L, Violante R, Arecco MA, De Isasi M, Rebesco M, Uenzelmann-Nebenf G, Nebeng S, Marshall P (2010) Giant mounded drifts in the Argentine Continental Margin: origins, and global implications for the history of thermohaline circulation. Mar Petrol Geol 27:1508–1530

Hervé F, Calderón M, Faúndez V (2008) The metamorphic complexes of the Patagonian and Fuegian Andes. Geologica Acta 6:43–53

Hinz K, Neben S, Schreckenberger B, Roeser HA, Block M, de Souza KG, Meyer H (1999) The Argentine continental margin north of 48 S: sedimentary successions, volcanic activity during breakup. Mar Petrol Geol 16:1–25

Isla FI, Angulo RJ (2015) Tectonic Processes along the South America Coastline Derived from Quaternary Marine Terraces. J Coastal Res

Kohn MJ, Spear FS, Harrison TM, Dalziel IWD (1995) 40Ar/39Ar geochronology and P-T–t paths from the Cordillera Darwin metamorphic complex, Tierra del Fuego, Chile. J Met Geol 13:251–270

Kraemer PE (1998) Structure of the Patagonian Andes: regional balanced cross section at 50°S. Argentina Int Geol Rev 40:896–915

Kraemer PE (2003) Orogenic shortening and the origin of the Patagonian orocline (56 S Lat). J S Am Earth Sci 15:731–748

Kraemer PE, Riccardi AC (1997) Estratigrafía de la región comprendida entre los lagos Argentino y Viedma (49°40–50°10 lat. S), Provincia de Santa Cruz. Rev Asoc Geol Argentina 52:333–360

Lagabrielle Y, Suarez M, Rossello EA, Hérail G, Martinod J, Régnier M, De la Cruz R (2004) Neogene to Quaternary tectonic evolution of the Patagonian Andes at the latitude of the Chile Triple Junction. Tectonophysics 385:211–241

Lagabrielle Y, Suárez M, Malavieille J, Morata D, Espinoza F, Maury RC, Bellon H (2007) Pliocene extensional tectonics in the Eastern Central Patagonian Cordillera: geochronological constraints and new field evidence. Terra Nova 19:413–424

Lagabrielle Y, Goddéris Y, Donnadieu Y, Malavieille J, Suarez M (2009) The tectonic history of Drake Passage and its possible impacts on global climate. Earth Planet Sci Lett 279:197–211

Likerman L, Burlando JF, Cristallini EO, Ghiglione MC (2013) Along-strike structural variations in the Southern Patagonian Andes: insights from physical modeling. Tectonophysics 590:106–120

Livermore RA, Nankivell AP, Eagles G, Morris P (2005) Paleogene opening of Drake Passage. Earth Planet Sci Lett 236:459–470

Lodolo E, Menichetti M, Bartole R, Ben-Avraham Z, Tassone A, Lippai H (2003) Magallanes-Fagnano continental transform fault (Tierra del Fuego, southern-most South America). Tectonics 22:1076

Lodolo E, Donda F, Tassone A (2006) Western Scotia Sea margins: improved constraints on the opening of the Drake Passage. J Geophys Res-Sol EA 111(B6)

Maffione M (2016) Kinematic evolution of the Southern Andean orogenic arc. In: Ghiglione MC (ed) Geodynamic Evolution of the Southernmost Andes. Springer Earth System Sciences, pp 173–200

Maffione M, Speranza F, Faccenna C, Rossello E (2010) Paleomagnetic evidence for a pre-early Eocene ($\sim$50 Ma) bending of the Patagonian orocline (Tierra del Fuego, Argentina): paleogeographic and tectonic implications. Earth Planet Sci Lett 289:273–286

Maffione M, Fernandez-Moreno C, Ghiglione M, Speranza F, van Hinsbergen DJJ, Lodolo E (2015) Constraints on deformation of the Southern Andes since the Cretaceous from anisotropy of magnetic susceptibility. Tectonophysics 665:236–250

Malumián N (2002) El Terciario marino. Sus relaciones con el eustatismo. In: Haller MJ (ed) Geología y Recursos naturales de Santa Cruz. Asociación Geológica Argentina, Relatorio XV Congreso Geológico Argentino, pp 237–244

Malumián N, Jannou G (2010) Los Andes Fueguinos: el registro micropaleontológico de los mayores acontecimientos paleooceanográficos australes del Campaniano al Mioceno. Andean Geol 37:345–374

Malumián N, Nañez C (2011) The Late Cretaceous-Cenozoic transgressions in Patagonia and the Fuegian Andes: foraminifera, palaeoecology, and palaeogeography. Biol J Linnean Soc 103:269–288

Malumián N, Olivero EB (2006) The Cabo Domingo Group, Tierra del Fuego: Biostratigraphy, paleoenvironments, and events of the marine Eocene-Miocene. Rev Asoc Geol Argentina 61:139–160

Malumián N, Scarpa R (2005) Foraminíferos de la Formación Irigoyen, Neogeno, Tierra del Fuego, Argentina: su significado paleobiogeográfico. Ameghiniana 42:363–376

Malumián N, Panza JL, Parisi C, Nañez C, Caramés A, Torre E (2000) Hoja Geológica 5172-III-Yacimiento Río Turbio, provincia Santa Cruz, 1:250.000. Servicio Geológico Minero Argentino, Boletín 247, p 108

Marenssi SA, Casadío S, Santillana S (2002) La Formación Man Aike al sur de El Calafate (Provincia de Santa Cruz) y su relación con la discordancia del Eoceno medio en la cuenca Austral. Rev Asoc Geol Argentina 57:341–344

Marenssi SA, Limarino CA, Tripaldi A, Net L (2005) Fluvial systems variations in the Rio Leona Formation: tectonic and eustatic controls on the Oligocene evolution of the Austral (Magallanes) Basin, southernmost Argentina. J S Am Earth Sci 19:359–372

Marshall LG, Sempere T, Butler RF (1997) Chronostratigraphy of the mammal-bearing Paleocene of South America. J S Am Earth Sci 10:49–70

Martínez OA, Kutschker A (2011) The 'Rodados Patagónicos' (Patagonian shingle formation) of Eastern Patagonia: environmental conditions of gravel sedimentation. Biol J Linnean Soc 103:336–345

Martínez OA, Rabassa J, Coronato A (2009) Charles Darwin and the first scientific observations on the Patagonian shingle formation (Rodados Patagónicos). Rev Asoc Geol Argentina 64:90–100

Mella PE (2001) Control Tectónico en la Evolución de la Cuenca de Antepaís de Magallanes, XII Región, Chile. Unpubl. Memoria para optar al título de Geólogo, Facultad de Ciencias Químicas, Departamento Ciencias de la Tierra, Universidad de Concepción, Concepción, p 149

Melnick D, Echtler HP (2006) Inversion of forearc basins in south-central Chile caused by rapid glacial age trench fill. Geology 34:709–712

Mercer JH, Sutter JF (1982) Late Miocene—earliest Pliocene glaciation in southern Argentina: implications for global ice-sheet history. Palaeogeog Palaeoclim Palaeoeco 38:185–206

Nelson EP (1982) Postectonic uplift of the Cordillera Darwin orogenic core complex: evidence for fission track geochronology and closing temperature time relationships. J Geol Soc London 139:755–761

Pankhurst RJ, Riley TR, Fanning CM, Kelley SP (2000) Episodic silicic volcanism in Patagonia and the Antarctic Peninsula: chronology of magmatism associated with the break-up of Gondwana. J Petrology 41:605–625

Panza JL (2002) La cubierta detrítica del Cenozoico superior. In Haller MJ (ed) Geología y Recursos Naturales de Santa Cruz, Relatorio XV Congreso Geológico Argentino, pp 259–284

Parras A, Griffin M, Feldmann R, Casadío S, Schweitzer C, Marenssi S (2008) Correlation of marine beds based on Sr-and Ar-date determinations and faunal affinities across the Paleogene/Neogene boundary in southern Patagonia, Argentina. J South Am Earth Sc 26:204–216

Parras A, Dix GR, Griffin M (2012) Sr-isotope chronostratigraphy of Paleogene-Neogene marine deposits: Austral Basin, southern Patagonia (Argentina). J South Am Earth Sc 37:122–135

Pedoja K, Regard V, Husson L, Martinod J, Guillaume B, Fucks E, Iglesias M, Weill P (2011) Uplift of Quaternary shorelines in Eastern Patagonia: Darwin revisited. Geomorphology 127:121–142

Perkins ME, Fleagle JG, Heizler MT, Nash B, Bown TM, Tauber A, Dozo MT (2012) Tephrochronology of the Miocene Santa Cruz and Pinturas Formations, Argentina. Early Miocene Paleobiology in Patagonia: High-Latitude Paleocommunities of the Santa Cruz Formation, Vizcaíno, SF, pp 23-40

Piola AR, Matano RP (2001) Brazil and Falklands (Malvinas) currents. In: Steele JH, Thorpe SA, Turekian KK (eds) Ocean currents: a derivative of the encyclopedia of ocean sciences. Academic Press, New York, pp 35–43

Ramos VA (1989) Andean foothills structures in Northern Magallanes Basin, Ar-gentina. Am Assoc Pet Geol Bull 73:887–903

Riccardi AC, Rolleri EO (1980) Cordillera Patagónica Austral. In Turner JCM (ed) Simposio de Geología Regional Argentina, vol 2, pp 1173–1306

Sachse VF, Strozyk F, Ankyr Z, Rodriguez JF, Primio R (2015) The tectono-stratigraphic evolution of the Austral Basin and adjacent areas against the background of Andean tectonics, southern Argentina, South America. Basin Res. 10.1111/bre.12118

Scalabrino B, Lagabrielle Y, de la Rupelle A, Malavieille J, Polvé M, Espinoza F, Morata D, Suarez M (2009) Subduction of an active spreading ridge beneath southern South America: a review of the Cenozoic geological records from the Andean foreland, central Patagonia (46–47 S). Subduction Zone Geodynamics. Springer, Berlin, pp 227–246

Scasso RA, McArthur JM, Del Río CJ, Martínez S, Thirlwall MF (2001) 87 Sr/86 Sr Late Miocene age of fossil molluscs in the 'Entrerriense'of the Valdés Peninsula (Chubut, Argentina). J South Am Earth Sci 14:319–329

Sue C, Ghiglione MC (2016) Wrenching tectonism in the Southernmost Andes and the Scotia Sea Constrained from Fault Kinematic and Seismotectonic Overviews. In: Ghiglione MC (ed) Geodynamic Evolution of the Southernmost Andes. Springer Earth System Sciences, pp 137–171

Thomas T, Livermore RA, Pollitz F (2003) Motion of the Scotia Sea plates. Geophys J Int 155:789–804

Thomson SN (2002) Late Cenozoic geomorphic and tectonic evolution of the Patagonian Andes between latitudes 42° and 46°S: an appraisal based on fission-track results from the transpressional intra-arc Liquine-Ofqui fault zone. Geol Soc Am Bull 114:1159–1173

Thomson SN, Herve F, Stockhert B (2001) The Mesozoic-Cenozoic denudation history of the Patagonian Andes (southern Chile) and its correlation to different subduction processes. Tectonics 20:693–711

Thomson SN, Brandon MT, Reiners PW, Tomkin JH, Vásquez C, Wilson NJ (2010) Glaciation as a destructive and constructive control on mountain building. Nature 467:313–317

Ton-That T, Singer B, Mörner NA, Rabassa J (1999) Datación de lavas basálticas por 40 Ar/39 Ar y geología glacial de la región del lago Buenos Aires, Provincia de Santa Cruz, Argentina. Rev Asoc Geol Argentina 54:333–352

Uliana VMA, Biddle KT, Cerdan J (1989) Mesozoic extension and the formation of Argentine sedimentary basins. In: Tankyrrd AJ, Balkwill HR (eds) Extensional tectonics and stratigraphy of the North Atlantic Margins. Am Assoc Pet Geol Bull, Memoirs 46:599–613

Wilson TJ (1991) Transition from back-arc to foreland basin development in the Southernmost Andes: stratigraphic record from the Ultima Esperanza District, Chile. Geol Soc Am Bull 103:98–111

Wrenching Tectonism in the Southernmost Andes and the Scotia Sea Constrained from Fault Kinematic and Seismotectonic Overviews

Christian Sue and Matías C. Ghiglione

Abstract The geodynamics of the study area includes subduction, orogen dynamics, and major transcurrent tectonics at a complex transform margin in the southern ocean. We present a synthesis of the fault system associated to major fault zones in the Fuegian Andes and the Scotia Sea and review the available kinematic databases from microtectonic measurements. Our synthesis of six independent fault kinematic studies is coherent with a very consistent shortening direction-oriented NE–SW in the Southernmost Andes. The stability of the stress pattern and orientation of the shortening axes on a more regional scale involving the Scotia Sea indicate a steady E–W to NNE–SSW $\sigma 1$/shortening direction since middle Eocene times. This observations reflect that the global left-lateral motion between Antarctica–Scotia–South America plates circuit is the main driving force for the entire area and in particular for the Southernmost Andes, over the last $\sim$45 Ma. Both the observed short-term geodetic and the long-term geological slip rates of the Magallanes–Fagnano fault system from Tierra del Fuego are moderate ($\sim$5 mm/year), and the expected time span between major M8 earthquakes would be around 10 kyr. Yet the time between the two most recent large earthquakes was about 70 years. Such a great discrepancy suggests a complex mechanics on the Magallanes–Fagnano fault system, leading to complex recurrence time history for the characteristic earthquake ($M \geq 7$).

Keywords Fuegian andes · Scotia plate · Fault kinematic · Seismotectonics · Wrenching · Magallanes–Fagnano fault system · North scotia ridge

C. Sue (✉)
Bourgogne Franche-Comté University, Laboratoire Chrono-environnement, Besançon, France
e-mail: christian.sue@univ-fcomte.fr

M.C. Ghiglione
Departamento de Ciencias Geológicas, Instituto de Estudios Andinos "Don Pablo Groeber" CONICET, FCEN—Universidad de Buenos Aires, Buenos Aires, Argentina

© Springer International Publishing Switzerland 2016

M.C. Ghiglione (ed.), *Geodynamic Evolution of the Southernmost Andes*, Springer Earth System Sciences, DOI 10.1007/978-3-319-39727-6_6

1 Introduction

Since the very first expedition in Tierra del Fuego and southern Patagonia, a mythic end-of-the-world in our collective unconscious (Fig. 1a) (Darwin 1845, 1846), a growing number of scientific and more specifically geoscientific researches have been led in this fascinating region from the twentieth century onward. Given the local geography, a lot of these studies approached their topic by sea, despite the rough weather conditions in this part of the world leading to navigation comparable to the drunken boat of Rimbaud (1871). Beyond its formal beauty (Fig. 2), the geodynamic complexity of the connection between the Southernmost Andes and the Scotia plate, through the Magallanes–Fagnano fault system (MFF) and the North Scotia Ridge (NSR), demands for proper geodynamic models. They concern the evolution of the overall system from 200 Ma onward (e.g., Vérard et al. 2012; Eagles et al. 2005), the development of the Southernmost Andes' orogenic arc (e.g., Ghiglione and Cristallini 2007; Maffione et al. 2010, 2015; see Maffione 2016 for a review) and related crustal rotations, the matters of triple junction between the major plates and plate boundaries, the dislocation in continental blocks of the southern tip of the South American plate along the NSR (Eagles 2016), and the concomitant opening of the Drake passage and Magallanes Strait (e.g., Ghiglione et al. 2013), which had serious implications in terms of global climatology (e.g., Livermore et al. 2004; Lagabrielle et al. 2009). Indeed, these questions are addressed as a whole in the present book. In this chapter, we focus on the most recent and current deformation of the region, including a synthesis on the fault

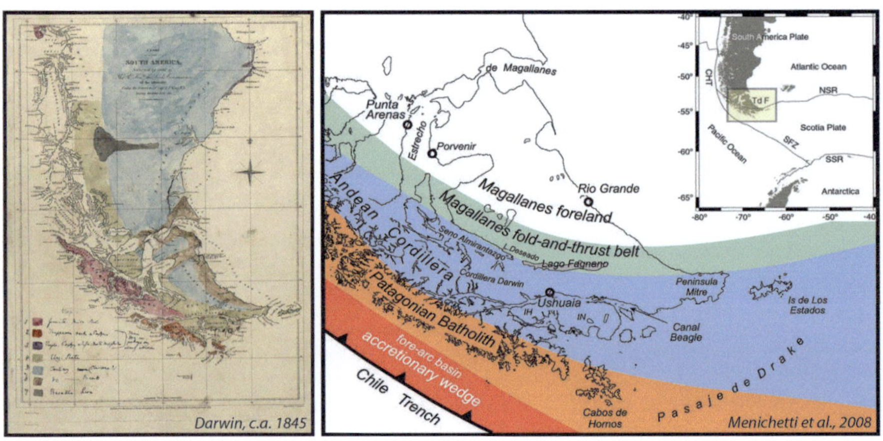

Fig. 1 *Left* Very first geological map of Patagonia and Tierra del Fuego drafted by Darwin (1845, 1846). *Right* Tectonostratigraphic provinces in the Southernmost Andes (modified from Menichetti et al. 2008)

system carving the area, together with a critical review of the available kinematic and microtectonic data, an overview of the seismicity and paleoseismicity leading to current fault mechanism and actual seismic deformation, which then could be compared to geodetically related deformation observed by GPS.

The geodynamics of the study area between 50°S–57°S and 55°W–75°W (Fig. 2) includes subduction, orogen and orogenic arc dynamics, and major transcurrent tectonism at a complex transform margin. The complexity of this system rises up from the variety of geodynamic processes still ongoing. Indeed, the main one is related to the left-lateral movement between the South America and Scotia plates (e.g., Pelayo and Wiens 1989).

The Andes is one of the longest orogenic belts on earth. They run N–S linearly from 18°S to 50°S over almost 4000 km and abruptly turn to E–W direction in the so-called Patagonian orocline between 50°S and 56°S. The transform plate limit south of the system acted and still acts as first-order tectonic feature in the area and is probably at least partly linked to the bending of the Southernmost Andes (e.g., Cunningham 1993; Ghiglione et al. 2010; Maffione 2016).

In this chapter, we will firstly present a synthesis of the fault system associated with the MFF, both from geometric and kinematic viewpoints. Then, we will review the available kinematic databases from microtectonic measurements and present a comprehensive model. Neotectonic observations on this fault system will then allow opening on the current tectonics.

Indeed, along the MFF and satellite faults, evidence of active left-lateral movement is described in response to the ongoing westward motion of Scotia plate relative to South America (Perucca et al. 2015). The overall area undergoes a diffuse and moderate seismicity (Febrer et al. 2000). Nevertheless, paleoseismological observation established the occurrence of large earthquakes along this fault system (e.g., Pedrera et al. 2014; Perucca et al. 2015), and major earthquakes stroke the area during the last century. The strongest sequence happened in the seventeenth of December 1949 (Forsyth 1975; Jaschek et al. 1982; Pelayo and Wiens 1989).

The overall MFF–NSR left-lateral strike-slip fault is actually quite active and also produced the recent M7.0 strike-slip earthquake, southeast of the Malvinas Island on November 25, 2013 (Fig. 3). It is clear that this fault system remains a source of actual seismic risk, as it may produce major events close to quite large cities such as Ushuaia, Rio Grande, or Punta Arenas. In terms of a geodetic approach, some recent works quantificate the current deformation (Smalley et al. 2003, 2007; Mendoza et al. 2011). They led to an estimation of the relative left-lateral velocity of about 4–6 mm/year between Scotia and South America plates, coherently with the focal mechanisms of the earthquakes in the region, and with structural studies. In this framework, we present hereafter a new synthesis of brittle deformation, fault kinematics and related seismicity, and deformation pattern in the Southernmost Andes and surrounding southern oceans.

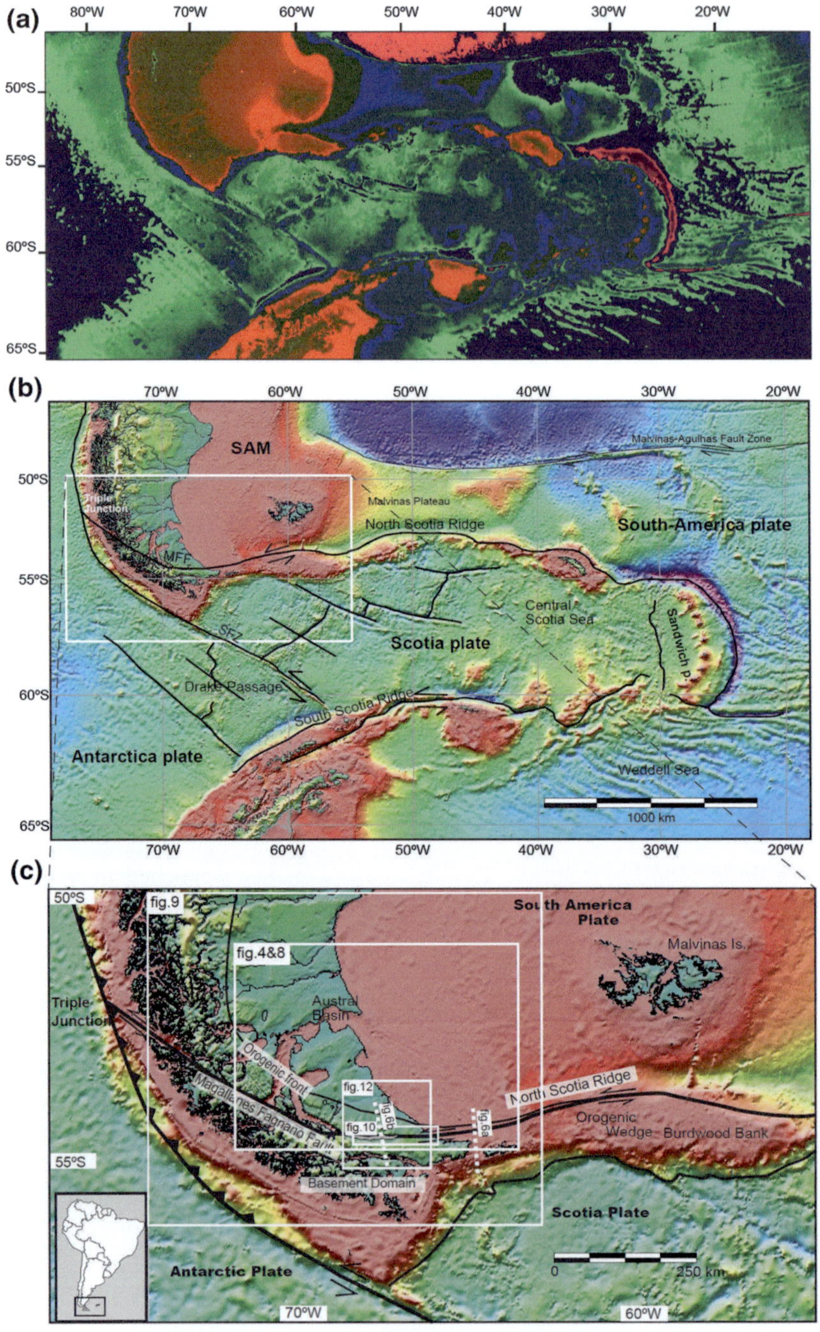

Fig. 2 **a** Morphological settings of the overall Scotia area (data from ETOPO1). **b** Major tectonic features enclosing the Scotia Sea region, **c** Location of main morphostructural units and location of figures in the following of this chapter. *MFF* Magallanes–Fagnano fault system; *SFZ* Shakleton fault zone

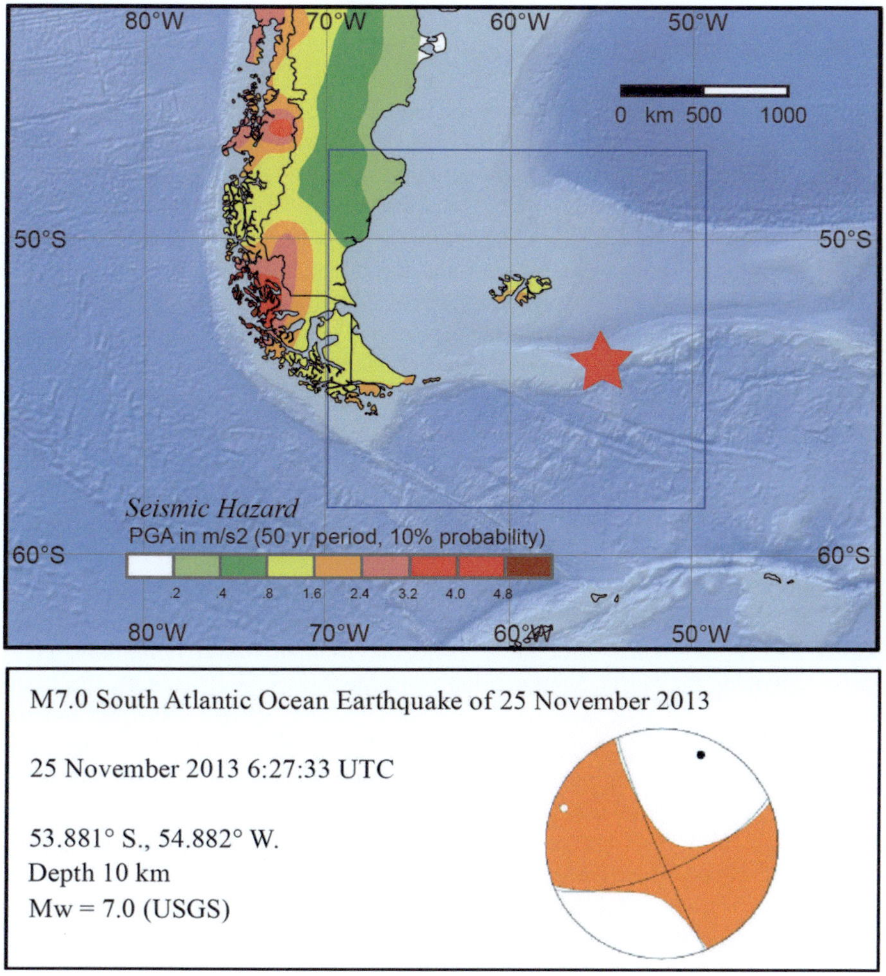

Fig. 3 Location of the November 25, 2013, M7.0 South Atlantic Earthquake (*red star*) together with the focal solution and the onshore seismic hazard map (from USGS Web site http://earthquake.usgs.gov/)

2 Brittle Deformation in the Southernmost Andes

2.1 Geodynamic Setting and Fault Systems

At continental scale, three main fault systems shape the study area (Fig. 2a, b) (see Eagles 2016 for a review on the geometry and evolution of this plate tectonic configuration): (i) the Chile trench to the west, which evolves southeastward into the Shackleton fracture zone, and the boundary between the South America and Scotia plates composed of (ii) the North Scotia Ridge (NSR) offshore to the east, which evolves into (iii) the continental Magallanes–Fagnano fault system (MFF), cutting through the Tierra del Fuego Island and its surroundings (Fig. 4).

In the studied region South of 50°S, the Pacific trench corresponds to the subduction of the Antarctica plate underneath South America (Fig. 2c), with a relative velocity of about 20 mm/year (Barker 2001). Therefore, intersection of the MFF with the Chile trench conforms a triple junction (Fig. 2c), defining one of the major boundaries of the overall study area. Southeastward of the active subduction, the geometry and kinematics of the plate boundary lead to a transcurrent motion: the Shackleton fracture zone, which forms the boundary between the Antarctica and Scotia plates in the Drake Passage (Fig. 2b) (Livermore et al. 2004). The SFZ is a left-lateral transform boundary with relative velocities of about 8 mm/year (Thomas et al. 2003), active since the formation of the Scotia plate at approximately 50 Ma (Eagles 2016).

The NSR bounding the Scotia and the South American plates is moving at a rate of about 7 mm/year (Thomas et al. 2003). It is a major left-lateral fault, which produced the recent Mw 7.0 Malvinas Island earthquake (November 25, 2013, Fig. 3). It is marked by an offshore system of shallow banks and ridges, which corresponds to continental dislocated blocks. It extends about 2000 km from the Tierra del Fuego's tip (Isla de los Estados) to South Georgia Island, in a slightly arcuated overall W–E linear shape (Fig. 2b) (Ludwig and Rabinowitz 1982). Northward, the Malvinas Trough, a Cenozoic foreland basin (Ghiglione et al. 2010), separates the North Scotia Ridge from the Malvinas Plateau on the South America plate. The NSR worked as a transpressional system with accretionary complex (Ludwig and Rabinowitz 1982) and a strong left-lateral component; although in its western segment toward the onshore, it has an overall transtensional component (Lodolo et al. 2003). The general evolution of the NSR is coeval with the development of the west Scotia Sea domain (Barker 2001; Thomas et al. 2003; Eagles et al. 2005). The lithologies observed along the NSR are those of Mesozoic South America rocks (Macfadyen 1933; Tyrrell 1945; Dalziel et al. 1975), which allow connecting these blocks with the Fuegian Andes. Actually, the transpressional deformation is well described up to the South Georgia Island, both in ductile and in brittle domains, and related to the mid-Cretaceous Andean tectonic phase (Curtis et al. 2010). Thus, the left-lateral component of the deformation appears as a very steady tectonic feature all along the development of the NSR domain in space and time. The cumulative deformation may reach 1500 km of transcurrent movement along the NSR (Cunningham 1993).

The western tip of the NSR connects with the MFF (e.g., Kranck 1932; Winslow 1982; Dalziel and Brown 1989), a unique onshore transform plate boundary in the world (Fuenzalida 1972, 1976; Lodolo et al. 2003). This left-lateral fault crosscuts the Tierra del Fuego Island, from the eastern tip of the continent to the Chile subduction trench, along a ~600-km-long lineament well visible on every remote sensing images (Fig. 4). Nevertheless, the exact path followed by the MFF from the Chile trench to the west toward the NSR to the east remains unclear (Lodolo et al. 2003). It follows part of the western Magallanes Strait and runs through the south of the Dawson Island, and through the Lago Fagnano to the Irigoyen river valley

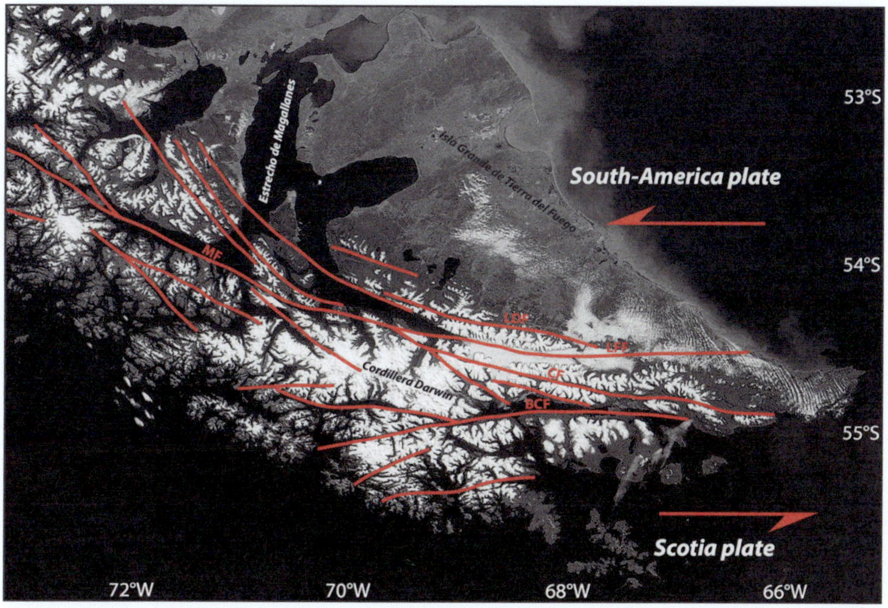

Fig. 4 Overall fault system corresponding to the Magallanes–Fagnano continental strike-slip plate boundary and related individual strike-slip faults. *BCF* Canal Beagle fault; *CF* Carbajal valley fault zone; *LFF* Fagnano fault (Eastern part of the MFF system); *LDF* Lago Deseado fault; *MF* Magallanes fault (Western part of the MFF system)

toward the Atlantic coast (Ghiglione 2002). Actually, most authors agree that left-lateral strike-slip deformations along the MFF system are active since the Oligocene (e.g., Ghiglione 2002; Lodolo et al. 2003; Menichetti et al. 2008).

The strong morphological expression, the size of the linear structure, and the connection with the NSR, together with signs of recent activity (Fuenzalida 1972; Winslow 1982; Lodolo et al. 2002, 2003; Ghiglione et al. 2013), support the interpretation of the MFF as the actual plate boundary between the continental crust of the Scotia and South America plates (Fuenzalida 1972; Cunningham et al. 1995; Eagles et al. 2005; Galindo-Zaldívar et al. 2006). Indeed, it accommodates the current left-lateral relative movement between the two plates (Pelayo and Wiens 1989), with some transtensive component affecting the Fuegian Andes (Ghiglione 2002; Lodolo et al. 2003). The left-lateral movement on the MFF is ongoing for at least 30 Myr (Cunningham et al. 1995; Ghiglione 2002) and developed in close relation to the Scotia plate oceanic floor opening (Barker and Burrell 1977; Eagles et al. 2005) and to the Shackleton fracture zone (Fig. 2b). Sinistral deformation along the MFF could have increase after the end of western Scotia's seafloor opening ($\sim$9 Myr, Eagles et al. 2005). Prior to the Scotia's history, strike-slip deformation has been suggested along the Canal Beagle fault zone since Late Cretaceous time, implying reactivation of inherited basement structures (Klepeis and Austin 1997; Grunow et al. 1992; Cunningham et al. 1995).

The quantification of the transcurrent deformation along the whole MFF system still remains a matter of debate. The displacement along the main MFF fault could reach up to a 55 km (Rossello 2005) or 80 km offset (Cunningham 1993; Lodolo et al. 2003), and satellite faults connected on the same system also exhibit large left-lateral displacements (e.g., 25 km for the north segment in the Deseado fault zone, Klepeis 1994). The overall displacement along the MMF system may reach ~100 km, but still remains much lower than the 1500 km proposed for the NSR (Cunningham 1993 and references herein). For a fault system active since ~30 Ma (Ghiglione 2002; Eagles 2016), that means low slip rates of around ~2.5 mm/year.

The present-day GPS-measured deformation also shows a slow slip rate of ~5 mm/year (Del Cogliano et al. 2000; Mendoza et al. 2011), while the movement across the MFF system as a whole reaches rates as high as 6.6–9.6 mm/year (Smalley et al. 2003; DeMets et al. 2010).

Beyond the major morphological expression of the 600-km-long lineament of its main branch, the MFF is made of several tectonic structures segmented in a wide system of subvertical faults. Kinematic indicators show left-lateral movement along major faults running from the front of the Magallanes fold and thrust belt to the north, to the Canal Beagle and the Navarino Island to the south (Fig. 4). In the Isla Grande de Tierra del Fuego, it comprises in particular the MFF itself running along the Magallanes Strait–Seno Almirantazgo–Lago Fagnano–Irigoyen river (Ghiglione 2002, 2003; Lodolo et al. 2003), the Deseado fault to the north (Winslow 1982), the Canal Beagle fault to the south (Cunningham 1993), and the Carbajal fault in between (Figs. 4 and 5) (Caminos et al. 1981).

The Lago Deseado fault is a linear structure about 100 km long, marked by a wide zone of fault breccias, which follows a steep valley across the external Magallanes fold and thrust belt (Klepeis 1994). Kinematic analysis of mesoscale faults measured in this area indicates a major left-lateral movement, with some extensional component, within the same overall tectonic scheme as the MFF. Sag ponds along the fault with clear neotectonic lineaments suggest that it could have been active during Quaternary times (Winslow 1982).

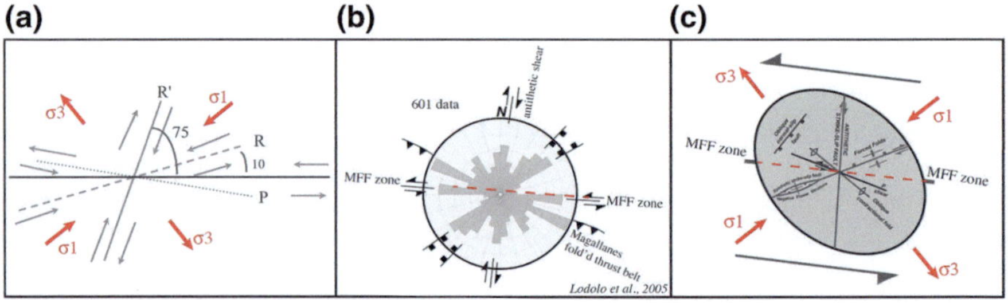

Fig. 5 **a** Theoretical Riedel fault system and its corresponding principal stress direction orientation, **b** Rose diagram summarizing the azimuth and length distribution of 601 lineaments identified in satellite images from Tierra del Fuego, and their tectonic interpretation. **c** Geometric relation of faults and folds to left-lateral wrench fault combined schematically with a strain ellipse and their equivalent principal stress directions. Modified from Lodolo et al. (2003)

The Carbajal fault is a ~100-km-long linear structure, which seems to connect the MFF with the Canal Beagle fault (Fig. 4). It strikes very slightly oblique on the MFF direction and is marked by a wide valley carved in the Central Cordillera. Left-lateral movement is also documented along strike-slip faults that cut through the fold and thrust belt (Caminos et al. 1981).

The Canal Beagle fault is a former extensional detachment, which constitutes an important tectonic and metamorphic discontinuity in the Fuegian Andes linked to the Andean orogenesis (e.g., Darwin 1846; Dalziel and Brown 1989). Strike-slip faulting plays an important role in the tectonic history of this fault, both in ductile and brittle conditions, probably from the Cretaceous onward (Cunningham 1993). It has evolved toward a left-lateral brittle strike-slip branched into MFF system as a whole. The most recent movements of this fault linked to the development of the MFF probably occurred in brittle conditions and were favored by inherited structures. Kinematic indicators in this area are numerous mesoscale left-lateral faults and en-échelon Riedel structures at mesoscale and regional scale (Cunningham 1993).

These main faults in Tierra del Fuego (MFF, Canal Beagle, Carbajal, and Deseado) have been carefully analyzed from a structural viewpoint, and fault kinematic data are available in their vicinities (Cunningham 1993; Klepeis 1994; Diraison et al. 2000; Lodolo et al. 2003; Menichetti et al. 2008; see below for an overview of microtectonic data). Associated to these main faults, the whole Tierra del Fuego presents a set of lineaments that could be related to the overall faulting. Indeed, Lodolo et al. (2002) proposed a map of 601 lineaments analyzed using remote sensing images (SPOT and SAR satelittes). The 2-D statistics of these lineaments (Fig. 5) (Lodolo et al. 2002, 2003) provides a straight interpretation of the whole system as a large-scale Riedel system perfectly coherent with the main transcurrent kinematics rising up from direct fault analysis, i.e., with NE–SW-oriented shortening axis, and NW–SE-oriented extensional axis. The left-lateral crustal wrenching is thus responsible for both the main morphotectonic features, directly related to major faults, and for most of the incipient morphologic markers of the area, which suggest its very recent activity.

Along the main branch of the overall fault system, and specially in the Lago Fagnano's vicinity, several geological and geophysical campaigns (bathymetry, seismic profiles, gravimetry) have been achieved to constrain the morphology, geometry, kinematics, and timing of the brittle deformation by Lodolo and collaborators (e.g., Lodolo et al. 2002, 2003; Tassone et al. 2008; Menichetti et al. 2008 and references therein). They pointed out in particular the asymmetric pull-apart geometry of the Lago Fagnano, which is also observed along most of the MFF system in seismic profiles (Fig. 6). The lake itself is interpreted as a large pull-apart basin (Lodolo et al. 2003). The Fig. 6 shows an overview of crustal cross sections across the MFF system (from Tassone et al. 2008; Menichetti et al. 2008) and underlines the transtensional behavior of the whole system, with several negative flower-like structures, which suggest coeval strike-slip and extensional components of the deformation. Actually, the deformation history of the area is quite complex, with first shortening phases (Late Cretaceous), predating the

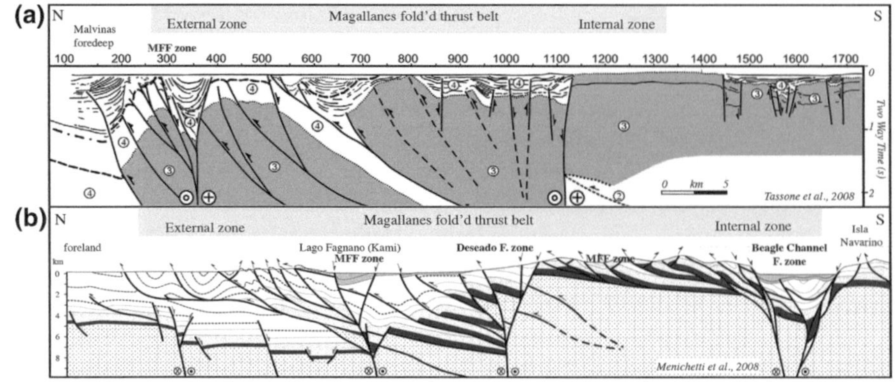

Fig. 6 Overview of crustal cross sections across the MFF system from **a** Tassone et al. (2008) and **b** Menichetti et al. (2008). See section's traces on Fig. 2

evolution of the MFF system toward a crustal wrenching, with an extensional component, potentially during Miocene times (Ghiglione 2002).

This scenario would imply potential reactivation of compressional structures in transtensive ones. Note that the extensional component is also locally observed on the field by kinematic indicators (see below, Sect. 2.2). In terms of geodynamic interpretation, the extensional part of the overall transcurrent deformation could be associated with rotational movement of Scotia plate with respect to South America.

2.2 *Kinematic Indicators and Paleostrain/Stress Determination*

Complementary to morphotectonic analyses and field mapping of fault systems, the systematic and statistical analysis of brittle deformation measured on the field at meso-/microscale allows quantifying the corresponding kinematic axes, namely either strain or stress axes, depending on the methodological and theoretical backgrounds. In terms of methodology, fault slickenside analysis is based on the Wallace and Bott principle (Wallace 1951; Bott 1959), which has long been discussed and still remains a matter of debate concerning stress/strain relations (e.g., Angelier and Mechler 1977; Angelier and Goguel 1979; Angelier 1990; Gephart 1990; Twiss and Unruh 1998; Yamaji 2000, 2003, 2006; Lacombe 2012 and references herein). Regardless of whether the stress or kinematic hypothesis is assumed, this principle states that fault slip parallels the direction of maximum resolved shear stress (on the considered plane) of the local spatially homogeneous stress tensor. Basic assumptions for its application are the following: (i) The rock volume should be isotropic and homogeneous; (ii) displacements on faults should be incremental and faults should slip independently; (iii) the stress/strain field should be uniform and steady during a given tectonic phase; (iv) the deformation is non-rotational; and (v) no post-faulting reorientation of the fault-slip data has occurred. In practice, the direction of movement is given by slickensides, striae, and grooves on the fault plane. The sense of movement is generally given by calcareous fibers, drag folds, steps impressed in the plane by striator objects, and Riedel criteria (Riedel 1929), among other less common indicators. Although the related methods have proved their robustness in many areas and various tectonic contexts (e.g., Sue and Tricart 2002, 2003; Champagnac et al. 2003; Sue et al. 2014), fault-slip data may depart from these basic assumptions, mainly because of strong local deformation and stress variations due to preexisting anisotropies and inherent rock heterogeneities (see for instance Homberg 2000).

The long-lasted discussions brought by these methods concerning strain versus stress are beyond the scope of this paper. A critical discussion on this point has been recently published by Lacombe (2012). We refer to this paper and references herein for further precisions. Several methods are available for fault kinematic quantification. We can quote for instance a non-exhaustive list: the geometrical right dihedra method (RD, Angelier and Mechler 1977; Pfiffner and Burkhard 1987), the pressure–tension method (PBT), the numerical dynamic analysis (NDA), the direct inversion Angelier's method (INVD, Angelier 1990) (all implemented in the TectonicsFP software from Sperner et al. 1993), the XYZ method (Marrett and Allmendinger 1990), the Etchecopar's method (Etchecopar et al. 1981), and the Yamaji's method (Yamaji et al. 2006), among others.

In the study area, several fault kinematic analyses have been achieved with various methods during the last 20 years. We synthesize hereafter the data provided by Cunningham (1993) in the Canal Beagle area, Klepeis (1994) along the Deseado fault, Klepeis and Austin (1997) in the Seno Almirantazgo fiord, Diraison et al. (2000) in the Magallanes fold and thrust belt, and Menichetti et al. (2008) and Branellec (2010) in the Lago Fagnano–Carbajal valley–Beagle channel area. Although these papers do not use all the same theoretical background, all of them present mesoscale to microscale fault-striae measurements.

We made a global synthesis of these data, in terms of kinematic axes, that is to say main strain/stress axes. Figure 7 shows an overview of the data published by these authors. We refer to each specific paper for a more precise discussion on fault measurement.

Cunningham (1993) studied the ductile and brittle kinematic of the Canal Beagle faults (corresponding to the SW and NW arms of the channel) (Fig. 4). This author measured about 400 single faults and reported them following their kinematics and orientation (Fig. 7a). The offsets range from millimeters to kilometers. The measured faults can be divided into two families: left-lateral faults oriented N45E to N135E and right-lateral faults oriented N30W to N35E. These two families correspond to conjugated fault system kinematically compatible. Although a proper inversion or geometrical reconstruction of the kinematic axes was not performed, an overall shortening axis-oriented N40E (red arrows in Fig. 7a), and a corresponding N130E-oriented extensional axis, is coherent with Cunningham (1993)'s data.

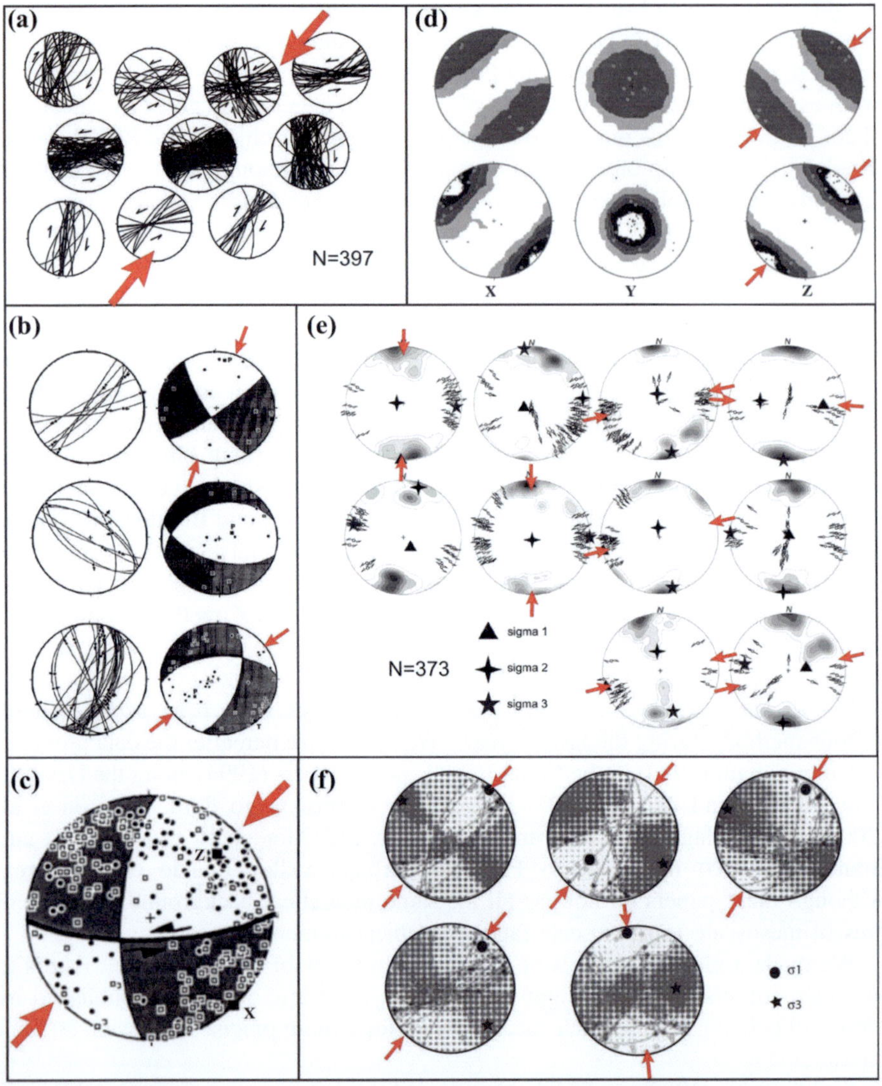

Fig. 7 Fault kinematic analyses achieved with various methods in the study area. *Red arrows* show the orientations of compressional stress axis σ1. Data are from: **a** Cunningham (1993) in the Canal Beagle area. **b** Klepeis (1994) along the Deseado fault. **c** Klepeis and Austin (1997) in the Seno Almirantazgo fiord. **d** Diraison et al. (2000) in the Magallanes fold and thrust belt and foreland and in the Lago Fagnano's vicinity. **e** Menichetti et al. (2008) in the Fagnano–Carbajal–Beagle area. **f** Branellec (2010) along the Lago Fagnano and Yehuin lakes shores

Klepeis (1994) investigated the Deseado fault zone, describing a major left-lateral system, with typical Riedel shear geometry, made of a set of late faults, cutting compressional structures. The fault measurements were subdivided into three families: (i) left-lateral faults oriented ~N75E and interpreted as R shear planes, (ii) a left-lateral fault family oriented N125E, corresponding to P shear planes, and (iii) a right-lateral fault family oriented N20E and interpreted as R' shear planes. The respective kinematics of these three families is represented in Fig. 7b, using the Marrett and Allmendinger's method (1990) method. As in Cunningham (1993), the fault was divided by sense of movement and orientation. The R family exhibits almost pure strike-slip movement, whereas the R' and P families show an oblique normal transcurrent component (Fig. 7b). In consequence, the Deseado fault zone seems to accommodate a transtensional overall deformation in brittle conditions. Indeed, the three kinematics (fault plane solutions) are compatible with a main shortening axis oriented in the ~N40°E direction, and an extensional axis in the ~N130 direction, and are largely comparable in direction with the results obtained by Cunningham (1993) to the south.

Klepeis and Austin (1997) applied the same approach and the method of Marrett and Allmendinger (1990), to the region of Seno Almirantazgo as Klepeis (1994) did on the Deseado fault. The Seno Almirantazgo area belongs to the central part of the MFF system, just east of the Deseado fault (Fig. 4). Indeed, a similar relation between three fault families in terms of a Riedel shear scheme was suggested. Beyond surface mapping showing that the transtensional deformation overprints the former compressional deformation along the MFF (Winslow 1982; Klepeis 1994), they constrained the corresponding strain axes. For the last transtensional phase, they computed an average kinematic solution using 136 measurements for the whole area (Fig. 7c), which exhibits an outstanding strike-slip solution, with a minor extensional component. A main left-lateral plane E–W-oriented symbolizes the orientation of the MMF in the area. It is related to a N45°E z-axis (shortening) associated with a N135°E-oriented x-axis (extension). Indeed, this kinematic solution provides a proper view to sum up the regional tectonics controlled by the transtensional deformation along the entire MFF system.

Diraison et al. (2000) led a widespread tectonic analysis in the Patagonian and Fuegian Andes on minor faults of Cenozoic age, mainly in the foothills of the belt. These authors reported that shortening directions vary in trend, from ENE in the Patagonian foothills with dominant thrusting (following the results of Coutand et al. 1999), to NE in the Fuegian cordillera, where wrenching is dominant. In the following, we will restrict our synthesis to the southern part of their work (named regions 3 and 4 in Diraison et al. 2000). They used the geometrical method of the right diedra to determine the strain axes (Angelier and Mechler 1977; Pfiffner and Burkhard 1987) computed with the code published by Marrett and Allmendinger (1990). The corresponding datasets are represented in Fig. 7d. In the Fuegian Andes, strike-slip faults dominate, either left lateral or right lateral in the same proportion. The two families trend about N–S for the dextral family and E–W for the sinistral one, with strain ellipsoids generally of plane-strain type. In the axial zone of the Magallanes–Austral Basin and in the foreland of eastern Tierra del Fuego, some localities display a transtensional stress regime. The average directions of shortening strike from N52°E for the axial zone in the west Magallanes Strait (zone 3 in Diraison et al. 2000), to N43°E for the Fuegian foothills (zone 4).

The synthetic paper on Fuegian Andes' structural geology from Menichetti et al. (2008) also includes a specific brittle deformation section, with new data in the Lago Fagnano–Canal Beagle area (Fig. 6b). They worked in terms of paleostress axes, using the Angelier's methods (Angelier 1990). Their data plotted in Fig. 7e show a combination of strike-slip ($\sigma 1$ and $\sigma 3$ horizontal, $\sigma 2$ vertical) and extensional stress regime ($\sigma 2$ and $\sigma 3$ horizontal, $\sigma 1$ vertical). Looking carefully at their dataset (Fig. 7e), one can see that some extensional tensors are computed with a majority of strike-slip faults. Indeed, such inversion may somehow lead to stress inversion between $\sigma 2$ and $\sigma 3$ (Sue et al. 2014). Even if such a discussion is not the scope of our paper, anyhow, the overall system exhibits a global transtensional regime, with six strike-slip tensors (over ten tensors) showing $\sigma 1$ oriented in the NE–SW direction, and four extensional tensors showing N–S-oriented $\sigma 2$ (namely shmax in this case). This study confirms the overall orientation of the main stress/strain axis ($\sigma 1$/shortening) in the NE–SW direction, as described above. Moreover, it quantifies the extensional part of the deformation, with noticeable sites showing E–W extension, in close relation to the pull-apart-like geometry described along the MFF (see above, and Lodolo et al. 2003).

Finally, we report the data from Branellec (2010) corresponding to a fieldwork during spring 2010, mainly along the north shore of Lago Fagnano (C. Sue Unpubl. data). Using the stress inversion method of Angelier (1990), we also found a combination of strike-slip tensors, with globally NE–SW-oriented $\sigma 1$, and extensional tensors, with roughly NW–SE-oriented $\sigma 2$ (shmax) axes. Our data are reported in Fig. 7f. Using this more local work, we can confirm the quantification of the overall stress/strain scheme obtained by the previous authors, that is to say the overall transtensional system along the MFF, with left-lateral movements along the globally E–W planes. It is quite important using such kinematic analyses of small or "minor" fault planes to cross-correlate and compare the results of different research teams, as the methodology itself remains quite controversial (see Lacombe 2012 for a review).

In an attempt of synthesis, we combined the results presented above in a single map (Fig. 8). The overall synthetized database includes 63 axes, with 40 axes from Diraison et al. (2000), 10 from Menichetti et al. (2008), and 8 from Branellec (2010). We also added the main axis that can be inferred from Cunningham (1993), the main transcurrent structures from Klepeis and Austin (1997), and the 3 from Klepeis (1994). We mapped the main shortening axes: Z/$\sigma 1$ for strike-slip kinematics and Y/$\sigma 2$ for extensional kinematics.

The density of data is much more important in the Lago Fagnano–Canal Beagle area and in the foreland than in the rest of the Fuegian Andes. Nevertheless, we obtain a very homogeneous map, with a quite regular field for the main shortening axes (Fig. 8). Wrenching appears largely dominant, not only in the MFF's realm, but also in the whole Tierra del Fuego. Indeed, strike-slip is the main signal obtained all along the Southernmost Andes cordillera (inner zones), the Magallanes fold and thrust belt, as well as the northeastern sector, in the foreland of the belt (Magallanes-Austral basin).

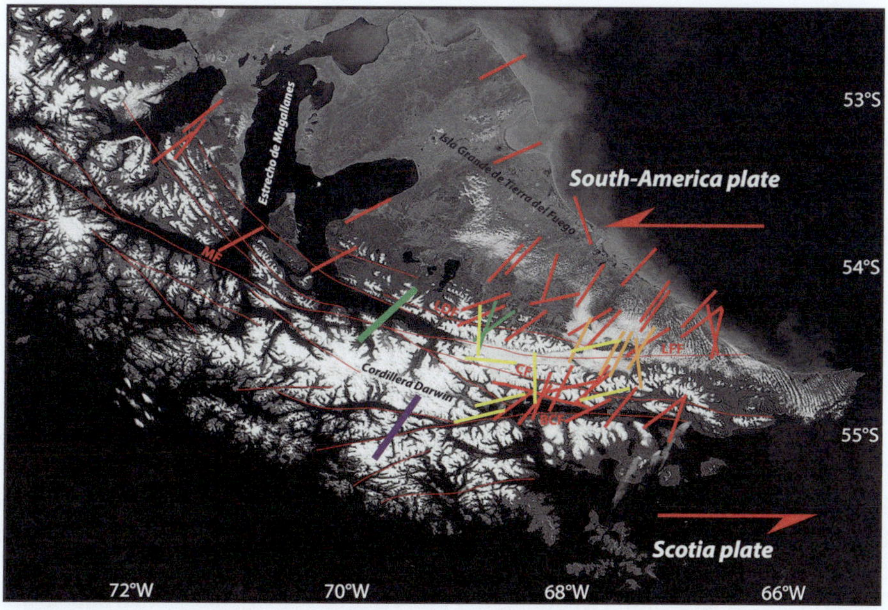

Fig. 8 Synthesized map of 63 principal stress axes (shmax) from fault kinematic analysis presented in Fig. 7. Z/σ1 for strike-slip kinematics and Y/σ2 for extensional kinematics are plotted. Data from Cunningham (1993) in *purple* (average axis); Klepeis (1994) in *green* (small symbols); Klepeis and Austin (1997) in *green* (large symbol; average axis); Diraison et al. (2000) in *red*; Menichetti et al. (2008) in *yellow*; and Branellec (2010) in *orange*. See text for discussion

The six independent studies are coherent with a main shortening axis-oriented NE–SW, associated with a main extensional axis-oriented NW–SE. Actually, we do not observe any significant rotation of the main stress axes along the MFF system. Although kinematic data are still missing in the westernmost part of the system, it seems that the overall strain/stress scheme is very stable in the whole domain. It could correspond to far field, directly linked to the boundary condition of the system, that is to say the wrenching kinematic between Scotia and South America plates. In this interpretation, the stress/strain field would stay very steady, whatever was the orientation of the main fault system (MFF) from E–W to ENE–WSW.

However, such a steadiness could also be interpreted in term of chronology and related to a late faulting or coeval deformation with respect to the bending of the Southernmost Andes (e.g., Maffione et al. 2010, 2015; Ghiglione et al. 2013; Maffione 2016). The precise timing of the overall brittle deformation remains a matter of debate (e.g., Cunningham 1993; Diraison et al. 2000; Menichetti et al. 2008) and could have last from the Late Cretaceous onward. Our synthetic map arises from faulting that more probably acted during the last 40 Myr, in close relation to the opening of the Drake Passage and Scotia plate's development (Eagles 2016). The problem of the beginning of this overall wrenching could hardly be addressed due to the lack of absolute fault dating in the area. However, this tectonics has controlled the deformation of the whole Fuegian Andes and surrounding area during at least Oligo–Miocene times. Beyond the matter of timing of the incipient wrenching tectonics, the question of its present-day activity is of first importance as it bears implications in terms of seismic risk assessment.

3 Seismotectonics at the Boundary Between South America and Scotia Plates

3.1 Seismicity in the Vicinity of the Magallanes–Fagnano Fault System

Among a large cluster of shallow earthquakes that originate on the plate boundary (Fig. 9a), the largest earthquake ever recorded in Tierra del Fuego stroked on December 17, 1949, between the Magallanes Strait and the western tip of the Lago Fagnano. Two successive events occurred at 6h53′ and 15h06′ (GMT time), with magnitudes estimated at 7.5 and 7.8, respectively (Lomnitz 1970; Goodstein et al. 1980; Febrer et al. 2000). According to Perucca et al. (2015), the second shock was stronger in magnitude than the first one, based on wave amplitudes recorded in Weston (USA). Indeed, the first event could be described as a foreshock for the largest one. Among the classical aftershock sequence (Scholz 1990), the largest event stroked on January 30, 1950. These 3 events are represented by red stars in Fig. 9b (modified from Perucca et al. 2015). The locations of the foreshock, mainshock, and aftershock have been reassessed by Bonorino et al. (2012), following Jaschek et al. (1982). A discussion on the precision of the location of these events has also been proposed by Adaros Cárcamo (2003). Bonorino et al. (2012) proposed the following locations: (53.89°S, 69.67°W) for the mainshock, (54.24°S, 69.03°W) for the foreshock, and (53.50°S, 71.50°W) for the aftershock. Indeed, these three major quakes seem organized in a direction subparallel to the MFF, with a rupture length of about 300 km, which is consistent with a M7.5 earthquake (Wells and Coppersmith 1994).

Prior to this major earthquake, another large one has been reported on February 2, 1879, which strokes the region of Ushuaia and Tierra del Fuego as a whole, with a MSK intensity estimated at VIII in Tierra del Fuego, and an estimated magnitude

Fig. 9 **a** Worldwide seismicity around the Scotia Sea realm, from IRIS database, www.iris.edu. ► **b** Epicentral location of the largest instrumentally recorded earthquakes on December 17, 1949, numbered as (1) and (2) (Jaschek et al. 1982), and the January 1950 (3) aftershock. *Red squares* represent the seismicity recorded in 1998–1999 by Febrer et al. (2000). *Colored circles* represent the seismicity recorded from local deployments in 1997–1998 studied by Adaros Cárcamo (2003). *Dark gray dots* represent the seismicity recorded by the 2007 campaign of Buffoni et al. (2009). Focal mechanisms are from Forsyth (1975), Pelayo and Wiens (1989), Adaros Cárcamo (2003), and USGS database. Map has been modified from Perucca et al. (2015). **c** seismic cross section across the MFF modified from Buffoni et al. (2009). See text for details

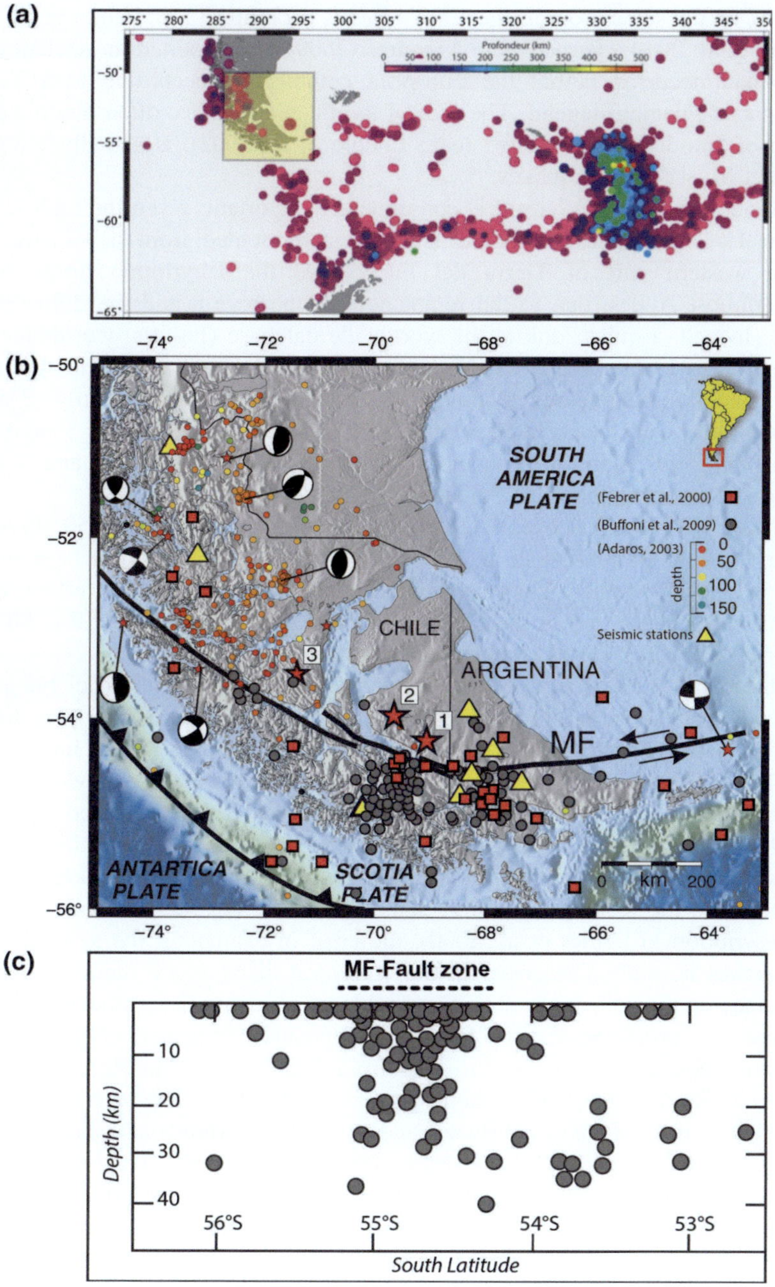
(a)
Profondeur (km)
0 50 100 150 200 250 300 350 400 450 500
275° 280° 285° 290° 295° 300° 305° 310° 315° 320° 325° 330° 335° 340° 345°
-50°
-55°
-60°
-65°
(b)
-74° -72° -70° -68° -66° -64°
-50°
-52°
-54°
-56°
SOUTH AMERICA PLATE
(Febrer et al., 2000)
(Buffoni et al., 2009)
(Adaros, 2003)
depth
0
50
100
150
Seismic stations
CHILE
ARGENTINA
MF
1
2
3
ANTARTICA PLATE
SCOTIA PLATE
0 km 200
(c)
MF-Fault zone
Depth (km)
10
20
30
40
56°S
55°S
54°S
53°S
South Latitude

M7, i.e., on the same order as the 1949's events (Lomnitz 1970; Febrer et al. 2000; Pedrera et al. 2014). Perucca and Moreiras (2009) also reported an ancient earthquake that occurred before the European colonization according to a Yahgán (indigenous Fuegian) legend. The area of the Fuegian Andes, often described as a region of "moderate seismicity" (e.g., Lodolo et al. 2002), is actually a zone of potential M $\sim$7–8 earthquakes.

Although the seismic hazard is consequently important, given the high level of historical seismicity, the overall area is poorly instrumented, from the Atlantic shore to the western part of Tierra del Fuego and the labyrinthic fiords of the Southernmost Andes, due to the remoteness of the region and the difficulties of access. Indeed, a quick look at the worldwide database (Fig. 9a; worldwide seismicity from IRIS database, www.iris.edu) and from the USGS site for the Ushuaia's region (http://earthquaketrack.com/ar-23-ushuaia/biggest) gives only 3 $M > 4$ recent events, well localized along the MFF system. A network of 5 permanent stations (Fig. 9b) is installed in the Ushuaia's vicinity (Argentina) and 3 in the western side of Southernmost Andes (Chile).

Beyond the historical events (1879, 1929, 1930, 1944, 1949, 1950, 1970; see Febrer et al. 2000; Buffoni et al. 2009), instrumental seismicity has been investigated in 3 recent studies. They correspond to temporary seismic recording, over a one year each at the end of the 1990s. We synthesized these data in Fig. 9b (after Febrer et al. 2000; Adaros Cárcamo 2003; Buffoni et al. 2009).

Febrer et al. (2000) analyzed the regional seismicity in Tierra del Fuego and surrounding areas using single station localization (SSL) technique with the data recorded at the broadband seismic station of Ushuaia (USHU) from June 1998 to June 1999. They detected and analyzed 39 events in this time window with magnitudes ranging from 2.0 to 5.4 (Table 3 in Febrer et al. 2000; red squares in Fig. 9b). This moderate seismicity recorded over one year of observation is largely consistent in terms of localization with the major historical earthquakes described above. Given the scarcity of seismic stations, and the location method used (SSL, see description in Febrer et al. 2000), the error in estimating the azimuth of the earthquakes is $\sim 3°$. The errors in the epicentral distance are mainly caused by inaccuracy on the velocity model. The final location errors are estimated ~ 5 % for the transverse position and 10 % for the radial position, in respect of the azimuth of the station-epicenter path. On the map from Fig. 9b (red squares), 2 groups of events cluster close the MFF system. One east of the Lago Fagnano and a second one in the vicinity of the Canal Beagle and near the Navarino Island show the actual activity on the MFF as a whole (Winslow 1982; Klepeis 1994). The positions of the 3 main earthquakes of the 1949s sequence are also located on the same fault system, taking into account the large uncertainty of their precise epicenters (Adaros Cárcamo 2003). Epicenters located eastward in the Atlantic Ocean could be linked to the continuation of the MFF toward the NSR. The southernmost event in the Pacific–Atlantic boundary could be associated with the Chilean trench and with its southward continuation in the Shackleton fracture zone. The relatively moderate size seismicity recorded during one year suggests a continuous release of seismic energy, in relation to the driving mechanism of plates wrenching at larger scale.

Anyhow, a specific study on the regional value, as investigated in various tectonic contexts (e.g., Sue et al. 2002), and taking into account the Febrer's database, together with the historical one, and the more recent ones by Adaros Cárcamo (2003) and Buffoni et al. (2009) could bring new constrain on the mechanism of faulting in this area. Indeed, a state-of-the-art work dedicated to the seismotectonics of the MFF system as whole is still missing.

A precise work on the recent seismicity of Southernmost Andes has been achieved in the PhD. of Adaros Cárcamo (2003), analyzing the data of the SEPA experiment. He worked on the seismicity of southernmost Chile and includes the connection with the MFF system to the south. His data have been reanalyzed by Cisternas and Vera (2008) and Perucca et al. (2015). Figure 9b shows the earthquakes recorded during the experiment used in this PhD. (colored dots, depending on the event's depth) and conducted from February 1997 to November 1998 using a local network of 5 seismometers installed in Southernmost Andes and in the foreland (see location in Adaros Cárcamo 2003). The recorded seismicity corresponds to 320 crustal earthquakes, which are registered within a depth lower than 40 km, with magnitudes below M_l4.5. On quality consideration, Adaros Cárcamo (2003) reduced the published catalogue to 250 events. Seismicity north of the MFF corresponds to the convergence between the South America and Antarctic plates, and mechanisms of the observed earthquakes show roughly E–W-oriented compression axes (Adaros Cárcamo 2003). The seismicity viewed by this network is less frequent south of Magallanes Strait and MFF. Indeed, some events seem to locate on the MFF system, with WNW–ESE elongated clusters. As a whole, the seismic activity is clearly greater in the frontal part of the Southernmost Andes (the fold and thrust belt), and north of the Evangelista lighthouse, which corresponds to the western end of the MFF. Cisternas and Vera (2008) also pointed out the potential active volcanic activity linked to the seismicity. In terms of plate motions, this greater activity to the north than in the southern part of the area (namely the MFF system) would rise up from the velocity of convergence between South America plate and Antarctic plate (20 mm/year), compared to the velocity in strike-slip between the Scotia and Antarctica plates ($\sim$7 mm/year). The low seismicity observed by this experiment along the MFF system has been interpreted by Adaros Cárcamo (2003) in terms of seismic gap: Such failure would be activated only by large earthquakes (e.g., the 1949's sequence), energy accumulated through being partly released by small events on lateral faults, organized as Riedel shear. This would imply an overall locked MFF system, in-between major earthquakes. This interpretation is at least partly invalidated by the data of Febrer et al. (2000), who observed a quite steady seismic activity along the MFF system. Much longer recording of local seismicity would be necessary to better constrain the fault mechanism along the MFF. The particular case of the diffuse seismicity located east of the frontal fold and thrust belt, in Santa Cruz Province of Argentina, would be associated with seismicity induced by oil and gas exploration and exploitation in southern Argentina (e.g., Grasso 1992), although it is likely that such activity is associated with the reactivation of preexisting Jurassic upper crustal faults (Ghiglione et al. 2013).

A third analysis of the regional (micro)-seismicity in the Magallanes Strait and overall Tierra del Fuego area has been recently published by Buffoni et al. (2009). They used the recording provided by 5 seismic stations located in eastern Isla Grande (yellow triangles in Fig. 9b), from January to December 2007, and provided a careful assessment of the event's location. Due to the scheme of the seismic network with respect to the main active zones, the overall precision of this database remains quite rough, with uncertainties most of the time over 5 km in horizontal and depth location (see discussion in Buffoni et al. 2009). However, there were 185 recorded events, with local magnitudes between 2 and 5 (dark gray dots in Fig. 9b, c). As for the previous studies, the small time window, together with the geometry of the network, prevents to obtain a long-term and regional analysis. Nevertheless, this work provides a good idea of the current activity in the MFF. The distribution of epicenters shows that earthquakes concentrate in several clusters: The main one seems located in the vicinity of the Cordillera Darwin, another one around the Lago Fagnano, and probably a third one close to the Canal Beagle. Indeed, this database clearly shows that the MFF system, and more generally speaking the actual limit between Scotia and South America plates, is currently seismically active, with earthquakes of low to medium magnitude, but steadily occurring. Consequently, the MFF system would not be seismically locked as suggested by Adaros Cárcamo (2003). The cross section (Fig. 9c) shows that the seismicity along the MFF system locates mainly in the first 30 km of the crust. On this N–S section, the activity on the MFF is clearly visible, with an elongated subvertical cluster at $\sim$55°S.

3.2 Paleoseismicity and Neotectonics Along the Magallanes–Fagnano Fault System

Complementary to actual seismic monitoring, several studies reported neotectonic and paleoseismologic features on the MFF system, which allow investigating its Quaternary tectonics. Evidences are found in the best outcrops of the MFF, along the Turbio and Irigoyen rivers (Fig. 10) (Costa et al. 2006). Several morphotectonic evidences are reported by Perucca et al. (2015), such as linear valleys, deflected river streams, shutter ridges, sag ponds, and pressure ridges. This kind of drainage anomalies can be followed downstream from Tolhuin to the Atlantic shore and could be quantitatively analyzed using river profile analyses (e.g. Rabin et al. 2015). According to Coronato et al. (2002), drainage diversion must have occurred during the late glacial period sometime after 12 kyr. For instance, the Turbio River flowed into the Atlantic Ocean before being diverted into Lago Fagnano by the Fagnano fault scarp. River drainages exhibit subtle topographic modifications concerning for instance the incision depth along the river. Another feature corresponds to the change in the river pattern (from anastomosing to meandering) or in the alluvial plain width, which would indicate uplift or sagging (Audemard 1999). Sag ponds are also recognized along the subsiding bedrock and glacial plains, with a W–

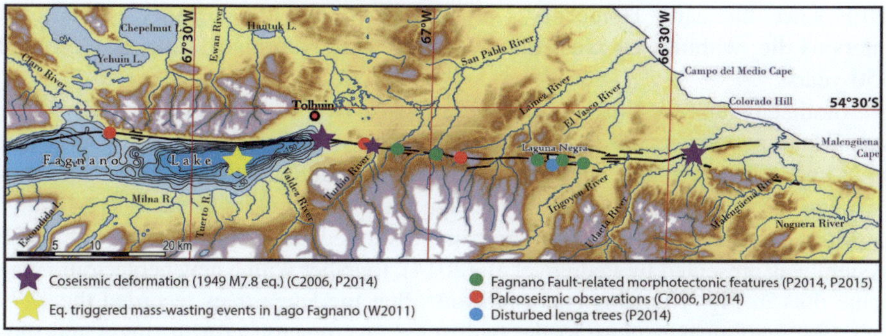

Fig. 10 Paleoseismological map of Lago Fagnano–Irigoyen river area including observations from C2006, Costa el al. (2006); W2011, Waldmann et al. (2011); P2014, Pedrera et al. (2014); and P2015, Perucca et al. (2015). Map modified after Pedrera et al. 2014 and Perucca et al. 2015

E-trending orientation. In particular, the area of Laguna Negra constitutes a pull-apart basin developed between two overlapping segments of the MFF (Fig. 10). Perucca et al. (2015) reviewed the tectonic-related anomalies (green dots in Fig. 10) and propose the left-lateral, and actually transtensive, movement of the MFF during quaternary times (probably post-LGM).

In terms of quantitative paleoseismology (see McCalpin 2009 for methodologies), several methods complementary to morphotectonics have been applied to the MFF system: trenching of fault scarp in Quaternary rocks (Costa et al. 2006), high-resolution seismic imaging of the sediments in the Lago Fagnano (Waldmann et al. 2011), geophysics (ground-probing radar, vertical electric sounding, seismic refraction; Bonorino et al. 2012), and disturbed tree analyses (Pedrera et al. 2014).

Costa et al. (2006) analyzed paleoseimic record associated with the ruptures of the twin 1949s earthquakes (M7.8 and 7.5) and recognized 3 or 4 events within the last 9 kyr, in which ruptures reached the surface (indeed M6 or more, Scholz 1990). They observed the fault scarp at several localities east of Lago Fagnano, and they dug a trench across a secondary fault of the MFF at Río San Pablo. The cumulative scarp reaches heights up to 11 m in late Pleistocene–Holocene deposits. The components of the 1949 events reach maximum 1 m vertically and 4 m horizontally. A maximum average recurrence interval of 2 kyr is thus estimated for major earthquake ruptures. Other palaeoseismic studies have proposed comparable recurrence intervals: 750 years for M7.8 earthquakes (Smalley et al. 2003) and 2–2.7 kyr for surface rupture earthquakes over the last 8 kyr (Schwartz et al. 2001).

Using high-resolution seismic imaging, coring, and dating methods (AMS-14C), Waldmann et al. (2011) provided a very accurate analysis of the sediment infill in the Lago Fagnano and thus quantified and dated a sequence of 19 Holocene mass-wasting events (subwater landslides) over the last 2 kyr. They interpreted slope failures and mega-turbidites as driven by large earthquakes occurring along

MFF since the early Holocene. They suggested that the majority of repeated intervals lie within two clusters of events with recurrence time of 350 and 850 years.

Another recurrence-time estimation proposed by Bonorino et al. (2012) using subsurface geophysics (radar, electric sounding, seismic refraction) indicates that strand plain deposits have been affected by rupture along the MFF at least six times between 0.9 and 6.4 ky, giving an average recurrence rate of about 1 kyr.

More recently, a study of seismically disturbed lenga trees across the MFF system was presented by Pedrera et al. (2014), together with a new neotectonic map (blue dots in Fig. 10). This analysis shows that the lenga trees recorded the 1879 and 1949 earthquakes close to the fault trace (asymmetric growth rings) and thus established the possible location of the earthquake on this fault.

To sum up the paleoseismological results, it appears that the MFF would produce major earthquakes comparable to the 1879 and 1949 events (i.e., with magnitude ~7 or more) with a recurrence time in the order of 1 kyr. Anyhow, the recent seismic activity, with major earthquakes in 1879 and 1949, plus on reported by the Yahgán people "before the colonization", that is to say probably in the eighteenth century, exhibits a noticeable lower recurrence time, in the order of one century or less. This apparent discrepancy rises up the matter of the tectonic coupling between the actual driving geodynamics and the regional fault mechanism (see discussion in Sect. 4).

3.3 Focal Mechanisms and Stress Inversion at the Scotia and South America Plate Boundaries

Few focal mechanisms are actually available in the literature in the vicinity of the MFF system (Fig. 9b). Forsyth (1975) published 3 focal mechanisms in the study area, together with a first complete analysis of Scotia's tectonics. Pelayo and Wiens (1989) reassessed one of the Forsyth's solutions. Adaros Cárcamo (2003) computed 2 focal mechanisms using the SEPA database (see above) and also reported 3 mechanisms from the USGS Harvard-CMT database (http://www.globalcmt.org). The parameters of these 8 solutions are given in Table 1. The scarcity of focal mechanisms in the overall Southernmost Andes rises up from the configuration of the seismological permanent networks, which have a quite low density of stations. Indeed, the available mechanisms have been generally computed with earthquakes strong enough to be detected by worldwide networks. The typical magnitudes of these events are about M5.5 (Forsyth 1975; Pelayo and Wiens 1989). The solutions computed by Adaros Cárcamo with the SEPA database correspond to M4.5 and M4.3 events detected during the SEPA experiment's data acquisition.

The synthesized seismotectonic map shown in Fig. 9b allows investigating the overall geodynamics of the Southernmost Andes' realm. To the west, one solution in reverse faulting with a flat focal plane is probably linked to the subduction in the

Table 1 Parameters of the focal mechanisms drawn in Fig. 9

Date	Lat (S)	lon (W)	Depth	Mag.	Strike	Dip	Rake	Ref
14/06/1970	52.0	73.8	33	6.0	315	12	14	F
15/06/1970	54.3	63.6	10	5.6	179	87	192	F-P&W
09/02/1972	51.8	74.0	33	5.5	305	27	27	F
06/06/1979	52.9	74.6	15	5.6	7	8	104	USGS-A
30/07/1992	51.1	72.6	15	5.4	31	25	114	USGS-A
31/08/1996	53.4	73.1	30	4.9	54	77	−148	USGS-A
27/05/1998	51.5	72.4	22	4.5	5	55	90	A
26/04/1997	52.2	71.6	19	4.3	2	50	90	A

Data after Forsyth (1975) (F), Pelayo and Wiens (1989) (P&W), Adaros Cárcamo (2003) (A) including three solutions he took from the USGS database (USGS-A)

Chile trench. In the northeastern part of the study area, 3 reverse faulting mechanisms oriented more or less N–S are related to active thrusting in the frontal fold and thrust belt. Concerning the MFF system, 3 focal mechanisms establish clearly its left-lateral transcurrent behavior. Eastward of the MFF, the left-lateral movement is transferred to the North Scotia Ridge, which behavior in terms of focal mechanisms has been well investigated by Thomas et al. (2003). This regional-scale analysis is integrated in the large-scale processes acting between Antarctica, South America, and Scotia plates (Eagles et al. 2005, Galindo-Zaldívar et al. 2006).

Focal mechanisms are classically used to determine the actual stress of a geodynamic system (see for instance Zoback et al. 1989 at worldwide scale). At the scale of the South America plate, the stress field has been investigated by Stefanick and Jurdy (1992) and Coblentz and Richardson (1996). A recent study including focal mechanism inversion, and numerical modeling at the scale of the Scotia plate (Sue et al. 2012; Sue et al. in prep.), allows to better characterize the stress field in the realm of the Scotia plate in general and in the Southernmost Andes in particular (Fig. 11 and Table 2). This study provides a new regionalization and quantification of the stress variations in the larger South America–Scotia–Antarctic plate system. It rises up the matter of regional evolution from compressional (Patagonian Andes–Southernmost Andes, Sandwich subduction front), to strike-slip (North and South Scotia ridges, Shackleton fracture zone), and extensional areas (Bransfield basin, Sandwich subduction zone).

By zooming on the northwestern part of the work of Sue et al. (2012) (Fig. 11a, b), we can propose a new map of the actual stress in the Fuegian Andes and surrounding areas (Fig. 11b). We applied stress inversion methods to the subsets of focal mechanisms extracted from the USGS Harvard-CMT database (http://www.globalcmt.org) from 1976 to 2009 (Fig. 11c, see Sue et al. 2012; Sue et al. in prep for details). We used the software ZMAP (Wiemer 2001) to perform the stress inversions from focal mechanisms. Stress inversion methods assume a uniform state of stress within the study area as well as newly formed faults. Furthermore, in contrast to the inversion of fault-striae data (see above), standard inversion of earthquake data (e.g., Michael 1987; Gephart 1990; Delvaux 1993) does not a priori

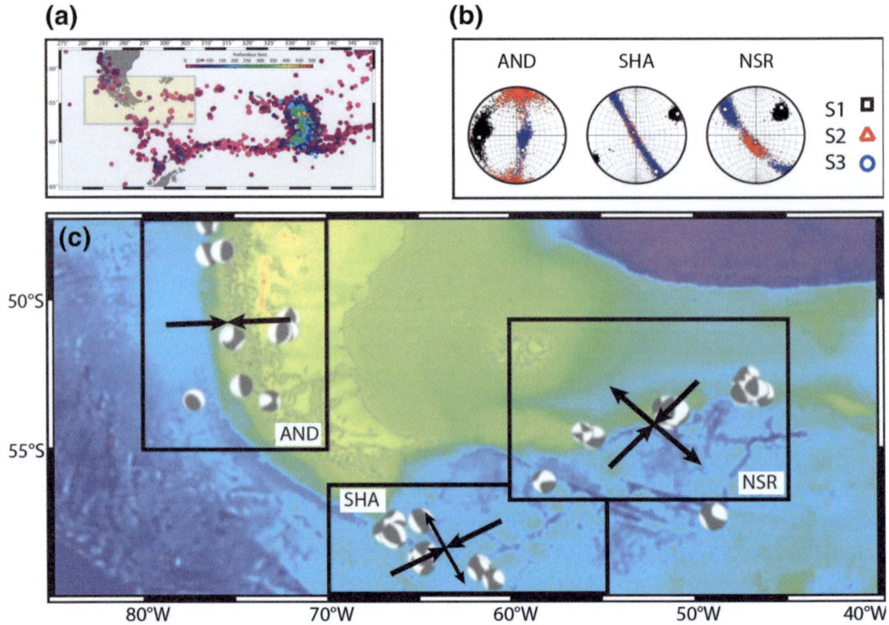

Fig. 11 **a** Location of the study area on the IRIS database (from Fig. 9a). **b** Stress tensors of the 3 inversions (AND, SHA, NSR). **c** Stress map obtained with the 3 inversions, with horizontal s1 and/or s3 (see parameters in Table 2)

Table 2 Parameters of the stress inversion performed in the Southernmost Andes (AND), Shackleton fracture zone (SHA), and North Scotia Ridge area (NSR)

Zone	*N*	s1 az	s1 pl	s2 az	s2 pl	s3 az	s3 pl	Var	phi ratio
AND	12	93.3	13.9	6.2	33.8	157.7	52.7	0.07	0.20
SHA	26	62.1	3.1	−47.3	80.6	152.5	8.8	0.07	0.26
NSR	17	42.3	14.5	173.9	72.1	−50.3	10.1	0.04	0.41

Data from sue et al. (2012, 2016 in prep.). *N* number of focal mechanisms used in the inversion; *Sx-az* and *Sx-pl* azimuth and plunge of the x stress axis. *Var* variance of the inversion; *phi* phi ratio of the stress tensor

discriminate between the active and the virtual nodal plane. Despite such restrictions, stress inversion has been shown to provide a powerful tool to analyze focal plane mechanism datasets (e.g., Sue et al. 1999; Kastrup et al. 2004; Delacou et al. 2004).

The aim of the inversion of focal plane data is the determination of a regional stress tensor that satisfies most, if not all, observed individual earthquakes in a given area. In contrast to the simple interpolation of isolated, projected p- and t-axis directions, inversion methods take the entire 3-D orientation of focal planes into account and search for a common stress tensor.

Our inversion is based on an objective, visual selection of large homogeneous zones, which are characterized by an apparently uniform type of deformation. Indeed, a synthetic analysis combining the various information coming from seismicity (active zones), focal mechanisms (type of deformation), DEM (topography plus bathymetry), structural setting, and geophysical data allows to differentiate regionally three tectonically active homogeneous sectors (Fig. 11b): (i) the southern Patagonian Andes and west of Tierra del Fuego, namely the Southernmost Andes (AND), (ii) the North Scotia Ridge (NSR), and (iii) the Shackleton fracture zone (SHA). These three sectors related to specific plate tectonic boundaries (Fig. 2b, c) have been used to decipher the current stress pattern and its variation in the overall area, using state-of-the-art inversion method of focal mechanism subdatasets (Table 2, see Wiemer 2001, Gephart 1990). The overall current stress pattern is surprisingly coherent at the scale of the selected sectors (Fig. 11c) and is comparable to the kinematic results from Giner-Robles et al. (2003). Indeed, the tectonic complexities of the Scotia plate's surroundings induce stress variation from pure compression (AND), to transpression (SHA), and pure strike-slip (NSR). Nevertheless, the principal stress directions show a good stability with $\sigma1$ oriented E–W to NNE–SSW and $\sigma3$ oriented NW–SE to NNW–SSE.

The stability of the stress pattern reflects the global steady left-lateral motion between the Antarctic, Scotia, and South America plates (Fig. 11c) and the consequent westward movement of the Pacific trench (Fig. 2b), as the main driving force for the entire area.

4 Crustal Deformation at South America–Scotia Plate Boundaries Inferred from GPS

Current motion of South America with respect to Antarctica is left lateral along the E–W strike. In between, the intermediate Scotia plate also moves left laterally with respect to both the South America and Antarctic plates (Fig. 2b). Plate interaction is mainly distributed along the North and South Scotia Ridges, transcurrent faults, and the Shackleton fracture zone (Fig. 2b, c). The largest Scotia plate onshore exposure is the portion of Tierra del Fuego south of the left-lateral MFF, which is the landward continuation of the NSR and the current and active continental plate boundary (Fig. 4) (Smalley et al. 2007).

At the beginning of the 1990s, the Scotia plate was not tied into the global spreading circuit and could not be included in the NUVEL-1 plate model (DeMets et al. 1990). Using closure arguments and refined models, it subsequently integrated the NUVEL-1A model (DeMets et al. 1994). At plate scale, the current deformation is described by Smalley et al. (2007), who proposed a reassessment of the Scotia's Euler poles relatively to South America and Antarctica, using GPS measurements (King et al. 1997; Smalley et al. 2003) and earthquakes slip vectors (Pelayo and Wiens 1989; Thomas et al. 2003). They also integrated a modeling of the MFF deformation and Antarctica subduction along the Chile trench. The Smalley et al. (2007)'s Euler vector for Scotia–South America relative movement is approximately parallel to that for Antarctic–South America movement, indicating that the overall motions of Scotia and Antarctica with respect to South America are subparallel and differ only in velocities. The poles or rotation previously proposed by Pelayo and Wiens (1989) and Thomas et al. (2003), together with the solution of Smalley et al. (2007), stays roughly on an N–S-oriented great circle perpendicular to the Scotia plate's along east–west axis.

Looking closer to the MFF system, two recent publications constrained the strain rates and crustal velocities in Tierra del Fuego, investigating the mechanical behavior of the eastern MFF (Smalley et al. 2003; Mendoza et al. 2011). Using repeated measurements of 20 GPS stations in Tierra del Fuego and southern Patagonia that spans the central portion of the MFF, Smalley et al. (2003) precisely constrained the velocity of the MFF at 6.6 ± 1.3 mm/year. This velocity is obtained by more regional databases, and somehow with less resolution, of 5 mm/year and 9.6 mm/year from Del Cogliano et al. (2000) and DeMets et al. (2010), respectively. Indeed, campaign measurement sites have been occupied between two and four times over total observation spans of two to six years between 1993 and 1998. The velocity field obtained by Smalley et al. (2003) (red arrows in Fig. 12a) shows an increase in velocity from north to south with respect to stable South America. The velocity gradient is concentrated in the region of the MFF and is associated with the plate boundary. Movements of the sites south of the fault are almost perfectly eastward, which is left lateral with respect to stable South America (Fig. 12a).

Beyond the GPS velocity quantification, this study also proposes a modeling of the MFF mechanism, on the basis of an E–W left-lateral strike-slip, assumed to be locked at 15 km. The formalism developed by Savage et al. (1975) for strike-slip modeling has been used (see discussion in Smalley et al. 2003), together with an inversion for the position, azimuth, and velocity of an E–W-striking fault. The position of the modeled fault is coincident with the surface trace of the MFF (54.54°S), defining a subvertical fault. The diagram in Fig. 12b, which fits the data with the modeling, clearly establishes the velocity of the deformation on the MFF. Assuming a 5 m offset for the M7.5 1949 event, the GPS rate of 6.6 mm/year results in a recurrence time in the order of 750 years. Although the uncertainty of this estimate is large, it is quite well coherent with the recurrence-time estimations proposed using paleoseismological observations. Actually, an additional 1–2 mm/year of relative plate motion may also be diffusely distributed across a

Fig. 12 **a** Positive strain rate principal components represented by strain tensors of each triangle ▶ cell are plotted (*black arrows*) and interpolated in the background (μstrain) from Mendoza et al. (2011). *Red arrows* are velocity field from Smalley et al. (2003). **b** Fit of GPS data to simple strike-slip model of Savage et al. (1975): cross section of eastward component of velocity (V_e). *Horizontal dashed green lines* show velocity asymptotes and relative velocity axis. *Vertical dashed green line* shows horizontal location of the down-dip end of the fault or location of the screw dislocation. Modified from Smalley et al. (2003)

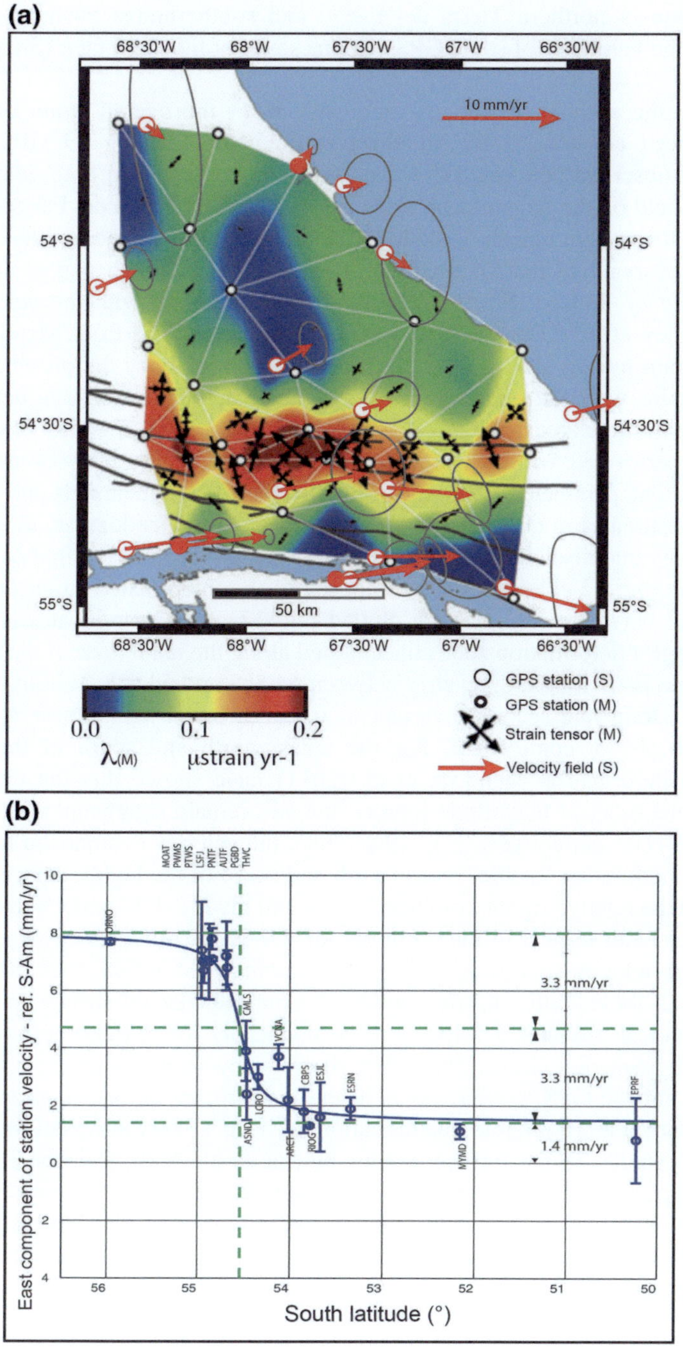
(a)
68°30'W
68°W
67°30'W
67°W
66°30'W
10 mm/yr
54°S
54°30'S
55°S
50 km
GPS station (S)
GPS station (M)
Strain tensor (M)
Velocity field (S)
0.0
0.1
0.2
λ(M) μstrain yr-1
(b)
East component of station velocity - ref. S-Am (mm/yr)
South latitude (°)
3.3 mm/yr
3.3 mm/yr
1.4 mm/yr

larger region in northern Tierra del Fuego and southernmost Patagonia, as suggested by the velocities of some GPS stations south of the main fault (Smalley et al. 2003).

Beyond the search for velocity along the MFF, the overall strain pattern has recently been constrained by Mendoza et al. (2011). Using 29 GPS stations, repeatedly observed between 1993 and 2008, they determined both velocity field and strain field on the Argentinean side of Isla Grande de Tierra del Fuego, that is to say across the eastern branch of the MFF. Some sites investigated by Mendoza et al. (2011) correspond to that of Smalley et al. (2003), and both papers are largely complementary. Indeed, Fig. 12a combines both of them, with the velocity field from Smalley et al. (2003) (red arrows) and the strain field from Mendoza et al. (2011) (strain tensors drawn with black arrows). To quantify the present-day horizontal strain, with the assumption of stationary and homogeneous deformation rates in small finite areas, a strain analysis rising up from the regional velocity field has been performed within 44 triangular cells dividing the network as a whole (Fig. 12a). The parameters resulting from this analysis are invariants and constitute unbiased estimates for the regional strain rate (see Mendoza et al. 2011 for methodology and discussion on the database). The strain rate principal components ($\lambda 1$, shortening, $\lambda 2$ extension) superimposed to the interpolated strain rate second invariant ($\lambda = (\lambda 12 + \lambda 22)1/2$) are plotted Fig. 12a. This map locates and quantifies a strongest deformation zone concentrated along the main trace of the MFF, in a 30-km-wide E–W longitudinal strip between 54.5°S and 54.6°S. Within this belt, a maximum strain rate second invariant of 0.253 ± 0.004 µstrain/year is observed (within 2σ, 95 % confidence). For the areas, north and south of the zone of maximum deformation Mendoza et al. (2011) have shown that the deformation rates are one order of magnitude weaker, but still remain significant with values of about 0.03–0.05 µstrain/year (Fig. 12a). Then, the velocity comparison for stations located in both rather "stable" areas (north and south of the highly deforming strip) confirmed the relative rigidity of these blocks and gives a differential velocity of the south block with respect to the north one of 4.4 ± 1.1 mm/year eastward, and an almost null velocity southward (0.3 ± 0.8 mm/year) (Mendoza et al. 2011).

Although these results nicely confirm the pure left-lateral strike-slip along the MFF system, it provides a velocity value noticeably lower than the 6.6 mm/year provided by Smalley et al. (2003). The discrepancy may rise up from the longer time span for the GPS measurements in the more recent work, which would converge toward a more reliable quantification, as seen in very slow deforming areas (e.g., GPS studies in the European Alps by Sue et al. 2000; Delacou et al. 2005, 2008; Walpersdorf et al. 2015), but also from differences in the data processing strategy. Indeed, the higher value of $\sim$6.6 mm/year from Smalley et al. (2003) could also integrate a significant part of the overall deformation diffusely distributed across a larger region.

5 Discussion and Concluding Remarks

It should be first highlighted that although the gathered information from previous works in the Southernmost Andes and the Scotia Sea is somehow scarce and scattered, such as fault kinematic data and focal mechanism, it was, however, very consistent. After data were filtered for minimum technical requisites, compiled in our regional databases, and reprocessed, they resulted in explanatory figures (e.g., Figs. 8 and 11b, c), illuminating some key discussion points in the geodynamic models of the Southernmost Andes.

Our synthesis of six independent fault kinematic studies is coherent with main shortening consistently oriented NE–SW, without significant rotation of the axes across the orogen (Fig. 8). The fault kinematic data present a very consistent shortening direction, i.e., on the lines of oblique convergence models presented by Ghiglione and Cristallini (2007). Instead of the two-stage orthogonal convergence model preferred by these authors, it seems that the oblique convergence of a basement indenter in one ~NW steady direction, presented in the same paper, probably better explains the shortening direction of steady patterns around the Southernmost Andes, as proposed by Maffione et al. (2015) (cf. Maffione 2016). These observations, together with the stability of the stress pattern and orientation of the shortening axes on a bigger scale (Fig. 11b, c), reflect a steady E–W to NNE–SSW $\sigma 1$/shortening direction since middle Eocene times (Ghiglione et al. 2013), reflecting that the global left-lateral motion between Antarctica–Scotia–South America plate circuit is the main driving force for the entire area and in particular for compression in the Southernmost Andes, over the last 50 or 40 Ma.

Elastic rebound theory predicts that the major earthquakes on a fault are time-dependent, as they are linked to a period of built-up energy (interseismic) with abrupt relaxation stages (coseismic). Both the short-term geodetic (Del Cogliano et al. 2000; Mendoza et al. 2011) and the long-term geological observation (i.e., Lodolo et al. 2003; Rossello 2005), slip rates of the MFF system are low (<5 mm/year). Therefore, the time span between major earthquakes should be larger than the one obtained over the last 2 or 3 centuries. Considering a simple tectonic setting of a pure left-lateral strike-slip fault with a constant 5 mm/year slip rate (geodetic slip rate, Mendoza et al. 2011) able to generate ~6 m of left-lateral maximum displacement at the surface during the largest earthquakes (effects of the 1949 Ms 7.8 earthquake; Costa et al. 2006), the expected time span between comparable earthquakes would be around 10 kyr. Yet the time between the two most recent large earthquakes on the eastern onshore fault section of the MFF was about 70 years. Such a great discrepancy could result (i) from the presence of locked sections along the MFF; (ii) variable slip rate history and tectonic loading; (iii) delayed and pulsed stress release due to steady tectonic loading and stored in the crust over millennium time scale; and (iv) the complexities of the extensional *en échelon* geometry of the MFF, which could rise up discrepancies from the simple strike-slip loading model.

Our results fill up the time gap between the 1949 event and the two, possibly three, paleoearthquakes detected by trenching over the last 9 kyr (Costa et al. 2006). The discrepancies between long-term and short-term velocities, that is to say between the evaluated tectonic loading and the actual recurrence time, suggest a complex mechanics on the MFF, leading to complex recurrence time history for the characteristic earthquake ($M \geq 7$).

Acknowledgements This work was supported by Buenos Aires and Besancon University (UBA and UBFC resp.), the Besançon Observatory (OSU-THETA), the French CNRS-INSU (project SCOPE), the Argentinean CONICET, and the ECOS-SUD comity (project A15U02). Many thanks to the people who came with us and helped us on the field and specifically to M. Branellec. Some maps were drawn using GMT code (Wessel and Smith 1991). Special thanks to Ombeline, Felix, and Simon.

References

Adaros Cárcamo RE (2003) Sismicidad y Tectónica del extremo sur de Chile. Tesis Magíster, Universidad de Chile, 82 pp

Angelier J (1990) Inversion of field data in fault tectonics to obtain the regional stress—III. A new rapid direct inversion method by analytical means. Geophys J Int 103:363–376

Angelier J, Mechler P (1977) Sur une méthode graphique de recherche des contraintes principales également utilisable en tectonique et séismologie: la méthode des dièdres droits. Bull Soc géol Fr 19:1309–1318

Angelier J, Goguel J (1979) Sur une méthode simple de détermination des axes principaux des contraintes pour une population de failles. C R Acad Sci Paris 282:307–310

Audemard FA (1999) Morpho-structural expression of active thrust fault systems in the humid tropical foothills of Colombia and Venezuela. Zeitschrift für Geomorphologie 118:1–18

Barker PF (2001) Scotia Sea regional tectonic evolution: implications for mantle flow and palaeocirculation. Earth Sci Rev 55:1–39

Barker PF, Burrell J (1977) The opening of Drake Passage. Mar Geol 25:15–34

Bonorino GG, Rinaldi V, del Valle Abascal L, Alvarado P, Bujalesky GG, Güell A (2012) Paleoseismicity and seismic hazard in southern Patagonia (Argentina-Chile; 50°–55°S) and the role of the Magallanes-Fagnano transform fault. Nat Hazards 61(2):337–349

Bott MH (1959) The mechanism of oblique slip faulting. Geol Mag 96:109–117

Branellec M (2010) Analyse morphotectonique de la terminaison des Andes Fuegones (Terre de Feu, Argentine). Master thesis, Brest University, 90 pp

Buffoni C, Sabbione NC, Connon G, Ormaechea JL (2009) Localización de hipocentros y determinación de su magnitud en Tierra del Fuego y zonas aledañas. Geoacta 34:75–86

Caminos R, Haller M, Lapido J, Lizuain O, Page A, Ramos VA (1981) Reconocimiento geológico de los Andes Fueguinos, Territorio Nacional de Tierra del Fuego. VIII Congreso Geológico Argentino (San Luis) Actas 1:759–786

Champagnac JD, Sue C, Delacou B, Burkhard M (2003) Brittle orogen-parallel extension in the internal zones of the Swiss Alps (south Valais). Eclogae Geol Helvet 96:325–338

Cisternas A, Vera E (2008) Sismos históricos y recientes en Magallanes, Magallania (Punta Arenas) 36(1):43–51

Coblentz DD, Richardson RM (1996) Analysis of the South American intraplate stress field. J Geophys Res 101:8643–8657

Coronato A, Roig C, Mir X (2002) Geoformas glaciarias de la región oriental del Lago Fagnano, Tierra del Fuego, Argentina. In: Cabaleri N, Cingolani C, Linares E, López de Luchi M, Ostera H, Panarello H (eds) XV Congreso Geológico Argentino Actas, pp 457–462

Costa CH, Smalley R, Schwartz D, Stenner H, Ellis M, Ahumada E, Velasco MS (2006) Paleoseismic observations of an onshore transform boundary: the Magallanes-Fagnano fault, Tierra del Fuego, Argentina. Rev Asoc Geol Argentina 61:647–657

Coutand I, Diraison M, Cobbold PR, Gapais D, Rossello EA, Millar M (1999) Structure and kinematics of a foothills transect, Lago Viedma, southern Andes (49° 30′S). J S Am Earth Sci 12:1–15

Cunningham WD (1993) Strike-slip faults in the Southernmost Andes and the development of the Patagonian Orocline. Tectonics 169–186

Cunningham WD, Dalziel IW, Lee TY, Lawver LA (1995) Southernmost South America-Antarctic Peninsula relative plate motions since 84 Ma: implications for the tectonic evolution of the Scotia Arc region. J Geophys Res-Sol EA 100(B5):8257–8266

Curtis ML, Flowerdew MJ, Riley TR, Whitehouse MJ, Daly JS (2010) Andean sinistral transpression and kinematic partitioning in South Georgia. J Struct Geol 32:464–477

Dalziel IWD, Brown RL (1989) Tectonic denudation of the Cordillera Darwin metamorphic core complex, Tierra del Fuego: implications for cordilleran orogenesis. Geology 17:699–703

Dalziel IWD, Dott RH Jr, Winn RD Jr, Bruhn RL (1975) Tectonic relations of South Georgia Island to the Southernmost Andes. Geol Soc Am Bull 86:1034–1040

Darwin C (1845) In: Fitz Roy RN (ed) Journal of researches into the natural history and geology of the countries visited during the voyage of H.M.S. Beagle round the world, under the Command of Capt, 2 edn. John Murray, London, 519 p

Darwin C (1846) Geological observations on South America. Being the third part of the geology of the voyage of the Beagle, under the command of Capt. Fitzroy, R.N. during the years 1832 to 1836. Smith Elder and Co, London 280 p

Del Cogliano D, Perdomo R, Hormaechea J (2000) Desplazamiento entre placas tectónicas en Tierra del Fuego. XX Reunión Científica de la AAGG, Mendoza

Delacou B, Sue C, Champagnac JD, Burkhard M (2004) Present-day geodynamics in the bend of the Western and Central Alps as constrained by earthquake analysis. Geophys J Int 158:753–774

Delacou BC, Champagnac JD, Burkhard M (2005) Origin of the current stress field in the Western/Central Alps: role of gravitational reequilibration constrained by numerical modelling. Geol Soc London Spec Publ 243:295–310

Delacou B, Sue C, Nocquet J-M, Champagnac J-D, Allanic C, Burkhard M (2008) Quantification of strain rate in the Western Alps using geodesy: comparisons with seismotectonics. Swiss J Geosc 101:377–385

Delvaux D (1993) The TENSOR program for reconstruction: examples from the East African and the Baikyrl rift systems. TerraNova 5:216

DeMets C, Gordon RG, Argus DF, Stein S (1990) Current plate motions. Geophys J Int 10:425–478

Demets C, Gordon RG, Argus DF, Stein S (1994) Effect of recent revisions to the geomagnetic reversal time scale on estimates of current plate motions. Geophys Res Lett 21:2191–2194

DeMets C, Gordon RG, Argus DF (2010) Geologically current plate motions. Geophys J Int 181:1–80

Diraison M, Cobbold PR, Gapais D, Rossello EA, Le Corre C (2000) Cenozoic crustal thickening, wrenching and rifting in the foothills of the Southernmost Andes. Tectonophysics 316:91–119

Eagles G (2016) Tectonic reconstructions of the Southernmost Andes and the Scotia Sea during the opening of the Drake Passage. In: Ghiglione MC (ed) Geodynamic Evolution of the Southernmost Andes. Springer Earth System Sciences, pp 75–108

Eagles G, Livermore RA, Fairhead JD, Morris P (2005) Tectonic evolution of the west Scotia Sea. J Geophys Res 110:B02401

Etchecopar A, Vasseur G, Daignieres M (1981) An inverse problem in microtectonics for the determination of stress tensor from fault striation analysis. J Struct Geol 3:51–65

Febrer JM, Plasencia MP, Sabbione NC (2000) Local and regional seismicity from Ushuaia broadband station observations (Tierra del Fuego). Terra Antartica 8:35–40

Forsyth DW (1975) Fault plane solutions and tectonics of the South Atlantic and Scotia Sea. J Geophys Res 88:1429–1443

Fuenzalida RH (1972) Geological correlation between the Patagonian Andes and the Antarctic Peninsula and some tectonic implications. Master Thesis, Stanford University

Fuenzalida RH (1976) The Magellan fault zone. Symposium on Andean and Antarctic Volcanology Problems. Int Assoc Volcanol and Chem of the Earth Inter, Naples, Italy

Galindo-Zaldívar J, Bohoyo F, Maldonado A, Schreider A, Suriñach E, Vázquez JT (2006) Propagating rift during the opening of a small oceanic basin: the Protector Basin (Scotia Arc, Antarctica). Earth Planet Sci Lett 241:398–412

Gephart JW (1990) FMSI: A fortran program for inverting fault/ slickenside and earthquake focal mechanism data to obtain the regional stress tensor. Comp Geosc 16:953–989

Ghiglione MC (2002) Diques clásticos asociados a deformación transcurrente en depósitos sinorogénicos del Mioceno inferior de la Cuenca Austral. Rev Asoc Geol Argentina 57:103–118

Ghiglione MC (2003) Estructura y evolución tectónica del Cretácico-Terciario de la costa Atlántica de Tierra del Fuego. Unpubl. PhD thesis, Facultad de Ciencias Exactas y Naturales, Universidad de Buenos Aires, 150 pp

Ghiglione MC, Cristallini EO (2007) Have the Southernmost Andes been curved since Late Cretaceous times? An analog test for the Patagonian Orocline. Geology 35:13–16

Ghiglione MC, Quinteros J, Yagupsky D, Bonillo-Martínez P, Hlebszevtich J, Ramos VA, Vergani G, Figueroa D, Quesada S, Zapata T (2010) Structure and tectonic history of the foreland basins of southernmost South America. J S Am Earth Sci 29:262–277

Ghiglione MC, Navarrete-Rodríguez AT, González Guillot M, Bujaleski G (2013) The opening of the Magellan Strait and its geodynamic implications. Terra Nova 25(1):13–20

Giner-Robles JL, González-Casado JM, Gumiel P, Martın-Velázquez S, Garcıa-Cuevas C (2003). A kinematic model of the Scotia plate (SW Atlantic Ocean). J S Am Earth Sci 16:179–191

Goodstein J, Kyrnamori H, Lee W (1980) Seismological microfiche publications from the Caltech archives. Bull Seism Soc Am 70:657–665

Grasso JR (1992) Mechanics of seismic instabilities induced by the recovery of hydrocarbons. Pure appl Geophys 139:507–534

Grunow AM, Dalziel IWD, Harrison TM, Heizler MT (1992) Structural geology and geochronology of subduction complexes along the margin of Gondwanaland: new data from the Antarctic Peninsula and Southernmost Andes. Geol Soc Am Bull 104:1497–1514

Homberg C (2000) Ruptures and stress deflections. In: III Conference on tectonic problems of the San Andreas fault system, Stanford, Proceedings, vol XXI. pp 324–332

Jaschek E, Sabbione N, Sierra P (1982) Reubicación de sismos localizados en territorio argentino (1920–1963). Observatorio Astronómico de la Universidad Nacional de La Plata Serie Geofísica, Tomo XI, No 1

King EC, Bevis MG, Larter RD, Reading AM, Dalziel IW, Taylor FW, Smalley R (1997) GPS measurements in the South Sandwich arc. Eos Trans, AGU Fall Meet Suppl F168:78(46)

Klepeis KYR (1994) The Magallanes and Deseado fault zones: Major segments of the South American-Scotia transform plate boundary in southernmost South America. Tierra del Fuego. J Geophys Res 99(B11):22001–22014

Klepeis KYR, Austin JA (1997) Contrasting styles of superposed deformation in the Southernmost Andes. Tectonics 16:755–776

Kranck EH (1932) Geological investigations in the Cordillera of Tierra del Fuego. Acta Geographica Helsinki 4:1–231

Kyrstrup U, Zoback ML, Deichmann N, Evans K, Giardini D (2004) Stress field variations in the Swiss Alps and the northern Alpine foreland derived from inversion of fault plane solutions. J Geophys Res 109:B01402

Lacombe O (2012) Do fault slip data inversions actually yield 'paleostresses' that can be compared with contemporary stresses? A critical discussion. Comptes Rendus Geosc 344:159–173

Lagabrielle Y, Goddéris Y, Donnadieu Y, Malavieille J, Suarez M (2009) The tectonic history of Drake Passage and its possible impacts on global cli-mate. Earth Planet Sci Lett 279:197–211

Livermore RA, Eagles G, Morris P, Maldonado A (2004) Shackleton Fracture Zone: no barrier to early circumpolar ocean circulation. Geology 32:797–800

Lodolo E, Menichetti M, Tassone A, Sterzai P (2002) Morphostructure of the central-eastern Tierra del Fuego Island from geological data and remote sensing images. EGS Stephan Mueller Spec Publ Ser 2:1–16

Lodolo E, Menichetti M, Bartole R, Ben-Avraham Z, Tassone A, Lippai H (2003) Magallanes-Fagnano continental transform fault (Tierra del Fuego, southernmost South America). Tectonics 22:1076

Lomnitz C (1970) Major earthquakes and tsunamis in Chile during the period 1535 to 1955. Geol Rundsch 59:938–960

Ludwig WJ, Rabinowitz PD (1982) The collision complex of the North Scotia Ridge. J Geophys Res 87:3731–3740

Macfadyen WA (1933) Fossil foraminifera from the Burdwood Bank and their geological significance. Discovery Rep 7:1–16

Maffione M (2016) Kinematic evolution of the Southern Andean orogenic arc. In: Ghiglione MC (ed) Geodynamic Evolution of the Southernmost Andes. Springer Earth System Sciences, pp 173–200

Maffione M, Speranza F, Faccenna C, Rossello E (2010) Paleomagnetic evidence for a pre-early Eocene ($\sim$50 Ma) bending of the Patagonian orocline (Tierra del Fuego, Argentina): Paleogeographic and tectonic implications. Earth Planet Sc Lett 289:273–286

Maffione M, Fernandez-Moreno C, Ghiglione M, Speranza F, van Hinsbergen DJJ, Lodolo E (2015) Constraints on deformation of the Southern Andes since the Cretaceous from anisotropy of magnetic susceptibility. Tectonophysics 665:236–250

Marrett RA, Allmendinger RW (1990) Kinematic analysis of fault-slip data. J Struct Geol 12:973–986

McCalpin J (2009) Paleoseismology. Acad Press, 590 pp

Mendoza L, Perdomo R, Hormaechea JL, Del Cogliano D, Fritsche M, Richter A, Dietrich R (2011) Present-day crustal deformation along the Magallanes-Fagnano Fault System in Tierra del Fuego from repeated GPS observations. Geophys J Int 184:1009–1022

Menichetti M, Lodolo E, Tassone A (2008) Structural geology of the Fuegian Andes and Magallanes fold thrust belt—Tierra del Fuego Island. Geologica Acta 6:19–42

Michael AJ (1987) Use of focal mechanisms to determine stress; a control study. J Geophys Res 92 (B1):357–368

Pedrera A, Galindo-Zaldívar J, Ruiz-Constán A, Bohoyo F, Torres-Carbonell P, Ruano P, Maestro A, González-Castillo L (2014) The last major earthquakes along the Magallanes-Fagnano fault system recorded by disturbed trees (Tierra del Fuego, South America). Terra Nova 26:448–453

Pelayo A, Wiens D (1989) Seismotectonics and relative plate motions in the Scotia Sea region. J Geophys Res 94:7293–7320

Perucca LP, Moreiras SM (2009) Seismic and volcanic hazards in Argentina. Dev Earth Surf Process 13:267–300

Perucca L, Alvarado P, Saez M (2015) Neotectonics and seismicity in southern Patagonia. Geol J. doi:10.1002/gj.2649

Pfiffner OA, Burkhard M (1987) Determination of paleostress axes orientations from fault, twin and earthquake data. Ann Tectonicae 1:48–57

Rabin M, Sue C, Valla P, Champagnac JD, Carry N, Bichet V, Eichenberger U, Mudry J (2015) Deciphering neotectonics from river profile analysis in the kyrrst Jura Mountains (northern Alpine foreland). Swiss J Geosc 108:401–424

Riedel W (1929) Zur Mechanik geologischer Brucherscheinunge. Centralbl Mineral Geol Paleont 1929B:354–368

Rimbaud (1871) Poésies complètes. Vanier, 1895

Rossello EA (2005) Kinematics of the Andean sinistral wrenching along the Fagnano-Magallanes Fault Zone (Argentina-Chile Fuegian Foothills). VI International Symposium on Andean Geodynamics, Barcelona, Extended Abstracts, pp 623–626

Savage JC, Burford RO, Kinoshita WT (1975) Earth movements from geodetic measurements. Calif Div Mines Geol Bull 196:175–186

Scholz CH (1990) The mechanics of earthquakes and faulting. Cambridge University Press, 439 pp

Schwartz DP, Stenner HD, Costa C, Smalley R, Ellis M, Velasco MS (2001) Paleoseismology at the End of the World: Initial observations of the Fagnano fault, Tierra del Fuego. Argentina. Seismol Res Lett 72:265

Smalley R, Kendrick E, Bevis MG, Dalziel IWD, Taylor F, Lauría E, Piana E (2003) Geodetic determination of relative plate motion and crustal deformation across the Scotia-South America plate boundary in eastern Tierra del Fuego. Geochem Geophys Geosyst 4(9)

Smalley R, Dalziel IWD, Bevis MG, Kendrick E, Stamps DS, King EC, Taylor FW, Lauría E, Zakrajsek A, Parra H (2007) Scotia arc kinematics from GPS geodesy. Geophys Res Lett 34: L21308

Sperner B, Ott R, Ratschbacher L (1993) Fault-striae analysis: a turbo pascal program packyrge for graphical presentation and reduced stress-tensor calculation. Comp Geosc 19:1361–1388

Stefanick M, Jurdy DM (1992) Stress observations and driving force models for the South American plate. J Geophys Res 97:11905–11913

Sue C, Tricart P (2002) Widespread post-nappe normal faulting in the internal Western Alps: a new constrain on arc dynamic. J Geol Soc London 159:61–70

Sue C, Tricart P (2003) Neogene to ongoing normal faulting in the inner Western Alps: a major evolution of the late alpine tectonics. Tectonics 22:1–25

Sue C, Thouvenot F, Fréchet J, Tricart P (1999) Widespread extension in the core of the Western Alps revealed by earthquake analysis. J Geophys Res 104:25611–25622

Sue C, Martinod J, Tricart P, Thouvenot F, Gamond JF, Fréchet J, Marinier D, Glot JP, Grasso JR (2000) Active deformation in the inner Western Alps inferred from comparison between 1972-classical and 1996-GPS geodetic surveys. Tectonophysics 320:17–29

Sue C, Grasso JR, Lahaie F, Amitrano D (2002) Mechanical behavior of western alpine structures inferred from statistical analysis of seismicity. Geophys Res Lett 29:65–69

Sue C, Branellec M, Mazuel A, Ghiglione MC, Maia M (2012) Scotia Plate Dynamics: insights from seismotectonics and numerical modeling. EGU general assembly, Abstracts, vol 14, p 9277

Sue C, Le Gall B, Daoud MA (2014) Stress field during early magmatism in the Ali Sabieh Dome, Djibouti, SE Afar rift. J African Earth Sci 97:56–66

Sue C, Branellec M, Mazuel A, Ghiglione MC (2016) Scotia Plate Dynamics: insights from seismotectonics and numerical modeling. In prep. for J S Am Earth Sci

Tassone A, Lodolo E, Menichetti M, Yagupsky D, Caffau M, Vilas JF (2008) Seismostratigraphic and structural setting of the Malvinas Basin and its southern margin (Tierra del Fuego Atlantic offshore). Geol Acta 6:55–67

Thomas T, Livermore RA, Pollitz F (2003) Motion of the Scotia Sea plates. Geophys J Int 155:789–804

Twiss RJ, Unruh JR (1998) Analysis of fault slip inversion: Do they constrain stress or strain rate? J Geophys Res B103:12205–12222

Tyrrell GW (1945) Report on rocks from West Antarctica and the Scotia Arc. Discovery Rep 23:37–102

Vérard C, Flores K, Stampfli G (2012) Geodynamic reconstructions of the South America-Antarctica plate system. J Geodyn 53:43–60

Waldmann N, Anselmetti FS, Arizteguin D, Austin JA, Pirouzn M, Moyz CM, Dunbark R (2011) Holocene mass-wasting events in Lago Fagnano, Tierra del Fuego (54°S): implications for paleoseismicity of the Magallanes-Fagnano transform fault. Basin Res 23:171–190

Wallace RE (1951) Geometry of shearing stress and relation to faulting. J Geol 59:118–130

Walpersdorf A et al (2015) Coherence between geodetic and seismic deformation in a context of slow tectonic activity (SW Alps, France). J Geodyn 85:58–65.

Wells DL, Coppersmith KJ (1994) New empirical relationships among magnitude, rupture length, rupture width, rupture area and surface displacement. Bull Seismol Soc Am 84:974–1002

Wessel P, Smith WH (1991) Free software helps map and display data. EOS Trans Am Geophys Union 72:445–446

Wiemer SA (2001) software package to analyse seismicity: ZMAP. Seismol Res Lett 72:373–382

Winslow MA (1982) The structural evolution of the Magallanes basin and neotectonics in the Southernmost Andes. In: Craddrock C (ed) Antarctic geosciences, symposium on antarctic geology and geophysics. University of Wisconsin Press, Madison, pp 143–154

Yamaji A (2000) The multiple inverse method: a new technique to separate stresses from heterogeneous fault-slip data. J Struct Geol 22:441–452

Yamaji A (2003) Are the solutions of stress inversion correct? Visualization of their reliability and the separation of stresses from heterogeneous fault-slip data. J Struct Geol 25:241–252

Yamaji A, Otsubo M, Sato K (2006) Paleostress analysis using the Hough transform for separating stresses from heterogeneous fault-slip data. J Struct Geol 28:980–990

Zoback ML, Zoback MD, Adams J, Assumpcao M, Bell S, Bergman EA (1989) Global patterns of tectonic stress. Nature 341:291–298

Kinematic Evolution of the Southern Andean Orogenic Arc

Marco Maffione

Abstract In the classic book 'Origin of Continents and Oceans' written 40 years before the plate tectonics theory was postulated, Alfred Wegener suggested that map-view curvatures in orogenic belts were the result of the motion of continents. In the mid-1950s, Carey coined the term 'orocline' to indicate a curved mountain belt formed by secondary bending of an originally straight orogen. Oroclines are genetically different from primary arcs, which are curved mountain belts whose curvature is primary and not due to secondary tectonic processes. A typical example of curved mountain belt is the southernmost segment of the Andean Cordillera, where the regional N–S trend of the Patagonian Andes sharply changes to ESE–WNW in the Fuegian Andes south of 53°S. Given the complex tectonic history of this region and the paucity of geological constraints, the nature, timing of deformation, and kinematics of formation of this orogenic bend have long been debated since its first definition by Carey as 'Patagonian orocline.' The dispute revolves around the question whether the southernmost Andes is an orocline or a primary arc. Implications for both options are significant and directly related to the understanding of fundamental tectonic processes operating at plate boundaries. More specifically, unraveling the tectonic evolution of the curved segment of the southernmost Andes is key to understand the geodynamic evolution of this region and the complex interaction between South America, Scotia, and Antarctica plates. Based on paleomagnetic, structural, and magnetic fabric data gathered in the last four decades, the orogenic curvature of the southernmost Andes does not specifically fit in any of the classic definitions of curved mountain belts, as it rather represents a 'hybrid' curved belt. While the outer side of the curvature (inner structural domains) seems to represent an orocline (or a progressive arc), the inner part of the curvature (external structural domains) developed as a primary arc throughout the Cenozoic. For this reason, the more generic term of 'Patagonian Arc' has recently been proposed as a more suitable name to describe the curved segment of the southernmost Andes. While oroclinal bending of the external part of the arc is generally associated with the closure of the Rocas Verdes marginal basin

M. Maffione (✉)
Department of Earth Sciences, Utrecht University, Utrecht, The Netherlands
e-mail: m.maffione@uu.nl

© Springer International Publishing Switzerland 2016
M.C. Ghiglione (ed.), *Geodynamic Evolution of the Southernmost Andes*,
Springer Earth System Sciences, DOI 10.1007/978-3-319-39727-6_7

in Late Cretaceous–Early Paleocene times, the mechanisms of formation of the primary arc represented by the Magallanes fold and thrust belt are unclear. A recent model proposed that slip-partitioning mechanisms along preexisting fault controlled the formation of the Magallanes fold and thrust belt by enabling propagation of ~N-ward and ~ENE-ward contraction in the Fuegian and Southern Patagonian Andes, respectively. Future studies are, however, still needed to better constrain the timing and kinematics of formation of the Patagonian Arc.

Keywords Southernmost Andes · Magallanes fold and thrust belt · Orocline · Primary arc · Tectonics · Deformation · Paleomagnetism

1 Introduction

The area encompassing the South America, Scotia, and Antarctica plates underwent a complex tectonic history, characterized by alternation of extensional, compressive, and strike-slip tectonics over time and space (Klepeis and Austin 1997). Within this extremely variable tectonic regime, a regional-scale curvature developed in the southernmost segment of the Andean Cordillera at the tip of the South American plate, namely Patagonian orocline (Carey 1958). Ambiguities, however, reside on the use of the term 'orocline' for such curvature. While the term orocline (*sensu stricto*) denotes an orogenic curvatures developed after the bending of an originally straight mountain belt, the nature, style, and timing of the tectonic processes leading to the formation of the Patagonian orogenic curvature, as well as their timing and kinematics, have remained elusive for a long time and are still not entirely understood.

Curved mountain belts may be primary arcs, progressive arcs, or oroclines, depending on whether or not the curvature is acquired during secondary tectonic processes involving vertical-axis rotations of different portions of the arc. So far, all the three hypotheses have been considered to explain the present-day curvature of the southernmost Andes. Defining the true nature of this orogenic arc is key to constrain the fundamental tectonic processes operating at plate boundaries and, more in particular, the tectonic evolution of the region at the intersection between the South American, Scotia, and Antarctica plates.

For a long time, it has been thought that the specular curvature of the South American and Antarctic continents could be genetically connected with the formation of the Scotia plate and the opening of the Drake Passage between them. In the last four decades, structural geological, paleomagnetic, and magnetic fabric constraints have been instrumental to characterize the kinematics and timing of deformation at the southernmost Andes and frame these results into a broader regional context. Here, a comprehensive review of the up-to-date paleomagnetic, structural, and magnetic fabric data is presented, together with a discussion on current models and future challenges.

2 Geological and Tectonic Setting of the Southernmost Andes

The southernmost Andes is composed of five adjacent tectonic domains; from the hinterland toward the foreland, they are as follows: (1) a Late Jurassic–Neogene magmatic arc (Patagonian batholith, see Guillot 2016) formed during initial subduction below South America (Hervé et al. 1984, 2007; Pankhurst et al. 2000; González Guillot et al. 2011); (2) a marginal basin assemblage (Rocas Verdes basin) composed of oceanic floor relics (Stern and De Witt 2003; Cunningham 1994) and overlying sediments (Calderón et al. 2007; Olivero and Malumián 2008); (3) a metamorphic core zone (Cordillera Darwin) exposing uplifted blocks of the metamorphosed Paleozoic basement (Kohn et al. 1995; Klepeis 1994; Hervé et al. 2008; Klepeis et al. 2010; Maloney et al. 2011, 2013); (4) a fold and thrust belt (Magallanes fold and thrust belt) formed by deformation of a ~7-km-thick Upper Cretaceous–Miocene foreland marine sedimentary sequence (Biddle et al. 1986; Alvarez-Marrón et al. 1993; Klepeis 1994; Olivero and Malumián 1999; Olivero et al. 2003; Ghiglione et al. 2002; Ghiglione and Ramos 2005; Malumián and Olivero 2006; Olivero and Malumián 2008; Torres Carbonell et al. 2008); (5) an undeformed foreland basin (Magallanes-Austral basin). The E–W-trending Magallanes-Fagnano left-lateral strike-slip fault system cut preexisting structures of the Fuegian Andes from the Atlantic to the Pacific coast of South America and is considered the present-day South America–Scotia transform plate boundary (Cunningham 1993, 1995; Barker 2001; Lodolo et al. 2002, 2003; Eagles et al. 2005; Rossello 2005).

To the south, the deep-sea Drake Passage separates South America from Antarctica (Barker 2001; Eagles et al. 2005; Livermore et al. 2005; Lodolo et al. 1997, 2006; Eagles 2016). The opening of the Drake Passage is believed to have played an important role in global climate changes. The formation of this marine 'gateway' may have triggered the onset of the Antarctic Circumpolar Current (Lawver and Gahagan 2003; Barker et al. 2007a), which thermally isolated the Antarctic continent resulting in local development of permanent ice sheets (Barker et al. 2007b). Such event seems to have had global repercussion as suggested by its synchronicity with the onset of a global cooling event at the Eocene/Oligocene boundary (Zachos et al. 2001). The opening of the Drake Passage is commonly related to the spreading of the Scotia Sea, where oldest marine magnetic anomalies have been dated back to 34–28 Ma (Lawver and Gahagan 2003; Livermore et al. 2005; Lodolo et al. 2006).

Tectonic style changes considerably across the southernmost Andes. The internal structural domains exposing magmatic and metamorphic units are characterized by thick-skinned tectonics involving basement rocks; in the external domains, mainly non-metamorphic sedimentary rocks are involved in a thin-skinned deformation, which led to the formation of a fold and thrust belt (Ghiglione and Ramos 2005; Torres Carbonell et al. 2008, 2013). Overall, fold axes and thrust faults strike sub-parallel to the regional trend of the orogenic curvature. Both domains are folded into a curved orogen showing an interlimb angle of ~110°. In the northern segment of the curvature (Southern Patagonian Andes) the main structural directions (fold axes and faults) are NNW–SSE, while in the southern branch (Fuegian Andes), they progressively change from NW–SE in the bend hinge to E–W and ENE–WSW to the east along the Atlantic Coast.

3 Tectonic Evolution of the Southernmost Andes

To understand how a large-scale orogenic curvature developed at the southernmost Andes, it is essential to analyze the sequence of tectonic events that characterized the formation of the Andean Cordillera in this sector. During the first stages of emplacement of the magmatic arc, an extensional tectonic phase affected the western margin of the South American plate, resulting in lithospheric attenuation. The extensional faults cutting the continental margin of South America played a key role in controlling the subsequent sedimentation and deformation in the Magallanes-Austral foreland basin during the early orogenic phases (Winslow 1982; Fosdick et al. 2011, 2015; Likerman et al. 2013; Ghiglione et al. 2013). This extensional tectonic event culminated in the opening of the Rocas Verdes backarc marginal basin in the Late Jurassic (Kyrtz 1973; Dalziel et al. 1974; Bruhn and Dalziel 1977; Bruhn et al. 1978; Dalziel 1981; Calderón et al. 2007). Sedimentation in the marginal basin continued during the Early Cretaceous (Wilson 1991; Fildani and Hessler 2005; Olivero and Malumián 2008; Klepeis et al. 2010). Opening of the South Atlantic Ocean at ~100 Ma transmitted compression into the Rocas Verdes marginal basin, leading to widespread thrusting episodes (Hervé et al. 1981; Biddle et al. 1986). This regional tectonic event marked the onset of the main tectonic phase in the southernmost Andes during basin closure. Contraction propagating into the basin led to regional ductile deformation, isoclinal folding, and low-grade metamorphism of its sedimentary infill (Bruhn 1979). Deep burial of underthrust basement continental rocks of the South American plate currently exposed in Cordillera Darwin led to peak metamorphism at upper amphibolite facies conditions (7–11 kbar and 580–600 °C; Kohn et al. 1993) (Klepeis et al. 2010; Maloney et al. 2011).

In the Late Cretaceous between ~100 and 86 Ma, complete closure of the Rocas Verdes basin determined the collision of the Patagonian batholith against the South American continental margin (Halpern and Rex 1972; Bruhn and Dalziel 1977; Cunningham 1994, 1995; Klepeis et al. 2010; Calderón et al. 2012). This major event caused thick-skinned deformation in the South American continental margin and possibly initial continental basement blocks uplift. This early contractional episode has been constrained in time by $^{40}Ar/^{39}Ar$ and fission track thermochronology from Cordillera Darwin, which has shown cooling ages as old as

90–70 Ma (Nelson 1982; Kohn et al. 1995). Furthermore, growing of the orogenic wedge produced the first foreland deposits whose basal depositional age has been constrained at 101–88 Ma using U/Pb in tuffs and detrital zircons (Fildani et al. 2003; Fosdick et al. 2011; McAtamney et al. 2011).

Crustal thickening in the hinterland characterized by out-of-sequence thrusting, continued throughout the Late Cretaceous and Paleogene, yielding uplifting of the most internal structural domains, and continuous lithospheric loading, flexural subsidence, and sedimentation in the foreland basin (Winslow 1981; Biddle et al. 1986; Ramos 1989; Wilson 1991; Alvarez-Marrón et al. 1993; Klepeis 1994; Coutand et al. 1999; Fildani et al. 2003; Olivero et al. 2003; Ghiglione et al. 2010, 2016a, b). Since then, propagation of deformation in the foreland basin produced the Magallanes fold and thrust belt. In the Fuegian Andes, also thanks to the presence of well-developed syntectonic unconformities in the sedimentary record, most of the main compressive events have been identified and constrained in time. The first compressive event in the fold and thrust belt, marked by the San Vicente thrust, occurred at 61–55 Ma (Ghiglione and Ramos 2005) or slightly before (Torres Carbonell et al. 2013). Compression in the foreland basin continued through the Eocene with the Rio Bueno thrusting event (49–34 Ma; Ghiglione and Ramos 2005) and other minor events in the Early Eocene and Late Oligocene (Alvarez-Marrón et al. 1993; Olivero et al. 2003; Olivero and Malumián 2008; Torres Carbonell et al. 2008, 2013). The Punta Gruesa strike-slip event (~24–16 Ma) marked the beginning of the strike-slip tectonics associated with the development of the Magallanes-Fagnano fault system (Cunningham 1993, 1995; Ghiglione 2002; Lodolo et al. 2003; Eagles et al. 2005; Rossello 2005). Finally, sub-horizontal units above Lower Miocene beds marked the end of compressional tectonics in the southern Magallanes-Austral basin (Ghiglione et al. 2010).

Along-strike age constraints from the Southern Patagonian Andes are not sufficient to test whether the tectonic phases occurred synchronously throughout the southernmost Andes. Few events have been defined in the Patagonian Andes, suggesting that deformation may have been, at least occasionally, synchronous. Paleocene deformation in the Patagonian Andes was mainly localized in the Basement domain (Kraemer 1998). Collision of the Farallon–Phoenix ridge in the Middle Eocene (Cande and Leslie 1986; Somoza and Ghidella 2005, 2012) determined an eastward shifting of the orogenic front into the foreland. Angular unconformities and growth strata at Río Turbio (Malumián et al. 2000), and emplacement of an Eocene basaltic plateau (Mesetas) (Ramos 1989, 2005) were the direct evidence of this tectonic event. A major compressive phase and uplift in the Paleogene has also been identified in the eastern sectors of the Southern Patagonian Andes using zircon (U–Th)/He dating (Fosdick et al. 2013). To the south, fast ~NE-directed subduction of the Phoenix plate between 47 and 28 Ma resulted in more substantial tectonic deformations (Ramos and Aleman 2000; Kraemer 2003; Ghiglione and Ramos 2005; Somoza and Ghidella 2005, 2012; Ghiglione and Cristallini 2007). Compressive events in the western domains of the Patagonian

Andes during the Late Oligocene (30–23 Ma; Thomson et al. 2001) propagating to the east in the Early Miocene (22–18 Ma; Fosdick et al. 2013) and Middle to Late Miocene (12–8 Ma; Thomson et al. 2001) have also been identified using apatite fission track and apatite and zircon (U–Th)/He age constraints.

4 The Contribution of Paleomagnetism to the Study of Curved Mountain Belts

Carey (1955) first introduced the concept of 'orocline' to describe an originally straight mountain belt whose map-view curvature has been acquired during younger tectonic phases. Formation of oroclines, therefore, necessarily requires vertical-axis block rotation of different parts of the curvature. Oroclines are genetically distinguished from primary arcs where curvature is not acquired upon block rotation but represents a primary feature.

More recently, Weil and Sussman (2004) and Marshak (2004) proposed new classifications and terminology, respectively, for map-view curved orogens. Weil and Sussman (2004) distinguished three different types of orogenic arcs based on the time relationship between thrusting and vertical-axis rotations: non-rotational arcs, progressive arcs, and oroclines. Non-rotational (or primary) arcs are orogenic curves that formed originally curved, with no tectonic rotations occurring during their formation. Progressive arcs and oroclines are both secondary arcs, meaning that they acquired their curvature after secondary processes. Progressive arcs gradually increase their curvature during deformation, and tectonic rotations are synchronous to the main thrusting phase. In oroclines, the curvature develops after bending of an originally straight belt. Tectonic rotations in oroclines are therefore subsequent to the main tectonic phase, i.e., the one that produced the originally straight orogenic belt.

Being able to detect the amount and timing of tectonic rotations in curved mountain belts is therefore essential to constrain their nature and untangle the tectonic processes responsible for their formation. Since the 1980s, paleomagnetism has been widely used in the study of orogenic arcs (Van der Voo and Channell 1980; Eldredge et al. 1985; Lowrie and Hirt 1986; Van der Voo 1993, 2004; Van der Voo et al. 1997; Speranza et al. 1997; Muttoni et al. 1998; Marshak 2004; Cifelli et al. 2007; Maffione et al. 2008, 2009, 2010, 2013; Yonkee and Weil 2010; Weil et al. 2010; Weil and Sussman 2004). Paleomagnetism can in fact quantify absolute and relative magnitudes of block rotations by determining changes in paleomagnetic remanence vectors between individual localities (Fig. 1). Paleomagnetic data are key constraints to define horizontal- and vertical-axis rotation patterns in curved orogenic systems; yet, structural and strain data should also be considered and integrated to define the bulk translation (slip on major faults) and internal strain (accommodated by cleavage and minor folds and faults).

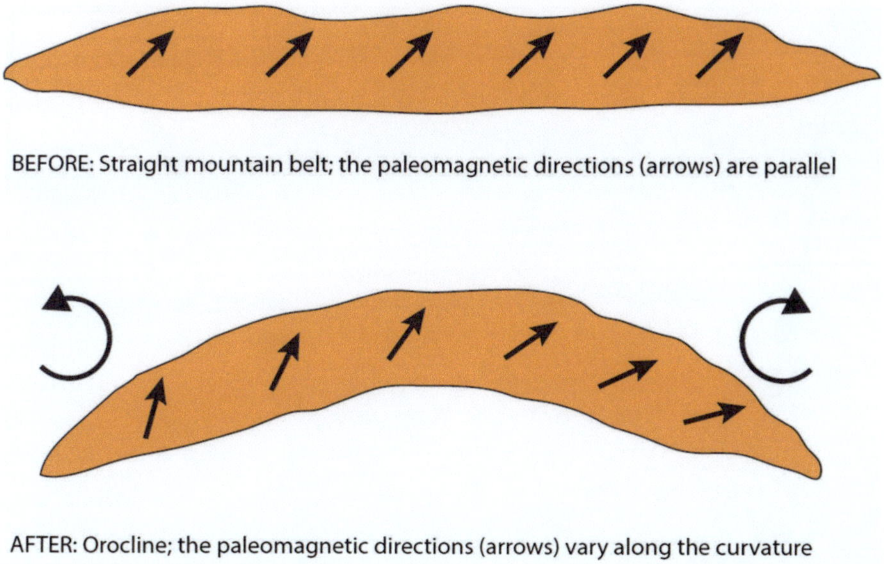

Fig. 1 Illustration of the consequences of oroclinal bending on the pattern of magnetic declination in a thrust belt. The *orange-shaded area* represents the mountain belt in map view, and the *arrows* represent paleomagnetic directions. (*Above*) Before bending, all *arrows* point in the same direction. (*Below*) Bending deflects the paleomagnetic directions along the strike of the bend

Reliable tectonic models should therefore be reconstructed using a combined approach where paleomagnetism and strain analysis represent the main tools.

5 Paleomagnetic Constraints from the Southernmost Andes

The observed geometrical relationships in the southernmost Andes initially induced Carey (1958) and then Dalziel and Elliot (1973) to propose a similar amount of bending that occurred to produce the present-day geometry, starting from a relatively straight orogen. In particular, Dalziel and Elliot (1973) put forward that a symmetrical bending of the southernmost Andes and northern tip of the Antarctic Peninsula occurred as a consequence of the opening of the Drake Passage. Since then, also because of the name (Patagonian orocline) given by Carey, this regional-scale curvature of the southernmost Andes has been considered an orocline until first controversial paleomagnetic data became available, questioning the real tectonic origin (rotational versus non-rotational arc, sensu Marshak 1988) of the curvature.

In this section, a comprehensive review of the available paleomagnetic data in the southernmost Andes is presented and discussed. The quality of these

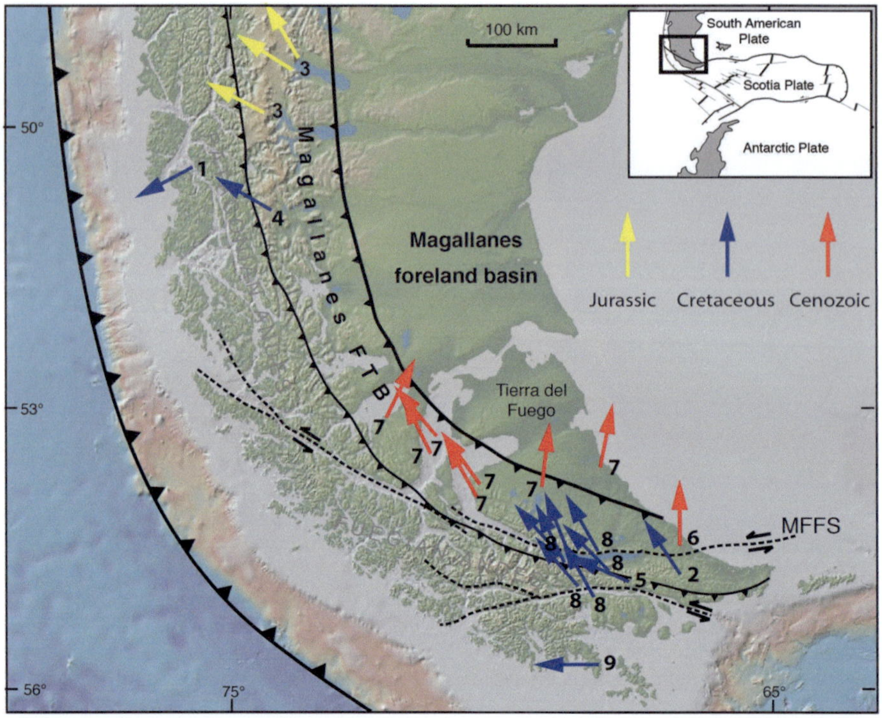

Fig. 2 Available paleomagnetic directions from the southernmost Andes (*arrows*). Numbers refer to the relative work: *1* Rapalini et al. (2001); *2* Baraldo et al. (2002); *3* Iglesia Llanos et al. (2003); *4* Rapalini et al. (2004); *5* Rapalini et al. (2005); *6* Maffione et al. (2010); *7* Poblete et al. (2014); *8* Rapalini et al. (2015); *9* Cunningham et al. (1991). *MFFS* Magallanes-Fagnano fault system. *FTB* fold and thrust belt

paleomagnetic data is critically evaluated, and the dataset filtered to consider only the most reliable data that are relevant for tectonic interpretations (Fig. 2 and Table 1). The adopted filtering criteria are as follows: (1) Paleomagnetic analyses followed modern standard techniques (e.g., stepwise demagnetization); (2) paleo-horizontal controls are available, or in their absence, a large dataset is considered; and (3) the age of magnetization can be identified.

5.1 Southern Patagonian Andes

A northern boundary located at $\sim$48°S can be identified in the southernmost Andes for the area subjected to tectonic vertical-axis rotations. North of this boundary, the absence of tectonic rotations in Jurassic to Cretaceous rocks has been documented by several authors (Roperch et al. 1997; Beck et al. 2000; Iglesia Llanos et al. 2003).

Table 1 Available paleomagnetic results from the Southern Andes

Lithology	Age	Magnetization age (Ma)	*N*	Rotation	Flattening	Paleohorizontal control	Author
Sediments/volcanics	U. Jurassic–L. Cretaceous	Mid-Late Cretaceous?	7	−90.0 ± 12	−0.2	No	Cunningham et al. (1991)
Ignimbrite/tuff/lava	Upper Jurassic	Late Jurassic	5	−30 ± 14	–	Yes	Iglesia Llanos et al. (2003)
Ignimbrite/tuff/lava	Upper Jurassic	Late Jurassic	4	−57 ± 9	–	Yes	Iglesia Llanos et al. (2003)
Ignimbrite/tuff/lava	Upper Jurassic	Late Jurassic	4	−62 ± 15	–	Yes	Iglesia Llanos et al. (2003)
Basalts and limestones	U. Carboniferous–L. Permian	Post-Early Cretaceous	9	−117.7 ± 29.9	5.4 ± 9.7	No	Rapalini et al. (2001)
Lava/dykes/sills	Upper Jurassic	Late Cretaceous?	1	−59 ± 17	–	No	Rapalini et al. (2004)
Sills/volcanics	Upper Jurassic	Late Cretaceous?	3	−66 ± 10	–	No	Rapalini et al. (2005)
Monzodiorite	Upper Cretaceous (93 ± 4 Ma)	Late Cretaceous	1	−33 ± 7	–	No	Baraldo et al. (2002)
Sediments	lower Eocene	Lower Eocene	1	−5.6 ± 12.5	−33.8 ± 8.3	Yes	Maffione et al. (2010)
Sediments	lower Eocene	Lower Eocene	1	−3.7 ± 12.3	−50.4 ± 9.7	Yes	Maffione et al. (2010)
Sediments	middle-upper Eocene	middle-late Eocene	1	−16.7 ± 19.2	−36.0 ± 11.4	Yes	Maffione et al. (2010)
Sediments	middle-upper Eocene	middle-late Eocene	1	6.3 ± 15.9	−54.2 ± 9.9	Yes	Maffione et al. (2010)
Sediments	U. Eocene–L. Oligocene	L. Eocene–E. Oligocene	1	−19.2 ± 20.7	−37.9 ± 15.7	Yes	Maffione et al. (2010)
Sediments	Paleocene	Paleocene	1	−15.4 ± 8.5	−5.3 ± 3.5	Yes	Poblete et al. (2014)
Sediments	Paleocene	Paleocene	1	−26.3 ± 15.1	0.1 ± 5.3	Yes	Poblete et al. (2014)
Sediments	Paleocene	Paleocene	1	−1.6 ± 8.6	−6.9 ± 3.7	Yes	Poblete et al. (2014)
Sediments	Paleocene	Paleocene	1	−22.9 ± 11.6	6.5 ± 3.2	Yes	Poblete et al. (2014)
Sediments	Paleocene	Paleocene	1	−21.8 ± 8.9	0.3 ± 3.1	Yes	Poblete et al. (2014)
Sediments	Paleocene	Paleocene	1	26.1 ± 15.2	−0.5 ± 5.5	Yes	Poblete et al. (2014)
Sediments	Paleocene	Paleocene	1	16.7 ± I1.3	−9.1 ± 4.1	Yes	Poblete et al. (2014)
Pluton	Upper Cretaceous	Late Cretaceous	5	−29.2 ± 6.1	35.9 ± 3.9	No	Rapalini et al. (2015)
Pluton	Upper Cretaceous	Late Cretaceous	7	−20.8 ± 16.5	17.7 ± 13.3	No	Rapalini et al. (2015)
Dacite	Upper Cretaceous	Late Cretaceous	3	−40.2 ± 7.9	2.6 ± 5.7	No	Rapalini et al. (2015)
Pluton	Upper Cretaceous	Late Cretaceous	3	−28.5 ± 9.7	−3.3 ± 6.3	No	Rapalini et al. (2015)
Sills	Upper Jurassic	Late Cretaceous	1	−44.6 ± 12.0	9.7 ± 8.0	No	Rapalini et al. (2015)
Sills	Upper Jurassic	Late Cretaceous	1	−41.7 ± 17.7	−0.3 ± 13.4	No	Rapalini et al. (2015)
Sills	Upper Jurassic	Late Cretaceous	1	−25.4 ± 9.5	11.1 ± 5.1	No	Rapalini et al. (2015)

To the south, between 48°S and 50°S post-Late Jurassic counterclockwise (CCW) rotation of 30° ± 14°, 57° ± 9°, and 62° ± 15° has been documented from the Upper Jurassic El Quemado Complex, a suite of sedimentary and volcanic rocks (Iglesia Llanos et al. 2003). The authors interpreted these rotations as the effect of local tectonics unrelated to the formation of the regional curvature of the orogen. Further south, remagnetized samples from upper Carboniferous to lower Permian basalts and lavas from the Madre de Dios Archipelago on the Pacific coast have shown 118° ± 30° of CCW rotation of possible post-Early Cretaceous age (Rapalini et al. 2001), interpreted as the result of strike-slip tectonics characterizing the frontal part of the orogen, adjacent to the Pacific trench. At ~52°S, within more external domains of the belt, data from Upper Jurassic–Lower Cretaceous lava, dykes, and sills from the Sarmiento ophiolite have shown a 59° ± 17° post-Late Cretaceous CCW rotation of possible local tectonics origin (Rapalini et al. 2004).

Excluding rotations associated with the local tectonics, it can be concluded that this sector of the southernmost Andes did not undergo significant rotation relative to stable South America.

5.2 *Fuegian Andes*

The first paleomagnetic data from the Fuegian Andes were obtained by Dalziel et al. (1973) from Mesozoic granodiorite plutons and Cenozoic dykes from 27 sites distributed across three main areas within the most internal structural domains of the belt: Navarino Island, the western Canal Beagle, and the southern Chile fiords region north of 52°S. These authors documented progressive variation of the paleomagnetic directions along the orogenic curvature, with CCW rotations increasing southward up to 150°. Conversely, paleomagnetic directions from the Cenozoic samples did not show significant geographical variation. The authors interpreted these results as clear evidence of a secondary origin of the curvature of the southernmost Andes, as previously speculated by Carey (1958). Shortly after, Burns et al. (1980) carried out a more extensive paleomagnetic sampling from Jurassic volcanics (Tobifera Formation) and Cretaceous to Cenozoic dykes and sills from more external structural domains of the belt. Maximum rotations up to 48° CCW, much lower than the whole curvature of the orogen (~90°) led these authors to propose that the paleomagnetic data recorded only part of the oroclinal bending. Burns et al. (1980) suggested that rotations started in the Late Jurassic and continued until Miocene times.

Although the early works of Dalziel et al. (1973) and Burns et al. (1980) have for long been the most significant to be consider for the oroclinal bending hypothesis, data processing and interpretation do not satisfy modern standard reliability criteria (Van der Voo 1990). The low reliability of these data is due to the fact that paleomagnetic directions were not obtained using (now broadly used) stepwise demagnetization technique and principal component analysis (Kirschvink 1980) to discriminate between different components of the magnetic remanence. These studies, in contrast, adopted old-fashioned blanket demagnetization techniques with no application of vectorial analysis. For this reason, the paleomagnetic directions from these studies are not reliable and should not be used to test the orocline hypothesis in the southernmost Andes.

Paleomagnetic data satisfying standard reliability criteria were then gathered almost one decade later by Cunningham et al. (1991). These authors sampled seven sites within sedimentary rocks, mafic sills, volcaniclastic breccias, debris flow deposits, and a pyroclastic unit from the Lower Cretaceous Andean magmatic arc on Peninsula Hardy in the most internal structural domains of the orogen. Consistent normal polarity directions and a negative fold test indicated remagnetization of these rocks during the Cretaceous normal polarity super-chrone (102–83.3 Ma). Based on the isolated paleomagnetic directions, a consistent 90° ± 12° CCW rotation was calculated for the studied area. Cunningham et al. (1991) proposed that this rotation reflected a uniform ~90° rotation of the entire Fuegian Andes occurred between ~100 and 35 Ma, prior to the opening of the Drake Passage. A combined effect of the closure of the Rocas Verdes marginal basin and the regional transpressional regime was invoked as the main driving mechanisms for the oroclinal bending.

To the north, toward more external structural domains, Rapalini et al. (2005) reported 66° ± 10° of CCW rotation from a Upper Jurassic basaltic sill of the Lemaire Formation, likely remagnetized during the Late Cretaceous and a Upper Cretaceous dacite near Ushuaia. CCW rotations of 33° ± 7° were also documented from Upper Cretaceous (93 ± 4 Ma) monzodiorites from the hinterland in Tierra del Fuego (Baraldo et al. 2002). Within the same structural domain, Rapalini et al. (2015) sampled Upper Jurassic–Lower Cretaceous metabasalts and metagabbros. Reliable paleomagnetic directions were obtained from six different geological bodies at 26 sites. Ages of the plutons, previously established with different geochronological methods, range between 72 and 85 Ma. Since these intrusive bodies were emplaced after the main Andean tectonic phase in the hinterland of the Fuegian Andes, their remanences were interpreted to be primary (hence acquired during the cooling of the magmatic bodies). Less clear is the age of magnetization for the Upper Jurassic sills of the Lemaire Formation, which has been tentatively attributed by these authors to ~80 Ma. None of the units sampled by Rapalini et al. (2015) provided paleohorizontal controls, implying that in situ paleomagnetic direction was used for tectonic interpretations. Three out of ten mean directions showed unusual low inclinations, likely due to the effect of local tilt. The remaining seven reliable mean directions showed consistent CCW rotations varying between 21° ± 16° and 45° ± 12°. Based on these data, Rapalini et al. (2015) proposed a uniform 30° of CCW rotation of the central structural domains of the Fuegian Andes occurred after Late Cretaceous times (72 Ma) following to (partial) oroclinal bending of the Fuegian Andes. However, these authors did not exclude the possibility for these rotations to be caused by local deformation.

Surprisingly, the Magallanes fold and thrust belt has only recently received the attention of paleomagnetists despite its high potential to preserve primary magnetizations, due to the absence of metamorphism, and provides paleohorizontal-controlled paleomagnetic directions. The most comprehensive paleomagnetic study

(85 sites), encompassing large part of the northern and southern branches of the Magallanes fold and thrust belt, is that of Poblete et al. (2014). Unfortunately, due to the low intensity and stability of magnetization, and the absence of paleohorizontal control for several sites, reliable paleomagnetic directions could be obtained from only seven sites from the Paleocene (~65 Ma) sedimentary units of the Magallanes fold and thrust belt in the Magallanes and Tierra del Fuego provinces. Excluding two sites that provided anomalous moderate clockwise (CW) rotations, and one site at the northern edge of the fold and thrust belt in Tierra del Fuego showing no significant rotation, the remaining four sites located in the Magallanes province and western Tierra del Fuego showed consistent CCW rotations ranging between 15° ± 9° and 26° ± 15°. The authors interpreted these rotations as resulted from the accommodation of differential shortening, increasing eastward, rather than oroclinal bending of the Fuegian Andes.

The most external domains of the Magallanes fold and thrust belt were sampled by Maffione et al. (2010). Due to the low intensity and stability of the remanence, only five sites out of 22 provided reliable paleomagnetic directions. Although somewhat noisy, the remanence isolated at these five sites was interpreted to be prefolding and likely primary, with opposite polarity directions that passed the fold test. The mean paleomagnetic directions at these five sites have an inclination that is substantially lower than expected, likely because of compaction of the clay-rich sediments during diagenesis. Despite the limitation due to the low number of sites and the not ideal quality of data, this study provided the first paleomagnetic constraints from the Cenozoic sedimentary sequences of the Magallanes-Austral basin. The results of this study have three major implications. First, the absence of rotations in the Fuegian Andes during the last ~50 Ma excluded that oroclinal bending occurred during the formation of the fold and thrust belt. Second, the curved Magallanes fold and thrust belt mainly developed during the Cenozoic; hence, the absence of tectonic rotations since 50 Ma classifies it as a non-rotational (primary) arc. Third, the present-day curvature of the southernmost Andes was already acquired prior to the opening of the Drake Passage in the Early Oligocene. This implies that the opening of the Drake Passage and growth of the Scotia plate did not play any role in the evolution of the curvature at the southernmost Andes.

5.3 *Antarctic Peninsula*

Data from the curved segment of the Antarctic Peninsula, mirroring the shape of the southernmost Andes, are scarce but sufficient to exclude any connection between the tectonic evolutions of the two continental margins. Early paleomagnetic data from the Antarctic Peninsula (Grunow 1993) documented in fact two distinct phases of earlier rotation: CW with respect to East Antarctica between 175 and 155 Ma due to the early opening in the Weddell Sea basin, and CCW with respect to West Antarctica in the 155–130 Ma time interval. Poblete et al. (2011) reported paleomagnetic results from Permo-Triassic and Jurassic rocks remagnetized during the Middle Cretaceous and primarily magnetized Cretaceous and Cenozoic rocks. These data indicated no relative rotations between the different sectors of the Antarctic Peninsula since the Cretaceous. Therefore, well before the beginning of the Andean orogeny in the Late Cretaceous, the curved shape of the Antarctic Peninsula was already developed.

6 Structural and Strain Data from the Southernmost Andes

The most prominent structural feature of the southernmost Andes is the change in direction of its structural lineaments. The main strike of faults and fold axes varies from 345° in the Southern Patagonian Andes to 320°–295° at the hinge of the bend and to 285°–260° in the Fuegian Andes. This geometry is reflected in a similar variation of the shortening directions along the curvature. Diraison et al. (2000) carried out an extensive structural geological study measuring 1600 fault planes in Mesozoic–Cenozoic sedimentary units of the Magallanes fold and thrust belt. Shortening directions in the southernmost Andes are sub-perpendicular to major folds and thrusts and progressively rotate from 075° in the Southern Patagonian Andes to 052° in the western part of the Fuegian Andes and to 043° in the eastern Fuegian Andes (Fig. 3) (see Sue and Ghiglione 2016 for a review). At the easternmost tip of the Fuegian Andes ~N–S, shortening directions have also been documented by Diraison et al. (2000) and Torres Carbonell et al. (2013).

These two studies also showed a progressive southeastward increase of the shortening in the Fuegian Andes from ~5 to ~22 %. Shortening estimates along the belt are quite variable. Regional balanced cross sections of the Fuegian Andes by Kraemer (2003) indicated a total shortening of 300–600 km. This author proposed that most of this shortening occurred in the Middle Cretaceous during the closure of the Rocas Verdes marginal basin, while minor shortening was accommodated during subsequent tectonic phases until the Neogene. Kraemer (2003) suggested that the amount of shortening in the southernmost Andes was consistent with a model of oroclinal bending.

Besides fault kinematic analyses, anisotropy of magnetic susceptibility (AMS) is a powerful tool to perform strain determination even when macroscopic kinematic constraints are either absent of difficult to identify (Tarling and Hrouda, 1993; Sagnotti et al. 1994; Parés et al. 1999). Clustering of the longest axis of the AMS ellipsoid defines a magnetic lineation, which tends to orient sub-parallel to the maximum axis of the strain ellipsoid during deformation. As a first-order approximation, magnetic lineation is parallel to the shortening direction in compressive settings, or to the stretching direction in extensional regimes. With AMS analyses, it is therefore possible to calculate the style of tectonics (compressive versus extensional) and the shortening or stretching directions.

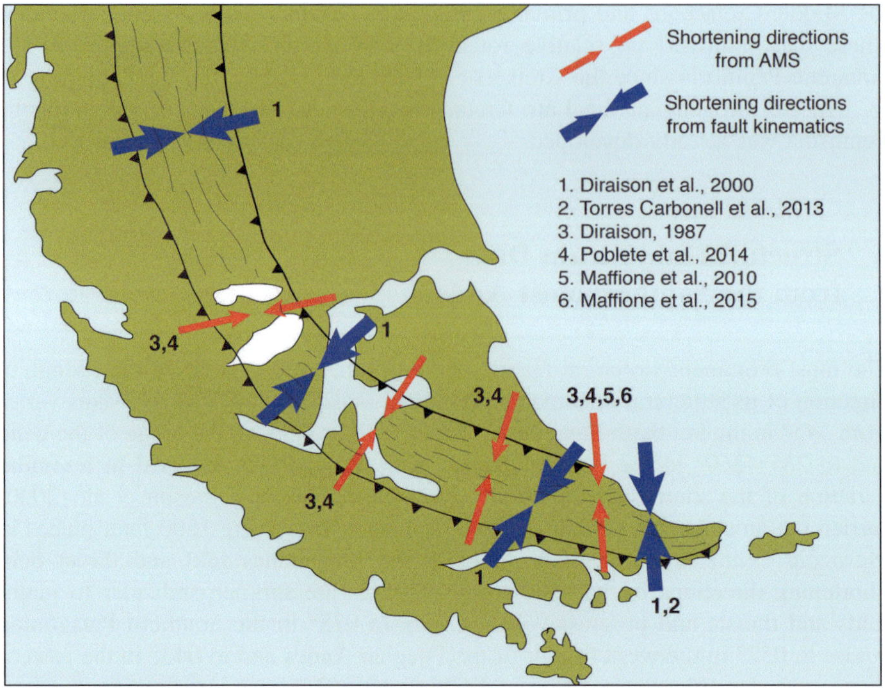

Fig. 3 Shortening directions in the southernmost Andes inferred from fault kinematics analyses (*blue arrows*) and anisotropy of magnetic susceptibility (AMS) analyses (*red arrows*)

Few AMS studies have been performed in the southernmost Andes (Diraison, 1998; Rapalini et al. 2005; Maffione et al. 2010, 2015; Esteban et al. 2011; Poblete et al. 2014). AMS analyses from the non-metamorphic sedimentary units of the Magallanes fold and thrust belt reveal the presence of well-developed magnetic lineations of clear compressive origin, which are sub-parallel to local faults and fold axes (see review by Maffione et al. 2015). These magnetic lineations change in direction following the curvature of the southernmost Andes and rotating from a ~N–S direction in the Southern Patagonian Andes to a ~NNE–SSW trend in the Fuegian Andes. Since the orientation of the AMS and strain ellipsoids are comparable, the lineations from the southernmost Andes identify a radial strain pattern where the shortening direction progressively rotate from ~ENE–WSW in the Southern Patagonian Andes to ~N–S in the eastern sectors of the Fuegian Andes (Fig. 3). Importantly, Maffione et al. (2015) show that this radial strain field was continuous in this region from the Early Cretaceous until at least the Oligocene (Fig. 4).

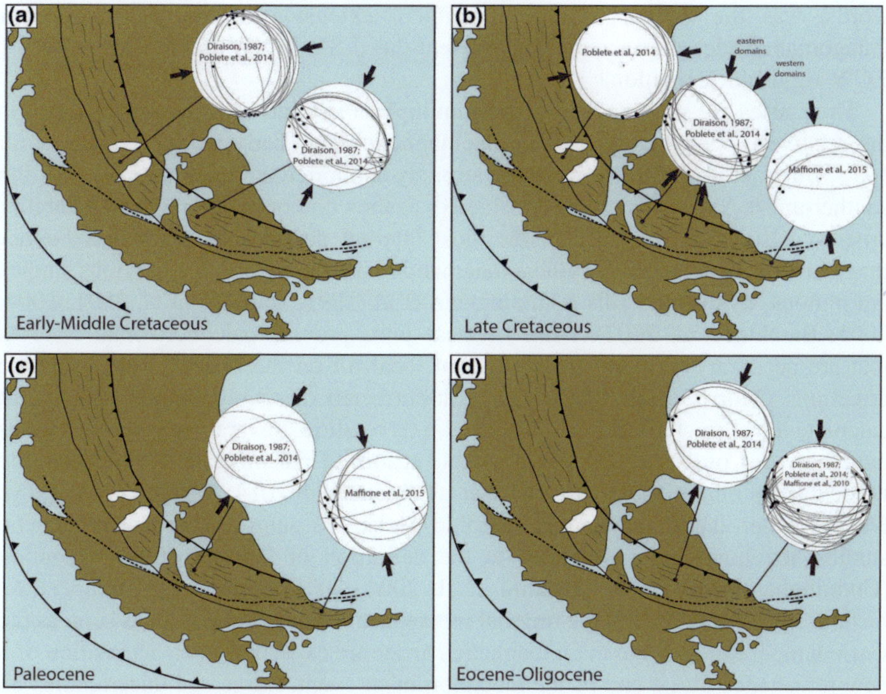

Fig. 4 Evolution of the paleostrain field (*black arrows* on the stereonets) within the Magallanes fold and thrust belt during four progressive time intervals, inferred from the published AMS data. Stereonets in each figure show the site mean lineations (*black dots*) and local bedding planes (*great circles*). Redrawn from Maffione et al. (2015)

7 Nature and Tectonic Evolution of the Curved Segment of the Southernmost Andes as Inferred from Available Paleomagnetic and Strain Constraints

Several hypotheses on the origin of the curvature of the southernmost Andes have been put forward since the 1950s in the absence of paleomagnetic and structural constraints. Early models considered the formation of the orogenic curvature as due to the drag associated with the left-lateral shearing between the South America and Scotia–Antarctic plates (Carey 1958; Winslow 1982). Paleomagnetic data obtained since the first study of Dalziel et al. (1973) have been crucial to provide a first-order reconstruction of the main tectonic events responsible for the present-day geometry of the southernmost Andes. However, although the amount and quality of data have recently increased, more than forty years of paleomagnetic investigations in the southernmost Andes have not yet been sufficient to provide a straightforward answer to the key question: Is the curved segment of the southernmost Andes (so far ambiguously called Patagonian orocline) a primary arc or an orocline? In other words, was the Fuegian Andes a southern extension of the ~N–S-oriented Patagonian Andes that acquired its present-day ESE–WNW trend after ~90° of CCW rigid block rotation?

The causes for such uncertainty are multiple. First and most important cause is the relatively poor quality of the paleomagnetic data. Early works (Dalziel et al. 1973; Burns et al. 1980), which more strongly supported an oroclinal model for the southernmost Andes, cannot be considered, as they do not satisfy modern reliability criteria. Also, rocks exposed in the most internal domains of the belt are either magmatic or metamorphic. Typical magmatic rocks sampled in the previous studies are plutons, dykes, and sills (Cunningham et al. 1991; Rapalini et al. 2001, 2004, 2005; Baraldo et al. 2002). None of these units provide paleohorizontal controls that are necessary to remove the effect of local tilt on the isolated paleomagnetic directions (Morris et al. 1998). Only tilt-corrected directions should be used to calculate the correct magnitude of vertical-axis rotation. Several more recent studies have however proposed that the inevitable scatter of paleomagnetic directions due to local tilt is minimized in large datasets (Poblete et al. 2014; Rapalini et al. 2015). On the other hand, where metamorphic rocks are sampled for paleomagnetic studies, the main issue is related to the definition of the age of magnetization (Cunningham et al. 1991; Rapalini et al. 2004, 2005). During high-temperature metamorphism (>500 °C), the original remanence is likely to be reset as soon as the Curie temperatures of the main magnetic carrier are exceeded, with acquisition of a new remanence at the end of the metamorphism event. Large age uncertainties on the remagnetization event can therefore bias the estimates on the timing of the tectonic rotations. Most of the remagnetized samples from the hinterland of the Fuegian Andes have probably been magnetically overprinted during the Late Cretaceous tectonic phase (Menichetti et al. 2008); still, a precise definition of the timing of the rotations cannot be achieved using metamorphic rocks.

A second issue even when high-quality paleomagnetic data are available is the distinction between regional and local tectonic rotations. The southernmost sector of the Fuegian Andes was affected by widespread left-lateral strike-slip tectonics (Cunningham 1993; Diraison et al. 2000; Menichetti et al. 2008). CCW vertical-axis rotations can be associated with sinistral strike-slip faults. The tectonic rotations documented in the southern Patagonian Andes by Iglesia-Llanos et al. (2003) and Rapalini et al. (2001) are a typical example of rotation associated with local faulting, and similar data should therefore not be used to test the orocline hypothesis (Rapalini 2007). Large strike-slip fault-related rotations have also been identified in the most internal domains of the belt along the southern shore of the Canal Beagle (Rapalini et al. 2015). Local deformation was also invoked as a possible cause of the ~20° CCW rotations in the southern sectors of the Magallanes fold and thrust belt (Poblete et al. 2014).

Finally, a third potential cause for the ongoing debate on the nature of the curvature of the southernmost Andes is the scarce and unevenly distributed paleomagnetic sampling coverage that still inhibits precise tectonic reconstructions by sheding uncertainties on the local versus regional character of the tectonic rotations.

So far, all three possible models have been proposed for the nature of the curvature of the southernmost Andes: primary or non-rotational arc (Diraison et al. 2000; Ramos and Aleman 2000; Ghiglione and Cristallini 2007), rotational arc or orocline sensu stricto (Dalziel and Elliot 1973; Burns et al. 1980; Cunningham et al. 1991; Kraemer 2003), and a 'hybrid' structure composed of a rotational and a non-rotational arc (Poblete et al. 2014; Maffione et al. 2015; Rapalini et al. 2015). What seems to emerge from the most recent paleomagnetic and structural geological data is that the hybrid model seems the most plausible possibility.

Although, as discussed, paleomagnetic data from Burns et al. (1980) cannot be used for tectonic interpretation, these authors were the first to put forward the hypothesis of a systematic pattern in the Fuegian Andes in which the magnitude of rotations decreased northward from a maximum value of ~90° comparable with the interlimb angle between the Patagonian and Fuegian Andes. The available paleomagnetic data seem to confirm this scenario with ~90° of CCW rotation of the southernmost (most internal) domains of the Fuegian Andes (Cunningham et al. 1991), ~30° of CCW rotations in the central belt where metamorphic sedimentary and magmatic rocks are exposed (Baraldo et al. 2002; Rapalini et al. 2015), and little (~20° CCW; Poblete et al. 2014) or no rotation (Maffione et al. 2010) affecting the non-metamorphic sedimentary sequences of the Magallanes fold and thrust belt in the most external domains of the belt. Furthermore, the timing of the rotations appears to decrease northward, from post-mid-Late Cretaceous (~90 Ma) in the south to post-Late Cretaceous (~72 Ma) in the central belt, and to post-Paleocene (~65 Ma) until ~50 Ma in the fold and thrust belt. Similar patterns of magnitude and timing of rotations are commonly found in fold and thrust belts like the Southern Apennines (Italy), where the magnitude of the rotations decrease from the firstly deformed nappes from the most internal paleogeographic domains to the youngest nappes from the frontal part of the belt (e.g., Maffione et al. 2013). This mechanism is typical of progressive arcs (sensu Weil and Sussman 2004) in which the tectonic rotations accompany the propagation of the orogenic front into the foreland. Based on the sparse data from the Fuegian Andes, it can only be speculated that a similar mechanism occurred here. According to this scenario, the largest rotations in Peninsula Hardy from the innermost domains of the belt (Cunningham et al. 1991) would be the effect of initial compressive deformation of the most internal paleogeographic domains.

Most of the tectonic models explaining the evolution of the curvature of the southernmost Andes invoked the closure of the Rocas Verdes marginal basin as a possible cause (Burns et al. 1980; Cunningham et al. 1991; Kraemer 2003; Maffione et al. 2010, 2015; Poblete et al. 2014; Rapalini et al. 2015). According to these reconstructions, at the beginning of the southern Andean orogeny (~100 Ma), the Jurassic magmatic arc (Patagonian Batholith) formed a relatively straight structure along the Pacific coast, bounding a triangle-shaped marginal basin to the west. Closure of the marginal basin occurred by continuous CCW rotation of the southern segment of the magmatic arc, leading to progressive nappe stacking in the basin's sedimentary cover until the magmatic arc collided with the South American continental margin in the Late Cretaceous. It is therefore plausible that during this event, tectonic slices originally rooted in different paleogeographic domains of the basin recorded different amounts of rotation. After continental collision, rigid rotation of what is today the Fuegian Andes might have continued for few million years progressively involving more external paleogeographic domains. Evidence against this reconstruction is the magnitude of shortening along the Fuegian Andes. Although shortening increases from west to east (Diraison et al. 2000), the magnitude of the gradient is not enough to explain a ~90° bending, as inferred from paleomagnetic data (Cunningham et al. 1991) and the overall geometry of the southernmost Andes. The total amount of oroclinal bending remains therefore unclear.

The beginning of a possible rigid block rotation of the Fuegian Andes is constrained to a general Late Cretaceous time due to remagnetization of the rocks exposed in this region (Cunningham et al. 1991). Conversely, the end of the rotations is more precisely constrained at ~50 Ma by paleomagnetic data from the Cenozoic Magallanes fold and thrust belt (Maffione et al. 2010). These data for the first time proved that no rotations were produced during the deformation of the Magallanes-Austral foreland basin since ~50 Ma. Minor (~20°) post-65 Ma CCW rotations documented by Poblete et al. (2014), of possible local origin, seem to confirm this trend. Assuming that the formation of the Magallanes fold and thrust belt started at ~61 Ma (Ghiglione and Ramos 2005) or slightly earlier (Torres Carbonell et al. 2013) and ended in the Early Miocene, the absence of tectonic rotations since 50 Ma (and possibly 65 Ma) classifies this curved belt as a primary (non-rotational) arc.

Combining all the available paleomagnetic data, a complex tectonic evolution of the curved segment of the southernmost Andes emerges. What has been so far implicitly defined as an orocline due to its more popular name of Patagonian orocline is more likely a hybrid orogenic arc composed of two parts: an older orocline or progressive arc in the outer side of the curvature, mainly formed in Cretaceous times upon closure of the Rocas Verdes marginal basin, and a younger primary arc in the inner side of the orogenic arc, developed in the Cenozoic (Poblete et al. 2014; Maffione et al. 2015). For this reason, Maffione et al. (2015) have recently proposed to use the more generic name of 'Patagonian Arc' to describe the curved segment of the southernmost Andes.

Precise discrimination between temporal and geographical patterns of oroclinal bending is still necessary to validate this model.

8 Mechanisms of Bending of the Southernmost Andes

Considering the hypothesis that the southernmost Andes form a hybrid orogenic arc, it is likely that different tectonic processes contributed to their formation. As previously mentioned, oroclinal bending of the outer part of the arc has been commonly associated with the closure of the Rocas Verdes marginal basin in the mid-Late Cretaceous. Reconstructing the evolution of the primary arc in the inner side of the orogenic curvature is much more complicated once strain data from fault kinematics, AMS, and analogue experiments are taken into account. The first high-resolution structural model for the Patagonian bending taking into account faults kinematics and Cenozoic brittle faulting was proposed by Cunningham (1993). This author proposed the Patagonian orocline to be the product of broad interplate shearing accommodated by strike-slip faulting, block rotation, and contraction, i.e., the strike-slip orocline model. However, this model implies a succession of E–W-oriented, left-lateral faults cutting the trend of folds and thrusts transversally to the orogen; the strike-slip faults recognized so far in the southernmost Andes are orogen-parallel; and faults and folds are continuous along strike (Figs. 2 and 3) (Ghiglione and Cristallini 2007).

Analogue sandbox experiments were performed by Ghiglione and Cristallini (2007) using both indentation of a curved rigid buttress and rotation of one arm of the indenter. Models using indentation of a rigid indenter provided the most consistent results with the geological evidence. In particular, these authors concluded that a two-phase indentation, NE-ward and then ESE-ward, was able to reproduce with a higher detail the deformation pattern observed in the fold and thrust belt. The deformation paths produced by this model are, however, not entirely consistent with the shortening directions from fault kinematic analysis reported by Diraison et al. (2000). While shortening directions show a progressive variation along the orogenic arc (Fig. 3), the models of Ghiglione and Cristallini (2007) indicate the formation of two distinct, overlapping families of shortening directions parallel to the motion of the indenter.

When shortening directions from both fault kinematic and AMS analysis are considered, a clear pattern emerges, defined by Maffione et al. (2015) as a radial strain field (Figs. 3 and 4). Different processes, other than oroclinal bending, need to be found to explain the presence of a radial strain field in a primary (non-rotational) arc such as the Magallanes fold and thrust belt. Although radial strain patterns have been documented in primary arcs, caused by, e.g., gravitational forces within the growing wedge (Ramsay 1981; Marshak 2004; Weil et al. 2010), extreme $\sim 90°$ variation of the shortening directions as observed in the southernmost Andes is likely caused by more complex tectonic processes. Paleomagnetic data from Rapalini et al. (2015) indicate a post-72 Ma CCW rotation of $\sim 30°$ in the central belt just south of the Magallanes fold and thrust belt. This result suggests that minor rotations could have occurred before the beginning of deformation in the foreland basin at ~ 60 Ma (Ghiglione and Ramos 2005). A rigid indenter model like that proposed by Ghiglione and Cristallini (2007), where an already curved proto-Andes converged during the Cenozoic into the Magallanes-Austral foreland basin to produce a thin-skinned fold and thrust belt, has been considered as a plausible scenario (Maffione et al. 2015).

Simple convergence of a rigid indenter cannot produce a radial strain field (Ghiglione and Cristallini 2007). A two-phase indentation mechanism, as already discussed, is inconsistent with both the pattern of shortening directions and the geological evidence that seems to indicate continuous deformation along both branches of the orogenic arc during the Cenozoic (Maffione et al. 2015).

Maffione et al. (2015) proposed that a radial strain field could be produced upon indentation of a curved rigid buttress if slip partitioning operated along preexisting faults oriented parallel to the margin of the indenter. Jurassic normal faults rooted in the basement of the South American continental margin played a key role in the evolution of the Magallanes-Austral foreland basin sedimentation and subsequent deformation (Menichetti et al. 2008; Likerman et al. 2013; Fosdick et al. 2011, 2014; Ghiglione et al. 2013). The occurrence of ~N–S-oriented normal faults in the Southern Patagonian Andes and ~E to NE oriented in the Fuegian Andes associated with the rifting of the Rocas Verdes basin has also been documented (Ghiglione et al. 2009, 2013; Fosdick et al. 2011; Likerman et al. 2013). Based on this evidence, Maffione et al. (2015) proposed that the preexisting Jurassic faults in the basement of the southernmost Andes controlled the Late Cretaceous to Cenozoic kinematic evolution and final geometry of the curved Magallanes fold and thrust belt (Fig. 5). Slip-partitioning mechanisms along these faulted blocks in the central belt allowed

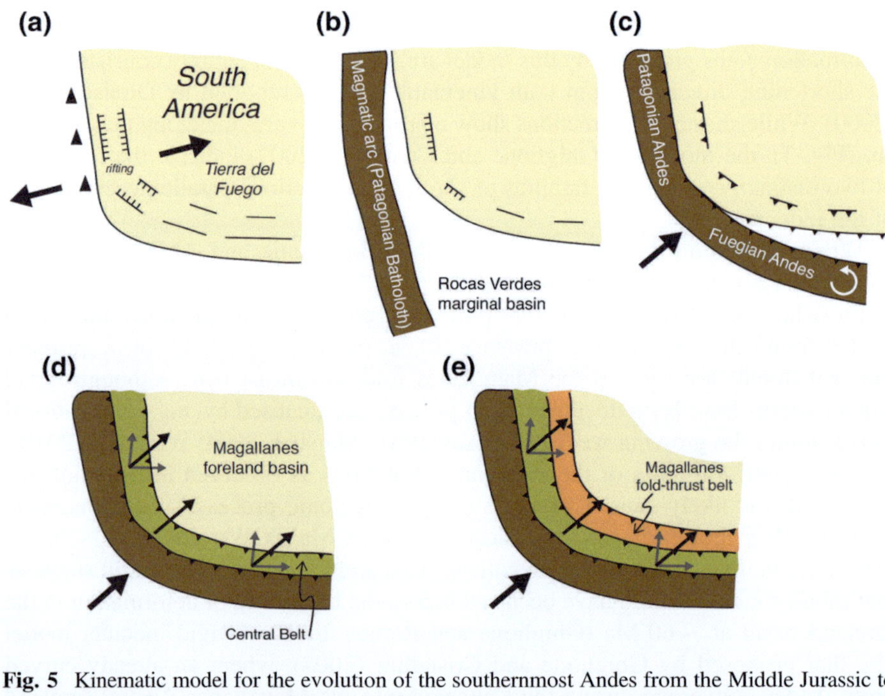

Fig. 5 Kinematic model for the evolution of the southernmost Andes from the Middle Jurassic to the Oligocene proposed by Maffione et al. (2015). *Thick black arrows* indicate the direction of the regional stress field. **a** *Black triangles* indicate active magmatism. **c** *Circular white arrow* indicates the regional vertical-axis rotation of the Fuegian Andes during the closure of the Rocas Verdes basin. **d**, **e** *Thin black arrows* indicate the displacement vectors of the basement-rooted faulted blocks within the central belt; *thin gray arrows* indicate the fault-parallel and fault-perpendicular components resulting from slip partitioning propagating into the Magallanes-Austral foreland basin and leading to the formation of the fold and thrust belt. Redrawn from Maffione et al. (2015). **a** Mid-Late Jurassic, **b** Late Jurassic, **c** Early Late Cretaceous, **d** Middle Late Cretaceous, **e** Paleocene–Oligocene

transference of fault-normal compression components into the Magallanes-Austral foreland ($\sim$N-ward and $\sim$E-ward contraction in the Fuegian and Patagonian Andes, respectively) throughout the Cenozoic, resulting in a radial strain field.

9 Conclusive Remarks

Since its first definition as Patagonian orocline by Carey (1958), the real nature and evolution of the regional-scale curvature of the southernmost Andes have been controversial. The debate revolved around the two conflicting hypotheses considering the curvature of the southernmost Andes either as a non-rotational (primary) arc, or a rotational arc (i.e., orocline or progressive arc). The curved shape of the Antarctic Peninsula and the growth during the Cenozoic of a new oceanic plate (Scotia plate) intervening between the South American and Antarctic plate were used as evidence at the basis of early tectonic models. With time, most of the early models resulted to be imprecise. Only with the acquisition of the first paleomagnetic data from the southernmost Andes, the debate on the nature of its curvature was resumed, continuing until today.

The mechanism proposed by Maffione et al. (2015) requires widespread sinistral and dextral strike-slip displacements on main thrusts from the internal domains of the Fuegian and Patagonian Andes, respectively. Further data identifying and quantifying such strike-slip tectonics in the southernmost Andes are, however, needed to fully test the proposed model.

Although more than 40 years of paleomagnetic investigation in the southernmost Andes has provided key constraints to define the timing, style, and kinematics of deformation in this region, their number and quality are still too low to allow detailed reconstructions. The available paleomagnetic data indicate that $\sim$90° of CCW vertical-axis rotation affected the most internal structural domain of the Fuegian Andes (i.e., the outer part of the curvature) since $\sim$90 Ma, $\sim$30° of which occurred after $\sim$72 Ma, and that rotations in the Magallanes fold and thrust belt (i.e., the inner part of the curvature) ended at $\sim$50 Ma. Although these data are affected by uncertainties related to the age of magnetization, the absence of paleohorizontal control, and a poor stability of the magnetic remanence, an overall pattern of CCW rotations in the internal sectors of the belt appear evident, while rotations did not affect the Magallanes fold and thrust belt at least since the Early Eocene. These data suggest that the outer part of the curvature represents an orocline or a progressive arc (sensu Weil and Sussman 2004). Definition of precise timing and total amount of the bending awaits further investigations. On the other side, the absence of tectonic rotations in the Magallanes fold and thrust belt indicates that it developed in the Cenozoic as a primary arc (sensu Weil and Sussman 2004).

Based on the available paleomagnetic constraints and the most recent tectonic models, the broadly used term of 'Patagonian orocline' appears to be outdated and inappropriate. According to formal definitions (e.g., Weil and Sussman 2004), an orocline is a large-scale curved mountain belt formed upon bending of a previously straight orogen. This definition, as demonstrated, may only apply to the outer side of the curvature of the southernmost Andes, being the inner side a primary arc. As recently proposed by several authors (Poblete et al. 2014; Maffione et al. 2015), the orogenic bend of the southernmost Andes is composed of two adjacent arcs: a rotational arc (orocline or progressive arc) and a non-rotational arc (primary arc) in the outer and inner part of the curvature, respectively. For this reason, Maffione et al. (2015) suggested that the old term 'Patagonian orocline' should be abandoned in favor of a more generic 'Patagonian Arc'.

A broad consensus exists that invokes the closure of the Rocas Verdes marginal basin in the Late Cretaceous as the main mechanism of oroclinal bending of the internal sectors of the southernmost Andes. More uncertain is the evolution of the primary arc of the Magallanes fold and thrust belt. Proposed scenarios to explain the formation of the curved fold and thrust belt, in the context of a primary arc model, considered indentation of a curved rigid buttress (represented by the previously bent internal domains of the southernmost Andes) into the foreland sediments of the Magallanes-Austral basin. The presence of highly variable (radial pattern) shortening directions along the Magallanes fold and thrust belt (Maffione et al. 2015), uncommon in primary arcs, has been explained by the possible effect of slip partitioning along preexisting Jurassic faults at the South American margin, which could have enabled ~N-ward contraction in the Fuegian Andes and ~E-ward contraction in the Patagonian Andes. A slip-partitioning model in the southernmost Andes requires widespread strike-slip tectonics with dextral and sinistral kinematics in the Patagonian and Fuegian Andes, respectively. Future investigations are needed to test the occurrence of such tectonics and validate this model.

References

Alvarez-Marrón J, McClay KR, Harambour S, Rojas L, Skyrrmeta J (1993) Geometry and evolution of the frontal part of the Magallanes foreland thrust and fold belt (Vicuna-Area), Tierra del Fuego, Southern Chile. Am Assoc Petr Geol Bull 77:1904–1921

Baraldo A, Rapalini A, Tassone A, Lippai H, Menichetti M, Lodolo E (2002) Estudio paleomagnético del intrusivo del cerro Hewhoepen, Tierra del Fuego, y sus implicancias tectónicas. XV Congreso Geológico Argentino (El Calafate) Actas 1:285–290

Barker PF (2001) Scotia Sea regional tectonic evolution: implications for mantle flow and palaeocirculation. Earth Sci Rev 55:1–39

Barker PF, Diekmann B, Escutia C (2007a) Onset of Cenozoic Antarctic glaciation. Deep-Sea Res II 54:2293–2307

Barker PF, Filippelli GM, Florindo F, Martin EE, Scher HD (2007b) Onset and role of the Antarctic circumpolar current. Deep-Sea Res II 54:2388–2398

Beck ME Jr, Burmester R, Cembrano J, Drake R, García A, Hervé F, Munizaga F (2000) Paleomagnetism of the North Patagonian Batholith, southern Chile. An exercise in shape analysis. Tectonophysics 326:185–202

Biddle KT, Uliana MA, Mitchum RM, Fitzgerald MG, Wright RC (1986) The stratigraphy and structural evolution of the central and eastern Magallanes Basin, southern South America. In: Allen A, Homewood P (eds) Foreland Basins (vol 8). Blackwell Scientific Publications, London (Int Assoc Sediment Special Publication), pp 41–61

Bruhn RL (1979) Rock structures formed during back-arc basin deformation in the Andes of Tierra del Fuego. Geol Soc Am Bull 90(11):998–1012

Bruhn RL, Dalziel IWD (1977) Destruction of the Early Cretaceous marginal basin in the Andes of Tierra del Fuego. Island Arcs, Deep Sea Trenches Back-Arc Basins 395–405

Bruhn RL, Stern CR, de Wit MJ (1978) Field and geochemical data bearing on the development of a mesozoic volcano-tectonic rift zone and back-arc basin in southernmost South America. Earth Planet Sci Lett 41:32–46

Burns KL, Rickyrrd MJ, Belbin L, Chamalaun F (1980) Further palaeomagnetic confirmation of the magallanes orocline. Tectonophysics 63:75–90

Calderón M, Fildani A, Hervé F, Fanning CM, Weislogel A, Cordani U (2007) Late Jurassic bimodal magmatism in the northern sea-floor remnant of the Rocas Verdes basin, Southern Patagonian Andes. J Geol Soc Lond 164:1011–1022

Calderón M, Fosdick JC, Warren C, et al (2012) The low-grade Canal de las Montañas Shear Zone and its role in the tectonic emplacement of the Sarmiento Ophiolitic Complex and Late Cretaceous Patagonian Andes orogeny, Chile. Tectonophysics 524–525:165–185. doi: 10.1016/j.tecto.2011.12.034

Cande SC, Leslie RB (1986) Late Cenozoic tectonics of the southern Chile Trench. J Geophys Res-Sol EA 91(B1):471–496

Carey SW (1955) The orocline concept in geotectonics. Proc R Soc Tasmania 89:255–288

Carey SW (1958) The tectonic approach to continental drift. Continental drift: a symposium: Hobart. Geology Department, University of Tasmania, Tasmania, pp 177–355

Cifelli F, Mattei M, Rossetti F (2007) Tectonic evolution of arcuate mountain belts on top of a retreating subduction slab: the example of the Calabrian Arc. J Geophys Res-Sol EA 112(B9)

Coutand I, Diraison M, Cobbold PR, Gapais D, Rossello EA, Miller M (1999) Structure and kinematics of a foothills transect, Lago Viedma, southern Andes (49°30′S). J S Am Earth Sci 12:1–15

Cunningham WD (1993) Strike-slip faults in the southernmost Andes and the development of the Patagonian Orocline. Tectonics 12:169–186

Cunningham WD (1994) Uplifted ophiolitic rocks on Isla Gordon, southernmost Chile: implications for the closure history of the Rocas Verdes marginal basin and the tectonic evolution of the Beagle Channel region. J S Am Earth Sci 7:135–147

Cunningham WD (1995) Orogenesis at the southern tip of the Americas: the structural evolution of the Cordillera Darwin metamorphic complex, sou-thernmost Chile. Tectonophysics 244:197–229

Cunningham WD, Klepeis KYR, Gose WA, Dalziel IW (1991) The Patagonian Orocline: new paleomagnetic data from the Andean magmatic arc in Tierra del Fuego. Chile. J Geophys Res-Sol EA 96(B10):16061–16067

Dalziel IWD (1981) Backarc extension in the southern Andes: a review and critical reappraisal. Philos Trans R Soc Lond 300:319–335

Dalziel IWD, Elliot DH (1973) The Scotia arc and Antarctic margin. In: Nairn AEM, Stehli FG (eds) The Ocean Basins and margins (vol 1). The South Atlantic Plenum Press, New York, pp 171–245

Dalziel IWD, Klingfield R, Lowrie W, Opdyke ND (1973) Paleomagnetic data from the southernmost Andes and the Antarctandes. In: Tarling DH, Runcom SK (eds) Implications of continental drift to the earth sciences. Academic Press, San Diego, pp 87–101

Dalziel IWD, De Wit MF, Palmer KF (1974) Fossil marginal basin in the southern Andes. Nature 250:291–294

Diraison M (1998) Evolution cénozoique du Bassin de Magellan et tectonique des Andes Australes. Mém Doc Géosciences-Rennes 85:332

Diraison M, Cobbold PR, Gapais D, Rossello EA, Le Corre C (2000) Cenozoic crustal thickening, wrenching and rifting in the foothills of the Southernmost Andes. Tectonophysics 316:91–119

Eagles G (2016) Tectonic reconstructions of the Southernmost Andes and the Scotia Sea during the opening of the Drake Passage. In: Ghiglione MC (ed) Geodynamic Evolution of the Southernmost Andes. Springer Earth System Sciences, pp 75–108

Eagles G, Livermore RA, Fairhead JD, Morris P (2005) Tectonic evolution of the west Scotia Sea. J Geophys Res-Sol EA 110:1–19

Eldredge S, Bachtadse V, Van der Voo R (1985) Paleomagnetism and the orocline hypothesis Tectonophysics 119:153–179

Esteban F, Tassone A, Menichetti M, Rapalini AE, Remesal MB, Cerredo ME, Lippai H, Vilas JF (2011) Fabrica Magnética y Microestructuras a través de los Andes de Tierra del Fuego, Argentina. Andean Geol 38:64–81

Fildani A, Hessler AM (2005) Stratigraphic record across a retroarc basin inversion: Rocas Verdes-Magallanes Basin, Patagonian Andes, Chile. Geol Soc Am Bull 117:1596–1614

Fildani A, Cope TD, Graham SA, Wooden JL (2003) Initiation of the Magallanes foreland basin: Timing of the southernmost Patagonian Andes orogeny revised by detrital zircon provenance analysis. Geology 31:1081. doi: 10.1130/G20016.1

Fosdick JC, Romans BW, Fildani A, Bernhardt A, Calderón M, Graham SA (2011) Kinematic evolution of the Patagonian retroarc fold and thrust belt and Magallanes foreland basin, Chile and Argentina, 51°30′S. Geol Soc Am Bull 123:1679–1698

Fosdick JC, Graham SA, Hilley GE (2014) Influence of attenuated lithosphere and sediment loading on flexure of the deep-water Magallanes retroarc foreland basin, Southern Andes. Tectonics 33:2505–2525. doi: 10.1002/2014TC003684

Fosdick JC, Grove M, Hourigan JK, Calderón M (2013) Retroarc deformation and exhumation near the end of the Andes, southern Patagonia. Earth Planet Sci Lett 361:504–517. doi: 10.1016/j.epsl.2012.12.007

Fosdick JC, Graham SA, Hilley GE (2015) Influence of attenuated lithosphere and sediment loading on flexure of the deep-water Magallanes retroarc foreland basin, Southern Andes. Tectonics 33:2505–2525

Ghiglione MC (2002) Diques clásticos asociados a deformación transcurrente en depósitos sinorogénicos del Mioceno inferior de la Cuenca Austral. Rev la Asoc Geol Argentina 57:103–118

Ghiglione MC, Cristallini EO (2007) Have the southernmost Andes been curved since Late Cretaceous times? An analog test for the Patagonian Orocline. Geology 35:13–16

Ghiglione MC, Ramos VA (2005) Chronology of deformation in the southernmost Andes of Tierra del Fuego. Tectonophysics 405:25–46

Ghiglione MC, Ramos VA, Cristallini EO (2002) Fuegian Andes foreland fold and thrust belt: structure and growth strata. Rev Geol Chile 29(1):17–41

Ghiglione MC, Suarez F, Ambrosio A, Da Poian G, Cristallini EO, Pizzio MF, Reinoso RM (2009) Structure and evolution of the Austral basin fold–thrust belt, Southern Patagonian Andes. Rev Asoc Geol Argentina 65:215–226

Ghiglione MC, Quinteros J, Yagupsky D, Bonillo-Martínez P, Hlebszevtich J, Ramos VA, Vergani G, Figueroa D, Quesada S, Zapata T (2010) Structure and tectonic history of the foreland basins of southernmost South America. J S Am Earth Sci 29:262–277

Ghiglione MC, Navarrete-Rodríguez AT, González Guillot M, Bujaleski G (2013) The opening of the Magellan Strait and its geodynamic implications. Terra Nova 25(1):13–20

Ghiglione MC, Sue C, Ramos ME, Tobal JE, Gallardo RE (2016a) The relation between Neogene Denudation of the southernmost Andes and sedimentation in the offshore Argentine and Malvinas basins during the opening of the Drake Passage. In: Ghiglione MC (ed) Geodynamic Evolution of the Southernmost Andes. Springer Earth System Sciences, pp 109–135

Ghiglione MC, Ramos V, Cuitiño J, Barberón V (2016b) Growth of the Southern Patagonian Andes (46–53°S) and its relation with subduction processes. In: Folguera A, Naipauer M, Sagripanty L, Ghiglione MC, Orts D, Giambiagi LB (eds) Growth of the Southern Andes. Springer Earth System Sciences, Berlin, pp 201–240

Guillot MG (2016) Magmatic evolution of the southernmost Andes and its relation with subduction processes. In: Ghiglione MC (ed) Geodynamic Evolution of the Southernmost Andes. Springer Earth System Sciences, pp 37–74

González Guillot M, Escayola M, Acevedo R (2011) Calc-alkyrline rear-arc magmatism in the Fuegian Andes: implications for the mid-Cretaceous tectonomagmatic evolution of southernmost South America. J S Am Earth Sci 31:1–16

Grunow AM (1993) New paleomagnetic data from the Antarctic Peninsula and their tectonic implications. J Geophys Res-Sol EA 98(B8):13815–13833
Halpern M, Rex DC (1972) Time of folding of the Yahgan Formation and age of the Tekenikyr beds, southern Chile, South America. Geol Soc Am Bull 83:1881–1886
Hervé F, Nelson E, Kyrwashita K, Suárez M (1981) New isotopic ages and the timing of orogenic events in the Cordillera Darwin, southernmost Chilean Andes. Earth Planet Sci Lett 55(2):257–265
Hervé M, Suárez M, Puig A (1984) The Patagonian Batholith south of Tierra del Fuego, Chile. Timing and tectonic implications. J Geol Soc Lond 141(5):909–917
Hervé F, Pankhurst R, Fanning C, Calderón M, Yaxley G (2007) The South Patagonian Batholith: 150 my of granite magmatism on a plate margin. Lithos 97:373–394
Hervé F, Calderón M, Faúndez V (2008) The metamorphic complexes of the Patagonian and Fuegian Andes. Geologica Acta 6:43–53
Iglesia Llanos MP, Lanza R, Riccardi AC, Geuna S, Laurenzi MA, Ruffini R (2003) Palaeomagnetic study of the El Quemado complex and Marifil formation, Patagonian Jurassic igneous province, Argentina. Geophys J Int 154:599–617
Kirschvink JL (1980) The least-squares line and plane and the analysis of palaeomagnetic data. Geophys J 62:699–718
Klepeis KYR (1994) Relationship between uplift of the metamorphic core of the southernmost Andes and shortening in the Magallanes foreland fold and thrust belt, Tierra del Fuego, Chile. Tectonics 13:882–904
Klepeis KYR, Austin JA (1997) Contrasting styles of superposed deformation in the southernmost Andes. Tectonics 16:755–776
Klepeis K, Betkyr P, Clarke G, Fanning M, Hervé F, Rojas L, Mpodozis C, Thomson S (2010) Continental underthrusting and obduction during the Cretaceous closure of the Rocas Verdes rift basin, Cordillera Darwin, Patagonian Andes. Tectonics 29:TC3014
Kohn MJ, Spear FS, Dalziel IWD (1993) Metamorphic P-T Paths from Cordillera Darwin, a Core Complex in Tierra del Fuego, Chile. J Petrol 34:519–542. doi: 10.1093/petrology/34.3.519
Kohn MJ, Spear FS, Harrison TM, Dalziel IWD (1995) 40Ar/39Ar geochronology and P-T-t paths from the Cordillera Darwin metamorphic complex, Tierra del Fuego, Chile. J Metamorph Geol 13:251–270
Kraemer PE (1998) Structure of the Patagonian Andes: Regional balanced cross section at 50°S. Argentina Int Geol Rev 40:896–915
Kraemer PE (2003) Orogenic shortening and the origin of the Patagonian orocline (56S). J S Am Earth Sci 15:731–748
Kyrtz HR (1973) Contrasts in tectonic evolution of orogenic belts in the southwest Pacific. J R Soc N Z 3:333–362
Lawver LA, Gahagan LM (2003) Evolution of Cenozoic seaways in the circum-Antarctic region. Palaeogeo Palaeoclim Palaeoecol 198:11–37
Likerman L, Burlando JF, Cristallini EO, Ghiglione MC (2013) Along-strike structural variations in the Southern Patagonian Andes: insights from physical modeling. Tectonophysics 590:106–120
Livermore RA, Nankivell AP, Eagles G, Morris P (2005) Paleogene opening of drake passage. Earth Planet Sci Lett 236:459–470
Lodolo E, Coren F, Schreider AA, Ceccone G (1997) Geophysical evidence of a relict oceanic crust in the southwestern Scotia Sea. Mar Geophys Res 19:439–450
Lodolo E, Menichetti M, Tassone A, Sterzai P (2002) Morphostructure of the central-eastern Tierra del Fuego Island from geological data and remote-sensing images. EGS Stephan Mueller Spec Publ Ser 2:1–16
Lodolo E, Menichetti M, Bartole R, Ben-Avraham Z, Tassone A, Lippai H (2003) Magallanes-Fagnano continental transform fault (Tierra del Fuego, southernmost South America). Tectonics 22:1076
Lodolo E, Donda F, Tassone A (2006) Western Scotia Sea margins: improved constraints on the opening of the Drake Passage. J Geophys Res-Sol EA 111(B6)
Lowrie W, Hirt AM (1986) Paleomagnetism in arcuate mountain belts. In: Wezel FC (ed) The origin of arcs, developments in geotectonics (vol 21). Amsterdam, Elsevier, pp 141–158

Maffione M, Speranza F, Faccenna C, Cascella A, Vignaroli G, Sagnotti L (2008) A synchronous Alpine and Corsica-Sardinia rotation. J Geophys Res 113(B3)

Maffione M, Speranza F, Faccenna C (2009) Bending of the Bolivian orocline and growth of the central Andean plateau: paleomagnetic and structural constraints from the Eastern Cordillera (22–24°S, NW Argentina). Tectonics 28:TC002402

Maffione M, Speranza F, Faccenna C, Rossello E (2010) Paleomagnetic evidence for a pre-early Eocene ($\sim$50 Ma) bending of the Patagonian orocline (Tierra del Fuego, Argentina): paleogeographic and tectonic implications. Earth Planet Sci Lett 289:273–286

Maffione M, Speranza F, Cascella A, Longhitano SG, Chiarella D (2013) A $\sim$125° post-early Serravallian counterclockwise rotation of the Gorgoglione formation (Southern Apennines, Italy): new constraints for the formation of the Calabrian arc. Tectonophysics 590:24–37

Maffione M, Fernandez-Moreno C, Ghiglione M, Speranza F, van Hinsbergen DJJ, Lodolo E (2015) Constraints on deformation of the Southern Andes since the cretaceous from anisotropy of magnetic susceptibility. Tectonophysics 665:236–250

Maloney KT, Clarke GL, Klepeis KYR, Fanning CM, Wang W (2011) Crustal growth during back-arc closure: cretaceous exhumation history of Cordillera Darwin, southern Patagonia. J Metamorph Geol 29(6):649–672

Maloney KT, Clarke GL, Klepeis KYR, Quevedo L (2013) The Late Jurassic to present evolution of the Andean margin: drivers and the geological record. Tectonics 32:1049–1065

Malumián N, Olivero EB (2006) The Cabo Domingo Group, Tierra del Fuego: biostratigraphy, paleoenvironments, and events of the marine Eocene-Miocene. Rev Asoc Geol Argent 61:139–160

Malumián N, Panza JL, Parisi C, Nañez C, Caramés A, Torre E (2000) Hoja Geológica 5172-III-Yacimiento Río Turbio, provincia Santa Cruz, 1:250.000. Servicio Geológico Minero Argentino, Boletín 247: p 108

Marshak S (1988) Kinematics of orocline and arc formation in thin-skinned orogens. Tectonics 7:73–86

Marshak S (2004) Salients, recesses, arcs, oroclines, and syntaxes—a review of ideas concerning the formation of map-view curves in fold-thrust belts. Thrust Tecton Hydrocarb Syst 82:131–156

McAtamney J, Klepeis K, Mehrtens C, Thomson S, Betkyr P, Rojas L, Snyder S (2011) Along-strike variability of back-arc basin collapse and the initiation of sedimentation in the Magallanes foreland basin, southernmost Andes (53–54.5°S). Tectonics 30(5):2010T. doi:10.1029/1107C002826

Menichetti M, Lodolo E, Tassone A (2008) Structural geology of the Fuegian Andes and Magallanes fold thrust belt—Tierra del Fuego Island. Geologica Acta 6:19–42

Morris A, Anderson MW, Robertson AHF (1998) Multiple tectonic rotations and transform tectonism in an intraoceanic suture zone, SW Cyprus. Tectonophysics 299:229–253

Muttoni G, Argnani A, Kent DV, Abrahamsen N, Cibin U (1998) Paleomagnetic evidence for Neogene tectonic rotations in the northern Apennines, Italy. Earth Planet Sci Lett 154:25–40

Nelson EP (1982) Postectonic uplift of the Cordillera Darwin orogenic core complex: evidence for fission track geochronology and closing temperature time relationships. J Geol Soc Lond 139:755–761

Olivero EB, Malumián N (1999) Eocene stratigraphy of Southeastern Tierra del Fuego Island, Argentina. Am Assoc Petr Geol Bull 83:295–313

Olivero EB, Malumián N (2008) Mesozoic-Cenozoic stratigraphy of the Fuegian Andes, Argentina. Geologica Acta 6:5–18

Olivero EB, Malumián N, Palamarczuk S (2003) Estratigrafía del Cretácico superior-Paleoceno del área de bahía Thetis, Andes Fueguinos, Argentina: acontecimientos tectónicos y paleobiológicos. Rev Geol Chile 30:245–263

Pankhurst RJ, Riley TR, Fanning CM, Kelley SP (2000) Episodic silicic volcanism in Patagonia and the Antarctic Peninsula: chronology of magmatism associated with the break-up of Gondwana. J Petrol 41:605–625

Parés JM, Van der Pluijm BA, Dinarès-Turell J (1999) Evolution of magnetic fabrics during incipient deformation of mudrocks (Pyrenees, northern Spain). Tectonophysics 307:1–14

Poblete F, Arriagada C, Roperch P, Astudillo N, Hervé F, Kraus S, Le Roux JP (2011) Paleomagnetism and tectonics of the South Shetland Islands and the northern Antarctic Peninsula. Earth Planet Sci Lett 302:299–313
Poblete F, Roperch P, Hervé F, Diraison M, Espinoza M, Arriagada C (2014) The curved Magallanes fold and thrust belt: tectonic insights from a paleomagnetic and anisotropy of magnetic susceptibility study. Tectonics 33:2526–2551
Ramos VA (1989) Andean foothills structures in northern Magallanes basin. Argentina. Am Assoc Petr Geol Bull 73(7):887–903
Ramos VA (2005) Seismic ridge subduction and topography: foreland deformation in the Patagonian Andes. Tectonophysics 399:73–86
Ramos VA, Aleman A (2000) Tectonic evolution of the Andes Tectonic Evolution of South America. In: Proceedings of XXXI International Geololgical Congress (Rio de Janeiro) Brazil, pp 453–480
Ramsay JG (1981) Tectonics of the Helvetic nappes. Geol Soc Spec Publ 9:293–309
Rapalini AE (2007) A paleomagnetic analysis of the Patagonian orocline. Geologica Acta 5:287–294
Rapalini AE, Hervé F, Ramos VA, Singer SE (2001) Paleomagnetic evidence for a very large counterclockwise rotation of the Madre de Dios Archipelago, Southern Chile. Earth Planet Sci Lett 184:471–487
Rapalini AE, Calderón M, Hervé F, Cordani U, Singer S (2004) First Paleomagnetic Results on the Sarmiento Ophiolite, Southern Chile: implications for the Patagonian Orocline. Bolletino di Geofisica Teorica ed Applicata 45:246–249
Rapalini AE, Lippai H, Tassone A, Cerredo ME (2005) An AMS and paleomagnetic study across the Andes in Tierra del Fuego. In: VI international symposium on Andean geodynamics (Barcelona), extended abstracts, pp 596–599
Rapalini AE, Peroni J, Luppo T, Tassone A, Cerredo ME, Federico E, Lippai H, Vilas JF (2015) Palaeomagnetism of Mesozoic magmatic bodies of the Fuegian Cordillera: implications for the formation of the Patagonian Orocline. Geol Soc Spec Publ 425:SP425–3
Roperch P, Chauvin A, Calza F, Palacios C, Parraguez G, Pinto L, Goguitchaivilli A (1997) Paleomagnetismo de las rocas volcánicas del Jurásico tardío al Terciario temprano de la región de Aysén (Coyhaique-Cochrane). In: VIII Congreso Geológico Chileno (Antofagasta) Universidad Católica del Norte, Actas 1, pp 236–240
Rossello EA (2005) Kinematics of the Andean sinistral wrenching along the Fagnano-Magallanes Fault Zone (Argentina-Chile Fuegian Foothills). In: VI international symposium on Andean geodynamics (Barcelona), extended abstracts, pp 623–626
Sagnotti L, Faccenna C, Funiciello R, Mattei M (1994) Magnetic fabric and structural setting of Plio-Pleistocene clayey units in an extensional regime: the Tyrrhenian margin of central Italy. J Struct Geol 16:1243–1257
Somoza R, Ghidella ME (2005) Convergence in the western margin of South America during the Cenozoic: subduction of Nazca, Farallon and Aluk plates. Rev Asoc Geol Argentina 60:797–809
Somoza R, Ghidella ME (2012) Late Cretaceous to recent plate motions in western South America revisited. Earth Planet Sci Lett 331–332:152–163
Speranza F, Sagnotti L, Mattei M (1997) Tectonics of the Umbria-Marche-Romagna Arc (central northern Apennines, Italy): new paleomagnetic constraints. J Geophys Res-Sol EA 102 (B2):3153–3166
Stern CR, de Witt MJ (2003) Rocas Verdes ophiolites, southern South America: remnants of progressive stage of development of oceanic-type crust in a continental margin back-arc basin. Geol Soc Spec Publ 218:665–683
Sue C, Ghiglione MC (2016) Wrenching tectonism in the southernmost Andes and the Scotia Sea constrained from fault kinematic and seismotectonic overviews. In: Ghiglione MC (ed) Geodynamic Evolution of the Southernmost Andes. Springer Earth System Sciences, pp 137–171
Tarling DH, Hrouda F (1993) The magnetic anisotropy of rocks. Chapman, Hall, Lond, 217 pp

Thomson SN, Herve F, Stockhert B (2001) The Mesozoic-Cenozoic denudation history of the Patagonian Andes (southern Chile) and its correlation to different subduction processes. Tectonics 20:693–711

Torres Carbonell PJ, Olivero EB, Dimieri LV (2008) Structure and evolution of the Fuegian Andes foreland thrust-fold belt, Tierra del Fuego, Argentina: Paleogeographic implications. J South Am Earth Sci 25:417–439

Torres Carbonell PJ, Dimieri LV, Martinioni DR (2013) Early foreland deformation of the Fuegian Andes (Argentina): constraints from the strain analysis of upper Cretaceous-Danian sedimentary rocks. J Struct Geol 48:14–32

Van der Voo R (1990) The reliability of paleomagnetic data. Tectonophysics 184:1–9

Van der Voo R (1993) Paleomagnetism of the Atlantic, Tethys and Iapetus oceans, 411 p

Van der Voo R (2004) Paleomagnetism, oroclines, and growth of the continental crust. GSA Today 14:4–9

Van der Voo R, Channell JET (1980) Paleomagnetism in orogenic belts. Rev Geophys Space Phys 18:455–481

Van der Voo R, Stamatakos JA, Parés JM (1997) Kinematic constraints on thrust-belt curvature from syndeformational magnetizations in the Lagos del Valle Syncline in the Cantabrian Arc, Spain. J Geophys Res-Sol EA 102(B5):10105–10119

Weil AB, Sussman AJ (2004) Classifying curved orogens based on timing relationships between structural development and vertical-axis rotations. Geol S Am S 383:1–15

Weil AB, Yonkee A, Sussman A (2010) Reconstructing the kinematic evolution of curved mountain belts: a paleomagnetic study of Triassic red beds from the Wyoming salient, Sevier thrust belt, USA. Geol Soc Am Bull 122:3–23

Wilson TJ (1991) Transition from back-arc to foreland basin development in the southernmost Andes: stratigraphic record from the Ultima Esperanza District, Chile. Geol Soc Am Bull 103 (1):98–111

Winslow MA (1981) Mechanisms for basement shortening in the Andean foreland fold belt of southern South America. Geol Soc London, Spec Publ 9:513–528. doi: 10.1144/GSL.SP.1981.009.01.46

Winslow MA (1982) The structural evolution of the Magallanes Basin and neotectonics in the southernmost Andes. Antarctic Geoscience, III symposium on Antarctic geology and geophysics, Madison, pp 143–154

Yonkee A, Weil AB (2010) Quantifying vertical axis rotation in curved orogens: correlating multiple data sets with a refined weighted least squares strike test. Tectonics 29(3):TC2312

Zachos J, Pagani H, Sloan L, Thomas E, Billups K (2001) Trends, rhythms, and aberrations in global climate 65 Ma to present. Science 292:686–693

Index

A
Accretionary complexes, 8, 14, 115
Active seismicity, 2
Adakitic magmatism, 55
Aftershock, 152
Alkali basalts, 89, 96
Analogue sandbox experiments, 191
Andean denudation, 113
Andean metamorphic complex, 9–11
Andean orogeny, 8, 43, 185, 189
Anisotropy of Magnetic Susceptibility (AMS), 185, 186, 191
Antarctica, 1, 2, 4, 76, 92, 96, 98, 100, 102, 115, 159, 161, 175
Antarctica–Nazca–South America triple junction, 116
Antarctic Circumpolar Current, 76, 86, 92, 128, 175
Antarctic Peninsula, 1–3, 10, 13, 14, 18, 25–27, 29, 40, 76, 78, 79, 88, 91, 98, 100, 102, 179, 184, 185, 193
Antarctic plate, 76, 83, 84, 98, 99, 155, 159, 161, 187, 193
Antarctic waters, 4, 110, 114
Apatite (U–Th)/He (AHe), 117
Aptian-Albian, 28, 78, 99, 103
Argentine basin, 4, 112, 119, 123–125, 127, 128
Argentine continental margin, 123, 126
Argentine continental shelf, 127
Asthenospheric slab window, 116
Asthenospheric window, 113, 121, 123, 127
Atlantic ocean, 76, 121, 154, 156, 176
Atlantic offshore basin, 110
Atlantic platform, 2, 113
Australia, 76
Austral Volcanic Zone (AVZ), 55

B
Bahía de la Lancha, 13
Barker/Aurora Bank, 85
Barros Arana basalts, 65
Barros Arana Formation, 52
Basement complexes, 10, 14
Basement domain, 113–118, 123, 128, 177
Beagle suite, 42–44, 48, 56, 57, 59, 61, 62
Beauvoir-La Paciencia Formations, 41
Bending, 41, 79, 94, 139, 151, 174, 178, 179, 190, 193
Bimodal magmatism, 22, 27
Bransfield basin, 159
Bransfield Strait, 77
Bruce Bank, 77, 88, 97, 101
Burdwood Bank, 77, 88, 90, 94, 97, 101, 117, 128
Burdwood transform fault, 94–96, 98

C
Calc-alkaline trench, 37
Cambrian, 15, 23, 25, 26
Canal Beagle, 22, 41–43, 48, 57, 110, 118, 144, 145, 154, 156, 182, 188
Canal Beagle fault zone, 143
Canal Bertrand Formation, 52
Canal de Hoces, 128
Canal de las Montañas Shear Zone (CMSZ), 24, 28
Cape Horn, 89
Capitán Aracena Complex (CAC), 19, 22
Carbajal valley fault zone, 143
Carboniferous, 13–16, 25, 27, 182
Carmen Silva Formation, 112, 123, 127
Centinela Formation, 119
Central Scotia Plate, 85, 90
Cerro Negro, 13

© Springer International Publishing Switzerland 2016
M.C. Ghiglione (ed.), *Geodynamic Evolution of the Southernmost Andes*, Springer Earth System Sciences, DOI 10.1007/978-3-319-39727-6

Chenque, 119
Chile ridge, 4, 116, 122
Chile seismic ridge, 116
Chon Aike, 10, 40, 90, 100, 101
Chon Aike Silicic Large Igneous Province, 10
Chonide orogeny, 27
Chonos Metamorphic Complex (ChMC), 9, 16
Continental extension, 27, 30, 95–97
Continental rifting, 17
Convergent margin, 27, 83
Cooper Bay shear zone, 78
Cordillera Darwin, 9, 22, 41, 42, 48, 57, 59, 114, 115, 118, 156, 175, 176
Cordillera Darwin Metamorphic Complex (CDMC), 9, 14, 15, 25, 27, 28
Cretaceous, 3, 4, 10, 17, 18, 22, 24, 28, 29, 38, 40, 41, 43, 44, 51, 53, 57, 59, 60, 63–66, 78, 79, 81, 83, 98, 99, 101, 103, 115, 126, 142, 145, 151, 175, 176, 180, 182, 183, 185, 186, 188–190, 192, 194
Crustal extension, 27, 30, 97, 100
Cumberland Bay Formation, 78, 101
Curved belt, 5, 190
Curved mountain, 5, 174, 178, 193
Curved orogen, 41, 176, 178

D
Darwin, Charles, 76
Darwin suite, 42, 58, 63, 64
Davis Bank, 77, 86, 88, 101
Dawson Island, 142
Deformation paths, 191
Denaro Complex (DC), 15, 16
Deseado Massif, 10, 13, 14, 25, 100
Detrital U-Pb zircon, 8
Detrital zircon, 11, 13, 15, 16, 19, 25–27, 52, 100, 177
Devonian, 13–16, 25–27
Diego de Almagro Metamorphic Complex (DAMC), 17–19, 24, 28
Diego Ramirez Island, 18
Diego Ramírez Metamorphic Complex (DRMC), 18
Discovery Bank, 77, 94, 98
Dove basins, 89, 97
Drake Passage, 1–4, 76, 87, 92, 115, 117, 123, 127–129, 138, 142, 151, 174, 175, 179, 183, 184
Duque de York Complex (DYC), 9, 15, 17, 24, 26, 27

E
Earthquake, 5, 139, 141, 142, 152, 154–159, 161, 165, 166
Eastern Andean Metamorphic Complex (EAMC), 9, 13–15, 17, 21, 25–27
East Scotia Sea, 83, 86, 87
El Chacay Formation, 119
Elephant, 93
El Quemado Complex, 115, 182
Endurance collision zone, 92, 98
Entrerriense sea, 113, 123
Estancia 25 de Mayo, 119
Euler vector, 162
Extra-Andean Patagonia, 4, 113, 114, 121, 123

F
Fagnano fault, 5, 9, 22, 38, 42, 116, 117, 128, 138, 140, 142, 143, 152, 156, 177, 180
Falkland Plateau, 97, 100, 101
Falkland Trough, 77, 88
Farallon/Nazca–Phoenix ridge, 65
Farallon–Phoenix ridge, 177
Faulting mechanisms, 159
Fault kinematic, 2, 139, 145, 147, 148, 151, 165, 186, 191
Fault mechanism, 139, 155, 158
Fault slickenside, 146
Fault slip data, 4, 147
Fault system, 4, 128, 139, 143, 144, 146, 147, 151, 154, 175
Fission-track, 117
Fold and thrust belt, 28, 38, 41, 113–115, 117, 118, 122, 123, 144, 145, 147, 148, 150, 155, 159, 175, 177, 180, 183–194
Foreland basin, 3, 4, 28, 142, 175–177, 190–192
Fuegian Andes, 1, 5, 10, 14, 38, 56, 60, 65–67, 113–116, 118, 126–128, 142, 143, 145, 149, 150, 152, 154, 159, 175–177, 182–186, 188–190, 192–194
Fuegian basement, 12
Fuegian Batholith (FB), 41–46, 48, 55–57, 59, 61, 63, 65, 66, 114
Fuegian orogeny, 8
Fuegian Potassic Magmatism (FPM), 38, 45, 46, 48–53, 56, 57, 59–61, 63, 65

G
Gabbroic assemblage, 43–45, 56
Gaiman formations, 119
Geochemistry, 4, 8, 23, 45, 51, 52, 54
Geothermobarometry, 8, 17
Glaciation, 76, 113, 114
Gondwana, 8, 24–27, 40, 43, 98, 101, 114
Gondwanaland, 15, 16
Gondwanan, 3, 24, 100
Gordon Island, 52

Graham Land, 98
Greenschist facies, 14, 17, 23, 24, 114

H

Hardy Formation, 40, 41, 43, 45, 56, 62

I

Indenter, 165, 191, 192
Irigoyen basin, 110, 128
Isla de los Estados, 91, 142
Isla Diego de Almagro, 24
Isla Navarino, 22, 79
I-type granites, 43

J

Jane Basin, 85–87
Jeu-Jepén, 38, 44–46, 48, 49, 51, 60, 61
Jurassic, 3, 8, 10, 12, 14–18, 21, 24, 25, 27–29, 40–42, 56, 57, 63, 64, 66, 78, 100, 103, 114, 115, 155, 175, 176, 180, 182–184, 189, 192, 194

K

Kinematic databases, 4, 139
Kinematic evolution, 3, 5, 192
Kinematic reconstructions, 2, 3, 82, 116
Kranck pluton, 44, 49–51, 60

L

Ladrillero beds, 121
Lago Argentino, 110
Lago Belgrano, 110
Lago Buenos Aires, 110
Lago Deseado Fault, 143, 144
Lago Fagnano, 110, 128, 142, 144, 145, 147, 148, 150, 152, 154, 156, 157
Lago General Carrera unit, 13
Lago Posadas, 110, 118, 121
Lago San Martín, 110
Lago Viedma, 110
La Pera Formation, 52
Lu-Hf isotopic, 8

M

Maastrichtian, 98
Madre de Dios Accretionary Complex, 15
Madre de Dios Terrane (MDT), 9, 15
Magallanes–Austral basin, 3, 10, 12, 28, 110, 113, 115, 118, 119, 123, 126, 149, 150, 175, 177, 184, 194
Magallanes–Central Scotia plate boundary, 85, 90, 91
Magallanes–Fagnano fault system (MFF), 5, 38, 42, 116, 117, 128, 138–140, 142–146, 149–152, 154–159, 161, 162, 164, 165
Magallanes-Fagnano fault zone (MFFZ), 9
Magallanes fault, 143
Magallanes fold and thrust belt, 144, 147, 148, 150, 175, 177, 183–194
Magallanes strait, 110, 117, 128, 138, 142, 144, 149, 152, 155, 156
Magallanes Thrust and Fold Belt (MTFB), 9
Magmatic arc, 13, 16, 25, 28, 38, 40, 41, 43, 48, 63, 66, 114, 175, 176, 183, 189
Magmatic belt, 3, 8, 10, 14, 25, 27
Magmatic growth, 4
Magmatic suite, 4, 56
Magnetic fabric data, 5, 174
Main arc, 4, 38, 42, 43
Main Range Metamorphic Complex (MRMC), 9
Malvinas, 112, 127
Malvinas basin, 113, 114, 117, 119, 123, 126–128
Malvinas current, 127–129
Malvinas island, 139, 142
Malvinas trough, 113, 123, 128, 142
Map-view curvatures, 178
Marginal basin, 3, 10, 18, 25, 27, 28, 38, 41, 59, 63, 64, 66, 67, 79, 175, 176, 189
Marginal basin inversion, 40, 41, 64, 66
María Cristina beds, 121
Maurice Ewing Bank, 97
Mesozoic, 8–10, 14, 15, 19, 24, 38, 40, 42, 45, 78, 98, 142, 182, 185
Mesozoic metamorphic complexes, 3, 10
Metamorphic complex, 3, 8, 9, 11–19, 25, 30, 114
Metamorphic core complex, 15
Metamorphic pressure, 10
Metamorphic P–T conditions, 15
Metasedimentary complexes, 27
Metasedimentary rocks, 8, 11, 13, 15, 18, 21, 22, 26
Mid-Oceanic Ridge Basalts (MORB), 15, 23, 45, 51
Moat pluton, 44, 49–51, 60, 61
Monte León, 119
Monzonites of Navarino Island, 52
Mylonitic Belts, 10, 24

N

Navarino Island, 45, 46, 52, 56, 144, 154, 182
Nazca–Phoenix ridge, 66

Nazca plate, 89, 116
Non-rotational arc, 178, 179, 189, 194
North-east Georgia Rise, 77, 85
Northern Patagonia, 14
North Patagonian Batholith, 41
North Scotia Ridge (NSR), 76, 79, 83, 85, 86, 88, 90, 91, 94, 95, 100, 110, 116, 117, 120, 128, 138, 139, 141–144, 154, 159–161

O
Oblique convergence, 41, 65, 165
Oceanic-type seafloor, 8
Offshore sedimentation, 2
Omond Land, 98
Ona Basin, 93
Opening of the Drake Passage, 1–4, 92, 128, 138, 151, 174, 175, 179, 183, 184
Ophiolitic complexes, 8, 10, 18, 20, 29, 78
Ordovician, 13–16, 25–27
Oroclinal bending, 4, 67, 78, 179, 182–185, 190, 191, 194
Orocline, 2, 3, 5, 41, 65, 78, 79, 102, 139, 174, 178, 179, 183, 187, 189, 190, 193, 194
Orogenic arc, 138, 139, 174, 178, 190, 191
Orogenic curvature, 5, 174, 176, 182, 187, 191
Orogenic growth, 2, 4, 113

P
Pacific-Atlantic boundary, 154
Pacific plate, 38, 63, 66, 76
Pacific trench, 110, 114, 116, 127, 142, 161, 182
Packsaddle Formation, 38, 41
Packsaddle volcanics, 41, 43, 59
Paleogeographic evolution, 10
Paleomagnetic, 5, 100, 174, 178–180, 182–184, 187–190, 193
Paleomagnetic constraints, 100, 179, 184, 187, 193
Paleomagnetic directions, 180, 182–184, 188
Paleomagnetism, 2, 178
Paleozoic, 1–3, 9, 10, 14, 19, 21, 24–26, 28, 42, 78, 114, 175
Patagonia, 1, 2, 4, 14, 21, 27, 28, 38, 40, 66, 100, 110, 113–115, 121–123, 126, 127, 138, 164
Patagonian Andes, 27, 30, 38, 42, 99, 110, 115, 118, 159, 178, 186, 188, 193, 194
Patagonian arc, 5, 190, 194
Patagonian batholith, 2, 10, 41, 62, 114, 117, 118, 175, 176, 189
Patagonian glaciations, 113
Patagonian orocline, 2, 3, 5, 41, 139, 174, 179, 187, 190, 191, 193, 194
Patagoniense, 112, 121
Patagoniense transgression, 119
Peninsula Hardy, 66, 183, 189
Permian, 13–16, 22, 23, 25–27, 182
Petrotectonic assemblages, 10
Phoenix–Nazca plate boundary, 89
Phoenix plate, 84, 177
Pillow, 16, 20, 22, 23
Pillow basalt, 15, 18, 20, 22, 23
Pirie Bank, 77, 88, 94, 101
Pirie Province, 90
Plutonic complex, 14, 25
Plutonic suite, 4, 55, 59
Potassic, 38, 42, 48, 59, 60
Potassic magmatism, 4, 38, 48, 59, 63, 64, 66
Powell Basin, 88, 91, 95, 97
Pre-Scotia times, 4
Pressure–Temperature, 10
Primary arc, 41, 174, 178, 187, 190, 191, 193, 194
Progressive arc, 41, 174, 178, 189, 190, 193
Protector Basin, 77, 87, 92, 95, 96
Provenance areas, 8
Provenance studies, 13
Puerto Edén, 14
Puerto Edén Igneous and Metamorphic Complex (PEIMC), 9, 17
Pull-apart basin, 145, 157
Punta Gruesa stage, 128

R
Radial strain field, 186, 191, 193
Rain shadow, 112
Rear-arc, 4, 62
Rear-arc suites, 44, 46, 56, 58, 59, 61, 62, 66
Recent Cook deposits, 41
Recent volcanism, 37
Riedel, 145, 147–149, 155
Rift, 8, 10, 18, 23, 25
Rio Bueno thrusting, 177
Río Chico high, 118, 126
Río Guillermo, 118
Río Lácteo Formation, 13
Río Mayer Formation, 115
Río Turbio, 110, 115, 177
Rocas Verdes basin, 3, 8–10, 15, 20, 22–25, 28, 29, 40, 41, 52, 63, 67, 80, 100, 102, 175, 176, 192
Rocas Verdes marginal basin, 2, 4, 8, 38, 40, 64, 66, 79, 176, 183, 185, 189, 190, 194
Rocas Verdes ophiolites, 3, 19
Rocas Verdes ophiolitic complexes, 18, 20
Rodados Patagónicos, 113, 121, 123
Rotational arc, 189, 194

S

Sandebugten Formations, 79, 99–101
Sandwich, 77, 86, 159
Sandwich plate, 83, 85, 86
San Pablo, 119, 157
Santa Cruz Formation, 112, 113, 116, 121, 126
San Vicente thrust, 177
Sarmiento Complex (SC), 19–22, 24, 28
Sarmiento Cordillera, 21
Satellite pluton, 4, 40, 42, 46
Scan Basin, 77, 85, 89, 91, 94
Scotia Arc, 2
Scotia Plate, 83, 85, 91, 116, 138, 139, 141–143, 146, 151, 159, 161, 162, 174, 184, 193
Scotia Ridge, 100, 110, 117, 120, 128, 138, 141, 142, 159, 161
Scotia Sea, 1, 2, 4, 76–93, 95–99, 102, 103, 110, 116, 121, 128, 140, 142, 152, 165, 175
Scotia–South America plate boundary, 88, 128
Seafloor-spreading, 79, 83, 84, 87, 90–93, 97, 98
Sediment dispersal, 3, 4
Seismic, 82, 86, 112, 116, 119, 123–126, 129, 139, 145, 154–158
Seismicity, 139, 152, 154–156, 161
Seismic risk, 139, 152
Seismotectonic, 3, 152, 155, 158
Seno Almirantazgo, 110, 128, 144, 147–149
Seno Año Nuevo suite, 43–45, 56, 59, 65
Seno Arcabuz Shear Zone (SASZ), 17, 18, 24
Shackleton fracture zone, 84, 92, 141–143, 154, 159–161
Shag Rocks, 76, 86
Shelf break, 113, 126
Shortening, 28, 63, 64, 88, 96, 101, 114, 115, 122, 128, 145, 147, 149–151, 164, 165, 184–186, 190, 191, 194
Shoshonites, 52, 56, 63
Slab flattening, 38, 64–66
Slab shallowing, 63
S-like granitoids, 42, 56, 66
Slip partitioning, 192
South America, 1, 4, 8, 9, 18, 19, 26, 29, 30, 38, 40–42, 66, 76, 79, 95, 96, 102, 115, 116, 139, 141, 142, 146, 155, 159, 161, 162, 174–176, 182, 187
South America–Antarctica, 117
South American-Antarctic Ridge, 86
South American plate, 78, 83, 91, 95, 96, 98, 99, 101, 116, 138, 142, 174, 176
South America-Scotia continental plate boundary, 128
South Atlantic, 79, 86, 98, 102, 128, 141, 176
Southern Andean orogenic arc, 138
Southern Atlantic margin, 114
Southernmost Andes, 1–5, 19, 38, 40, 41, 59, 65, 76, 110, 115, 126, 138, 139, 151, 154, 155, 158–161, 165, 174–177, 179, 180, 182–187, 188, 190–194
Southern ocean, 1, 76, 139
Southern Patagonia, 3, 4, 8, 10, 13, 14, 18, 25–28, 58, 64, 138, 162
Southern Patagonian Andes, 1, 10, 14, 18, 24, 48, 59, 63, 65, 66, 78, 110, 113–118, 121, 123, 161, 176, 177, 180, 185, 186, 188, 192
South Georgia, 4, 40, 76, 78–83, 86, 91, 94, 95, 97, 99–103, 142
South Orkney islands, 76, 78, 99
South Orkney Microcontinent, 77, 79, 91, 96–98
South Patagonian Batholith (SPB), 17, 21, 28, 40–42, 52, 58, 59, 65
South Sandwich islands, 76
South Sandwich subduction zone, 82, 98
South Sandwich Trench, 77, 82, 85, 90, 91, 94–96, 98
South Scotia Ridge, 1, 2, 76, 83, 86, 88, 91, 97, 159, 161
South Shetland Islands, 76, 78, 84
South Shetland Trench, 77
Springhill, 115
Staines Complex (SC), 9, 13
Strain analysis, 164, 179
Stress pattern, 4, 161, 165
Strike-slip orocline model, 191
Strike-slip tectonics, 174, 177, 182, 188, 193, 194
Subantarctic current, 127, 128
Supra-subduction, 17, 27, 28, 30
Supra-subduction zone, 10
Synorogenic, 112, 115, 116
Synorogenic sequences, 118
Synrift, 21, 27, 115

T

Tarlton Limestones (TL), 9, 15
Tectonic reconstructions, 4, 188
Tectonics, 80, 83, 87–89, 139, 152, 156, 158, 175, 177, 182
Terra Australis orogen, 2, 8
Terror Rise, 77, 88, 92, 97, 98, 101
Thermochronology, 8, 176
Tholeiitic magmas, 23, 24
Tierra del Fuego, 1, 4, 9, 10, 12, 38, 40, 44, 48, 56, 58, 59, 62–66, 76, 78, 79, 81, 82, 88, 91, 95–99, 102, 103, 110, 112–116, 119,

121, 123, 126–128, 138, 142, 144, 145, 149, 150, 152, 154, 161, 162, 164
Tierra del Fuego Igneous and Metamorphic Complex (TFIMC), 12–15, 25, 26
Tobífera Formation, 21, 22, 25, 43
Tobífera/Lemaire, 115
Tortuga ophiolite, 21–23
Transform margin, 139
Trapecio Hill, 38, 53
Triassic, 13, 14, 16, 24, 27, 40, 114, 184
Trinity Peninsula Group (TPG), 27
Triple junction, 122, 138, 142
Turbiditic successions, 13, 15, 114

U
Última Esperanza, 115, 118
Ushuaia Peninsula Andesites (UPA), 38, 48, 53–56, 59, 61–63, 65
Ushuaia pluton, 44, 49, 50, 60, 61

V
Vertical-axis rotations, 174, 180, 188, 192, 193
Volcanic front, 4, 38, 57, 59, 60, 62, 63, 65, 66
Volcaniclastic sequences, 48

W
Weddell Sea, 40, 86, 98, 100, 184
West Antarctica large igneous, 40
Western Andes Metamorphic Complex, 114
West Scotia Ridge, 84–86, 88, 90–92
Widening of the arc, 4, 59, 66
Wrenching, 41, 145, 146, 149–151, 154

Y
Yahgán Formation, 19, 22, 41, 78–80

Z
Zapata Formation, 21
Zeballos Group, 112

MIX
Papier aus verantwortungsvollen Quellen
Paper from responsible sources
FSC® C105338

If you have any concerns about our products,
you can contact us on
ProductSafety@springernature.com

In case Publisher is established outside the EU,
the EU authorized representative is:
Springer Nature Customer Service Center GmbH
Europaplatz 3, 69115 Heidelberg, Germany

Printed by Libri Plureos GmbH
in Hamburg, Germany